# Statistics for Biology and Health

## Statistics for Biology and Health

*Borchers/Buckland/Zucchini:* Estimating Animal Abundance: Closed Populations.
*Everitt/Rabe-Hesketh:* Analyzing Medical Data Using S-PLUS.
*Ewens/Grant:* Statistical Methods in Bioinformatics: An Introduction.
*Hougaard*: Analysis of Multivariate Survival Data.
*Klein/Moeschberger:* Survival Analysis: Techniques for Censored and Truncated Data.
*Kleinbaum/Klein:* Logistic Regression: A Self-Learning Text, 2nd ed.
*Lange:* Mathematical and Statistical Methods for Genetic Analysis, 2nd ed.
*Manton/Singer/Suzman:* Forecasting the Health of Elderly Populations.
*Salsburg:* The Use of Restricted Significance Tests in Clinical Trials.
*Sorensen/Gianola:* Likelihood, Bayesian, and MCMC Methods in Quantitative Genetics.
*Stallard/Manton/Cohen:* Forecasting Product Liability Claims: Epidemiology and Modeling in the Manville Asbestos Case.
*Therneau/Grambsch:* Modeling Survival Data: Extending the Cox Model.
*Zhang/Singer:* Recursive Partitioning in the Health Sciences.

Eric Stallard
Kenneth G. Manton
Joel E. Cohen

# Forecasting Product Liability Claims

## Epidemiology and Modeling in the Manville Asbestos Case

Foreword by The Honorable Jack B. Weinstein

Eric Stallard
Kenneth G. Manton
Center for Demographic Studies
Duke University
Durham, NC 27708
USA

Joel E. Cohen
Laboratory of Populations
Rockefeller and Columbia Universities
New York, NY 10021
USA



Library of Congress Cataloging-in-Publication Data
Stallard, Eric.
Forecasting product liability claims: epidemiology and modeling in the Manville asbestos case / Eric Stallard, Kenneth G. Manton, Joel E. Cohen.
p. cm — (Statistics for biology and health)
Includes bibliographical references and index.

1. Products liability—Asbestos—United States—Mathematical models. 2. Asbestos—Epidemiology—Mathematical models. I. Manton, Kenneth G. II. Cohen, Joel E. III. Title. IV. Series.
KF1297.A73S73 2003
346.7303'8—dc21 2003045445

ISBN 978-1-4419-2860-3 Printed on acid-free paper.



Printed in the United States of America. (HP)

9 8 7 6 5 4 3 2 1

springeronline.com

# Foreword

I write this foreword for two reasons: first, to acknowledge the gratitude of our court system to scientists willing to lend their talents to forensic tasks, and of myself, in particular, for the pathbreaking work of Eric Stallard, Kenneth G. Manton, and Joel E. Cohen in the Manville Asbestos Case; and second, because their work suggests both great strength and utility in their statistically based design and its limitations in predicting events strongly affected by political and social choices that are difficult to foretell as well as by demographic and epidemiologic factors that can be prophesied with somewhat more confidence – at least in the short term.

It is by now almost axiomatic that almost every important litigation in the United States requires experts to help judges and juries arrive at an understanding of the case sufficient to permit a sensible resolution within the flexible scope of our rules of law. The Supreme Court has laid down useful rough criteria for the courts in assessing the capability of proffered experts beginning with the *Daubert* line of cases.[1] It has also allowed the courts to appoint experts to supplement those designated by the parties.[2] Dr. Joel E. Cohen and Professor Margaret E. Berger were appointed by me in the *Manville* asbestos cases pursuant to Rule 706 of the Federal Rules of Evidence to help project future claims. Discovery provisions have improved utilization of experts by requiring advance reports and depositions.[3]

The result of acknowledgments by the judiciary of the critical role experts play in our work has resulted in an explosion of academic and other writings on ways to improve the relationship of science and the law; it is too extensive

[1] See *Daubert v. Merrell Dow Pharmaceuticals, Inc.*, 509 U.S. 579, 113 S.Ct. 2786, 125 L.Ed.2d 469 (1993); *Fed. R. Evid.* 702. There has been greatly increased concern with training judges in the use of scientific data and experts. See, e.g., Federal Judicial Center, *Reference Manual on Scientific Evidence* (2d ed. 2000); Science for Judges, Conference at the Brooklyn Law School Center for Health Law and Policy (Mar. 28 & 29, 2003).

[2] See *Fed. R. Evid.* 706.

[3] See *Fed. R. Civ. P.* 26(a)(2), (b)(4).

to review here. It is enough to acknowledge my own gratitude for the help of experts in a few of my own cases and publications.[4]

In the *Manville* and other asbestos cases one of the problems is attempting to predict future claims, and the future value of assets set aside in bankruptcy proceedings to pay for them, so that an equitable method of deciding the amount of payment to present and future claimants can be devised.

From the scientist's point of view, forecasting future claims requires the definition of diseases, causation and medical proof. In the case of some diseases such as mesothelioma, an asbestos signature disease, the definition, cause and proof are fairly precise and epidemiology is adequate to predict future diseases, although attribution to particular defendants is problematic. In the case of other diseases attributable to asbestos exposure, causation is not as clear, definition tends to be somewhat ambiguous, analysis of X-rays and other indicia highly subjective, and epidemiology soft.

From the legal point of view, there is the question of what claims should be recognized – for example, fear of cancer, nondisabling lung obstructions and the like. Courts and legislatures can change the universe of claims for reasons hard to predict. Attorneys' efficiencies in finding clients, sympathetic doctors and efficient methods of filing claims at little expense may enormously affect the number of supportable claims made. The economic problem of predicting the future value of assets set aside for payment is brought home by the recent huge changes in stock market values.

The nature of the almost insuperable hurdles to accurate prediction of claims for personal injury by asbestos is suggested by the fact that Manville sought bankruptcy in the early 1980s, first projections were made in the mid-eighties, first payments were made in the late eighties, and by the early

[4] See, e.g., Jack B. Weinstein & Margaret E. Berger, *Weinstein's Federal Evidence* §§ 702.01–702.08; 706.01–706.06 (Joseph M. McLaughlin ed., 2002); *In re Simon II Litig.*, 211 F.R.D. 86 (E.D.N.Y. 2002) (tobacco); *In re Breast Implant Cases*, 944 F.Supp. 958 (E.D. & S.D.N.Y. 1996); *United States v. Shonubi*, 895 F.Supp. 2d 460 (E.D.N.Y. 1995), vacated on other grounds, 103 F.3d 1085 (2d Cir. 1997); *In re "Agent Orange" Product Liability Litig.*, 603 F.Supp. 239 (E.D.N.Y. 1985); Jack B. Weinstein, *Individual Justice in Mass Tort Litigation* 3, 115-17 (1995); Jack B. Weinstein, Introduction to Hans Zeisel & David Kaye, *Prove It with Figures: Empirical Methods in Law and Litigation* (1997); Jack B. Weinstein & Catherine Wimberly, *Secrecy in Law and Science*, 23 Cardozo L.R. 1 (2001); *Science, and the Challenges of Expert Testimony in the Courtroom*, 77 Or. L.R. 1005 (1998); *Improving Expert Testimony*, 20 U. Rich. L.R. 473 (1986); *Expert Witness Testimony: A Trial Judge's Perspective*, in *Medical–Legal Issues Facing Neurologists*, 17 Neurologic Clinics No. 2, 355 (May 1999); *The Effect of Daubert on the Work of Federal Trial Judges*, 2 Shepard's Expert & Sci. Evid. Q. No. 1 (1994); *Litigation and Statistics: Obtaining Evidence Without Abuse*, Bureau of National Affairs, Dec. 24, 1986; Scientific Evidence in the United States Federal Courts: Opening the Gates of Law to Science, Address Before the First Worldwide Common Law Jurists Conference (May 30, 1995); Trial Judge's View of *Daubert*, Presentation (Aug. 9, 1993), *Scientific Evidence Review*, Monograph No. 2.

nineties, the trust set up to make the payments was already running out of funds because claims and amounts paid per claim had ballooned. Yet the last claim for mesothelioma was not expected, because onset of the disease is so delayed after exposure, until almost 2050. Thus the bankruptcy court in the 1990s needed reliable projections for claims more than 60 years into the future. The Manville Personal Injury Settlement Trust history explaining the present problem is partially set out in the court's opinion of December 27, 2002. That opinion is available on the publisher's World Wide Web site associated with this book. All-in-all, the asbestos cases present difficult problems for the courts that can be reduced, but not eliminated, by works such as this one of Stallard, Manton, and Cohen.

A brief review of where the asbestos litigation now stands might be useful.[5]

## Overview of Current Asbestos Problems

I approach the asbestos problem with sharp memories of the some seventy asbestos cases I tried with juries in the 1970s and 1980s. Those injured by asbestos worked as young men during World War II in the Brooklyn Navy Yard, building battleships and aircraft carriers urgently needed to defeat the Germans and Japanese; they labored in clouds of asbestos. Many then went on to serve as sailors, soldiers, and airmen. Years later they were dead or dying from cancer, or frightened about asbestos in their lungs reported after X-rays. Navy doctors and others in government, private industry, and the unions knew of the dangers of asbestos in the thirties and forties, but had done nothing to protect workers exposed to asbestos dust.

### The Asbestos Litigation Crisis

The history of asbestos in this country has been marked by a failure to deal promptly and comprehensively with a series of related problems. The dangers of asbestos were known at least as early as the 1930s, but millions of workers and other Americans continued to breathe asbestos without adequate warnings.

Few agencies or people merit medals for their role in this country's asbestos disaster.

The federal government did not act in time to regulate asbestos or to protect workers adequately. It failed to act promptly through an Environmental

[5] The following sections are based upon my remarks at a symposium, "Asbestos: What Went Wrong," held at the Bar Association of the City of New York on October 21, 2002, expanded by my and Katherine L. Aschenbrenner's article in the Bureau of National Affairs' *Product Safety & Liability Reporter* of November 25, 2002 at p. 1053ff. See also *In re Joint E. & S. Dists. Asbestos Litig.*, 237 F.Supp.2d 297 (E.D.N.Y. 2002); *Norfolk & Western Ry. Co. v. Ayers*, 123 S.Ct. 1210 (2003).

Protection Agency or an Occupational Safety and Health Administration until the 1970s.[6] State legislatures and agencies were largely quiescent.[7] So, too, were the relevant unions. Most of them were uninvolved in protecting their at-risk members.

A handful of doctors under Dr. Irving J. Selikoff at Mount Sinai conducted the devastating epidemiological studies and sounded warnings. They are free from all blame. Other doctors did little even as they observed that increased worker diseases resulted from work with asbestos and that adequate ventilation was almost nonexistent.

Except for minimal employee worker compensation, little if any attention was paid to reparation for those who had suffered, and would suffer, long-delayed injuries as a result of their unwitting exposure to asbestos. As a result, those injured by asbestos have relied primarily on the courts in tort-based suits against manufacturers of products containing asbestos.[8]

The plaintiff bar has been extensively criticized for its conduct in asbestos litigation. In my opinion, plaintiffs' lawyers have, on the whole, acted responsibly and honorably. These lawyers are frequently blamed for concocting the current and on-going crisis in asbestos litigation, but had it not been for the plaintiff bar no one would have acted. Plaintiffs' lawyers deserve credit for having done more than anyone else to protect and compensate workers injured by asbestos and to deter further dangerous use. Nor can defense counsel be seriously faulted; in the main, they acted ethically on behalf of their clients.

There are obvious problems inherent in leaving the asbestos crisis to be dealt with by general tort law in the court system. The deficiencies of our private tort-court system have been the subject of extensive critical discussion.[9]

---

[6] See Clean Air Act, 42 U.S.C. §§ 7401, 7412 (2002) (Congress designated asbestos as a hazardous air pollutant in the Clean Air Act as passed in 1970); 40 C.F.R. § 61 (2002) (Environmental Protection Agency emission standards for hazardous air pollutants, including asbestos); 29 C.F.R. § 1926.1101 (2002) (Occupational Safety and Health Administration ["OSHA"] standards on asbestos in construction); 29 C.F.R. § 1915.1101 (2002) (OSHA standards on asbestos in shipyards); 29 C.F.R. § 1910.1001 (2002) (OSHA standards on asbestos in general industry); see also Mark A. Behrens, *Some Proposals for Courts Interested in Helping Sick Claimants and Solving Serious Problems in Asbestos Litigation*, 54 Baylor L. Rev. 331, 336 n.25 (2002).

[7] See, e.g., Edmund J. Ferdinand, III, *Asbestos Revisited*, 5 Tul. Envtl. L.J. 581, 588 (1992) (commenting on the asbestos industry's manipulation of state legislatures).

[8] See, e.g., Stephen J. Carroll, et al., RAND Institute for Civil Justice, *Asbestos Litigation Costs and Compensation: An Interim Report* (2002) ("RAND Report 2002").

[9] See, e.g., Griffin B. Bell, *Asbestos Litigation and Judicial Leadership: The Courts' Duty to Help Solve the Asbestos Litigation Crisis*, National Legal Center for the Public Interest, June 2002. In Great Britain and France where a different compensation system prevails, a large increase in claims now threatens these countries' insurance and industrial entities. See, e.g., Laurie Kazan-Allen, *Foreign Perspectives on the U.S. Asbestos Crisis*, Mealey's Litigation Report: Asbestos, May 7,

Yet plaintiffs' claims in asbestos have been accurately described as continuing like a perpetually "unrolling carpet. At any moment ... some are filed, some are resolved and some are yet to come."[10]

By the end of 2000, claims for compensation for asbestos-related injuries had already cost industry and insurers well over 50 billion dollars. It is estimated that such claims ultimately will cost businesses as much as $210 billion[11] – though this seems high.

The total number of asbestos lawsuits and the rate at which those lawsuits are being filed have increased rapidly in recent years, even though most industrial use of asbestos ended more than thirty years ago. Defendants have expanded from a core group of companies who processed asbestos or used it extensively to over 6000 companies in virtually every type of industry in the United States.[12] Claimants have increased from those who are seriously ill to those with noncancerous conditions to those who are functionally unimpaired but show minimal clinical signs of asbestos exposure such as pleural thickening barely evident in X-rays.[13] These recent developments have given rise to renewed questions of whether there will be enough money to pay all claims likely to be filed.

As more than three score companies have been driven into bankruptcy by these claims, leading plaintiff lawyers, financed by substantial fees, have moved aggressively to bring in peripheral players – in particular, (1) plaintiffs, by using mass X-rays and B-reader[14] pro-plaintiff diagnoses paid for by lawyers, as well as by mass media advertising for new clients, and (2) defendants with little asbestos connection, such as automobile manufacturers, who used asbestos in brake linings.[15]

---

2003 at 45; Andrew Murray, et al., *Asbestos: Too Hot to Handle for European Insurers?*, Fitch Ratings, March 31, 2003; Laurie Kazan-Allen, *French Supreme Court Supports Asbestos Victims*, March 11, 2002, at www.ibas.btinternet.co.uk.

[10] Record at 31, *Ortiz v. Fibreboard Corp.*, 527 U.S. 815 (1999) (No. 97-1704) (Oral argument of Elihu Inselbuch, Esq., on behalf of respondents). Probably less than one-third of those who could sue, actually do so. See statement of Dr. Mark A. Peterson, Senate Judiciary Committee, June 4, 2003.

[11] *RAND Report 2002*, supra, at 78.

[12] *Ibid* at 49-50.

[13] *Ibid* at 19-21.

[14] A "B" reader is a physician who has been specially trained and certified by NIOSH in the use of the ILO system for classifying X-rays in the presence of pneumoconioses.

[15] See, e.g., *Jackson v. Anchor Packing Co.*, 994 F.2d 1295 (8th Cir. 1993); *Clutter v. Johns-Manville Sale Corp.*, 646 F.2d 1151 (6th Cir. 1981); *United States v. Midwest Suspension and Brake*, 824 F.Supp. 713 (E.D. Mich. 1993); *In re Tire Worker Asbestos Litig.*, 1991 WL 195557 (E.D. Pa. 1991); *Covington v. Abex Corp.*, 1990 WL 204688 (D.D.C. 1990); *DiSantis v. Abex Corp.*, 1989 WL 150548 (E.D. Pa. 1989); *Lowie v. Raymark Industries*, 676 F.Supp. 1214 (S.D. Ga. 1987).

Today, electronic filing permits multiple claims at almost no cost to the plaintiffs' bar. Legal fee generation in the asbestos litigation business is extremely efficient and remunerative to plaintiffs' and to claimants' lawyers.

The federal Multidistrict Litigation ("MDL") judge for all federal court asbestos cases has had his fingers in the dyke for some years in an attempt to slow the flood of cases. He has had extensive opportunity to observe recent developments in asbestos cases. He writes:

> Since the original assignment of the matter in July, 1991, the number of cases [in the federal courts, with many times that number in the state courts] has grown to approximately 105,000. We have closed over 75,000 cases which represents more than 9,900,000 claims. ... The new defendants are not necessarily manufacturers, distributors, or installers of asbestos products but are peripherally involved. With 20 to 100 or more defendants named in every case, and frequently with more than one plaintiff, this translates into a myriad of new claims. ...
>
> From the onset, we have given priority to the claims of the very sick and to the victims with malignancies. Groups settlements were and are critical to the movement of these cases. We have attempted to establish registries or deferred lists of cases to husband the resources that are available in order to protect future plaintiffs who may yet become victims. ...
>
> Since the inception of the MDL consolidation, we have worked with all of the attorneys to encourage them to concentrate on the cases which involve malignancies and severe asbestosis claimants, and to defer the claims of those who at present have no evidence of impairment. Many of the attorneys have followed this path. ... We have been able to close approximately 60% of the claims and the remaining number of cases is in the area of 30,000. Agreements to close the cases have included both federal, state, and unfiled cases which grew out of a waiver of the statute of limitations by all the parties.[16]

Courts have taken varying approaches in response to the rapid rise in claims brought by unimpaired individuals. For example, one judge has held inactive without prejudice, with the tolling of applicable statutes of limitations, nonmalignant cases until the claimants show evidence of an asbestos-related debilitating condition.[17] Such rulings have diverted many cases from the federal to state courts since plaintiffs have the option of bringing their cases in more receptive courts. (In some locales courts are said to be particularly con-

[16] Letter from Charles R. Weiner, MDL Judge, Eastern District of Pennsylvania, to Jack B. Weinstein and other judges (Oct. 15, 2002), filed and docketed in *In re Joint E. & S. Dists. Asbestos Litig.* (No. 90-3973) (E.D. & S.D.N.Y. 2002).

[17] See Administrative Order No. 8, *In re Asbestos Prods. Liab. Litig.* (No. MDL 875) (E.D. Pa. Jan. 16, 2002).

siderate of asbestos plaintiffs.) The federal MDL system has not staunched the flow.

One federal judge has suggested denying voting rights in bankruptcy to individuals not presently demonstrating objective medical evidence of impairment.[18] In effect this cuts nondisabling cases off from effective claims in bankruptcy.

State court judges have put nondisabling claims on an ultra slow track, moving forward only with the most serious cases.[19] One court has construed state law to reduce the burden on jointly liable defendants.[20] Still others have construed state substantive laws to exclude claims of those not functionally disabled.[21]

Plaintiffs' bar has been split on how nondisabled plaintiffs should be handled. Some responsible plaintiffs' and defendants' lawyers would stop recognizing them so that available funds could be used for mesothelioma or other serious diseases. The problem with any approach denying or indefinitely postponing claims brought by functionally unimpaired claimants who show some clinical or X-ray symptoms of asbestos exposure and disease is that, under federal law,[22] state tort law determines whether the claims are actionable. Although a matter of some controversy, many of these claims are substantively valid under the law of many states.[23] In my opinion, nonimpaired claims can-

[18] See, e.g., Debtors' Motion for a Declaration with Respect to Voting Rights of Certain Putative "Claimants," *In re USG Corp. et al.* (No. 01-2094) (Bankr. D. Del., Aug. 21, 2002).

[19] See generally, e.g., Mark A. Behrens & Monica G. Parham, *Stewardship for the Sick: Preserving Assets for Asbestos Victims Through Inactive Docket Programs*, 33 Tex. Tech L. Rev. 1 (2001); Paul F. Rothstein, *What Courts Can Do in the Face of the Never-Ending Asbestos Crisis*, 71 Miss. L.J. 1 (2001).

[20] See Cerisse Anderson, *Asbestos Ruling Favors Solvent Defendants*, N.Y.L.J., Nov. 4, 2002, at 1 (Manhattan Supreme Court Justice Helen Freedman construes apportionment of liability to protect solvent defendants).

[21] Professor Aaron D. Twerski, Newell DeValpine Professor of Law, Brooklyn Law School, Statement at Symposium, "Asbestos: What Went Wrong," held at the Bar Association of the City of New York (October 21, 2002). See also n.16, infra.

[22] *Erie R. Co. v. Tompkins*, 304 U.S. 64 (1938).

[23] Compare Unofficial Committee of Select Asbestos Claimants Supplemental Memorandum on SCB Proposed Trust Modifications, *In re Joint E. & S. Dists. Asbestos Litig.* (90 CV 3983) ( E.D. & S.D.N.Y., Sep. 25, 2002) *with* SCB's Supplemental Response to Memoranda Regarding Amendments to the Manville TDP, *In re Joint E. & S. Dists. Asbestos Litig.* (90 CV 3983) ( E.D. & S.D.N.Y., Oct. 4, 2002). See also James A. Henderson, Jr. & Aaron D. Twerski, *Asbestos Litigation Gone Mad: Exposure-Based Recovery for Increased Risk, Mental Distress, and Medical Monitoring*, 53 S.C. L. Rev. 815 (2002) (arguing that preinjury claims brought by plaintiffs exposed to asbestos for increased risk, negligent infliction of emotional distress, or medical monitoring are "radical departures from long-standing norms of tort law" and should not be recognized); *RAND Report 2002*, supra, at 19-20.

not be ignored where state law recognizes them. It is still possible, however, to reduce the ratio of compensation paid to claimants who are not functionally impaired to better reflect the relative gravity of injury.

As funds run out, the percentage paid of the agreed value of any claim is reduced. In *Manville*, claimants have been receiving only 5% to 10% of the agreed value of their claims.

In general, the plaintiffs' bar has not been irresponsible. Even those who do not want to exclude nondisabling claims have now agreed to allow much higher relative payments for more serious diseases. In the Manville Personal Injury Settlement Trust, for example, the ratio has gone up from less than 10 to 1 to 30 to 1 with the agreement of most of the plaintiffs' bar, the Trust, and the Representative of Future Claimants. There was also some tightening of the criteria for demonstrating an asbestos-related disease, although the opportunity for fraud or exaggeration of dubious claims still remains substantial.

## Possible Future

There are a number of possible legal reactions to the asbestos problems.

### *Leaving Matters As They Are*

First, if the acceleration and expansion of asbestos lawsuits continues unaddressed, it is possible that almost every company with even a remote connection to asbestos may be driven into bankruptcy or at least suffer serious financial difficulties. This risk is increasingly imminent as more obvious defendants become bankrupt and aggressive plaintiff lawyers move to fresh prey.

Although bankruptcy does have its advantages, it is not an entirely satisfactory solution. Reliance on it will cause unnecessary business failures and result in many who will suffer from asbestos-related injuries being un- or under-compensated. The continued spread of asbestos litigation leaves a cloud hanging over the American economy. There will be a threat of loss of jobs in defendant companies and unnecessary harm to creditors and equity holders of stock.

A "litigation as usual" outcome is not satisfactory. It results in the inability of many workers and other claimants to receive adequate compensation for their asbestos-related injuries. It will result in excessive transaction costs. It will inhibit useful corporate consolidations with companies that have the least hint of asbestos relatedness.

Dealing successfully with asbestos litigation through the court system presents difficulties. Any comprehensive solution would require substantial reliance on the Stallard, Manton, and Cohen projection system.

### *Plaintiff Class Action*

One new litigation-based strategy is a global class action brought by all claimants exposed to asbestos against all defendants connected to the asbestos industry. The class could be certified as a limited fund, non–opt-out compensatory damage class on the ground that the massive threat to industry as well as claimants warrants an exercise of equitable power.[24] A non–opt-out punitive damage class could be simultaneously prosecuted.[25] Such a class action would be able to utilize national demographic and epidemiological data that could place realistic boundaries on total recoveries, provide for sensible distribution, and cut fees and transactional costs.

Except in limited special circumstances, a plaintiff class action is unlikely to be brought because the continued possibility of large attorney's fees in individual litigation provides an adverse incentive to reducing total transaction costs. Mass cases, individually settled in bulk, provide much higher fees for plaintiffs' lawyers. In class actions the court regulates fees. Total regulation of fees even in non–class actions would provide the condition for a rational disposition reasonably protective of all those injured; this appropriate result would probably require empowering legislation.

The Supreme Court's decisions in *Amchem*[26] and *Ortiz*[27] leave a cloud over any global settlement.[28] In my view the size of this cloud has been much exaggerated. A well designed class settlement or trial could succeed in providing closure to asbestos litigation in this country.

*Defendant Class Action*

A second litigation-based possibility is an action, based on a theory of interpleader or an action to quiet title, brought by the industrial firms and other potential defendants against a non–opt-out class of possible claimants. It would use a class action to reverse the litigation roles of plaintiffs and defendants.[29]

After a determination of total liability, the court would determine percentage liability and allocate funds through an administrative scheme. The data necessary to make these determinations could be obtained in part through an analysis of claims filed with existing bankruptcy trusts and projections of

[24] See *Fed. R. Civ. P.* 23.

[25] See *In re Simon II Litigation*, 211 F.R.D. 86, 2002 WL 31323751 (E.D.N.Y. 2002).

[26] *Amchem Products, Inc. v. Windsor*, 521 U.S. 591 (1997).

[27] *Ortiz v. Fibreboard Corp.*, 527 U.S. 815 (1999).

[28] On asbestos litigation after *Amchem* and *Ortiz*, see generally Deborah R. Hensler, *As Time Goes By: Asbestos Litigation After Amchem and Ortiz*, 80 Tex. L. Rev. 1899 (2002); Samuel Issacharoff, *"Shocked": Mass Torts and Aggregate Asbestos Litigation After Amchem and Ortiz*, 80 Tex. L. Rev. 1925 (2002).

[29] Cf. *Ortiz*, 527 U.S. at 834-36, 848 (Rule 23(b)(1)(B) class action appropriate to resolve claims made against a fund facing limits other than those imposed solely by agreement of the parties which render it insufficient to satisfy all claims); *In re Simon II Litigation*, 211 F.R.D. 86, 2002 WL 31323751 at *106 (E.D.N.Y. 2002); *In re The Exxon Valdez*, No. A89-0095-CV (HRH) (Consolidated), Order No. 180 Supp., at 8-9 (D. Alaska, Mar. 8. 1994), vacated as to amount of punitive award, *In re Exxon Valdez*, 270 F.3d 1215 (9th Cir. 2001).

future disease manifestations calculated by these trusts and others. Demographic and epidemiological data are available.

Defensive actions by prospective defendants are unlikely to be brought because of fear. Plaintiffs' attorneys are perceived to be too powerful and too threatening to allow such a sensible resolution that would limit future fees.

*Bankruptcy*

One variation on this possibility of a defendants' initiative is to use bankruptcy proceedings and Section 524(g) of Title 11 of the United States Code (the special asbestos provision) to attempt to achieve a global resolution without destruction of equity and the debt structure of defendants. This is a fairly radical circumvention of traditional bankruptcy rules. Before new defendants would consider this route they would want legislation to protect their investors. A prepackaged bankruptcy may be available in limited circumstances to safeguard parents of asbestos-tainted subsidiaries.[30]

*Legislation*

The best resolution to the asbestos problem may be through federal legislation. The hope of uniform state legislation to eliminate nonmalignant claims does not appear realistic. The Supreme Court, in *Amchem Products, Inc. v. Windsor*[31] and in *Ortiz v. Fibreboard Corp.*,[32] emphasized the need for legislative intervention in the "elephantine mass of asbestos cases."[33]

The desirability of legislation to deal with the problem of asbestos in a comprehensive manner and to ensure that those who suffer from injuries caused by exposure to asbestos are fairly compensated has become more apparent with the rapid rise in claims brought by functionally unimpaired claimants. Senator Leahy responded to the Supreme Court's call for congressional action, holding a full Senate Judiciary Committee hearing on Asbestos Litigation on September 25, 2002.[34] The Senate Judiciary Committee held additional hearings on asbestos litigation on March 5 and June 4, 2003.

Congress has the authority under the Commerce Clause and its Bankruptcy Power to deal with asbestos. It could set up a compensation scheme as the exclusive remedy for injuries resulting from exposure to asbestos on the theory that it was regulating events affecting the economy of the country and protecting the health and welfare of the people. It has acted in mass

---

[30] Alison Langley, "ABB Proposes a Settlement for All Asbestos Lawsuits," *N.Y.Times*, Nov. 2, 2002, at B2 (prepackaged bankruptcy of subsidiary to protect parent).

[31] 521 U.S. at 597-98.

[32] 527 U.S. 815.

[33] *Ibid* at 821.

[34] *Asbestos Litigation, Hearing Before the Senate Committee on the Judiciary*, 107$^{th}$ Cong. (2002) (statement of Sen. Patrick Leahy, Chairman, S. Comm. on the Judiciary).

situations on occasion when traditional forms of litigation alone would have provided only an incomplete and unsatisfactory solution to harm suffered by large numbers of people.[35] The national legislature has the power to preempt state statutes and common law if necessary to achieve its purpose.[36]

The "tax" imposed for clean-up of major polluting sites offers some experience.[37] Black Lung and industry-funded pollution clean-up programs have, however, proved very expensive. Congress probably will shy away from providing an ultimate deep pocket. Any legislation will have to sharply limit taxpayer impact.

Federal asbestos legislation should not affect prior completed bankruptcies. Trusts in most of those cases already have been set up and the rights of claimants and industry have vested. The legislation should be designed to deal with those whose claims have not yet been adjudicated and with future asbestos claimants. Whether those whose suits or claims are pending but not decided should be included in any new federal scheme presents an open question; my preference would be to fold within the scheme all claims which have not been resolved by judgments. Partial past resolutions could be handled through set-offs.

A legislative scheme could effectively be set up under a federal administrative system as follows:

1. The federal legislative system would establish an exclusive remedy; there would be no right for either plaintiffs or defendants to opt out to enter the tort system. Allowing the system to be voluntary would risk the federal statutory alternative both becoming underfunded and being rejected by plaintiffs' lawyers.
2. An adjudicative administrative board would allocate administrative and compensation costs against prospective defendants and their insurers. Allocation would be based on a finding of percentage of probable culpability.

---

[35] See, e.g., September 11th Victim Compensation Fund of 2001, Pub.L. 107-42, Title IV, 115 Stat. 237 (2001); National Childhood Vaccine Injury Act of 1986, 42 U.S.C.§§300aa-1 to 34 (2002) (establishing the National Vaccine Injury Compensation Program); Black Lung Benefits Act, 30 U.S.C. § 901 (2002); Price-Anderson Act (Atomic Energy Damages Act), Pub. L. 85-256, 71 Stat. 576 (1957); see also Jack B. Weinstein, *Individual Justice in Mass Tort Litigation: The Effect of Class Actions, Consolidations, and Other Multiparty Devices* 119-21 (1995).

[36] Cf., e.g., *Lorillard Tobacco Co. v. Reilly*, 533 U.S. 525 (2001); *Freier v. Westinghouse Electric Corp.*, 303 F.3d 176 (2d Cir. 2002) (holding that the Comprehensive Environmental Regulation, Compensation, and Liability Act's uniform standard for determining the accrual date for state law personal injury claims arising out of exposure to hazardous substances, although it effectively overrode state statutes of limitations, did not exceed Congress's Commerce Clause Authority).

[37] See Comprehensive Environmental Regulation, Compensation, and Liability Act, 42 U.S.C. §§ 9601-9675 (2002), amended by Superfund Amendments and Reauthorization Act of 1986, Pub. L. No. 99-499, 100 Stat. 1613 (Oct. 17, 1986).

3. Compensation for claimants would be determined according to a social security-like scheme or be based upon a number of current bankruptcy administrative plans, with allocations to be predicated on a reasonable extrapolation from the tort system. Changing conditions would be monitored, permitting continuous adjustments of awards and liability for payment into a fund.
4. The government would guarantee payments in any individual year, up to a set (relatively modest) sum, with total payments out of prospective defendants' payments not to exceed another set sum per year. Since the federal government bears a large responsibility because of its failure to act to protect workers, particularly during World War II, its contribution would be appropriate.
5. Those who are not functionally impaired, those who are impaired primarily because of other exposures, and those who did not suffer occupational exposure should be excluded. The statute of limitations would be tolled until impairment could be shown.
6. The preemptive control of the federal government based upon the Supremacy Clause would trump state law as in tobacco, children's vaccines, and other matters.
7. Attorneys' fees could be based on criteria similar to those used by the Social Security Administration – a 25% maximum of the claimant's award or such other lower figure as the administrative tribunal determines to be appropriate.

Obviously there are many other acceptable legislative solutions. The details set out above are illustrative.

## Screening Programs

A prerequisite to any successful resolution of the asbestos problem is an efficient and effective method for determining those who have an asbestos-related "disease." A comprehensive screening program in and of itself is not objectionable, provided that screening is for real asbestos-caused problems.[38] Spending large sums for widespread medical monitoring tends, except in unusual circumstances, to be wasteful of funds better spent on those who are impaired.[39]

We need more objective tests. We require an opinion from an impartial and reliable national organization concerning better methods for determining asbestos-related injuries with potential alternatives and their feasibility. One

---

[38] Cf. Breast and Cervical Cancer Mortality Prevention Act of 1990, 42 U.S.C. § 300k (2002) (authorizing the grant of funds to provide breast cancer screening for underserved women).

[39] But cf. Pankaj Venugopal, *The Class Certification of Medical Monitoring Claims*, 102 Colum. L. Rev. 1659 (2002) (assuming viability of medical monitoring, opt-out right is unnecessary because of the equitable nature of the remedy under a Rule 23(b)(2) of the Federal Rules of Procedure class action).

possibility is to ask the National Academy of Sciences or a similar national organization with great prestige to undertake a study.

We also need a reliable opinion from such a group on when clinical evidence of asbestos-related exposure such as pleural thickening will become debilitating, the percentage of individuals showing such evidence who eventually will probably become functionally impaired as a result of their exposure to asbestos, and timelines for discovery of future disabilities. Such an understanding would help in a number of different ways. It would aid in making proper awards by existing bankruptcy trusts. It would provide a basis for future state and federal legislation. It would help courts determine liability and damages. It would permit insurers and insured to better plan for future payments.

## Projections of Claims

No rational solution can be adopted without a projection of future diseases and, perhaps, claims. Congress, the courts, industry, insurers, workers, and the public need the most reliable estimates that scientists can provide.

## Conclusion

The time to resolve the asbestos problem and litigation crisis has come and is fast going. We need to now address the failure in our medical-legal-entrepreneurial-political systems to deal with asbestos-caused diseases effectively. In doing so the law and our political and economic systems will have to rely heavily on the pathbreaking work of Eric Stallard, Kenneth G. Manton, and Joel E. Cohen.

Jack B. Weinstein
Senior Judge
U.S. District Court
Eastern District of New York

# Preface

When the Manville asbestos product-liability litigation was settled in December 1994, the question was raised about an appropriate avenue for publishing the scientific analyses conducted for the Court. A follow-up research monograph was required to provide a comprehensive review of epidemiology and modeling involved in the analyses for two reasons. First, substantial prior analyses were not included in the trial record but were necessary for a full understanding of the issues. Second, the efforts at model building and claim forecasting were iterative, with a first report in the summer of 1993 being superceded by a final report in the spring of 1994. The Court record contains only the summary of these changes. The details are contained herein.

In preparing this volume, we set three goals that would help the general reader more easily understand the logic and methods of the analyses we conducted for the court. First, although our analyses were focused on asbestos, the presentation of our methods should be of interest and applicable to forecasting problems in other product-liability areas. Second, our presentation should emphasize the interaction between epidemiology and modeling while recognizing the context of the U.S. legal system and its unique methods for assessing the liability of tort-feasors. Third, our presentation should be scientifically rigorous, yet not so complex and technically demanding that it would fail to appeal to a broad audience.

Throughout the preparation of this volume, we were cognizant of these goals and attempted to broaden the scope of our analyses and assessments to appeal to a diverse audience. In so doing, we were guided by two basic principles. First, because the Manville litigation was among the largest class-action cases in which the Federal Court relied on a panel of neutral experts to develop statistical models to ensure reliable projections of future claims, our task included disseminating the findings in a format accessible to other parties to asbestos litigation, in particular, and to product-liability cases, in general.

Second, because we viewed our models and methods as having applicability beyond product liability, our task included recording the methodology and findings in a format accessible to scientists outside of this area. This, we believe, provides the best opportunity for applying these methods to evaluating effects of occupational and other exposures for both short- and long-term disease outcomes.

Eric Stallard
Kenneth G. Manton
Joel E. Cohen

# Acknowledgements

This project benefited from the efforts of many people. We wish to thank Margaret A. Berger, Chair of the Rule 706 Panel, who brought the project to our attention and consistently encouraged us in our efforts to develop a fully satisfactory forecasting model, and The Honorable Jack B. Weinstein, who oversaw the Rule 706 Panel and reviewed its numerous reports. We acknowledge an intellectual debt to other researchers in this area, especially Alexander M. Walker, Irving J. Selikoff, William J. Nicholson, George Perkel, Herbert Seidman, and Julian Peto. We acknowledge the efforts of the Manville Personal Injury Settlement Trust to provide us with timely data and to ensure accurate interpretation and processing. We especially thank Mark E. Lederer, David T. Austern, and Patricia G. Houser. We acknowledge the efforts of the Resource Planning Corporation, and their willingness to share with us the extensive details of their alternative forecasting model; our thanks go to B. Thomas Florence and Daniel Rourke. We acknowledge the efforts of the other members of the Rule 706 Panel, Burton H. Singer and Alan M. Ducatman, whose oversight and comments ensured the high quality of the intermediate and final reports to the Court. We acknowledge the efforts of the legal representatives of the various parties to the case for reviewing and commenting on various draft reports and for ensuring that our forecasting model was a realistic representation of a complex process. Special thanks go to Mark A. Peterson. We acknowledge the efforts of the legal representative for the Future Claimants in preparing the authors (Stallard and Cohen) for oral court testimony on the forecasts; special thanks go to Leslie G. Fagen, Beth Levine, and Jean H. McMahon. We acknowledge the efforts of our own staff to ensure that this project reached various milestones on time and without error. We especially thank David L. Straley, who did the computer programming, and Betsy A. Smith and Stella Cole, who jointly did the word processing and technical typesetting. We thank Gene R. Lowrimore for his effort in developing the data for the project. We thank Elaine M. McMichael and Sue P. Hicks for able and competent administration of the project. We thank John Kimmel, senior editor of Springer-Verlag, who provided expert guidance throughout the publication process.

The research was supported by the Rule 706 Panel using funds provided on court order by the Manville Personal Injury Settlement Trust. Our ability to conduct this research in an effective and efficient manner greatly benefited from prior and concurrent research support from the National Institute on Aging, through grants R01AG01159 and P01AG17937, and the National Science Foundation, through grants BSR92-07293 and DEB99-81552, as well as from the hospitality (to J.E.C.) of Mr. and Mrs. William T. Golden.

# List of Abbreviations

| Term | Definition |
|---|---|
| AMR | absolute mortality risk |
| AR | attributable risk |
| $AR_H$ | attributable risk in the high-risk cohort |
| ATSDR | Agency for Toxic Substances and Disease Registry |
| BLS | Bureau of Labor Statistics |
| CAGE | age at claim filing |
| CDEC | cancer decrement model |
| CEQ | Council on Environmental Quality |
| CHR | claim hazard rate |
| CMDIS | cancer-among-multiple-diseases criterion |
| CWHS | Continuous Work History Survey |
| d.f. | degrees of freedom |
| DHEW | Department of Health, Education, and Welfare |
| DNA | deoxyribonucleic acid |
| DOFE | date of first exposure |
| EPA | Environmental Protection Agency |
| f | fiber |
| g | gram |
| ICD | International Classification of Diseases |
| ILO | International Labor Office |
| IWE | insulation-worker equivalent |
| kDa | kilodalton — 1000 atomic mass units |
| kg | kilogram |
| MDEC | mesothelioma decrement model |
| MDIS | multiple diseases criterion |
| ml | milliliter |
| MLE | maximum likelihood estimation |
| $\mu$m | micron or micrometer |

| Term | Definition |
|---|---|
| mppcf | million particles per cubic foot |
| MPIST | Manville Personal Injury Settlement Trust |
| NCI | National Cancer Institute |
| NCMDIS | no-cancer-among-multiple-diseases criterion |
| NIEHS | National Institute for Environmental Health Sciences |
| NIOSH | National Institute for Occupational Safety and Health |
| NRC | National Research Council |
| OR | odds ratio |
| OSHA | Occupational Safety and Health Administration |
| PEL | permissible exposure limit |
| Pr | probability |
| POC | proof of claim |
| PTS | propensity to sue |
| RPC | Resource Planning Corporation |
| RR | relative risk |
| SDIS | most severe disease criterion |
| SE | standard error |
| SEER | Surveillance, Epidemiology, and End Results Program |
| SIC | Standard Industrial Classification |
| SMR | standardized mortality ratio |
| SSA | Social Security Administration |
| SV40 | simian virus 40 |
| TDEC | total decrement model |
| TDP | Trust Distribution Process |
| TNCS | Third National Cancer Survey |
| TOFU | time of follow-up |
| TSFE | time since first exposure |
| TSS | time since separation |
| TWA | time-weighted average |
| VADEC | validated asbestos-related disease decrement model |
| VCDEC | validated cancer decrement model |
| VDIS | validated disease criterion |
| VTDEC | TDEC projection with disease counts converted from the SDIS to VDIS criterion using the transition matrix in Table 6.1 |
| WWII | World War II |

# Contents

# 1 Overview

## 1.1 Introduction

This chapter provides background to the Manville asbestos case, an overview of our modeling task, the results, the range of uncertainty, the setting of payout percentages, the need to monitor the future claim process, and the implications of our results for asbestos product liability litigation. The chapter concludes with a discussion of the monograph's organization, indicating the following chapters' contents, and interrelations of the chapters. The organizing principle is that our forecasting model combined elements of two different approaches – one due to Selikoff and his collaborators, and the other due to Walker and his collaborators. In selecting the "best" parts of these two approaches, we rejected other parts. The evidence and rationale for doing so are an integral part of the description of our hybrid model's development.

## 1.2 Asbestos and Health

Asbestos is the commercial name for silicate fibers from the serpentine and amphibole mineral groups – two groups that are distinguished by their crystalline structure (NRC, 1984). The most common forms of asbestos in the United States are chrysotile (a white serpentine), amosite (a brown amphibole), and crocidolite (a blue amphibole). Chrysotile accounts for 90-95% of historical use, and amosite and crocidolite rank a distant second and third. Three other amphiboles – actinolite, anthophyllite, and tremolite – were rarely used in the United States and are generally found as contaminants in other minerals.

Important physicochemical properties of asbestos include its high tensile strength, flexibility, resistance to heat and fire, corrosion resistance, resistance to chemical degradation, and high electrical resistance. These properties made asbestos commercially important and ideally suited for eventual use in over 5000 products including cement pipes, flooring, friction and roofing products, cement sheets, packing and gaskets, and coatings and compounds (NCI, 1996). Unfortunately, other related physicochemical properties of asbestos make it dangerous to human health. Asbestos is recognized as a causal agent in several long-term chronic disease processes, including mesothelioma, lung cancer, gastrointestinal cancer, asbestosis, and pleural plaques/thickening.

The primary pathway for human asbestos exposure is through the inhalation of airborne asbestos dust fibers. The inhalation process includes deposition of fibers in the respiratory tract, retention of fibers at critical locations in the respiratory airways, and clearance and transportation of the fibers to other organs – upward toward the mouth through normal mucus clearance mechanisms, leading to a secondary ingestion pathway that could account for laryngeal, pharyngeal, esophageal, stomach, and colon/rectum cancers, or outward through the lung tissue or through the lymphatic channels, which could account for the pleural diseases. Mineral fiber content analyses of tissue samples from patients, by disease type and occupation, provide strong evidence that amphibole fibers exhibit significantly greater persistence than chrysotile fibers in the lung and suggestive evidence that a significant fraction of chrysotile fibers migrate to the pleura, where they accumulate (Roggli et al., 1992b; Smith and Wright, 1996).

The central role of the lung and pleura in this process is evidenced by the finding that these sites manifest both neoplastic and non-neoplastic asbestos-associated diseases. Asbestosis is a non-neoplastic fibrotic lung disease, frequently asymptomatic, that may produce dyspnea (labored breathing) on exertion or at rest, dry nonproductive coughing, crackles, digital clubbing (widening/thickening of ends of fingers and toes), cyanosis (discoloration of skin/mucous membranes), and tachypnea (rapid respiration) (Roggli and Pratt, 1992). Radiographically, the disease is characterized by small irregular opacities in the lower lung. The presence of asbestos fibers and asbestos bodies (iron-coated asbestos fibers) in the lung can help distinguish asbestosis from similar nonasbestos-related diseases such as idiopathic interstitial fibrosis, interstitial pneumonia, and fibrosing alveolitis (Garrard et al., 1992).

Pleural plaques and pleural thickening are distinct benign manifestations of asbestos exposure (Greenberg, 1992). The pleurae are the double-walled serous membranes that enclose the lungs. The right and left pleurae are entirely distinct from each other, as are the right and left lungs. Pleural plaques are areas of dense elevated fibrous tissue, generally found on the outer pleural wall (parietal pleura), generally asymptomatic, often calcified, with diameters up to 12 cm. Pleural plaques may be due to infections, trauma, blood in the chest, and empyema (pus), in addition to asbestos exposure (Garrard et al., 1992). Pleural thickening includes diffuse pleural fibrosis and rounded atelectasis (folded lung syndrome), both of which may be asymptomatic, but may also result in reduced vital capacity. The former may be due to rheumatoid arthritis and systemic lupus erythematosus, in addition to asbestos exposure. Benign asbestos-related pleural effusion is a form of pleurisy that may occur alone or in combination with pleural plaques/thickening.

Lung cancer (carcinoma of the lung) and mesothelioma are the two most frequent and serious neoplastic diseases associated with asbestos exposure. Both diseases are generally lethal and resistant to treatment. Lung cancer includes proximal bronchogenic carcinoma and peripheral carcinoma of the small airways. Currently, 85-90% of lung cancers in the United States are attributable to cigarette smoking; only about 2% are attributable to asbestos exposure (Roggli et al., 1992b). Relative risks for lung cancer mortality as functions of cigarette smoking and asbestos exposure levels exhibit independent and multiplicative dose-response patterns. In the general population of cigarette smokers, the primary cancer occurs in an upper lobe of the lung about two-thirds of the time, whereas among asbestos workers, the primary cancer occurs in a lower lobe of the lung about two-thirds of the time. However, there are no distinctive pathological, morphological, or histological features of lung cancer that permit the causative agent to be unambiguously identified as tobacco or asbestos (Greenberg and Roggli, 1992).

Malignant mesothelioma is a cancer of the mesothelium, the single layer of flat cells that forms the surfaces of each of the body's three major serous membranes: the pleura (which lines the lungs), the peritoneum (which lines the abdominal viscera), and the pericardium (which lines the heart and associated vessels). A benign form of mesothelioma, not asbestos related, exists but will not be considered further in this monograph. Human malignant mesothelioma (henceforth referred to as "mesothelioma") is primarily due to asbestos exposure (Spirtas et al., 1994). In contrast to lung cancer, cigarette smoking is not a risk factor for mesothelioma. Other potential risk factors include erionite mineral fibers, radiation, chronic inflammation/scarring, chemical carcinogens, viruses (e.g., SV40), chronic empyema, therapeutic pneumothorax, and hereditary predisposition (Roggli et al., 1992c).

The most common site of mesothelioma is the pleura, accounting for about 90% of cases, followed by the peritoneum, accounting for most of the remaining 10% of cases; however, it has been hypothesized that these proportions may change with changes in exposure intensity and duration (Roggli et al., 1992c). Primary pericardial mesothelioma is rare and may be difficult to distinguish from pleural mesothelioma. In turn, pleural mesothelioma may be difficult to distinguish from other common tumors, including adenocarcinoma metastases from the lung, breast, stomach, colon, or other primary sites (Garrard et al., 1992). Symptoms of pleural mesothelioma include chest pain, persistent cough, dyspnea, weight loss, pleural effusion, and digital clubbing. The tumor generally spreads over the surface of the lung, eventually encasing it (Roggli et al., 1992c).

In a similar manner, peritoneal mesothelioma generally spreads over the surface of the abdominal viscera, eventually encasing the abdominal organs. Peritoneal mesothelioma may be difficult to distinguish from metastatic carcinoma from other primary sites such as the stomach, pancreas, and ovaries, papillary tumors in women, and reactive mesothelial hyperplasia (Roggli et al., 1992c). Symptoms of peritoneal mesothelioma include abdominal pain, weight loss, abdominal distension, constipation, and ascites (effusion of serous fluids). The prognosis for all forms of mesothelioma is extremely poor and survival times of 7-15 months beyond diagnosis are typical (Roggli et al., 1992c).

Currently, the Manville Trust recognizes asbestos-related disease claims for mesothelioma, lung cancer, colon/rectum cancer, stomach cancer, laryngeal, pharyngeal, and esophageal cancers, asbestosis, and pleural plaques/thickening (Weinstein, 2002). Additional information on the pathology of these diseases can be found in the volume by Roggli et al. (1992a).

## 1.3 History of Asbestos

Asbestos has been used since ancient times. The earliest written reference to asbestos was by Theophrastus, a student of Aristotle. Later references describe uses of asbestos across the centuries by such historical figures as Pliny the Elder, Charlemagne, Marco Polo, and Benjamin Franklin (Alleman and Mossman, 1997).

Georgius Agricola published the first scientific study of asbestos in the 16th century. A series of scientific papers on asbestos was published in England by the Royal Society during the period 1660-1700. The first full scientific volume on asbestos was published in 1727 by the German mineralogist Franz Bruckmann (Alleman and Mossman, 1997). Modern treatments of this subject can be found in Michael and Chissick (1979) and the references therein.

The first U.S. patent for an asbestos product was issued in 1828 for insulation material in steam engines. This was followed by a series of U.K. and U.S. patents over the period 1834-1885 for asbestos products for safes, lubricants, fireboxes, electrical insulation, roofing felt, construction boards, and food filters. Eventually, thousands of products were based on innovative applications of asbestos (NCI, 1996).

Mining of substantial amounts of chrysotile began in Quebec in 1878, crocidolite in South Africa in 1910, and amosite in South Africa in 1916 (Roggli and Coin, 1992). Today, Russia, China, and Canada are the world's leading producers of asbestos (Virta, 2001).

The rise and fall of the asbestos industry in the United States occurred over a period of more than 100 years. Data from the U.S. Bureau of Mines and the U.S. Geological Survey indicate that the growth of asbestos consumption began in 1895 (NRC, 1984), with annual consumption reaching 46,000 metric tons in 1910 and 209,000 metric tons in 1925, dropping below 90,000 metric tons in 1932 during the Great Depression, and then resuming a period of rapid increase during WWII and thereafter, reaching 613,000 metric tons in 1948 and 723,000 metric tons in 1951 (Buckingham and Virta, 2002). This was followed by a period of stability in the 1950s and slow growth through the 1960s, with annual consumption peaking at 803,000 metric tons in 1973, at which time evidence of the adverse health consequences of asbestos was beginning to lead to increased regulation and searches for safer substitute materials (Buckingham and Virta, 2002). This was followed by periods of gradual and then precipitous declines, with annual consumption dropping from 619,000 metric tons in 1978 to 356,000 metric tons in 1980, 247,000 metric tons in 1982, 41,000 metric tons in 1990, 22,000 metric tons in 1995, and 15,000 metric tons in 2000 (Buckingham and Virta, 2002).

As of 2003, asbestos use in the United States continued in a highly regulated environment at levels far below those of 1910. Furthermore, most major manufacturers and users of asbestos products have been subjected to large volumes of lawsuits for personal injuries attributable to the adverse health effects of asbestos. Asbestos-related bankruptcy filings in the United States accelerated from 16 in the 1980s and 18 in the 1990s to 28 from January 2000 through December 2002 (Biggs, 2003).

The health hazards of asbestos are recognized outside the United States and the substance has been subjected to increasing levels of control and regulation. For example, 21 European countries, including France, Germany, Italy, and the United Kingdom, have imposed a near-total ban on the use of asbestos, allowing exceptions only where there is no safe substitute.

## 1.4 Epidemiological Discovery

The epidemiological databases and analyses supporting our current understanding of the health risks associated with asbestos fibers were developed over a period of 80 years. The earliest reported death due to an asbestos-related disease occurred in London in 1900 in a man whose asbestos exposure occurred over a 12-year period (Murray, 1907). The first case of asbestosis reported in the medical literature was in Cooke (1924, 1927). The first cases of asbestos-related pleural plaques were reported by Sparks (1931). The first cases of asbestos-related lung cancer were reported in 1935 (Lynch and Smith, 1935; Gloyne, 1935) and the first cases of asbestos-related malignant mesotheliomas were reported in the early 1950s (Weiss, 1953; Leicher, 1954). While U.S. companies were doing nothing to protect American workers against the health risks of asbestos exposure, Nazi Germany mounted a campaign against asbestos and included lung cancer as a compensable occupational disease of asbestos workers (Proctor, 1999, pp. 107-113). Quantitative epidemiological analyses of excess asbestos-related lung cancer risk were conducted for asbestos factory workers by Doll (1955) and for asbestos insulation workers by Selikoff et al. (1964). The connection between asbestos exposure and malignant mesothelioma was firmly established by Wagner et al. (1960). These initial studies have been validated in dozens of follow-up studies (NRC, 1984; EPA, 1986).

Other cancer sites have also been identified. The first cases of asbestos-related gastrointestinal cancer (i.e., primary cancer of the esophagus, stomach, colon, or rectum) were reported by Selikoff et al. (1964), based on their epidemiological analyses of relative mortality risks among asbestos insulation workers. Selikoff et al. (1979) reported increased cancer death rates among asbestos insulation workers for the larynx, buccal cavity, oropharynx, and kidney. Selikoff and Seidman (1991) indicated that gallbladder/bile ducts and pancreas should be added to this list. However, Greenberg and Roggli (1992), after reviewing results from other epidemiological and experimental studies, concluded that the existing evidence for an association with asbestos was inconclusive for cancers of the pancreas, stomach, colon, and rectum, and was negative for cancers of the kidney.

## 1.5 Johns-Manville Corporation

The rise and fall of the Johns-Manville Corporation closely tracks that of the asbestos industry in the United States (Macchiarola, 1996). Founded in 1858 as the H.W. Johns Manufacturing Company, its main product was fire-resistant asbestos roofing material. The Johns-Manville Corporation was established in 1901 when Charles Manville bought the business from the Johns family, 3 years after H.W. Johns had died from pulmonary fibrosis - an unrecognized result of asbestos exposure. By 1925, the company was producing over 200 different asbestos products. The company grew to become the world's largest manufacturer of asbestos, earning a Fortune 500 listing. Johns-Manville's share of the asbestos products market has been estimated in the range 25-40%, far exceeding that of major competitors Owens Corning (10-15%), Owens-Illinois (5%), and Armstrong World (<5%) (Hersch, 1992, p. 17). Of these four companies, only Owens-Illinois remained solvent in 2003.

Johns-Manville executives were initially informed of health hazards associated with asbestos in 1924 at the time of Cooke's first paper connecting asbestos and asbestosis. In the 1930s, Johns-Manville quietly settled a number of asbestos-related personal injury lawsuits (Macchiarola, 1996). Following Selikoff et al.'s (1964) report, a new stream of lawsuits against Johns-Manville began, growing slowly at first, with 159 cases filed in 1976, and then more rapidly, with about 6000 cases filed in 1982. At the time Johns-Manville filed for bankruptcy protection under Chapter 11 of the U.S. Bankruptcy Code in August 1982, the company had settled 3570 lawsuits at an average cost of about $20,000 per claim; yet nearly 17,000 claims remained to be settled and thousands more were being filed each year. The company's gross assets were worth $2.25 billion; its net worth was $830 million (Macchiarola, 1996).

## 1.6 Manville Trust

The Manville Plan of Reorganization under Chapter 11 was authorized by Judge Burton R. Lifland (Chief Bankruptcy Judge, Southern District of New York) in December 1986 and given final approval by the U.S. Court of Appeals (Second Circuit) in October 1988. This plan established two distinct trusts, the first to provide compensation for personal injuries (the Manville Personal Injury Settlement Trust – MPIST) and the second to provide compensation for property damage (the Manville Property Damage Settlement Trust). Under the Plan of Reorganization, claims could be filed against either trust but not against the reorganized (and renamed) Manville Corporation.

The personal injury trust had assets in excess of $2 billion and an 80% ownership interest in the Manville Corporation. In contrast, the property damage trust was relatively small, with assets of about $125 million. Our interest in this monograph will be restricted to the experience of the personal injury trust (henceforth referred to as the "Trust" or the "Manville Trust," for brevity).

At the time of its inception in November 1988, the Trust had a backlog of nearly 17,000 claims filed prior to the Johns-Manville bankruptcy filing in August 1982 and a projected total of 83,000-100,000 new claims to be filed over the life of the Trust. The Plan of Reorganization required that these 100,000-117,000 claims be paid on a first-in/first-out basis at 100% of the liquidated value of the victims' claims (Macchiarola, 1996).

By the end of 1989, the Trust had received 140,000 claims and new claims were arriving at a rate of 17,000 per year. Not only had the existing number of claims already exceeded the upper bound of the projected total for the life of the Trust, but the settlement costs were averaging $41,150 per claim – 65% higher than the $25,000 cost assumed in the projections (Macchiarola, 1996). The Trust was running out of cash, the Trust Plan was failing, and the Trust had been named in 89,000 new lawsuits and tens of thousands more lawsuits were expected. Given these circumstances and its inability to carry out its mandate in an orderly manner, the Trust petitioned the Court for legal protection.

## 1.7 Manville Trust Litigation

In July 1990, jurisdiction over the Trust was assigned to Judge Jack B. Weinstein (Senior U.S. District Judge, Eastern District of New York), who issued a stay temporarily suspending all Trust payments, except those for exigent health circumstances or extreme economic hardships. Testimony in early November 1990 by Special Master Marvin E. Frankel indicated that the Trust was "deeply insolvent" (Weinstein, 1994). Shortly thereafter, on November 19, 1990, a class-action suit against the Trust, captioned *Findley v. Blinken*, was filed on behalf of all Trust beneficiaries jointly in the Eastern and Southern Districts of New York under the jurisdiction of Judges Weinstein and Lifland. The suit sought a restructuring of the Trust and a fair and equitable distribution of its assets among its beneficiaries. On February 13, 1991, the suit was certified as a mandatory non–opt-out class action under Rule 23(b)(1)(B) of the Federal Rules of Civil Procedure.

Rule 23(b)(1)(B), governing class actions, provides that:

> An action may be maintained as a class action if the prerequisites of [Rule 23 (a)] are satisfied, and in addition: (1) the prosecution of separate actions by or against individual members of the class would create a risk of adjudications with respect to individual members of the class which would as a practical matter be dispositive of the interests of the other members not parties to the adjudications or substantially impair or impede their ability to protect their interests.

The prerequisites of Rule 23(a) are as follows:

> One or more members of a class action may sue or be sued as representative parties on behalf of all only if:
> (1) the class is so numerous that joinder of all members is impracticable,
> (2) there are questions of law or fact common to the class,
> (3) the claims or defenses of the representative parties are typical of the claims or defenses of the class, and
> (4) the representative parties will fairly and adequately protect the interests of the class.

The case was settled, appealed, remanded, and ultimately refiled as an amended complaint, captioned *Findley v. Falise*, on October 8, 1993. Trial commenced on March 15, 1994 and the proceedings continued through July 25, 1994, at which time, a negotiated settlement was reached among the Trust and the six certified subclasses:

1. Present claimants
2. Future claimants
3. Claimants with settlements and judgments dated prior to November 19, 1990
4. Codefendant asbestos manufacturers
5. Manville asbestos distributors
6. The MacArthur Company (a former Manville distributor)

Pursuant to Rule 706 of the Federal Rules of Evidence, in April 1991, and on the recommendation of Professor Margaret A. Berger, Judge Weinstein ordered that a panel of independent neutral experts be assembled to develop a statistical model that would ensure reliable projections of the number, timing, and nature of future claims against the Trust. Such projections were needed to ensure that a fair, adequate, and equitable distribution of the Trust's limited assets could be developed that would balance the interests of all present and future claimants.

Rule 706, governing the use of court appointed experts, states:

> The court may on its own motion or on the motion of any party enter an order to show cause why expert witnesses should not be appointed, and may request the parties to submit nominations. The court may

appoint any expert witnesses agreed upon by the parties, and may appoint expert witnesses of its own selection. An expert witness shall not be appointed by the court unless the witness consents to act. A witness so appointed shall be informed of the witness' duties by the court in writing, a copy of which shall be filed with the clerk, or at a conference in which the parties shall have opportunity to participate. A witness so appointed shall advise the parties of the witness' findings, if any; the witness' deposition may be taken by any party; and the witness may be called to testify by the court or any party. The witness shall be subject to cross-examination by each party, including a party calling the witness.

The members of the Rule 706 Panel were the following:

Margaret A. Berger, at that time Associate Dean of Brooklyn Law School, Head of Panel

Joel E. Cohen, Professor of Populations, Rockefeller University

Alan M. Ducatman, Professor of Medicine, West Virginia University

Kenneth G. Manton, Research Professor of Demographic Studies, Duke University

Burton H. Singer, at that time Professor of Epidemiology and Public Health, Yale University

Eric Stallard, at that time Associate Research Professor of Demographic Studies, Duke University

The panel's projections were developed by Stallard and Manton (1993, 1994), with input and review by Cohen. Further internal peer review and comments on the epidemiological assumptions and statistical/demographic methods employed were provided by Ducatman and Singer. Additional reviews and comments on the methods and assumptions were provided by outside experts advising parties to the Trust class-action suit. Following the final settlement of the case on December 15, 1994, it was agreed that Stallard, Manton, and Cohen would be responsible for the present research monograph.

## 1.8 Project History

From 1982 to 1984, Cohen conducted an independent reconstruction and assessment of uncertainty in Walker's (1982, edited slightly and reissued as Walker et al., 1983) model, under a request from the Committee of Unsecured Creditors in the Manville bankruptcy case. The results of that analysis were contained in an unpublished report (Cohen et al., 1984).

In 1983, the Congressional Research Service of the Library of Congress asked Manton to review projections of asbestos-related disease prepared by Walker (1982) and Selikoff (1981, reissued in 1982). As part of that effort (see Manton, 1983), Stallard implemented a computer program to replicate as closely as possible Walker's projections of U.S. mesothelioma incidence. A number of alternative projections were also run to test the sensitivity of Walker's model to various assumptions.

In 1991, at the request of Margaret Berger, the Head of the Rule 706 Panel, Stallard and Manton agreed to work with Cohen to evaluate, revise, and extend their projection model and software to project the number, timing, and nature of future claims against the Trust using data that had become available since the time of their first report. Most of these new data came from two sources: (1) the claim experience of the Trust recorded over its several years of operation and (2) a national cancer registry system maintained by the National Cancer Institute. After an extensive review of the epidemiological literature, it was determined that by updating the data and assumptions used in the earlier projections, new projections could be made that improved upon Walker's, Selikoff's, and Stallard and Manton's own prior projections.

In 1992, Stallard and Manton revised and extended the computer model and software. Those efforts were reviewed in Stallard and Manton (1993). Intermediate versions of the projection model, software specifications, and drafts of that report were examined by the other members of the Rule 706 Panel. The report and a separate executive summary (collectively cited as Stallard and Manton, 1993) were presented by Stallard and Cohen at a meeting of all parties in *Findley v. Blinken* on September 7, 1993 at the Trust offices in New York.

Stallard and Manton (1993) presented 45 different projections designed to respond to questions raised at various meetings and in written comments submitted by Rule 706 Panel members and other reviewers advising the Court and the Trust. Although it was not feasible to deal with all permutations of assumptions in the computer model, it was possible to respond to most questions. Limitations were reviewed at the September 7, 1993 meeting. Some projection alternatives were included to explore the range of sensitivity of the model assumptions – even though they represented scenarios judged to be highly unlikely.

On September 23, 1993, the Resource Planning Corporation (RPC), the technical advisor to the Trust in its own claim forecasts, issued its report (RPC, 1993). Preliminary results had been presented by RPC at meetings open to the Rule 706 Panel members. Although RPC's purpose was the same as ours, their projection model was substantially different. Consequently, we compared results to identify areas of agreement and disagreement and to gain insight into the specific assumptions responsible for the different results.

Following the September 7, 1993 meeting, Cohen asked Stallard and Manton to further evaluate the joint effects of occupation and date of first exposure to asbestos on the projections. Because different occupations had different types, levels, timings, and durations of exposure to asbestos, there was concern that stratification by occupation might project substantially different outcomes. Using additional stratifications also provided additional tests of the established uncertainty bounds for the projections. Although the new projections were within the established bounds, there was an unexpected significant downward shift in the preferred estimate of the total future claims.

These projections were presented in Stallard and Manton (1994), which was entered into the Court record during Stallard's testimony on March 15, 1994. Because of the prominent role of occupation and date of first exposure, these projections provided the framework that combined the approaches of Walker and Selikoff. The hybrid model depended critically on the assumption that the Trust's claim experience over the selected calibration period (1990-1992) was a scaled-down version of the national mesothelioma experience; that is, the calibration period was assumed to be characterized by a stable rate at which injured persons decide to file a claim against the Trust for asbestos-related damages.

## 1.9 Results

Our research focused on claims against the Trust arising from the manufacturing of and/or exposure to products containing asbestos produced by the Johns-Manville Corporation. Because Johns-Manville was the largest asbestos producer – with 25-40% market share – it was generally believed that the Manville Trust claim projections serve as a reasonable proxy for industrywide claims (Hersch, 1992). This was important because our projections, which were based on the most recent claim data, were substantially higher than most prior projections. For example, when the Trust was created in 1986, the projected total number of claims (for the period 1988 to 2050) ranged from 83,000 to 100,000 (Weinstein, 1994, p. 20). In fact, the upper limit of this range (100,000) was surpassed by claims filed before November 1989. In January 1992, Lehman Brothers projected that there would be 210,000 future claims filed industrywide, in addition to their estimated 210,000 claims filed up to that date, based on Selikoff's results (Hersch, 1992). Our count of the number of valid claims filed through 1991 against the Manville Trust alone was 16.5% lower (i.e., 175,000), but our preferred projection of the number of future claims against the Manville Trust alone was about 63% higher than the 1992 Lehman Brothers projection for the entire industry.

Forecasts of the timing and nature of claims are as important as forecasts of the total number of claims in determining the fiscal liability they represent. Different types of disease are awarded different monetary amounts. Because the Trust was insufficiently funded to carry out its mandate, a ruling was issued in July 1990 preventing any additional payments by the Trust until a fair, adequate, and equitable method of compensating bona fide beneficiaries could be implemented that reflected the interests of current and future claimants. Such a distribution mechanism was approved in December 1994, with the Trust paying claims at a pro rata rate of $0.10 on the dollar (Weinstein, 1994). This rate was believed to be conservative and could have been increased to as high as $0.134 in the future, with supplemental payments for prior settlements.

By the middle of the 21st century, we projected that the Trust would accumulate 517,000 claims for asbestos-related disease/injury, due to exposures occurring before 1975. About 194,000 claims were filed by the end of 1992. The remaining 323,000 were projected to be filed at the rate of 18,000 per year in 1993-1994, 16,000 per year in 1995-1999, and continuing at a declining rate throughout the first half of the 21st century. At an average estimated total cost of $120,000 per claim, these projections implied a total liability of about $62 billion; if legal defense fees were added, the total cost could approach $75-90 billion. The Trust's share was projected at $34.4 billion, with a present value in 1994 of about $16.0 billion, under our preferred projection (Lederer, 1994). With assets valued in the range $1.8-2.5 billion in 1994, it was obvious that the Trust could not pay 100% of its share. These considerations supported the decision to set the initial pro rata rate at $0.10 per dollar.

The Trust settlement recognized the need to monitor future claim experience. At least every 3 years, and as often as necessary, the Trust had to reestimate the value of its total assets and its total liabilities to determine if the pro rata distribution rate should be revised. By comparing the actual versus projected number and nature of claims early in the forecast period, the Trust could ensure that it would not be subject to an unanticipated rapid depletion of its assets. Thus, the generation of a set of projections was not a once-and-for-all undertaking.

The pro rata distribution rate was held fixed at $0.10 per dollar through the first 6 years following the settlement. However, by early 2001, it became apparent that changes in the external litigation environment were highly unfavorable. On June 19, 2001, the Trust decided to reduce the rate to $0.05 per dollar (Austern, 2001). On August 28, 2002, the Trust announced that it was implementing a revised payment process that addressed the changes in the numbers and types of claims being filed.

Changes in the external asbestos litigation environment had implications that reached far beyond the Johns-Manville case.

At the end of 1982 (the year of the Johns-Manville bankruptcy), about 300 U.S. companies had been named as defendants in asbestos-related personal injury claims filed by 21,000 distinct claimants; by the end of 2000, about 6000 U.S. companies had been named in similar claims filed by about 600,000 claimants; and by the end of 2002, about 8000 U.S. companies had been named in similar claims filed by about 750,000 claimants (Carroll et al., 2002, p. 51; Associated Press, 2003).

By the end of 2000, at least one company from each major U.S. industry was a defendant in asbestos litigation (Carroll et al., 2002, p. 49).

The number of named defendant companies increased from an average of 20 per claimant in the early 1980s to 60-70 per claimant by the mid-1990s (Carroll et al., 2002, p. 41).

At the end of 2002, about 250,000 claims were pending in the federal and state court systems (Netherton and Harras, 2003).

At the end of 1982, four companies had filed for bankruptcy due to asbestos claims. By the end of 1994, this number had increased to 27-29 companies; by the end of 2000, to 43 companies; and by the end of 2002, to 65 companies (Biggs, 2003). Carroll and Hensler (2003) reported similar but not identical counts of bankrupt companies, with a cumulative total of 67 companies by the end of 2002. Both reports were constructed so that a bankrupt parent corporation and all bankrupt subsidiaries of the same parent corporation would be tallied only once in the count.

Some companies tried to avoid bankruptcy by negotiated settlements with legal representatives of present and future claimants. However, the U.S. Supreme Court (1997, 1999) rejected negotiated global settlements in two high-profile cases (*Amchem Products Inc. v. Windsor*, June 25, 1997; and *Ortiz v. Fibreboard Corp.*, June 23, 1999), stating that "this litigation defies customary judicial administration and calls for national legislation."

Bills to establish such uniform national procedures for asbestos-related personal injury claims were introduced in the U.S. Congress in May 1998 and again in March 1999, with public hearings in July and October 1999. An amended form of the House bill was reported back to the House of Representatives in July 2000 with little chance of further consideration.

Hearings were restarted in the Senate in September 2002 and March 2003. During February-May 2003, five separate bills were introduced in Congress to provide for "fair and efficient" resolution of asbestos-related personal injury claims. Although the details of these bills differ substantially, the chance of resolution of asbestos litigation through national legislation increased as these bills were considered.

The legal principles under which liability is determined in the United States permit part of the liability of insolvent defendants to be assigned to other defendants. Thus, the upward revision of the total liability of any one insolvent defendant implies a downward revision of their payout rate and an upward revision of the liability of the remaining solvent defendants. To the extent that Manville Trust claims are a proxy for industrywide claims, our projections implied an upward revision in the number of claims against every other asbestos defendant and, more importantly, an upward revision to both the number of claims and the share of liability for all solvent defendants. This effect could be magnified in future years given the recent increases in claim filings against the Trust and the increasing frequency at which solvent asbestos defendants are currently seeking Chapter 11 bankruptcy protection.

Given the number and cost of claims projected, the number of companies affected, and the long time period during which claims will be filed, similar analyses will most probably be needed for other asbestos defendants or for other nonjudicial approaches to asbestos compensation.

Our models are most directly applicable to the forecasting needs of companies currently undergoing reorganization under Chapter 11 of the U.S. Bankruptcy Code, especially in cases where a trust similar to the Manville Trust, as described in Section 524(g)(2)(B) of the U.S. Bankruptcy Code, is established. However, our models can also be used by solvent companies to estimate their future liabilities and establish reserve funds to meet those claims. Our analyses supported the conclusion that the Manville Trust claim experience was a scaled-down version of the entire asbestos industry, so that our models can also be adapted to forecast industrywide asbestos liability.

## 1.10 Organization of Monograph

The monograph has 10 chapters. Chapters 2-5 provide background on asbestos epidemiology and modeling that supports the main analyses of the Manville Trust data in Chapters 6-9.

Chapter 2 reviews epidemiologic studies of the health risks of asbestos. Chapters 3 and 4 present detailed assessments of the models used by Selikoff (1981) and Walker (1982) – models that provide conceptual building blocks for our work. In both models, the disease incidence functions are derived from the same epidemiologic analyses of North American insulation workers. The projections in Chapter 3 are based on direct estimates of past exposure to asbestos, by age, date, and occupation. The projections in Chapter 4 are based on indirect estimates of past exposure by age and date, but not occupation. The latter projections were the basis of the original settlement of the Johns-Manville bankruptcy case, in part because Walker included asbestosis in his projections, whereas Selikoff did not. Chapter 5 presents sensitivity analyses of Walker's projections and their evidentiary bases, as reviewed in Manton (1983) and Cohen et al. (1984).

Chapter 6 presents a projection model based on indirect estimation of past exposure which performed well on the more reliable and more extensive claim database from the Manville Trust available to the Rule 706 Panel. This model was the basis of Stallard and Manton's (1993) report to the Court. Chapter 7 presents sensitivity analyses of this model and establishes plausible bounds to the uncertainty of the projections.

A limitation of the indirect estimation model is its inability to deal explicitly with depletion of the population eligible to file a claim due to prior claim filing. If the eligible population is large compared to the number with prior claims (say, at least 20 times larger), then this effect likely can be ignored. Otherwise, alternative models should be considered. Chapters 8 and 9 present the results of a hybrid model that uses elements of both the direct and indirect estimation approaches. The impact of depletion of the eligible claim population was unexpectedly large and resulted in a significant downward shift in the preferred estimate of the total future claims. This model generated the projections that the Rule 706 Panel found most credible and was the basis of Stallard and Manton's (1994) report to the Court. The model employed indirect estimation of past exposure for occupational groups defined by Selikoff, using his estimates of relative risk and average duration of exposure for those groups. A hybrid model could be estimated because information on occupation was recorded in the Manville Trust files for individuals filing a claim. The analyses in Section 6.8 showed that claim projections based on the Trust claims for mesothelioma are essentially the same as claim projections based on national estimates of mesothelioma incidence, using cancer registry data collected by the National Cancer Institute. The sensitivity analyses in Chapter 9 are similar to those in Chapter 7 – primarily scenario testing with parameters set at their plausible extremes.

Chapter 10 compares the projected Trust experience for the periods 1990-1994 and 1995-1999 with the observed experience. These comparisons provide further insight into the modeling and forecasting, given that 5 years was sufficient for the vagaries of the claim filing and Trust distribution processes to emerge. Consideration is given to issues involved in new applications of these methods and results in forecasts of asbestos-related injuries, including the impact of recent changes in the external litigation environment.

Throughout the monograph, we stress the iterative nature of model building and the uncertainty generated by lack of complete knowledge of the injury and litigation processes.

# 2

# Epidemiology of Asbestos-Related Diseases

## 2.1 Introduction

Forecasts of asbestos-related diseases typically rely on epidemiological studies that establish the connection between asbestos exposure in the workplace and disease. These studies report increased risks of cancer among workers who have been exposed to asbestos. In particular, lung cancer and mesothelioma risks are increased. Asbestos workers are also at risk of contracting noncancer diseases such as asbestosis, a pulmonary disease characterized by fibrosis and caused by protracted inhalation of asbestos particles. Experimental animal studies have described the physiological mechanisms that account for the relationship between asbestos exposure and these illnesses (Roggli and Brody, 1992).

The Occupational Safety and Health Administration (OSHA) began setting permissible exposure limits (PELs) on the amount of asbestos in the workplace environment in 1971. In May 1971, the PEL was set at 12 fibers per milliliter (f/ml). In December 1971, this was reduced to 10 f/ml, with an 8-hour time-weighted average (TWA) PEL of 5 f/ml. In July 1976, the 8-hour TWA PEL was reduced to 2 f/ml; in July 1986, to 0.2 f/ml; and in October 1995, to 0.1 f/ml – the current PEL. In conducting air monitoring under these standards, OSHA (e.g., 1986, p. 22,739) mandated that asbestos exposure samples must be collected on mixed cellulose ester filter membranes, that fiber counts must be made by positive phase-contrast optical microscopy at a total magnification of 400×, and that the count must include fibers with a length of 5 $\mu$m or greater and an aspect ratio (length-to-width ratio) of 3:1 or greater.

Environmental studies have established that historical workplace exposure concentrations of airborne asbestos fibers for many workers exposed to asbestos were 1000-100,000 times higher than the nonoccupational or environmental exposures faced by the general population (EPA, 1986, p. 162). This differential explains why most epidemiological studies of asbestos-related diseases focus on or identify workers with high levels of asbestos exposure and why most claims against the Manville Trust and other asbestos defendants are based on occupational exposures. Such claims are generally limited to occupational exposures because the proof of claim must identify exposure to the defendant's asbestos products and this is most easily done for the workplace environment where specific brand-name asbestos products were well known to the workers. In contrast, in the case of disease due to environmental exposures to low levels of asbestos in the ambient air, it would be difficult to identify Johns-Manville or any other asbestos defendant as the source, and the low levels of asbestos fiber content in the lungs following such exposures would make it difficult to confirm that asbestos was the causal agent.

These considerations lead us to expect occupational exposures to account for virtually all of the claims against the Manville Trust. Thus, forecasting the number, timing, and nature of future claims against the Trust requires that we can forecast these same factors for persons who were exposed to asbestos in the workplace, and this requires a firm understanding of the epidemiology of asbestos-related diseases.

In this chapter, we examine a range of epidemiological studies, including those used by Walker (1982) and Selikoff (1981, reissued in 1982) in their projections. This review will be conducted in five parts. First, we discuss design and data quality issues that are specific to epidemiological studies of the occupational health hazards of asbestos. Second, we review studies of health risks of occupational exposures to asbestos. Third, we examine the variation in estimates of the relation of disease to the level of asbestos exposure produced in different studies. Fourth, we consider evidence on the effects of different types of asbestos fiber on different disease risks. Fifth, we consider evidence on the potential role of simian virus 40 as a causative agent and cocarcinogen with asbestos in inducing human mesothelioma.

## 2.2 Design Issues in Studying Occupational Exposure

There are two main types of epidemiological study: the prospective cohort study and the retrospective case-control study. The interpretation of the results of a specific study requires that we know whether the study is of the cohort or case-control type and are aware of issues in applying each type of design to the health outcomes, exposure factors, and populations of interest (Liddell et al., 1977). The use of epidemiologic data in projections must be consistent with the properties of data determined by the study design and the particular characteristics of the study population and its exposure.

The first design involves collecting data prospectively on a cohort of workers followed over a period of time. The essential characteristic of this design is that a group of persons (the "cohort") is identified on the basis of some exposure of interest and followed to determine when and how many of them become ill.

In the retrospective case-control design, the researcher starts with a group of people who are afflicted by the disease (the "cases"). A second independent group is also selected from the population of persons who do not manifest the disease. This second group is selected to match certain characteristics of the cases. Consequently, it is referred to as the "control" population. Typically, cases and controls are "matched" on the basis of selected factors (e.g., age, sex, and smoking) to account for the effects of the variables used in matching. The goal is to identify differences in the distribution of exposure factors between those with and without the disease. People who do and do not have the disease are compared to see if the group with the disease has a higher exposure to a suspected cause even when other factors are taken into account.

### 2.2.1 Measures of Risk

The underlying logic of the two approaches can be clarified with a simple numerical example. Consider the following two-way table generated from prospective follow-up of two cohorts:

| | Outcome | | |
|---|---|---|---|
| Cohort | Disease | No Disease | Total |
| High Risk | 100 | 10,000 | 10,100 |
| Low Risk | 10 | 100,000 | 100,010 |
| Total | 110 | 110,000 | 110,110 |

The probabilities of the disease in the two cohorts are estimated as

$$\mathrm{Pr}_H(D) = 100/10{,}100 = 0.009901,$$
$$\mathrm{Pr}_L(D) = 10/100{,}010 = 0.000100,$$

and the relative risk as

$$\mathrm{RR} = \mathrm{Pr}_H(D)/\mathrm{Pr}_L(D) = 99.01,$$

or 99.01 to 1. Relative risks are often approximated by the odds (cross-product) ratio:

$$\mathrm{OR} = \frac{100 \times 100{,}000}{10 \times 10{,}000} = 100,$$

or 100 to 1. The odds ratio provides the essential link between the two study designs. Consider the following two-way table generated from retrospective case-control sampling of the outcomes above:

| | Outcome | | |
|---|---|---|---|
| Cohort | Case=Disease | Control=No Disease | Total |
| High Risk | 100 | 10 | 110 |
| Low Risk | 10 | 100 | 110 |
| Total | 110 | 110 | 220 |

Here, 100% of the cases are retained, but only 0.1% of the controls (people with no diseases). For simplicity, we assumed that the relative distribution of the controls in the sample was identical to the original table, so that the selection of controls is independent of the indicator of high-low risk. In addition, we assumed that each case was matched with one control. We compute the odds ratio as

$$\mathrm{OR} = \frac{100 \times 100}{10 \times 10} = 100,$$

which is the same as earlier. The odds ratio is the same no matter how many controls are sampled if controls are sampled independently of the indicator of risk. Cases may also be sampled independently of the indicator of high-low risk without changing the odds ratio.

The case-control design does not permit calculation of the disease probabilities $\mathrm{Pr}_H(D)$ or $\mathrm{Pr}_L(D)$, because the row totals are arbitrary functions of the number of controls selected to match each case. For example, with two controls for each case the above table becomes

| | Outcome | | |
|---|---|---|---|
| Cohort | Case=Disease | Control=No Disease | Total |
| High Risk | 100 | 20 | 120 |
| Low Risk | 10 | 200 | 210 |
| Total | 110 | 220 | 330 |

Here, the row total depends on the sampling fraction. However the odds ratio,

$$\mathrm{OR} = \frac{100 \times 200}{10 \times 20} = 100,$$

is the same as earlier. The odds ratio approximates the relative risk if the disease outcome is rare; however this cannot be confirmed from the case-control data.

A related concept is that of the attributable risk – the fraction of the disease that can be uniquely attributed to the risk factor. Fleiss (1981) defined this fraction as

$$\begin{aligned} \mathrm{AR} &= [\mathrm{Pr}_H(D) \cdot \mathrm{Pr}(H) - \mathrm{Pr}_L(D) \cdot \mathrm{Pr}(H)]/\mathrm{Pr}(D) \\ &= \frac{\mathrm{Pr}\,(H)\,[\mathrm{RR} - 1]}{1 + \mathrm{Pr}\,(H)\,[\mathrm{RR} - 1]}. \end{aligned}$$

The second expression derives from Fleiss (1981, p. 76, Eq. 5.76). Here, $\mathrm{Pr}\,(H)$ is the marginal probability of exposure to the "high risk," and $\mathrm{Pr}\,(D)$ is the marginal probability of manifesting the selected disease. Continuing the above numerical example,

$$\begin{aligned} \mathrm{Pr}\,(H) &= 10{,}100/110{,}110 = 0.09173, \\ \mathrm{Pr}\,(D) &= 110/110{,}110 = 0.0009990, \end{aligned}$$

and

$$\mathrm{AR} = \frac{0.09173 \times 98.01}{1 + 0.9173 \times 98.01} = 0.900.$$

Fleiss (1981, p. 94) provided an alternative expression for retrospective data:

$$\mathrm{AR} = \frac{\mathrm{Pr}_D(H) - \mathrm{Pr}_{ND}(H)}{1 - \mathrm{Pr}_{ND}(H)},$$

where $\mathrm{Pr}_{ND}(H)$ is the probability of "high-risk" exposure in the No-Disease group, where it is assumed that (a) $\mathrm{Pr}\,(D)$ is low enough that OR $\approx$ RR and (b) the control group (ND = No Disease) is a random sample of the ND population.

Continuing the above numerical example,

$$\mathrm{AR} = \frac{\frac{100}{110} - \frac{10}{110}}{1 - \frac{10}{110}} = 0.900.$$

The identical result obtains in the example with two controls per case. In these examples, 90.0% of the disease outcomes are attributable to the risk factor associated with the high-risk cohort. This may be compared with estimates that 85-90% of lung cancers are attributable to cigarette smoking (Roggli et al., 1992b, p. 325) and that 85% of mesotheliomas among men (23% among women) are attributable to asbestos exposure (Spirtas et al., 1994).

Two other calculations are important. First, the attributable risk in the high-risk cohort is the fraction of the disease in that cohort uniquely attributed to the risk factor:

$$\begin{aligned}\mathrm{AR}_H &= [\mathrm{Pr}_H(D) - \mathrm{Pr}_L(D)]\,/\mathrm{Pr}_H(D)\\ &= [\mathrm{RR} - 1]\,/\mathrm{RR},\end{aligned}$$

so that based on the above example,

$$\mathrm{AR}_H = 98.01/99.01 = 0.990,$$

which shows that virtually all disease in the high-risk cohort is due to the risk factor. Later, the assumption that all mesotheliomas among asbestos workers is due to asbestos exposure will be justified as an approximation based on $\mathrm{AR}_H$ values close to unity.

Second, the relationship between AR and $\mathrm{AR}_H$ is

$$\begin{aligned}\mathrm{AR} &= \mathrm{AR}_H \cdot \mathrm{Pr}_H(D) \cdot \mathrm{Pr}(H)/\mathrm{Pr}(D)\\ &= \mathrm{AR}_H \cdot \mathrm{Pr}_D(H)\\ &\approx \mathrm{Pr}_D(H)\end{aligned}$$

when $\mathrm{AR}_H$ is close to unity, where $\mathrm{Pr}_D(H)$ is the probability of high-risk exposure among persons manifesting the disease. In the above example,

$$\Pr_D(H) = \frac{100}{110} = 0.909,$$

which is just 1% higher than the above AR estimates. This approximation can be used in retrospective analyses of occupational exposure to asbestos among mesothelioma cases to estimate the risk fraction attributable to this exposure route.

The distinction between AR and $\text{AR}_H$ is important when reviewing epidemiological analyses in the context of product liability modeling. The population focus of epidemiology leads to consideration of AR (and RR) to measure risk and to guide primary prevention activities. The targeted subpopulation focus of product liability modeling leads to consideration of $\text{AR}_H$ as a fundamental risk measure for the cohort or group designated by "$H$". The inequality $\text{AR} < \text{AR}_H$ may yield vastly different estimates of attributable risk. For example, Roggli et al. (1992b, p. 325) indicated an AR of about 2% for lung cancer in the United States attributable to asbestos exposure. In contrast, results from Hammond et al. (1979; see Section 2.3.1c) imply an $\text{AR}_H$ of about 80% for insulation workers, with no differences between smokers and nonsmokers. Other occupations with lower levels of asbestos exposure would have $\text{AR}_H$ values in the range 2-80%. McDonald et al. (1980; see Section 2.3.4) found differences in relative risks of smokers among chrysotile miners and millers in Quebec that implied $\text{AR}_H$ values ranging from 50% to 90% for smokers and nonsmokers, respectively, supporting arguments that the lower compensation offered to smokers by the Manville Trust for lung cancer injuries among asbestos-exposed workers is justified (Weinstein, 1994).

### 2.2.2 Design Issues

Each type of study has its advantages. The cohort design is not subject to conscious or unconscious biases in criteria for participation in the study to the same degree as the case-control design because disease outcomes are not known ahead of time in cohort studies. The results of cohort studies can be expressed in terms of population incidence rates and the absolute risk attributed to a given level of exposure can be evaluated for a target population.

The effects of competing risks on the duration of exposure must be considered in cohort studies because termination of exposure may be associated with the diseases under study (Liddell et al., 1977). An inaccurate assessment of the risk of an exposure may result precisely where those risks are highest. If the risks from exposure are high, no one may live long enough to achieve a long duration of exposure. As a consequence, there may be little evidence of an increase in risk with longer exposure. Furthermore, the total duration and intensity of exposure are often not known until the exposure has ended.

The case-control method has several advantages over the cohort approach, perhaps the most important being its lower cost. This is because the cohort design may require a very large cohort to get adequate numbers of affected persons. The case-control method can also explicitly control for sources of variation such as age or sex through matching on the appropriate variables. A disadvantage of the case-control method is that incidence rates and dose-response functions cannot be estimated.

Liddell et al. (1977) identified six design issues for prospective cohort studies:

First, is the cohort grouped into appropriate exposure categories?

Second, has one selected an appropriate population for comparison as a standard? Such "standard" populations may be either external (e.g., state or national populations) or internal (e.g., groups of nonexposed workers).

Third, is the duration of exposure appropriately measured? The study interval over which duration of exposure is measured should start at the same point relative to entry to employment for each subject in the study. The definition of study interval becomes problematic when follow-up is continuous and the duration of exposure for a worker changes over the course of the study.

Fourth, is the measure of health outcome appropriate (e.g., is the rate of onset or the frequency of death from the disease of interest assessed against some index of the size of the population at risk)? The measure most generally accepted is based on person-years of observation (i.e., the number of years each person in the study remains disease-free). The number of cases of disease expected if there is no effect of exposure is calculated by applying incidence/death rates specific to age, year, and disease from the standard population to the corresponding numbers of person-years lived, by age and year, in the study cohort, where person-years for individual cohort members are accumulated from the start of the study to the point at which incidence/death, loss from follow-up, or the end of the study occurs.

Fifth, has one selected an appropriate summary measure of the cohort morbidity/mortality experience and an appropriate statistical model to determine the quantitative relation between the duration of exposure and the measure of morbidity/mortality? The summary measure most often employed is the standardized mortality ratio (SMR). The SMR is the ratio of two quantities. The first is the observed number of deaths at all ages in the study population. The second is the number of deaths expected to occur if their age-specific mortality rates were the same as those in the standard or unexposed population. Thus, the ratio of the observed to the expected number of deaths indicates whether the exposure has increased the risks of the study population (i.e., SMR $> 1.0$), whether it has no effect (SMR $= 1.0$), or whether the frequency of death is smaller in the exposed population (SMR $< 1.0$). The SMR is frequently multiplied by 100 to express the observed number of deaths as a percentage of the expected number. When the SMR is less than 100%, epidemiologists often search for factors which might cause only healthy persons to be drawn into the exposed population. This actually happened in Selikoff et al.'s (1979) study of asbestos insulation workers.

Sixth, are subcohorts properly defined? They should be as follows:

- Mutually exclusive and comprehensive
- Approximately equal in size
- Large enough to produce stable estimates of morbidity/mortality
- Small enough to be fairly homogeneous
- Detailed enough to provide estimates of a dose-response relationship (usually at least three categories of exposure are required)

Liddell et al. (1977) raised a different set of design issues for retrospective case-control studies. The most critical issue is whether an appropriate non-exposed control group has been selected. For example, in studying exposure characteristics of persons with mesothelioma, it would be inappropriate to select a control group of farmers (i.e., a population with little or no exposure to industrial concentrations of chemical dusts or vapors in a closed work environment). In the McDonald and McDonald (1980) study of mesothelioma deaths (to be reviewed in Section 2.3.2), the control group consisted of persons who died in the same hospital as the mesothelioma cases and who had pulmonary metastases from nonpulmonary primary tumors (i.e., the primary site of their disease was not the lung, but the disease had spread secondarily to the lung). Controls should be as similar as possible to cases except for manifestations of the disease under study.

After selecting an appropriate control group, two further issues must be addressed. First, a strategy is needed for matching cases and controls. Once the control population is identified on the basis of some characteristic which all controls must possess, each case must be paired with a control so that they are matched as closely as possible on factors that may be relevant to disease risks (e.g., age and sex). Because it is difficult to find an exact match on certain variables, auxiliary analyses may be required to make the matches as similar as possible.

Second, there can be gains in relative efficiency when more than one control is selected for each case. This may be necessary when the number of cases is small, and, in general, is a way of increasing statistical power.

Finally, one may select one of two basic approaches to analyzing case-control data. The first approach (e.g., Miettinen, 1969) analyzes the data in tabular form. Alternately, hazard-rate regression strategies have been developed for analyzing case-control data (e.g., Prentice and Breslow, 1978). An important difference between the two strategies is that hazard-rate regression permits the use of continuous variables in the analysis.

## 2.3 Studies of Health Risks of Occupational Exposures

In this section, we review studies of the health risks of occupational exposures to asbestos that can be used in developing projections.

### 2.3.1 Health Risks of a Cohort of Insulation Workers Occupationally Exposed to Asbestos

The first study, described by Selikoff et al. (1979) and extended by Selikoff and Seidman (1991), contains what Walker (1982, p. 18) argued to be the most extensive and complete data on the health risks of high levels of occupational exposure to asbestos. This study is based on the mortality experience of two groups of U.S. and Canadian insulation workers:

- A cohort of 632 asbestos insulation workers in the New York-New Jersey metropolitan area registered as members of the International Association of Heat and Frost Insulators and Asbestos Workers as of January 1, 1943, who were followed from January 1, 1943 to December 31, 1962
- A cohort of 17,800 members of the International Association of Heat and Frost Insulators and Asbestos Workers union who were listed as members on January 1, 1967, who were followed from January 1, 1967 to December 31, 1976.

The 17,800 insulation workers followed from 1967 to 1976 yielded the most extensive data on the health implications of occupational exposure to asbestos. That cohort suffered 995 cancer deaths, including 486 from lung cancer and 175 from mesothelioma, and 168 deaths from asbestosis. Selikoff and Seidman (1991) extended the follow-up to December 31, 1986, with a 20-year total of 2295 cancer deaths (1168 lung; 458 mesothelioma) and 427 asbestosis deaths. The extended follow-up data are used in Chapter 7 in our sensitivity analysis of the updated forecasts.

All workers in both cohorts were on the active union enrollment list on the date of start of follow-up. Thus, the onset of exposure to asbestos occurred at some earlier date, and this date was recorded and included in the calculation of time from first exposure to onset of asbestos-related disease. The duration of employment in an asbestos-related job was not reported for these cohorts. In the following, we will describe the experience of the 17,800 member cohort over the periods 1967-1976 and 1977-1986. The results are summarized in Table 2.1.

#### *2.3.1a Basic health effects*

Selikoff et al. (1979) found a considerable delay between the start of the exposure and the time at which the disease was diagnosed. They concluded that a person would have to be observed for at least 20 years before the adverse health effects of exposure could be reasonably expected to be manifest.

They further argued that for up to 20 years after the first occupational exposure to asbestos, a "healthy worker" effect kept any adverse health effects from being noticed. Persons who were accepted for employment were selected for good health. They did not find significant excesses in total mortality until 20-34 years after the start of occupational exposure to asbestos. For this

**Table 2.1: Observed and Expected Deaths Among 17,800 North American Insulation Workers, 1967-1976 and 1977-1986**

| | Number of Deaths | | | | | |
|---|---|---|---|---|---|---|
| | 1967-1976 | | | 1977-1986 | | |
| | Observed | | | Observed | | |
| Cause of Death | Best Evidence | Death Certificate | Expected [1] | Best Evidence | Death Certificate | Expected [1] |
| Total deaths, all causes | 2,271 | 2,271 | 1,658.9 | 2,680 | 2,680 | 1,794.6 |
| Total deaths, all cancers | 995 | 922 | 319.7 | 1,300 | 1,205 | 441.7 |
| Lung cancer | 486 | 429 | 105.6 | 682 | 579 | 163.1 |
| Mesothelioma | 175 | 104 | — | 283 | 77 | — |
| Colon/rectum | 59 | 58 | 38.1 | 62 | 67 | 50.4 |
| Larynx/buccal-cavity/oropharynx/esophagus | 50 | 43 | 21.9 | 46 | 41 | 28.5 |
| All other cancers | 225 | 288 | 154.1 | 227 | 441 | 199.8 |
| Total deaths, all noncancer causes | 1,276 | 1,349 | 1,339.2 | 1,380 | 1,475 | 1,352.9 |
| Noninfectious respiratory diseases | 212 | 188 | 59.0 | 295 | 277 | 85.8 |
| Asbestosis | 168 | 78 | — | 259 | 123 | — |
| All other noncancer causes | 1,064 | 1,161 | 1,280.2 | 1,085 | 1,198 | 1,267.1 |

Note 1: Expected deaths are based on U.S. white male age-specific death rates, 1967-1976 or 1977-1986.

(Continued)

**Table 2.1 (Continued)**

| Cause of Death | Standardized Mortality Ratios (SMR) and z-Score Statistics[2] | | | | | | | |
|---|---|---|---|---|---|---|---|---|
| | 1967-1976 | | | | 1977-1986 | | | |
| | SMR (%) | | z-Score | | SMR (%) | | z-Score | |
| | Best Evidence | Death Certificate | Best Evidence | Death Certificate | Best Evidence | Death Certificate | Best Evidence | Death Certificate |
| Total deaths, all causes | 137 | 137 | 15.03 | 15.03 | 149 | 149 | 20.90 | 20.90 |
| Total deaths, all cancers | 311 | 288 | 37.77 | 33.69 | 294 | 273 | 40.84 | 36.32 |
| Lung cancer | 460 | 406 | 37.02 | 31.47 | 418 | 355 | 40.64 | 32.57 |
| Mesothelioma | — | — | | | — | — | | |
| Colon/rectum | 155 | 152 | 3.39 | 3.22 | 123 | 133 | 1.64 | 2.34 |
| Larynx/buccal-cavity/oropharynx/esophagus | 228 | 196 | 6.00 | 4.51 | 161 | 144 | 3.28 | 2.34 |
| All other cancers | 146 | 187 | 5.71 | 10.79 | 114 | 221 | 1.93 | 17.07 |
| Total deaths, all noncancer causes | 95 | 101 | -1.73 | 0.27 | 102 | 109 | 0.74 | 3.32 |
| Noninfectious respiratory diseases | 359 | 319 | 19.92 | 16.79 | 344 | 323 | 22.58 | 20.64 |
| Asbestosis | — | — | | | — | — | | |
| All other noncancer causes | 83 | 91 | -6.04 | -3.33 | 86 | 95 | -5.11 | -1.94 |

Note 2: z-Score = (Observed – Expected)/Expected$^{0.5}$.

Source: Derived from Selikoff et al. (1979, Table 12) and Selikoff and Seidman (1991, Tables 2 and 3).

time interval from first exposure, they calculated that there were increased relative risks of death for all types of cancer (SMR $> 1$), except mesothelioma. The reason for not calculating relative risks for mesothelioma deaths was that Selikoff viewed mesothelioma as a "signal" disease (Selikoff, 1981, p. 26) whose presence is prima facie evidence for asbestos exposure. Thus, no mesothelioma would be expected in an unexposed population, and the SMR would be undefined.

Selikoff et al. (1979) also argued that the expected number of mesothelioma deaths cannot be computed for the general population because mesothelioma is not a distinct category in the various revisions of the International Classification of Diseases (ICD). However, it would have been possible to calculate an expected value from either the Third National Cancer Survey (TNCS), 1969-1971, or from the Surveillance, Epidemiology, and End Results Program (SEER), 1973 onward (see Hinds, 1978).

Selikoff et al. (1979) also found that, except for mesothelioma and asbestosis, the death certificate diagnoses of asbestos-related diseases were reasonably accurate. Death certificate diagnoses in the study cohort were generally consistent with diagnoses based on the "best available" evidence (i.e., in order of preference: autopsy findings, pathological information derived from surgical evidence, and clinical and roentgenological observations made during life). In the absence of any additional medical evidence, findings were based only on death certificates. This occurred in only 28 of 995 cancer deaths.

All 175 diagnoses of mesothelioma were supported by autopsy or surgical findings. This was particularly important because only 104 of the 175 mesothelioma cases were correctly diagnosed on the death certificate. Mesothelioma was so poorly reported on the death certificate because it was not an explicit diagnostic entry in the ICD. Many cases of mesothelioma were also diagnosed as other types of neoplasia; in particular, 15 of 49 pancreatic cancer cases were reassigned to mesothelioma upon review of the best medical evidence.

Asbestosis also was not well diagnosed on the death certificate. Only 78 cases were identified on the death certificate; 168 cases were identified from the "best evidence."

The primary substantive result from this study was the determination of the risk of a wide range of diseases for a heavily exposed occupational cohort. The results for 1977-1986 were based on 134,740 person-years of observation for the 15,529 survivors over the second 10-year period.

The results for 1967-1976 were based on 166,853 person-years of observation for the 17,800 asbestos insulation workers over the first 10-year period. At the onset of observation in 1967, most men were below age 40 (10,101 of 17,800) and most had not yet been followed for 20 years from the time of their first occupational exposure to asbestos (12,683 of 17,800). By the end of the first observation period in 1976, 12,051 men had been observed for 20 or more years after their first exposure. Over the first observation period, there were 89,462 person-years of exposure at less than 20 years after the start of insulation employment (presumed onset of asbestos exposure) and

77,391 person-years of exposure 20 or more years after the start of insulation employment. Significant numbers of excess deaths were noted for total mortality, asbestosis, total cancer mortality, mesothelioma, lung cancer, esophageal cancer, cancer of colon and rectum, cancer of the larynx, oropharynx, and buccal cavity, and kidney cancer. Stomach cancer had marginally significant elevation in the first 10-year period but not in the full 20-year period (Selikoff and Seidman, 1991). Cancers of the pancreas and gallbladder/bile ducts had significant elevations in the full 20-year period but not in the first 10-year period (Selikoff and Seidman, 1991).

For our analyses, we retabulated Selikoff's detailed tables to show the relative risks for lung cancer, mesothelioma, colon/rectum cancer, and a combined category representing the larynx and upper digestive tract (buccal cavity and oropharynx, and esophagus). These four specific cancer categories corresponded to the compensable categories recognized by the Manville Trust during 1995-2002, except that the Trust dropped buccal cavity but accepted all types of pharyngeal cancer, not just the oropharynx site (Weinstein, 1994); additionally, the Trust began paying for stomach cancer claims in January 2003 (Weinstein, 2002). A residual category was defined for all other cancers, including cancers of the stomach, pancreas, kidney, and gallbladder/bile ducts.

Table 2.1 presents the retabulated summary counts for observed and expected deaths for the major cancer and noncancer diseases associated with asbestos exposure, stratified by observation period. For 1967-1976, almost half (46.4%) of the person-years of observation were 20 or more years after first exposure; for 1977-1986, most (81.5%) were 20 or more years after first exposure.

The SMRs for the second observation period reflect the joint impact of longer times since first exposure and older attained ages. The SMRs (best evidence) increased for all causes of death and for all noncancer causes, decreased slightly for lung cancer, and dropped sharply for colon/rectum cancer, cancer of the larynx and upper digestive tract, and all other cancers. The absolute death counts increased sharply for mesothelioma and asbestosis, diseases for which an SMR was not defined. The SMRs for noncancer causes other than respiratory diseases (primarily asbestosis) were 83% and 86%, respectively, indicating that these workers were generally healthier than the standard reference population of U.S. white males over the period 1967-1986. The impact of lung cancer, mesothelioma, and asbestosis is evident in both observation periods. The impact of cancer of the larynx and upper digestive tact is relatively much lower, although still significant.

For the second period (but not the first), under the best evidence criterion, the SMRs for colon/rectum cancer and all other cancers were not significantly elevated. This loss of significance was not noticed by Selikoff and Seidman (1991) but it is consistent with Greenberg and Roggli's (1992) conclusion that the evidence for increased risk is inconclusive for colon/rectum cancer and several other cancer sites (i.e., pancreas, stomach, and kidney). Nonetheless, claims for colon/rectum cancer are compensable under the Trust Distribution Process (TDP) (Weinstein, 1994).

### *2.3.1b Time to onset of disease*

The 10-year follow-up data for 1967-1976 were evaluated in a second article (Selikoff et al., 1980). A more detailed examination of the increase in incidence of asbestosis, mesothelioma, and lung cancer with time since onset of occupational exposure to asbestos confirmed that there was little increase in deaths before 15 years from onset of exposure. Beginning at 15-19 years from onset of exposure, there was a superlinear (i.e., accelerating) increase in the absolute risks of death from mesothelioma, asbestosis, and lung cancer. The lung cancer risks turned to sublinear (i.e., decelerating) increases at 30-35 years, a point at which their relative risks (i.e., compared to the expected risks in the U.S. white male population) peaked at a ratio of 6.1 to 1.0. Asbestosis risks exhibited a downturn at 45-59 years after onset of exposure, but this was reversed at 50+ years (with 73 deaths) in the 20-year follow-up data of Selikoff and Seidman (1991). Mesothelioma risks turned to sublinear increases at 40-44 years, but this also was reversed in Selikoff and Seidman (1991), where a peak was found at 45-49 years and a decline at 50+ years.

These reversals in the trends at the longest time intervals from first exposure suggest that there may be an interaction between date of initiation of exposure and time since onset of exposure. This could result if the type or amount of exposure changed over time. For example, Selikoff et al. (1979, p. 92) noted that only one type of asbestos (chrysotile) was used in the United States until the early 1940s, when a second type (amosite) became much more common. This could account for anomalies in the risk functions above 40 years since first exposure in the 1967-1976 follow-up. In addition, with 10 or 20 years follow-up, person-years of exposure at the longest durations are about one-tenth those of the shorter durations and are not for the same people. Selection effects may be operating on these groups (e.g., effects of cohort differences in cigarette smoking).

These results indicate that data for at least 40 years after first exposure are necessary for the full health implications of asbestos exposure to become manifest.

### *2.3.1c Impact of cigarette smoking on asbestos-related disease*

A third study of the 1967-1976 follow-up data (Hammond et al., 1979) is the primary source of our current understanding of the effects of smoking on asbestos-related mortality. In this study, attention was restricted to the 12,051 men who, by 1976, had at least 20 years elapsed since onset of occupational exposure. This provided 77,391 person-years of observation. The average age during observation was 53.8 and the number of deaths observed was 1946.

Of the 12,051 men in the study, 8220 answered the smoking questionnaire, of whom 83% were current or ex-smokers.

This study had data that allowed the authors to select a control group where smoking history was available. Because smoking was not recorded on death certificates, national vital statistics were not suitable for this task. Fortunately, data from a prospective American Cancer Society (ACS) study, begun in 1959, of over one million persons were available. These persons were traced through September 30, 1972. Smoking information was recorded. From this group, a subset of 73,763 male subjects was selected as a comparison population who were white, not farmers, had no more than a high school education, and had a history of occupational exposure to dust, fumes, vapors, gases, chemicals, or radiation. It was expected that this group was likely to be physically active (to match the physical activity required by insulation work). Because deaths were observed for the control group only through 1972, the experience of the controls was extrapolated from cause-specific mortality changes observed in the national population over the period 1972-1976.

The number of deaths expected in the study population based on the mortality experience of the control group was calculated in two ways. First, the mortality rates from the ACS study were applied to the person-years of insulation workers to calculate an expected number of deaths. Second, the mortality rates of the U.S. white male population were applied to the person-years of insulation workers to calculate another expected number of deaths. The calculation of the expected number of deaths using the ACS study was the preferred method because education, work activity, and smoking could be controlled in those computations.

Hammond et al. (1979) found significant excess mortality among insulation workers for all causes of death and for cancer from all sites when compared to the mortality expected using either standard population. Among deaths due to specific types of cancer determined from the best medical evidence, cancers of the lung, larynx, buccal cavity and oropharynx, esophagus, and colon/rectum were found to be significantly elevated, with smoking controlled, when compared to the mortality experience of the ACS population.

Another comparison was between smoking and nonsmoking insulation workers. For insulation workers, smoking elevated both the risk of total mortality and the risk of lung cancer. Insulation workers who were current heavy smokers had a lung cancer mortality risk 10.4 times greater than expected on the basis of the nonsmoker insulation worker mortality rates. The level of risk was lower if a person had quit smoking more than 5 years previously. Thus, the lung cancer risk was greatly increased among insulation workers who smoked.

Fewer data were available to assess risks of death from other diseases among insulation workers who smoked, so only a few general observations were made. First, the risk of asbestosis mortality was 2.8 times higher among smoking insulation workers than expected from the mortality experience of nonsmoking insulation workers. Of the insulation workers who never smoked regularly, none died of cancer of the esophagus, larynx, or buccal cavity and oropharynx. This was interpreted as evidence that asbestos exposure, in the absence of smoking, may have no effect on the risks of these diseases.

Next, insulation workers who were smokers were compared with nonsmokers from the control group. The observed number of lung cancer deaths among insulation workers who were smokers was 46.2 times higher than expected for nonsmokers in the ACS subpopulation (for current heavy smokers, the ratio was 87.4; for ex-smokers, it was 36.6).

The conclusion was that a strong interaction existed between asbestos exposure and smoking for lung cancer risks. Specifically, if the lung cancer risks of nonsmokers in the control group were taken as a baseline, then nonsmoking insulation workers had a risk 5.2 times greater. Among smokers in the control population, the relative risk was higher (10.9) than for nonsmoking insulation workers. For smoking insulation workers, the relative risk was 53.2 to 1 compared with nonsmoking insulation workers, and 4.9 to 1 compared with smokers (i.e., 53.2/10.9). The nearly equal estimates of asbestos relative risk for smokers and nonsmokers (4.9 vs. 5.2) is consistent with a multihit/multistage model of carcinogenesis, with asbestos and smoking affecting different "hits" or "stages" of the process (see Section 2.3.1d). In this case, the attributable risk ($AR_H$) for asbestos induced lung cancer among insulation workers is approximately 80%, compared to 2% for the general population. This suggests that much of the total excess lung cancer mortality among insulation workers who smoked cigarettes was attributable to the interaction of asbestos with smoking.

There was no evidence of an elevation of mesothelioma risks among smokers, in distinct contrast to the strong elevation of lung cancer risks. This finding has been confirmed in other studies (Lemen et al., 1980; McDonald and McDonald, 1980; Peto et al., 1982; Tagnon et al., 1980; Muscat and Wynder, 1991).

#### *2.3.1d Biologically motivated models of mesothelioma risks*

A fourth study of the 1967-1976 follow-up data (Peto et al., 1982) estimated the parameters of a mathematical model of the increase of the risk of mesothelioma with the time since first exposure. Important findings were established by Peto et al. (1982) through the application of this model to the experience of the insulation workers. First, it was demonstrated that the *absolute* risk of mesothelioma was dependent on time since first exposure but independent of age. Similarly, Nicholson et al. (1981a), using the same data, showed that the *relative* (but not absolute) risk of lung cancer was dependent on time since first exposure but independent of age.

Peto et al. (1982) had to demonstrate that the incidence of mesothelioma was dependent upon the time since the onset of asbestos exposure but not age before they could legitimately apply a multihit/multistage model of carcinogenesis to the exposure experience over the period 1922-1946 for the insulation workers. The multihit/multistage model of carcinogenesis is a model of the biology of tumor initiation developed by Armitage and Doll (1954, 1961). The model suggests that a tumor initiates when $k + 1$ errors occur in the genetic code of a single cell, leading to the loss of the mechanisms regulating cell reproduction. The model is a widely accepted model of carcinogenesis for the following reasons: (Whittemore and Keller, 1978)

- It is based on a plausible biological mechanism.
- It leads to a very simple computational form for predicting the increase in the risk of tumor onset as a function of time since the initiation of exposure to agents that might cause the genetic errors.
- It fits a wide range of data (e.g., Cook et al., 1969).

The mathematical form of the multihit/multistage model is

$$I_t = bt^k,$$

where $I_t$ is the incidence rate (equivalently, hazard rate) of the tumor (in this case, mesothelioma) $t$ years after initiation of exposure to the risk factor (asbestos), $b$ is a proportionality constant, and $k + 1$ is the number of cellular errors that are required for a tumor to start. Mathematically, this expression for the incidence rate is identical to the hazard rate of the Weibull distribution – a distribution frequently used in reliability analysis in engineering applications. This distribution also arises in extreme value theory as the distribution of the smallest extreme of a set of independent and identically distributed times to failure of independent components of a multicomponent system. In a biological system, individual cells are the components and the transformation of any one of up to a billion or more cells in a given organ (pleura or peritoneum, in the case of mesothelioma) is sufficient to generate the disease. The multihit/multistage model explains the parameter $m = k + 1$ as either the number of stages or hits, depending on whether or not a specific fixed order of cellular errors is required. The choice of hit versus stage affects the interpretation of the parameter $b$, but not its estimated value.

This model was fitted to mesothelioma mortality data from the insulation workers by Peto et al. (1982), who obtained estimates of 3.20 for $k$ and $4.37 \times 10^{-8}$ for $b$. Mortality data, rather than incidence data, were used because the time from diagnosis to death for mesothelioma is typically under 1 year and because incidence data were unavailable. The parameter estimates were obtained by minimizing the chi-squared statistic used to measure the goodness-of-fit of the observed and expected deaths under the model. The parameters were reestimated using the maximum likelihood method and the results were virtually identical (e.g., $k$ = 3.17 vs. 3.20 under the minimum chi-squared method).

To assess the generalizability of the model, it was also fitted to data from four studies with different levels of asbestos dust exposure and fiber types (i.e., Newhouse and Berry, 1976; Peto, 1980; Hobbs et al., 1980; Seidman et al., 1979). The value of $k$ was fixed at 3.20 for each of these studies, but $b$ was allowed to vary (yielding estimates of $4.95\times10^{-8}$, $2.94\times10^{-8}$, $5.15\times10^{-8}$, and $4.91 \times 10^{-8}$, respectively). In each case, the model fits reasonably well to the mesothelioma mortality data. Thus, the increase of mesothelioma mortality could be well described by the 3.2 power of the time since first exposure (i.e., $I_t = bt^{3.2}$), over variations in asbestos fiber type, site [different mixes of pleural (lung) and peritoneal (abdominal) tumors were observed across the studies], and exposure levels. Variation in all of these factors could be modeled by changes in $b$.

The estimate of $k$ obtained from the insulation worker data had a very broad confidence interval (standard error of 0.36) so that any value between 2.5 and 4.0 would provide an adequate fit. Peto et al. (1982) suggested that lack of precision of $k$ would not greatly alter predictions of future mortality trends. However, if one employs a value of 4.0 instead of 3.2, the predicted lifelong mesothelioma risk for men first exposed at age 20 (with $b$ reestimated to account for the change in $k$ – a necessary step due to a correlation of the sample estimates of $k$ and $b$ on the order of $-0.998$) would be 19% instead of 15%, a relative difference of 27%. If one uses $k = 2.5$, then the lifelong risk would be 12% instead of 15%. The overall uncertainty (i.e., going from $k = 2.5$ to 4.0) is 58% (i.e., with estimates of lifelong mesothelioma risks ranging from 12% to 19%). For the purposes of projecting future mesothelioma mortality, this degree of uncertainty is noteworthy.

Peto et al. (1982) warned against attributing spurious precision to the estimate of $k$ and recommended that a value of 3.5 be used to imply a value between 3 and 4. The lack of precision in the estimate of $k$ cited by Peto et al. (1982) and the large effect that the variation in $k$ has on the projection of mesothelioma mortality suggest that long-term projections will be sensitive to this parameter.

Peto et al. (1982) examined the risk of the two subtypes of mesothelioma – peritoneal and pleural mesothelioma – and concluded that fiber type was a primary determinant of anatomical site. Amphiboles (i.e., amosite or crocidolite) were argued to be largely responsible for peritoneal tumors. They also showed that the lifelong risk of mesothelioma was very sensitive to the assumed distribution of age at first exposure. For example, for the insulation worker data, the lifelong mesothelioma risk was 15% for persons first exposed at age 20, 7% for persons first exposed at age 30, and only 3% for persons first exposed at age 40. This suggests that projections will be sensitive to variations in the distribution of age at first exposure.

Peto et al. (1982) considered that the low mortality 10-15 years after first exposure could be a result of a lengthy tumor growth time; that is, under the multihit/multistage assumptions, the Weibull incidence rate is actually the rate at which a single cell gains status as a bona fide cancer cell. However, a tumor does not generally become detectable until about a billion or more daughter cells have been generated by mitosis from the original transformed cell, and this takes time. Peto et al. (1982) tested a modified model,

$$I_t = b(t - w)^k$$

with $k = 2$ and $w = 10$ years, and found, with suitable adjustment to $b$, that this model fit better than the first model for the first 15 years since first exposure and fit equally well thereafter.

This modified form of the model was adopted by both the Occupational Safety and Health Administration and the U.S. Environmental Protection Agency in their risk assessment models (OSHA, 1983, 1986; EPA, 1986). These agencies made additional adjustments, however, to account for the fact that union insulation workers tended to have continuous career-long exposure histories (35 years or more), whereas other workers typically had shorter durations of exposures at lower intensities. We discuss these modifications in Section 2.4.

### 2.3.2 A Case-Control Study of Asbestos Risks in the United States and Canada

McDonald and McDonald (1980) conducted a large and frequently cited retrospective case-control study of occupational exposure to asbestos. This study provided Walker's (1982) projections with the proportion of the total number of mesothelioma deaths that were likely to result in lawsuits. The study also provided Selikoff's (1981) projections with occupation-specific measures of relative risks that could be multiplied by estimates of the number of workers in each occupational category to produce the projected number of mesothelioma cases.

McDonald and McDonald (1980) identified groups of diseased and nondiseased persons and examined retrospectively their differences in exposure. The retrospective design differs from Selikoff et al.'s (1979) study of insulation workers where the population was defined on the basis of exposure and prospectively followed to determine who got the disease. The retrospective design allows for better control of confounding factors by closely matching cases with controls. However, it cannot be used to produce estimates of the incidence rate of mesothelioma. The results of the retrospective and prospective studies complement each other.

McDonald and McDonald (1980) contacted nearly all U.S. and Canadian pathologists (7400 in number) to determine how many cases of mesothelioma they had observed. For the period 1960-1975 in Canada and for the year 1972 in the United States, the pathologists contacted reported a total of 668 cases (557 recorded through the end of 1972 and selected for detailed analysis). For each mesothelioma case, a staff physician visited the hospital where the case was recorded, reviewed the diagnostic evidence, and selected a control matched for sex, age, and year of death, and in which pulmonary (lung) metastases were present from a nonpulmonary malignant tumor. After the selection of cases and matched controls, interviews were conducted (generally with relatives) to determine occupational and residential histories and smoking habits. For each occupation recorded, respondents were questioned about occupational dust exposure.

For the 557 cases selected for detailed analysis, 71% (395) were male and 29% (162) were female. Among males, 78% of cases were pleural and 22% peritoneal mesotheliomas. Among females, the corresponding figures were 61% and 39%, respectively.

Occupation coding was conducted for 344 male cases and 344 controls using a list of occupations associated with asbestos exposure provided by Selikoff. Jobs were independently assessed for the likelihood of asbestos exposure by four research centers specializing in occupational health studies. The agreement among the four centers was quite good for exposures categorized as "definite" and "unlikely," and for the cases of "possible" and "probable" exposure taken together. In the United States, on average, 73.6% of male cases were classified as possibly-definitely exposed; in Canada, the corresponding average was 58.3%.

For females, only 2 of 162 cases had worked with asbestos, so it was not possible to carry out a similar analysis. However, six additional cases were spouses of an asbestos worker, suggesting that about 5% of female cases could be linked to occupational exposures.

From the 344 male cases and 344 controls, it was also possible to calculate the relative risks of asbestos exposure for five occupational groups with an established association with mesothelioma. Recall that the relative risk is the ratio of the probability of dying from mesothelioma in one of the occupational categories with identified exposure to asbestos to the probability of dying from mesothelioma in occupational categories without identified exposure to asbestos (in this case, all occupations other than the five selected groups). The relative risks were calculated using the odds ratio approximation described in Section 2.2.1. The odds ratios were as follows:

- 46.0 for insulation workers
- 6.1 for asbestos production and manufacturing
- 4.4 for heating trades (excluding insulators)
- 2.8 for shipyard workers
- 2.6 for construction workers

These were evaluated for consistency with risk estimates made from prospective cohort studies and found to be in substantial agreement (Selikoff, 1981; see Section 3.3, Task 3 and Table 3.1).

In the United States, 64.8% of cases had worked in one of the five occupations; in Canada, the corresponding figure was 45.9%. These figures are 8.8% and 12.4% lower, respectively, than the figures based on the possible-definite exposure classifications.

In evaluating the two methods of analysis, McDonald and McDonald concluded that:

> The list of occupations provided by the Environmental Sciences Laboratory, Mount Sinai School of Medicine, proved a satisfactory method of classifying occupations thought to entail asbestos exposure. With minor modifications, the list could improve the comparability of case-control surveys in different regions and countries. Greater discrimination was achieved between case and controls by selecting occupations reported to have been associated with mesothelioma than by assigning probabilities of asbestos exposure to all occupations listed. (McDonald and McDonald, 1980, pp. 1654-1655)

In addition, McDonald and McDonald (1980) indicated that the likelihood of asbestos exposure for an occupation may be underestimated by both methods because it was determined from interviews conducted after the subject had died. They noted that interview data, especially from secondary sources such as relatives, may not yield complete occupational or exposure histories.

McDonald and McDonald (1980) noted a tendency for male workers in higher-risk occupations to have relatively more peritoneal (abdominal) mesothelioma. Combined with the results of Peto et al. (1982; see Section 2.3.1d), this finding suggested that asbestos fiber exposures in high-risk occupations may include greater relative amounts of amphiboles than in low risk occupations. This interpretation would also be consistent with the finding that females have relatively more peritoneal mesotheliomas than do males, even though their asbestos exposure is much lower. Moreover, the finding of no association of mesothelioma risk with smoking was confirmed for males (see Section 2.3.1c).

Walker (1982) used the data provided by McDonald and McDonald (1980) to divide the total projected number of mesothelioma cases (see Section 4.4, Task 1b):

- Into a group with a plausible (i.e., "definite" or "probable") occupational exposure history and a group without such a history
- For workers with a plausible occupational exposure history, into subgroups that were heavily and less heavily exposed

### 2.3.3 Short-Term Amosite Exposure Among Factory Workers in New Jersey

Seidman et al. (1979) considered the long-term effects of short-term exposures: 933 men employed in an amosite asbestos factory during the period 1941-1945 were followed in cohort studies for 35 years. As Seidman et al. (1979, p. 62) state, "This resulted in a unique experience; men with a very limited duration of intense work exposure to amosite asbestos followed by long observation." Thus, it was possible to determine if very limited exposures (e.g., 1 month) increased the risk of cancer, whether cancer risks increased with greater exposure duration, and if the exposure duration was correlated with the length of the latency period. There were no direct observations of dust counts for this cohort, although measurements made in 1971 suggested average exposure levels as high as 23 fibers/ml, for fibers longer than 5 $\mu$m.

One hundred thirteen men were eliminated from the original 933 men, 20 because of prior asbestos work experience, 14 because during the first 5 years after employment, they took up asbestos work elsewhere, 41 died, and 38 were lost to follow-up. This left 61 workers who worked less than 1 month, 90 for 1 month, 82 for 2 months, 149 for 3-5 months, 125 for 6-11 months, and 313 for 12 or more months. The mortality experience of these workers was compared on an age- and date-specific basis for the period 1946-1977 with the mortality experience of New Jersey white males (New Jersey having some of the highest cancer rates in the United States). Total mortality, mortality from specific causes, lung cancer, and an "all-asbestos" disease category were analyzed. The "all-asbestos" disease category represented asbestosis, chronic pulmonary disease, lung cancer, mesothelioma, and cancers of the esophagus, stomach, colon, rectum, larynx, buccal cavity, pharnyx, and kidney.

The study yielded several conclusions. First, the lower the dose the longer it took for excess mortality to become evident and the smaller the magnitude of the effect. Second, the length of the latency period decreased with increasing age at exposure. It had been suggested that if asbestos-related diseases had long latent periods, then older workers, because of their age, would not live long enough to manifest those diseases. Unfortunately, Seidman et al. (1979) found that high levels of exposure for older persons (e.g., aged 50 to 59) produced increased mortality very quickly (i.e., within 5 to 14 years). Third, it was demonstrated that mortality risks increased with time, even after exposure had ceased, apparently due to the effects of permanently retained asbestos in lung tissue and other sites. Fourth, for light exposure, it was determined that the follow-up period would have to be lengthy to identify health effects.

### 2.3.4 Effects of Chrysotile Exposure Among Miners and Millers in Quebec

McDonald et al. (1980) followed until 1975 a cohort of 11,379 workers (10,939 men and 440 women) born 1891-1920 and exposed to chrysotile in the mines and mills of Asbestos and Thetford, Quebec. Data were analyzed using two cohort methods, using male mortality in Quebec as a standard, and a case-control method employing internal controls. Cumulative measures of exposure to asbestos were available.

In the first cohort analysis, the male cohort was subdivided into four groups on the basis of length of service (i.e., less than 1 year, 1-4 years, 5-19 years, and 20 or more years). The workers in each length of service category were further divided into four subgroups on the basis of cumulated dust concentrations for all kinds of airborne particles, not just airborne asbestos, measured as the number of millions of particles per cubic foot (mppcf) to which the worker was exposed weighted by the number of years he was exposed at that level (mppcf-yr)). The subgroups were defined so that there was little variation in the average daily level of exposure in each of the four sets of four accumulated exposure categories (i.e., "low" accumulated exposure groups had been exposed to a concentration of 2.5 to 4.2 mppcf on average, "medium" accumulated exposure groups experienced dust concentrations that varied from 4.3 to 9.4 mppcf, "high" exposure groups experienced dust concentrations that varied from 14.4 to 23.6 mppcf, and "very high" groups varied from 46.8 to 82.6 mppcf).

There was little association between exposure level and cause of death for gross service of less than 5 years. For service of 5-19 years, there were consistent trends across exposure levels for total mortality, asbestosis (pneumoconiosis), heart disease, and stroke. SMRs were elevated in the highest-exposure group for lung cancer and other respiratory diseases. For workers with 20 or more years service, the most severely exposed category had the highest SMRs for total mortality and for all listed causes other than laryngeal cancer and accidents. Furthermore, there was a relatively consistent gradient for asbestosis, heart disease, total mortality, lung cancer, respiratory tuberculosis, and other respiratory diseases.

McDonald et al. (1980) conducted a second analysis using exposure categories based on the dose accumulated by age 45 (three categories: less than 30 mppcf-yr, 30-299 mppcf-yr, and 300+ mppcf-yr). There were clear trends in the SMRs for total mortality, asbestosis, lung cancer, cancer of the colon and rectum, respiratory tuberculosis, other respiratory diseases, and stroke. At age 45, lung cancer risks increased linearly at a rate of 0.16% per mppcf-yr accumulated exposure to asbestos (with exposure of 30 mppcf-yr or more divided into four categories).

McDonald et al. (1980) also analyzed their data retrospectively, using the case-control method. Multiple controls and four exposure categories (i.e., less than 30 mppcf-yr, 30-299 mppcf-yr, 300-999 mppcf-yr, and 1000+ mppcf-yr) were employed with persons with less than 30 mppcf-yr exposure used as internal controls. Clear increases in risk were found for asbestosis, lung cancer, esophageal and stomach cancer, and colon/rectum cancer. For these four diseases persons who had accumulated 1000+ mppcf-yr asbestos exposure had risks respectively 30.6, 3.16, 4.69, and 5.26 times greater than expected based on the mortality experience of persons with less than 30 mppcf-yr accumulated exposure.

When the analysis was stratified by smoking status, lung cancer risk increased 10-fold for nonsmokers with the highest level of accumulated exposure, compared to internal controls (i.e., nonsmokers with the lowest level of accumulated exposure). For persons with undifferentiated (i.e., unknown or doubtful) smoking habits, the risk ratio for persons with high levels of asbestos exposure compared to those with low levels was nearly 14-fold but only 2-fold for definite smokers. In this case, the attributable risk ($AR_H$) for asbestos-induced lung cancer at the highest levels of exposure were 90% for nonsmokers versus 50% for smokers – compared to estimates of 80% for both groups in Hammond et al. (1979).

In summary, both retrospective and prospective analyses of the data showed the following:

- There was essentially a linear response to dose of risk for lung cancer, asbestosis, and total deaths based on accumulated exposure.
- Both an additive model of smoking interaction with asbestos exposure and a multiplicative model (found in Hammond et al., 1979) are consistent with the data.
- Because of the difficulty in identifying excess risks at lower exposure levels, the fitting of linear dose-response forms are essential to the task of setting standards for acceptable environmental exposure levels.

Perhaps the most important conclusion from this study is that chrysotile asbestos fibers appeared to be less potent in increasing mesothelioma risks than amphiboles (amosite or crocidolite):

> The incidence of malignant mesothelial tumors, especially of the peritoneum, is so very much higher after exposure to amphiboles (and amphibole-rich mixtures) than after exposure to chrysotile alone, that differences in dust concentrations are unlikely to explain it. (McDonald et al., 1980, p. 22).

Nonetheless, McDonald et al. (1980) suggested that the available evidence on the aggregate health implications of fiber type was not conclusive because (a) no comparable (i.e., as statistically reliable) studies had been made of crocidolite or amosite production and (b) for the available reports on single-fiber exposure, exposure was expressed only in terms of duration [e.g., no direct exposure measures were available in Seidman et al. (1979)].

### 2.3.5 Mesothelioma Risks Among World War II Shipyard Workers

Important evidence about the health implications of asbestos was provided by studies of mesothelioma and lung cancer risks among World War II shipyard workers. Because this workforce was so large (i.e., 4-5 million workers), a significant elevation of risk in this group served to raise concern for the magnitude of the total health effect of occupational exposure to asbestos. Early evidence of this effect was derived from cancer maps for the period 1950-1969 (Mason et al., 1975). Several areas of excess lung cancer mortality risk were noted in coastal counties. One hypothesis to explain this elevation was that increased lung cancer mortality risk was due to shipyard exposure to asbestos. Eventually, case-control studies were conducted in a number of areas observed to have elevated lung cancer mortality risks on the maps (e.g., Blot et al., 1978, Georgia; Tagnon et al., 1980, Virginia; Blot et al., 1982, Florida; see also Blot and Fraumeni, 1981). The study by Tagnon et al. (1980) of coastal Virginia illustrates the general design and results of those case-control studies.

Sixty-one cases of mesothelioma diagnosed 1972-1978 were identified among white males from discharge diagnoses, pathology files, and tumor registries at major hospitals in coastal Virginia and from records of the Virginia Tumor Registry. Pathological specimens were sought for all cases for independent review. Mesothelioma incidence rates were calculated for each sex, race, and age group. The observed numbers of cases were compared to the numbers expected based on national estimates of mesothelioma derived from the Surveillance, Epidemiology, and End Results Program (SEER) results (Hinds, 1978). The case-control study was limited to white males – the only group with elevated rates. Controls consisted of 320 local residents who died from 1972 to 1976 from causes other than chronic respiratory diseases and were similar with respect to age at death and county of residence. Personal interviews of 4 surviving cases and the next-of-kin of 52 deceased cases and 236 controls were conducted using a standard questionnaire to obtain data on (a) place, type, and length of employment for all jobs held for more than 6 months and (b) information on smoking habits and residential history.

The mesothelioma incidence rates were four times the national estimates from SEER, with the excess concentrated among white males. Shipyard employment was reported for 77% of the cases. The risk of mesothelioma was 15.7 times higher for shipyard workers who had reported contact with asbestos than for the controls – implying that $\mathrm{AR}_H = 93.6\%$. Among shipyard workers reporting no contact with asbestos, the risk of mesothelioma was 4.9 times higher than among controls. Because mesothelioma risks were significantly elevated among shipyard workers who were not identified as having contact with asbestos, it was suggested that the determination of asbestos exposure from the interviews may have been incomplete. Cigarette smoking was not associated with an increased risk of mesothelioma.

Tagnon et al. (1980) also reported results from a parallel study of the same population in which lung cancer risks of shipyard workers were 1.7 times greater than those of the controls. Furthermore, shipyard workers developing lung cancer tended to have shorter durations of exposure to asbestos than those developing mesothelioma.

Because latencies of 35 years were often noted for mesothelioma, it was suggested that the full impact of mesothelioma had not yet been felt. The authors concluded:

> Assuming that the Tidewater rate of 10 cases/year/100,000 white males ages 50 to 70 years ... is composed of a 15-fold increased risk among 12% (the percentage of the 236 controls) of this population who worked in shipbuilding prior to 1950 and either handled asbestos or were career employees, and assuming that the risk was usual among the remaining 88%, then the annual incidence of mesothelioma among former shipyard employees would be 56/100,000. This rate exceeds that for all cancers except those of the lung, prostate, colon, and bladder. Furthermore, since survival is poorer for mesothelioma than

for the other neoplasms, mesothelioma may claim as many or more deaths among shipyard workers than does any cancer except lung cancer. (Tagnon et al., 1980, p. 3878).

### 2.3.6 Effects of Asbestos Exposure Among a Cohort of Retired Factory Workers

Henderson and Enterline (1979) reported the mortality experience of 1348 men aged 65+ who had "completed their working life times as production or maintenance-service employees with a U.S. asbestos company and retired with a company pension" (p. 117). Of the 1348 men, 273 were excluded whose only known employment was in Canada. For the remaining 1075 men, 781 deaths were recorded. For these 781 deaths, death certificates could be located for 749. The cohort was composed of three types of retiree for the period 1941-1967:

- Normal retirees at age 65
- Those who retired before age 65 for nonmedical reasons but who lived to 65
- Those who retired due to disability before age 65 but who lived to 65

The mortality experience of this cohort was compared with that expected assuming that U.S. white male mortality rates for the same ages and dates applied to the study population. Although this is a cohort study, one must be aware of the implications of selecting a group of persons who must survive to age 65 and who must have adequate service to qualify for a pension. The health effects of intense exposure to asbestos may have already been manifest before age 65. Therefore, workers who succumbed to asbestos-related diseases before age 65 were excluded from consideration by the study design.

Despite the selectivity of their cohort, Henderson and Enterline (1979) found that total mortality, cancer mortality, and mortality from chronic respiratory diseases were elevated, although perhaps not as high as in other studies. The authors calculated cumulative dosages and studied the dose-response relations. Previously, with Canadian data and 4 fewer years of follow-up, it was speculated that the mathematical function describing the dose-response relation was nonlinear. The later data (with 4 more years of follow-up and with the Canadian data excluded) were found by Henderson and Enterline (1979) to be consistent with a linear dose-response form. The estimated dose-response equation for respiratory cancer was $\text{SMR} = 100.0 + 0.658 \times \text{mppcf-yr}$, where the SMR is the standardized mortality ratio (percent form) based on U.S. white male respiratory cancer mortality and the dust levels were estimated by job and time period.

The authors obtained information on the type of asbestos fiber to which each worker was exposed (i.e., amosite, chrysotile, crocidolite, or some combination thereof). The effect of exposure to specific types of asbestos fibers on disease risk could be adjusted for cumulative dose. Although the numbers were small, the 112 men exposed to both chrysotile and crocidolite asbestos had an SMR that was 94.3% higher than expected on the basis of cumulative dust exposure alone. In contrast, the 754 men exposed only to chrysotile had an SMR that was 5.3% lower than expected on the basis of cumulative dust exposure alone. Taken together, these results imply that the excess risk induced by the chrysotile-crocidolite mixture could be 99.6% higher than that of chrysotile alone. Because the majority of the men who had mixed exposures worked in asbestos cement pipe manufacturing, it was difficult to draw firm conclusions about the different effects of chrysotile and crocidolite. The SMRs for amosite were elevated, but the sample sizes were too small for those SMRs to achieve statistical significance.

Although useful information was generated from the study, the study design made it impossible to draw meaningful conclusions about the relation of asbestos exposure and mesothelioma. The authors had previously reported only 1 mesothelioma death during 1941-1969 for the 1348 men in the study. This was surprising given that the study summarized the experience of a group with typically long durations of employment (3-51 years; 25 year average), high exposure levels, and lengthy times since onset of exposure.

As noted by Henderson and Enterline (1979), this finding is frequently compared with a study conducted near the Manville, NJ plant, where 72 cases of mesothelioma were identified (Borow et al., 1973). Henderson and Enterline (1979) provided a table to indicate the status of 58 of 72 of Borow's cases. No explanation was given by Henderson and Enterline for the difference between the 58 cases reviewed and Borow's total of 72. Furthermore, as Henderson and Enterline (1979, p. 124) explain:

> Of the 58 cases, there were records of work at the plant for 41. Thirty-one of these men were not included in our cohort, however. That is, they did not, according to our records, retire during the period 1941-1967. Most of these men were too young or had too little service to retire. Of the 10 on whom we did have records, seven died at ages under 65 and were not part of our study, because we studied deaths only at ages 65 and over.

Thus, a major portion of the effect of asbestos exposure on health (i.e., deaths due to mesothelioma) was lost because of the requirement of the study design that persons be over age 65 and have adequate service to retire.

Since a major portion of the total health consequences of occupational exposure to asbestos was excluded by the study design, the data from this study cannot be directly employed in projections of the total health consequences of occupational exposure to asbestos. Using these data in such projections could grossly understate future mesothelioma incidence. It seems likely that the same limitations of the study design that caused mesothelioma mortality to be grossly underestimated could also lead to underestimation of other health consequences of asbestos exposure.

## 2.4 Increases in Disease Risk Associated with Exposure to Asbestos

In all of the above-cited studies, the risks of certain diseases and causes of mortality increased for persons with significant occupational exposure to asbestos. However, it was not possible in all studies to estimate a dose-response function because the level of asbestos exposure was not measured in all studies. The dose-response function is a mathematical expression indicating the exact magnitude of the increase in risk associated with a unit dose increase in exposure to asbestos. The coefficient applied to the measured level of asbestos exposure is called the "dose-response coefficient." In this section, we will examine dose-response coefficients estimated for studies where asbestos exposure was measured (Selikoff, 1981; EPA 1986).

In addition to quantitative measures of asbestos exposure, a second requirement must be satisfied before a dose-response function can be estimated. This requirement is that the mathematical form of the dose-response function be known. In general, one lacks adequate data to prove that a dose-response function is of a particular form. Consequently, one is required to (a) specify a theoretically acceptable dose-response function and (b) make sure that the form specified is consistent with the data. The specification of a particular dose-response function is important in projections because it determines the level of disease risk that can be expected for persons exposed to a given level of asbestos.

The most common type of dose-response function used in the analysis of the risks of asbestos exposure is the linear dose-response function, a function that derives from the multihit/multistage model under the assumption that asbestos affects only one "hit" or "stage" of the process of carcinogenesis (Whittemore and Keller, 1978). This function has the property that the increase in risk associated with a unit increase in asbestos exposure is the same at all levels of asbestos exposure. The EPA found this assumption to be plausible for mesothelioma and strongly indicated by the evidence for lung cancer (EPA, 1986, p. 30). Both Walker (1982) and Selikoff (1981) assumed that the dose-response function is linear. This assumption was important for both of their projection strategies in that it permitted them to treat the duration of exposure as equivalent to dose. Thus, exposing 2000 persons to a given level of asbestos for 1 year would produce the same amount of disease as exposing 1000 persons to the same level of asbestos for 2 years. This equivalence holds only for the linear form of the dose-response function and only for moderate variations of the exposure duration.

Although such an assumption has not been proven, most findings are consistent with a linear dose-response form. Furthermore, none of the supporting data suggest the existence of a threshold level required for disease response. As a result, a linear dose-response form is usually accepted for practical reasons and because no epidemiological study can give accurate risk estimates at the lower dosage levels (McDonald et al., 1980).

Given the linear form of the dose-response relation, a number of technical issues remain. First, what measure of cumulative dosage should be employed? A common measure is the number of asbestos fibers greater than 5 $\mu$m in length found in 1 ml of air to which a worker is exposed, for 40 hours per week, over some standard time unit like a year. This measure is frequently abbreviated as f-yr/ml. An alternate cumulative measure, the mppcf-yr (see Section 2.3.4), may be related to this measure by the simple approximation 1 mppcf-yr = 3 f-yr/ml (Selikoff, 1981, p. 211; or Selikoff, 1982, p. 124).

Actually, this conversion is more complex than it appears. Direct conversion from U.S. customary units to metric units yields 1 mppcf-yr = 35.3 f-yr/ml assuming that 1 fiber = 1 particle. The discrepancy occurs because the mppcf measure was typically used to measure the total dust concentration for all kinds of airborne particle – not just airborne asbestos fibers. Selikoff's approximation is equivalent to the assumption that 1 asbestos fiber = 11.8 airborne particles. Selikoff warned that the conversion factor for 1 mppcf could plausibly range from 1 to 8 f/ml, so that the assumed conversion factor 3 f/ml may be grossly in error. For additional discussion, see Dement et al. (1983a) and EPA (1986, pp. 42-46).

A second technical issue to consider is what measure of response to use. In Table 2.2, the dose-response coefficients from a range of studies are presented for two measures: (a) the change in lung cancer deaths due to each 1 f-yr/ml change in exposure, as a percent of the expected lung cancer deaths, and (b) the change of all asbestos-related deaths due to each 1 f-yr/ml change in exposure, as a percent of observed deaths. Selikoff (1981) summarized these studies to show how the estimates changed with study design and condition. To maximize comparability, the asbestos-related deaths were restricted to include only deaths from asbestosis, lung cancer, mesothelioma, and gastrointestinal cancer [defined by Selikoff (1981) to include cancers of the esophagus, stomach, and colon/rectum].

According to Table 2.2, in the study of Seidman et al. (1979), where factory workers were exposed to amosite fibers, there was a 9.1% increase in lung cancer risk for each f-yr/ml. The lowest estimate was 0.06% (McDonald and Liddell, 1979) where miners and millers were exposed to chrysotile fibers. The ratio of the largest dose-response coefficient to the smallest coefficient was 151 to 1 (i.e., 9.1/0.06). This variation is large and probably reflects unidentified systematic differences in study design, study population, and study conditions, or combinations of these factors. For example, amosite fibers may be more toxic than chrysotile fibers. Thus, one might expect the dose-response coefficient to be higher in studies where the primary fiber type is amosite, as in Seidman et al. (1979). The ratio of the largest and smallest dose-response coefficients relating the risk of all asbestos-related deaths to asbestos exposure was also large: 108 to 1 (i.e., 0.65/0.006).

The ranges of dose-response estimates for both lung cancer and all-asbestos-related diseases are so broad that it may be hazardous to pool such estimates. Instead, Selikoff (1981) recommended that one examine the esti-

**Table 2.2: Estimated Risk Increases Associated with Asbestos Exposure**

| | | Increase in Risk per f-yr/ml of Cumulative Exposure | | |
|---|---|---|---|---|
| Study Group | Asbestos Type | Lung Cancer as a % of Expected Lung Cancer Deaths | All Asbestos-Related Deaths as a % of Observed Deaths | Reference |
| Factory Employment | | | | |
| Insulation Manuf. – Paterson, N.J. | Amosite | 9.10 | 0.650 | Seidman et al., 1979 |
| Asbestos Products Manuf. – London, U.K. | | | | |
| Males | Crocidolite, chrysotile, and amosite | 1.30 | 0.140 | Newhouse and Berry, 1979 |
| Females | Crocidolite, chrysotile, and amosite | 8.40 | 0.140 | Newhouse and Berry, 1979 |
| Asbestos Manuf. Retirees – U.S. | Amosite and chrysotile; some crodidolite | 0.30 | 0.020 | Henderson and Enterline, 1979 |
| Asbestos Products Manuf. – Manville, N.J. | Chrysotile; some amosite and crocidolite | 1.10 | 0.080 | Nicholson et al., 1981b |
| Insulation Application | Chrysotile and amosite | 1.70 | 0.170 | Selikoff et al., 1979 |
| Textile Production | | | | |
| U.S. | Chrysotile | 5.30 | 0.290 | Dement et al., 1982 |
| Rochdale, U.K. | | | | |
| Before 1951 | Chrysotile | 0.07 | 0.020 | Peto, 1980 |
| After 1950 | Chrysotile | 0.80 | 0.080 | Peto, 1980 |
| Chrysotile Mining and Milling | | | | |
| Quebec | Chrysotile | 0.06 | 0.006 | McDonald and Liddell, 1979; McDonald et al., 1980 |
| Thetford Mines, Quebec | Chrysotile | 0.15 | 0.030 | Nicholson et al., 1979 |

Source: Selikoff (1981, Table 3-14) and NRC (1984, Table 7-3).

mates in terms of data quality and systematic differences in the exposure and setting in order to select plausible dose-response estimates for specific forecasting or risk assessment applications.

Selikoff pointed out that the three highest estimates (5.3%, 8.4%, and 9.1%) suggested that even very low exposures (e.g., 0.5 f/ml for workers employed for 40 years = 20 f-yr/ml) may produce twice the risk of lung cancer and 4-13% higher total mortality. The six highest estimates implied that an exposure of 2.5 f/ml would produce, after 40 years, at least a doubling of lung cancer risk and 10% higher total mortality. This exposure level is half the U.S. standard permissible exposure limit of 5.0 f/ml existing in 1972-1976 and is just above the 2.0-f/ml standard existing in 1976-1986 (OSHA, 1986).

Despite Selikoff's (1981, p. 219; 1982, p. 134) admonition that "it is not appropriate to average or otherwise combine the data from the various investigations," this is precisely what was done in the National Research Council (NRC) (1984) study which relied on the same set of nine estimates for lung cancer as reported in Table 2.2. The NRC (1984, p. 214) computed the median dose-response coefficient (1.1%) and rounded the result upward to 2.0% for computing lifetime risks of lung cancer for nonoccupational environmental exposures.

Likewise, the EPA (1986) used an averaging of the risk coefficients obtained in their review of 14 studies that permitted estimation of the OSHA (1983) form of the lung cancer model. This model extended previously developed SMR models to explicitly introduce a 10-year latency period during which asbestos exposure would have no observable impact. The SMR (percent form) at age $a$, exposure level $f$, duration $d$, and time since first exposure $t$, is represented as

$$\mathrm{SMR}\,(f,d,t) = 100 \; + \; K_L \times f \times d_{t-10},$$

which is independent of age; and where $K_L$ is the dose-response coefficient (percent form), $f$ is the exposure intensity in f/ml, and $d_{t-10}$ is the completed duration of exposure 10 years in the past (i.e., as of age $a - 10$), where

$$d_{t-10} = \begin{cases} d & (t \geq d+10) \\ t-10 & (d+10 \geq t \geq 10) \\ 0 & (t < 10)\,. \end{cases}$$

The EPA (1986) evaluated 14 studies that allowed the estimation of dose-response coefficients and confidence intervals for the OSHA (1983) lung cancer model. The results in Table 2.3 indicate that there were significant differences among the estimates.

The EPA (1986, p. 82) computed the geometric mean $K_L$ of the 14 studies as 0.65%. However, for assessing the impact of environmental exposures, they excluded the three studies of mining and milling workers and recomputed the geometric mean as 1.0% – nearly identical to the initial result of 1.1% in the NRC (1984) study. A 95% confidence interval from 0.4% to 2.7% was derived from an analysis of variance of the 11 separate estimates.

**Table 2.3: Estimates of Dose-Response Coefficients for Asbestos-Related Lung Cancer in 14 Epidemiologic Cohort Studies**

| Industrial Process / Location | Asbestos Type | Estimated Dose-Response Coefficient $K_L$ (%) | 95%-Confidence Interval (%) | | Reference |
|---|---|---|---|---|---|
| | | | Low | High | |
| Textile Production | | | | | |
| Charleston, SC | Chrysotile | 2.80 | 1.70 | 5.60 | Dement et al., 1983b |
| Charleston, SC | Chrysotile | 2.50 | 1.00 | 3.70 | McDonald et al., 1983a |
| Rochdale, U.K. | Chrysotile | 1.10 | 0.30 | 2.40 | Peto, 1980 |
| Lancaster, PA | Chrysotile; some amosite and crocidolite | 1.40 | 0.36 | 1.70 | McDonald et al., 1983b |
| Friction Products | | | | | |
| U.K. | Chrysotile | 0.06 | 0.01 | 0.80 | Berry and Newhouse, 1983 |
| Connecticut | Chrysotile | 0.01 | 0.01 | 0.55 | McDonald et al., 1984 |
| Mining and Milling | | | | | |
| Quebec | Chrysotile | 0.06 | 0.02 | 0.11 | McDonald et al., 1980 |
| Thetford Mines, Quebec | Chrysotile | 0.17 | 0.06 | 0.32 | Nicholson et al., 1979 |
| Turin, Italy | Chrysotile | 0.08 | 0.01 | 0.89 | Rubino et al., 1979 |
| Insulation Manufacturing | | | | | |
| Paterson, NJ | Amosite | 4.30 | 0.84 | 7.40 | Seidman, 1984 |
| Insulation Workers | | | | | |
| U.S. and Canada | Chrysotile and amosite | 0.75 | 0.60 | 1.10 | Selikoff et al., 1979 |
| Mixed Products | | | | | |
| U.S. | Amosite and chrysotile; some crocidolite | 0.49 | 0.24 | 0.91 | Henderson and Enterline, 1979 |
| Cement Products | | | | | |
| U.S. | Chrysotile, crocidolite, and amosite | 0.53 | 0.14 | 1.10 | Weill et al., 1979 |
| Ontario, Canada | Chrysotile and crocidolite | 6.70 | 3.50 | 11.20 | Finkelstein, 1983 |

Source: EPA (1986, Table 3-10 and Figure 3-7).

Following essentially the same logic, OSHA (1986, p. 22,637) derived an identical estimate of $K_L$ (1.0%) with an uncertainty interval of 0.3-3.0%. Neither confidence interval includes the three estimates in Table 2.3 for mining and milling (0.06%, 0.17%, and 0.75%), whose geometric mean is 0.091% – smaller than the 1.0% pooled estimate by a factor of 11.0. To deal with this, the EPA recommended an uncertainty factor of 10 in applications to new exposure situations.

Part of Selikoff's (1981) concern about pooling the risk coefficients was that it may lead to underestimation of the risk faced by certain classes of workers. Conversely, it may lead to overestimation of the risk faced by others. For example, Camus et al. (1998) evaluated the EPA parameterization of the OSHA lung cancer model using mortality data for women from two chrysolite mining areas of Quebec for the period 1970-1989. The estimated average cumulative exposure was 25 f-yr/ml, which was relatively high given that 95% of the exposure was nonoccupational. The predicted relative risk was 2.05 to 1. The observed relative risk was 0.994 or 1.101, depending on the method used in the calculation. On this basis, Camus et al. (1998, p. 1568) concluded that "the EPA's risk-assessment model overestimated the mortality attributable to asbestos by a factor of at least 10."

The authors offered six possible reasons for overestimation by the EPA's model:

- Overestimation of risk at low doses
- Inadequacy of cumulative exposure in measuring risk
- Overestimation of the exposure-risk gradient
- Lower risk for chrysotile versus amphibole asbestos
- Lower relative risk of lung cancer due to asbestos among nonsmokers than smokers
- Overestimation of the dose-response gradient

These reasons included no mention of the uncertainty of the dose-response coefficient due to pooling, nor of the large confidence intervals recommended by the EPA (1986). The EPA (1986, p. 82) stated that application of their model to new exposure situations should allow for a risk differential as large as a factor of 10 from their 1% dose-response coefficient. This would include the risk level found by Camus et al. (1998) at its lower bound.

Alternatively, the results of Camus et al. (1998) may be reinterpreted as providing validation of the EPA (1986) model. This requires that we view the Quebec exposures not as a new exposure situation, but as one similar to the mining and milling exposures in the three studies (including Quebec) excluded from EPA's pooled estimate (see Table 2.3). The pooled dose-response coefficient for these three studies is 0.091%, which implies a predicted relative risk of 1.096 – a value in-between the two observed values 0.994 and 1.101 provided by Camus et al. (1998). [*Note*: $1.096 = 1 + (2.05 - 1) \times 0.091$.]

This explanation is more plausible than any of the six explanations proposed by the authors and it provides additional support for the OSHA (1983) form of the lung cancer model used by the EPA (1986). The question remains: Why is the lung cancer risk coefficient for the mining and milling of chrysotile so much lower than for other asbestos processing activities, especially for textile production in the United States and the United Kingdom?

The EPA (1986) used 4 of the 14 studies to estimate risk coefficients for mesothelioma using the OSHA (1983) form of the mesothelioma model. This model extended the latency form of the multihit/multistage model used by Peto et al. (1982; see Section 2.3.1d) to explicitly represent (a) the permanent increase in risk associated with each fiber that is inhaled and retained and (b) the reduced rate of increase in risk following the cessation of exposure. The absolute mortality risk at exposure level $f$, duration $d$, and time since first exposure $t$, for $t \geq 10$, is represented as

$$\mathrm{AMR}(f,t,d) = K_M \times 10^{-8} \times f \times \left[(t-10)^3 - (t-10-d_t)^3\right],$$

where

$$d_t = \begin{cases} d & (t \geq d+10) \\ t-10 & (t < d+10), \end{cases}$$

where the final condition is set to zero out the second term in brackets in the expression for $\mathrm{AMR}(f,d,t)$ during the 10-year latency period following cessation of exposure. The constant $10^{-8}$ is extracted from the constant $K_M$ to simplify the scaling. The 10-year latency assumption is implemented by setting $\mathrm{AMR}(f,d,t) = 0$ for $t < 10$.

The EPA's (1986) estimates of dose-response coefficients for mesothelioma are presented in rows 2, 3, 4, and 6 of Table 2.4. These estimates range from 1.0 to 12.0 with a geometric mean of 2.75. However, when the two OSHA (1986) estimates are included, the geometric mean drops to 0.98. The EPA (1986) was concerned about bias in their estimates, noting that two of the four studies included the two highest lung cancer dose-response estimates. To deal with this concern, the EPA evaluated the ratios of the mesothelioma and lung cancer coefficients, noting that the ratios for the four selected studies were in much closer agreement, ranging from 0.74 to 2.00, with a geometric mean of 1.25. Following this, the EPA (1986, p. 95) developed a series of adjustments that incorporated mesothelioma death counts from the other 10 studies listed in Table 2.3 and concluded that the best estimate of the dose-response ratio was 1.00, so that $K_M = K_L$=1.0, with an approximate 95% confidence interval from 0.2 to 5.0 and an uncertainty factor of 20 in applications to new exposure situations.

This estimate of $K_M$ is almost identical to the geometric mean (0.98) of the six studies in Table 2.4. However, the EPA's uncertainty bounds are large, suggesting that use of the OSHA-EPA model may lead to serious errors of underestimation of the risk faced by some workers and overestimation of the risk faced by others.

OSHA (1986, p. 22,640) followed a similar logic in their assessment of the four studies in Table 2.4, arriving at the same final estimate: $K_M = 1.0$. They

**Table 2.4: Estimates of Dose-Response Coefficients for Asbestos-Related Mesothelioma in Six Epidemiologic Cohort Studies**

| Industrial Process/Location | Asbestos Type | Mesothelioma Dose-Response Coefficient $K_M$ ($10^{-8}$) | Lung Cancer Dose-Response Coefficient $K_L$ (%) | Mesothelioma-Lung Cancer Ratio $K_M/K_L$ (%) | Reference |
|---|---|---|---|---|---|
| Textile Production | | | | | |
| Charleston, SC | Chrysotile | 0.22 | 2.80 | 8 | Dement et al., 1983b |
| Rochdale, U.K. | Chrysotile | 1.00 | 1.10 | 91 | Peto, 1980 |
| Insulation Manufacturing | | | | | |
| Paterson, NJ | Amosite | 3.20 | 4.30 | 74 | Seidman, 1984 |
| Insulation Workers | | | | | |
| U.S. and Canada | Chrysotile and amosite | 1.50 | 0.75 | 200 | Selikoff et al., 1979; Peto et al., 1982 |
| Cement Products | | | | | |
| U.S. | Chrysotile, crocidolite, and amosite | 0.07 | 0.53 | 13 | Weill et al., 1979 |
| Ontario, Canada | Chrysotile and crocidolite | 12.00 | 6.70 | 179 | Finkelstein, 1983 |

Source: Mesothelioma parameters in rows 2, 3, 4, and 6 are from EPA (1986, Table 3-30); mesothelioma parameters in rows 1 and 5 are from OSHA (1986, p. 22640-22641). Lung cancer parameters are from EPA (1986, Table 3-10).

also reported the $K_M$ estimates for the data of Dement et al. (1983b) and Weill et al. (1979) included in rows 1 and 5 of Table 2.4. In our model development in Section 8.4, we used the updated insulation worker data in Selikoff and Seidman (1991) to obtain a $K_M$ value of 1.45, 45% higher than the EPA (1986) value, but only 3.3% lower than the insulation worker value in Table 2.4. The uncertainty in these estimates motivated us to develop additional constraints on our forecasting model that will be discussed in Chapters 6-10.

## 2.5 Effects of Fiber Type on Disease Risks

Two aspects of the OSHA (1983) mesothelioma model are important to our modeling applications. First, the fact that the absolute risk is proportional to the asbestos exposure level, $f$, means that no nonasbestos-related causes of mesothelioma are represented. In the general population, where 80-90% of mesotheliomas are attributable to asbestos exposure, this assumption is clearly only an approximation, and the approximation could be improved by better accounting of the rate for the remaining 10-20% not due to asbestos exposure. In the exposed worker population, however, where the attributable risk ($AR_H$) is on the order of 99% or higher, the approximation is much better and there would be little gain in modeling the nonasbestos-related risk.

Second, the exponent $k = 3$ in the formula for absolute risk implies a four-stage or four-hit multistage/multihit model, consistent with the mechanisms proposed by Hahn et al. (1999; see Section 2.6 for discussion). This is 1 unit higher than the estimate $k = 2$ obtained by Peto et al. (1982) for the fitted model with a 10-year latency (see Section 2.3.1d). However, when we re-fitted that model to the updated data in Selikoff and Seidman (1991), we found that $k = 3$ was the best integer estimate for the exponent, and $k = 2.8$ was the best overall estimate (see Table 8.7). Thus, the fixed parameters of the OSHA model are consistent with the mesothelioma experience of the insulation worker cohorts and with the biological evidence on the mechanisms underlying the disease.

The variability in risk coefficient estimates from the various cohort studies of workers exposed to asbestos has yet to be fully explained. At several points in the preceding sections, it was suggested that there may be a gradient in carcinogenicity across the different types of asbestos fiber, with the lowest risks for chrysotile, increased risks for amosite, and the highest risks for crocidolite. However, the variability in risk coefficients for chrysotile in Tables 2.3 and 2.4 indicates that consideration should also be given to risk gradients according to the type of industrial process.

OSHA reviewed evidence on risk differentials by asbestos fiber type and concluded "that epidemiological and animal evidence, taken together, fail to establish a definitive risk differential for the various types of asbestos fiber"

(OSHA, 1986, p. 22,628). OSHA further stated that there exists "a clear relationship between fiber dimension and disease potential" (1986, p. 22,629). OSHA (1994) reviewed additional evidence relating to its earlier analysis and determined that it would stand by that analysis. Three reasons were offered:

1. Similar risk potencies for chrysotile and amphiboles were found for both lung cancer and asbestosis; evidence for lower chrysotile risk was presented only for mesothelioma.
2. Chrysotile presents a significant risk of cancer, even if it is accepted that its risk is lower than for amphiboles.
3. Most occupational exposures involve mixed fiber types.

The EPA (1986, p. 106-117) reviewed evidence on the relative carcinogenicity of different asbestos fiber types. Based on the 14 epidemiological studies of lung cancer risk identified in Table 2.3, it concluded "that factors other than mineral types substantially influenced the studies reviewed" (EPA, 1986, p. 108). For example, it was pointed out that chrysotile textile production exhibited lung cancer risks significantly larger than chrysotile mining or friction products manufacturing. Based on the four epidemiological studies of mesothelioma risk identified in Table 2.4, it concluded "that the same factors affect the variability of mesothelioma risk as affect lung cancer risk" and "it appears impossible to separate the effect of mineral type from other factors contributing to the variability of potency" (EPA, 1986, p. 110).

Using a more extensive set of 41 epidemiological studies, the EPA developed a series of adjustments that allowed it to compute ratios of pleural and peritoneal mesothelioma to excess lung cancer incidence in each of the studies. Assuming that excess lung cancer incidence is a proxy for cumulative asbestos exposure, the mesothelioma ratios could be interpreted as measures of relative carcinogenicity of the asbestos fibers in a given study. Several conclusions were reached (EPA, 1986, p. 114-115):

1. Amphibole exposures produced comparable numbers of pleural and peritoneal mesothelioma; chrysotile exposures rarely produced peritoneal mesothelioma.
2. For pleural mesothelioma, the ratios for chrysotile, amosite, and mixed exposures were roughly comparable, whereas the ratios for crocidolite were two to three times greater.
3. For peritoneal mesothelioma, the ratios for pure chrysotile exposures were significantly lower than for amphiboles or mixed exposures.
4. On average, pure amosite exposure has a risk about twice that of pure chrysotile exposure, whereas pure crocidolite exposure has a risk about four times that of pure chrysotile exposure.
5. Within fiber type, significant differences appear to be related to the type of processing conducted (e.g., chrysotile mining versus textile production; amosite mining versus insulation manufacturing).

The EPA (1986, p. 116) considered differences in fiber size distributions and industrial processes in different work environments to be major factors in accounting for risk differentials in the various epidemiological studies. This is consistent with Stanton and Wrench (1972), who evaluated the carcinogenicity of amosite, chrysotile, crocidolite, and other fibers by direct application to the pleura of 1200 Osborne-Mendel rats and concluded that the carcinogenicity of asbestos was primarily related to its structural shape rather than to its physicochemical properties. In contrast to the human epidemiological results obtained by the EPA (1986), Stanton and Wrench (1972) found that the incidence of pleural mesothelioma in their experiments on rats did not differ significantly among the three types of asbestos fiber.

The EPA's findings (conclusions 1 and 3) that chrysotile exposure rarely produces peritoneal mesothelioma may explain an anomalous result that is often cited but never adequately explained – that peritoneal mesothelioma appears to be associated with heavier cumulative exposure intensities (e.g., Lemen et al., 1980; Antman, 1980; Walker, 1982; Browne and Smither, 1983; Roggli et al., 1987, 1992c). Roggli et al. (1987, 1992c, p. 112) noted that about 50% of peritoneal cases had concurrent asbestosis compared with 20% of pleural cases; whereas Roggli et al. (1992b, p. 312) reported a correlation of 0.46 between asbestosis scores and the lung fiber burden in 36 autopsied cases. Walker (1982) cited results from 11 epidemiological studies to support the association of peritoneal mesotheliomas with heavier exposure. Given both epidemiological and tissue burden evidence, the anomaly is that the relative amount of peritoneal versus pleural mesothelioma is significantly higher among females than among males (McDonald and McDonald, 1980; SEER, 2000) – exactly the opposite of what one would expect if the association were real.

We review Walker's (1982) evidence in Section 4.4.1, Task 1b where we find an alternative interpretation of no association to be more plausible. Roggli et al.'s (1992c) finding of a correlation with asbestosis was based on autopsied cases and is subject to three important limitations (Stayner et al., 1996): (1) Asbestosis is indicative of heavy fiber concentration in the lungs, not the mesothelium; (2) chrysotile asbestos is cleared more rapidly than the amphiboles from the lungs – it differentially migrates to the pleura, frequently leaving tremolite contaminants in the lungs as the only persistent evidence of its presence; and (3) the lung tissue fiber distribution at the time of death may not be representative of the distribution at the time of exposure 20-50 years earlier. Combined with the finding of higher relative frequencies of peritoneal mesothelioma for females, these considerations suggest that the alternative explanation of no association is more plausible.

Evidence in favor of this explanation is provided by comparing the ratios of male to female counts of peritoneal versus pleural mesothelioma in the SEER data for 1973-1997 (SEER, 2000): 1.19 versus 4.50. Given that chrysotile accounts for 90-95% of asbestos consumption in the United States, and that chrysotile rarely produces peritoneal mesothelioma, one would expect only a modest increase in peritoneal cases for males (with $AR_H \approx 16\%$) and a substantial increase for pleural cases (with $AR_H \approx 78\%$), where the "$H$" factor is associated with being male.

Lippmann (1988, 1990) reviewed the literature relating fiber characteristics to disease in animals and humans in an attempt to establish critical fiber parameters for the three main asbestos-related diseases: asbestosis, lung cancer, and mesothelioma. He concluded the following:

1. Asbestosis risk is related to the surface area of asbestos fibers longer than 2 $\mu$m with diameters in the range 0.15-2.0 $\mu$m.
2. Lung cancer risk is related to the number of asbestos fibers with lengths in the range 10-100 $\mu$m with diameters greater than 0.15 $\mu$m, especially diameters in the range 0.3-0.8 $\mu$m.
3. Mesothelioma risk is related to the number of asbestos fibers with lengths in the range 5-10 $\mu$m and diameters less than 0.1 $\mu$m.

Interestingly, there is no overlap between the mesothelioma and lung cancer fiber parameters, with respect to either length or diameter, nor between mesothelioma and asbestosis, with respect to diameter. These results could account for variability in the ratios of mesothelioma to excess lung cancer incidence in the studies reviewed by the EPA (1986). These results could also account for the variability of risk associated with different industrial processes, if those processes changed the lengths, diameters, or surface areas of asbestos fibers. Lippmann (1988, p. 103) noted that the phase-contrast optical method was recommended for counting fibers with diameters between 0.25 and 3 $\mu$m. However, Mossman et al. (1990, p. 299) and Gaensler (1992, p. 234) commented that the phase-contrast microscopy mandated by OSHA (1986, 1994) actually has a resolution only to 0.5 $\mu$m, more than three times the lower bound for diameters of fibers causing asbestosis and lung cancer and more than five times the upper bound for diameters of fibers causing mesothelioma. Lippmann (1988, p. 103) noted that fibers with diameters below the resolution limit cannot be counted using the methods mandated by OSHA (1986, 1994); he recommended electron microscopy or magnetic alignment and light scattering techniques.

The role of chrysotile asbestos as a causal agent in human mesothelioma has been challenged. Churg (1988) surveyed the literature on chrysotile-induced mesotheliomas and concluded that at most 53 cases could be accepted as valid, and he argued that the causal agent in most chrysotile-induced mesothelioma was actually tremolite asbestos contaminants. Mossman et al. (1990, p. 247) argued that the lower carcinogenicity of chrysotile combined with the high proportion of chrysotile in asbestos-containing materials in buildings and schools suggest that most environmental exposures to asbestos will not lead to asbestos-associated malignancy or functional impairment. Furthermore, they suggested that "exposure to chrysotile at current occupational standards does not increase the risk of asbestos-associated diseases" (Mossman et al., 1990, p. 247).

Counterarguments to Mossman et al. (1990) were provided by Nicholson (1991) and Dement (1991). Nicholson (1991, p. 82) concluded that there was "no difference in the potency of chrysotile and amosite for producing mesothelioma." He accepted that there was two to three times greater risk for crocidolite. Dement (1991) compared data on asbestos fiber distributions in human lung tissues from Quebec chrysotile miners and millers with South Carolina chrysotile textile workers. These two groups exhibited the largest risk differentials for lung cancer in Table 2.3. Dement (1991, p. 18) noted that the South Carolina workers had lower total fiber deposition rates and lower proportions of tremolite, leading to the conclusion that tremolite was not the principal causal agent for lung cancer among these chrysotile workers. Rall (1994a, 1994b) and Mossman (1994) continued the debate with a series of points and counterpoints. Stayner et al. (1996) reviewed lung burden studies, epidemiologic studies, toxicologic studies, and mechanism studies that provided evidence on the relative carcinogenicity of chrysotile and amphibole fibers and concluded that tremolite contamination is not the explanation of mesothelioma incidence among chrysotile asbestos workers. Smith and Wright (1996) reviewed evidence from animal and human studies of pleural mesothelioma, including analyses of the asbestos fiber content of pleural tissue, and concluded that the potency of chrysotile was comparable to that of amosite, with crocidolite 2-4 times more potent. However, given that chrysotile accounted for about 95% of asbestos usage, they also concluded that chrysotile was the main cause of pleural mesothelioma in the United States.

Liddell et al. (1997, 1998) and McDonald et al. (1997) completed follow-up on the cohort of 11,000 Quebec chrysotile miners and millers discussed in Section 2.3.5. For overall mortality, they concluded that exposure to less than 1000 f-yr/ml was essentially innocuous. For lung cancer and mesothelioma, analysis of the geographical variation in risk correlated with the geographical distribution of fibrous tremolite as a contaminant in chrysotile asbestos. This correlation was investigated further by McDonald and McDonald (1997), who suggested that the greater durability and biopersistence of amphiboles in lung tissue may be of critical importance. McDonald (1998) noted that the very high risk of lung cancer, but not of mesothelioma, among chrysotile textile workers remains unexplained.

Cullen (1998) attempted to provide some perspective to these divergent findings. In particular, he noted that the high lung cancer rates among South Carolina chrysotile textile workers were not explained by the tremolite contamination hypothesis, but, instead, required additional explanation and explication of the risks associated with fiber length, diameter, and other physical characteristics.

## 2.6 Simian Virus 40 and Mesothelioma

Bocchetta et al. (2000) noted that (1) 5-10% of asbestos workers get mesothelioma, (2) 10-20% of mesotheliomas are not associated with asbestos exposure, and (3) 60% of human mesotheliomas contain simian virus 40 (SV40 – a macaque polyomavirus that is tumorigenic in rodents and inactivates p53 and pRb tumor suppressor proteins) DNA fragments. The first point suggested to them that additional factors may be involved; the second point suggested that alternative factors may cause mesothelioma; and the third point suggested a potential causative role for SV40 in mesothelioma development. To test this latter hypothesis, Bocchetta et al. (2000) conducted a series of in vitro experiments that established that SV40 infection of human mesothelial cells was different from the lytic pattern seen in almost all other types of cells, that the difference was related to increased levels of p53 in mesothelial cells, and that infected mesothelial cells underwent tumorigenic transformation to immortal phenotype. In addition, they demonstrated that the rate of transformation increased when the cells were exposed to increasing concentrations of crocidolite asbestos. This led Bocchetta et al. (2000) to conclude that asbestos and SV40 are cocarcinogens in vitro and may be cocarcinogens in vivo. One anomalous result, however, was the finding that crocidolite alone, without SV40, did not produce tumorigenic transformations of mesothelial cells in vitro.

Several comments are in order. First, the fact that only 5-10% of asbestos workers get mesothelioma does not mean that additional factors must be involved. Under the multistage model of carcinogenesis, the tumor develops only after several tumorigenic transformations have occurred. Hahn et al. (1999) argued that changes are needed in at least four distinct intracellular signaling pathways and cited SV40 large tumor antigen, oncogenic *ras*, and the catalytic subunit of human telomerase as candidates for study. The identity of the fourth pathway was left unspecified, except that it was related to some fundamental difference in the biology of rodent and human cells. The plausibility of this conjecture was boosted by Killian et al. (2001), who found that primates have two functional copies of the IGF2R tumor suppressor gene, whereas virtually all nonprimate mammals (including rodents) have only one functional copy due to a process of "genomic imprinting." Damage to the IGF2R gene is associated with cancer development at multple sites. Humans, however, would need one additional tumorigenic transformation (to the second copy of the IGF2R gene) to reach an equivalent stage to that of rodents undergoing tumor development. Thus, Hahn et al.'s (1999) argument appears credible. Furthermore, the identification of four stages in the process of carcinogenesis is significant for our modeling effort because that number exactly matches the number of stages implied by the OSHA model of mesothelioma mortality in Section 2.3 (assuming a 10-year latency period).

Nonetheless, this does not mean that Hahn's model is the only mechanism underlying mesothelioma. Murthy and Testa (1999) identified a range of possible pathways to mesothelioma, including mutational deletions on chromosomes 1p, 3p, 6q, 9p, 13q, 15q, and 22q. Gene IGF2R is on chromosome 6q at a site adjacent to the deletions noted by Murthy and Testa (1999). Murthy and Testa (1999) concluded that multiple tumor suppressor genes are lost or inactivated in mesothelioma but that it was not currently possible to determine their identity or sequence.

Second, the fact that 10-20% of mesotheliomas are not associated with asbestos exposure means that the attributable risk (AR) for asbestos is in the range 80-90%. The attributable risk among exposed workers ($AR_H$) could be substantially higher (e.g., 99% or more). Thus, the fraction of cases among exposed workers not due to asbestos must be on the order of 1% or less and it would be difficult to segregate these cases for separate treatment in our models.

In addition, the meaning of the term "asbestos exposure" varies from one study to the next. Generally, the term includes occupational exposures; it may also include environmental exposures, some of which are known and documentable, with others unknown and undocumentable. Roggli et al. (1992b, p. 316) estimated the distribution of asbestos-body (coated asbestos fibers) counts from the lungs of 100 mesothelioma patients and found a bimodal distribution, with about 25% of cases overlapping the general population with a mean value approximately 1/1000 that of the high-count group. The ratio 1/1000 is consistent with estimates of environmental exposures for the general population (EPA, 1986, p. 162). The distribution for the general population had a mean of about 1.6 asbestos-bodies per gram of wet lung tissue, suggesting that there is a significant amount of asbestos fibers in the lungs of "nonexposed" persons. Consequently, it may be impossible to rule out asbestos as a causative agent in any mesothelioma.

Third, the finding that 60% of human mesotheliomas contain SV40 DNA fragments is somewhat tentative. Butel and Lednicky (1999, p. 128) noted that the 60% figure is the median of seven published estimates ranging from 0% to 86%, with a pooled mean of 48% (= 95/196). However, the 0% estimate was obtained in the largest data series with 50 tumors tested. Pilatte et al. (2000) tested six mesothelioma cell lines and found no evidence of SV40 DNA. However, they did find that commercially available mouse monoclonal antibodies are contaminated with a 90kDa protein of similar size to SV40 large tumor antigen and this may lead to false-positive results in some test series.

Butel and Lednicky (1999) reviewed the evidence on the cellular and molecular biology of SV40, noting that it is tumorigenic in rodents, that it is potentially tumorigenic in humans, and that it may have been a contaminant in polio vaccines given to 10-30 million children vaccinated in the United States between 1955 and 1963.

Strickler et al. (1998) used SEER data 1973-1993, Connecticut Tumor Registry data 1950-1969, and national mortality statistics 1947-1973 to evaluate cohort differentials in relative risks for three types of cancers linked to SV40 – mesothelioma, osteosarcoma, and ependymomas. No evidence of increased risk for cohorts exposed to SV40 via polio vaccinations was found. The possibility of effects becoming manifest in future years was recognized, but at least through 1993, no effect was detected. More recent data for 1994-1997 (SEER, 2000) indicate that the annual numbers of mesothelioma deaths have plateaued or declined slightly since reaching a peak in 1992, a pattern consistent with the diminution of asbestos exposure beginning in the early 1970s. To indicate the potential size of the effect, we considered the SEER (2000) report of the number of mesotheliomas in 1997 for the age group 45-49, the cohort identified by Strickler et al. (1998) as the group at highest risk of SV40 exposure. In total, there were seven mesothelioma cases (three males, four females) in the SEER data approximately 34-42 years after SV40 exposure. Five years earlier, there were four mesothelioma cases in the SEER data for this cohort. SEER represents about 10% of the U.S. cases, so that the national incidence in 1992 and 1997 was about 40 and 70 cases, respectively. These estimates do not support the hypothesis that SV40 exposure will result in large numbers of new mesothelioma cases.

The SEER (2000) data indicate that the male/female ratio of mesothelioma cases continues at about 3.5 to 1 – consistent with the hypothesis that the main cause is occupational exposure to asbestos. This ratio is consistent with attributable risks (AR) of 80% for males and 30% for females for asbestos-induced mesothelioma, which compares well with estimates of 85% and 23%, respectively, from Spirtas et al. (1994).

Fourth, the finding that crocidolite asbestos does not produce tumorigenic transformations of mesothelia cells in vitro, combined with the strong in vitro effect of SV40, must be interpreted in the context of the overwhelming amount of epidemiologic evidence in support of an asbestos effect and the lack of similar evidence for an SV40 effect on human mesothelioma incidence. Bocchetta et al. (2000) speculated that asbestos, in vivo, may act as an immunosuppressant that permits the SV40 infection to proceed without cell lysis in mesothelioma cells. Alternatively, asbestos may induce the production of oxygen free radicals that lead to gene alterations and carcinogenesis in vivo. Klein (2000) commented that both alternatives are possible mechanisms that should be further studied, but that the results to date do not prove that SV40 has a causative role in human mesothelioma.

We observe that neither mechanism nor the in vitro experiments conducted by Bocchetta et al. (2000) is consistent with the hypothesis that the simple physical presence of asbestos fibers in contact with mesothelial cells is sufficient to induce tumorigenic transformations. There is a large and growing literature on the molecular biology of asbestos-induced fibrogenesis and carcinogenesis that suggests a complex series of pathways through which the health effects of asbestos are mediated. Kamp and Weitzman (1999) reviewed this literature and concluded that free radicals, especially the iron-catalyzed hydroxyl radical and reactive nitrogen species, are important mediators of asbestos genotoxicity. They noted, however, that the precise mechanisms by which asbestos leads to DNA damage, disrupted signaling mechanisms, altered gene expression, mutagenicity, apoptosis, and altered immune responses are not firmly established.

Jaurand (1997) reviewed the literature on asbestos-induced genotoxicity and concluded that the mechanisms depended jointly on the fiber dimensions (length, diameter, and aspect ratio), its chemical composition, and the cell environment, with a critical role assigned to the process of phagocytosis.

Mossman and Churg (1998) reviewed the literature on asbestos-induced fibrogenesis, including lung burden studies, again finding a critical role for phagocytosis, with the fiber dimensions and chemical composition governing the cellular reactions. Additional details are provided in Robeldo and Mossman (1999).

# 3

# Forecasts Based on Direct Estimates of Exposure

## 3.1 Introduction

This chapter provides a broad overview of the models used by Selikoff (1981, 1982) and his collaborators in making direct estimates of past occupational exposure to asbestos and in projecting the current and future numbers of workers with various asbestos-related diseases (mesothelioma, lung cancer, and other cancers). Following this overview, we review in detail the methods and assumptions used by Selikoff (1981, 1982) in generating his forecasts and then discuss the sensitivity of these forecasts to various assumptions. The chapter concludes with a brief review of alternative projections of the impact of asbestos-related diseases.

## 3.2 Selikoff's Study: General Description

Selikoff (1981, 1982) estimated the size and composition of the asbestos-exposed labor force directly from labor force data. These estimates were independent of estimates of risk levels and disease incidence. The labor force estimates, specific to industrial and occupational categories, were applied to risk measures developed from a variety of case-control and cohort studies of the cancer risks of occupational exposure to asbestos. No effort was made to apportion health effects on the basis of inferred exposure levels. Exposure, to the degree it was represented in these projections, was represented by the stratification of the exposed labor force into occupational and industrial categories.

### 3.2.1 Data

Selikoff (1981, 1982) used three types of data. The first type of data was derived from government statistics on the structure and size of the U.S. labor force. Selikoff used Bureau of Labor Statistics (BLS) data to generate estimates of the labor force specific to occupation, industry, and date of employment. These data were adjusted by rates of job turnover derived from the Social Security Administration's (SSA) Continuous Work History Survey (CWHS).

The second type of data was derived from case-control and cohort studies of groups of workers exposed to asbestos. Selikoff used this type of data to develop direct estimates of typical exposure levels within industrial and occupational classifications and to derive indices of disease risk and dose-response from a wide range of studies.

The third type of data describes compensation patterns and tort litigation of the insulation worker cohort. These data provide insight into behavioral aspects of the decision to file suit. However, because they did not figure in Selikoff's disease-specific projections, we do not discuss these data further.

### 3.2.2 Model and Methods

The logic of Selikoff's projections differs from that used by Walker (see Chapters 4 and 5 for details). Selikoff used much more information on the structure of the exposed labor force, whereas Walker (1982) relied on a specific mathematical model of the disease mechanism (Peto et al., 1982) and epidemiological data (e.g., McDonald and McDonald, 1980) to infer many of the features of the exposed labor force. Selikoff's (1981, 1982) projection methodology is logically simpler, but Selikoff needed to make a number of specific assumptions about exposure levels in specific occupations. Thus, the simplicity of the projection model was, to a degree, offset by the complexity of assumptions made about data.

## 3.3 Selikoff's Six Tasks

Selikoff's (1981, 1982) methodology was complex. We will organize our review into a discussion of the six tasks necessary to carry out his projections. The *first* task involved an evaluation of data on the typical levels of asbestos exposure found in specific industries and occupations. The *second* task involved developing a model to estimate the turnover of the workforce in specific occupations and industries to determine the total number, timing, and duration of employment of exposed workers. The *third* task involved reviewing evidence from case-control and cohort studies to determine the level of cancer risks for specific occupational and industrial groups. The *fourth* task involved an assessment of two models for projecting the age, date, and exposure-duration specific health effects of occupational exposure to asbestos. Because Selikoff assumed that mesothelioma is so rare that there is virtually no nonasbestos-related incidence among occupationally exposed workers (equivalently, $AR_H \approx 1.0$), one projection model, based on absolute risks, was developed for mesothelioma and another, based on relative risks, for all other tumor types. The *fifth* task involved the computation of Selikoff's projections and the presentation of results. The *sixth* task involved simple strategies to project estimates of death due to asbestosis. Although asbestosis was a major component of Walker's projections, relatively little effort was expended by Selikoff on the asbestosis mortality projections.

### 3.3.1 Task 1: Identify the Industries and Occupations Where Asbestos Exposure Took Place

The first task Selikoff faced was to identify industrial categories [using Standard Industrial Classification (SIC) codes] where asbestos exposure occurred and, where possible, to identify typical exposure levels. In this examination, *mining and milling workers* were not included due to the relatively small (600) numbers of such workers. Selikoff identified 11 industrial or occupational categories whose workers likely had significant exposure to asbestos. These groups were identified on the basis of the following:

1. A qualitative evaluation of work tasks involving asbestos
2. Studies of direct and indirect exposure to asbestos in the workplace yielding quantitative measurements of the intensity of exposure
3. Evidence of increased risk of asbestos-related disease (e.g., mesothelioma)
4. "Subjective" judgments

The extensive use of subjective estimates and assumptions in Selikoff's review of occupational and industrial categories should be emphasized.

#### *3.3.1a Primary asbestos manufacturing*

Three major industrial groups were identified in the primary manufacturing sectors using information from sources such as the Asbestos Information Association (AIA).

First, the *asbestos products industry* (SIC 3292) produced a wide range of materials involving asbestos. For this group, it was possible to characterize early exposure levels on the order of 25 to 35 f/ml. More recent exposures (e.g., in 1975) were much lower (e.g., 1.5 to 4.0 f/ml). Within this category, all production and maintenance workers were assumed at risk. The information on exposure, although often taken from data and provided in quantitative terms, was based on a few studies. Furthermore, subjective assessments of exposure patterns and levels were often used. One major problem was that individual exposures to asbestos often were characterized on the basis of industrywide measures. This produced crude approximations of the actual exposures of any given individual.

The second group of workers was employed in the *gaskets, packing, and sealing devices industry* (SIC 3293). Prior to 1972, asbestos was the predominant raw material. The industrial classification was expanded in 1972 so that after that date, it was assumed that only half of the workers in this category worked with asbestos. Thus in the projections all workers up to 1972 were counted as exposed, and after 1972, only half were.

The third group of workers was employed in *building paper and building board mills* (SIC 2661). Because about half of the workers in 1972 were employed in plants where asbestos was the raw material, half of the production and maintenance workers in this group were assumed to be exposed to asbestos.

For primary manufacturing industries, quantitative measures of exposure were available in the Asbestos Information Association-Weston Report (Daly et al., 1976). This report contained information on the range and typical levels of exposure in various industrial components in 1975. The typical exposure in such industries varied from 0.5 to 4.0 f/ml.

### *3.3.1b Secondary asbestos manufacturing*

A second major component of asbestos exposure was in industries involving secondary manufacture of asbestos products (i.e., industries where asbestos products were used in the manufacture of other products). There were four major secondary manufacturing categories. The first, *heating equipment - except electric and warm air furnaces* (SIC 3433), produced such devices as boilers and heating furnaces, which were often constructed using asbestos. One-half of production and maintenance employees were assumed to be in the population at risk. *Fabricated plate workers* (*Boiler Shops*) (SIC 3443) were assumed to have half of their production and maintenance employees at risk. In the *industrial process furnaces and ovens* (SIC 3567) group, it was assumed that all production and maintenance employees were exposed. For *electric housewares and fans* (SIC 3634), it was assumed that 10% of the production and maintenance employees were at risk.

Although asbestos was used in a variety of other secondary manufacturing industries, it was impossible for Selikoff to estimate the number of exposed individuals in all secondary manufacturing using BLS data. In order to assess the estimates obtained from the four secondary manufacturing categories, Selikoff compared those numbers with estimates obtained by AIA-Weston (Daly et al., 1976) for the asbestos industry.

AIA-Weston categorized the secondary manufacturing industries in 1975 according to the primary source of asbestos materials used therein: asbestos paper (158,400 employees), friction products (27,600 employees), asbestos cement sheets (19,200), gaskets and packings (12,000), reinforced plastics (8400), asbestos textiles (6000), and other miscellaneous sources (8400). In total, AIA-Weston estimated that there were 240,000 persons exposed in secondary manufacturing in 1975 compared to Selikoff's estimate of only 38,000 for the 4 industries (i.e., SIC 3433, 3443, 3567, and 3634). It appears that Selikoff's and AIA-Weston's definitions of primary and secondary manufacturing differed because, in the industry analyses, only 23,000 were exposed in primary manufacturing as opposed to Selikoff's estimate of 31,000. The inconsistencies for secondary manufacturing were not directly resolvable (with the asbestos industry's figures much higher). As a compromise (based on the assumption that it was "unlikely" that 158,000 employees would have had significant exposure in the manufacture of products containing asbestos paper and with the number of other categories appearing "reasonable"), Selikoff selected "a number equal to twice the four groups specified by SIC numbers" as his estimate of workers involved in the secondary manufacture of asbestos: 76,000 in 1975.

As for primary manufacturing, an effort was made to determine typical asbestos exposure levels in secondary manufacturing. Only data on fiber concentrations in later years were available, and these data were from industry sources. These data suggested levels of exposure ranging from 0.2 to 6.5 f/ml in secondary manufacturing.

### 3.3.1c Shipbuilding and repair

Another industrial category with significant exposure to asbestos was *shipbuilding and repair* (SIC 3731). No direct measurements of the level of asbestos exposure could be determined for shipyard workers. As a consequence, Selikoff's rationale for including this group in the exposed workers category was twofold. First, the importance of asbestos in the shipbuilding process was argued by Selikoff to be well known. Second, there were many studies of shipyard workers (e.g., Tagnon et al., 1980) which demonstrated elevated mesothelioma risks even when exposure was intermittent or indirect. All production and maintenance employees of private and naval shipyards were included in the estimates of the number of exposed shipyard workers. Estimates for naval shipyards came from the U.S. Department of the Navy.

### 3.3.1d Construction

Construction industries accounted for 70-80% of the U.S. consumption of asbestos fiber. Based on this high level of consumption of asbestos, Selikoff assumed that many construction workers would have significant levels of exposure. Industrial categories which Selikoff judged to have significant asbestos exposure included *general contractors for multifamily residential buildings* (SIC 1522), *general contractors for nonresidential buildings* (SIC 154), *water, sewer, pipe line, communication, and power line construction* (SIC 1623), and *special trade contractors* (all workers in SIC 17 except 1771, 1781, 1791, 1794, and 1796).

The primary sources of direct exposures included installation and removal of asbestos-cement pipes, asbestos-cement sheets, architectural panels, built-up roofing, asbestos drywall, and asbestos roofing felts, asbestos insulation of pipes, tubings, heating units, and generators, and use of asbestos-containing paints, coatings, and sealants. In addition to direct exposure to building products, significant exposure was felt to exist from the practice (1958-1972) of spraying asbestos insulation. Selikoff felt that such spraying could increase the risk for all workers on a construction site.

For workers in the SIC codes involved in the construction industry, Selikoff attempted to analyze the extent and level of exposure using a variety of insights about the use of asbestos in specific construction tasks, limited exposure measurements, and the occurrence of mesothelioma as prima facie evidence of asbestos exposure among potentially exposed workers. For example, Selikoff assumed that all construction workers in SIC 1522 and 154 should be included in the at-risk group. Because 30% of the water distribution pipe sold in the United States in 1974 was made of asbestos cement, it was assumed that 30% of the workers in SIC 1623 were exposed. In addition, 5% of the workers in SIC 16 (*construction other than building construction*) were argued to be exposed through working on the brakes of heavy construction equipment. All workers in SIC 171 [*plumbing, heating (except electrical), and air conditioning*] and 172 (*painting, paperhanging, and decorating*) were assumed to have been exposed. For SIC 172, studies were cited of exposure of drywall taping workers employed in New York City with exposures ranging from 5.3 to 47.2 f/ml for various operations. Selikoff argued that other construction workers had a significant indirect exposure to asbestos dust from those operations. For the remainder of SIC 17 (except the five excluded groups), he argued that the proportion exposed was 50% during 1958-1972 and 20% during 1940-1957 and 1973-1979.

#### 3.3.1e Utility services

Selikoff also identified workers in *electric, gas, and combination utility services* (SIC 491, 492, and 493) as having significant exposure to asbestos. To substantiate this claim, he cited English and French studies of elevated risks of asbestos-related disease among persons engaged in maintenance work at power stations. Consequently, 25% of physical workers in electric and gas utilities were included in the population at risk to asbestos-related disease.

#### 3.3.1f Selected occupational categories

In addition to identifying asbestos exposure by industrial groups, Selikoff argued that there were occupational groups with significant asbestos exposure not included in the above-discussed industrial groupings (i.e., primary and secondary asbestos manufacturing, shipbuilding and repair, construction, and utility workers). The industry-based estimates were augmented (after eliminations for double counting) by six specific occupational groups at risk:

1. *Asbestos and insulation workers* – evidenced by both elevated disease risks and measurements of high dust concentrations for certain activities

2. *Automobile body repairers and mechanics* – evidenced by both measurements of asbestos exposure and radiological evidence of abnormalities compared to blue collar controls
3. *Engine room personnel, seagoing vessels, U.S. Merchant Marine* – evidenced by X-ray films showing a high proportion (16-20%) of pleural abnormalities among merchant marine engineers
4. *Maintenance employees: chemical and petroleum manufacturing* (including all maintenance workers in SIC 28 and 29) – evidenced by chest X-ray abnormalities among petroleum and chemical maintenance workers
5. *Steam locomotive repair employees* – evidenced by mesothelioma among such workers; all employees of railroad repair "back shops" are included for the 1940s and, for the 1950s, the annual number of workers was held proportional to the ratio of steam to the total of steam and nonsteam locomotives in service
6. *Stationary engineers, stationary firemen, and power station operators* – evidenced by a high proportion (60%) of chest X-ray abnormalities among a sample of 34 stationary engineers with over 20 years experience in this trade

### 3.3.2 Task 2: Estimate the Number, Timing, and Duration of Employment of Exposed Workers

Beyond identifying 11 occupational and industrial categories in which workers were likely to be exposed to asbestos, Selikoff (1981, 1982) aimed to project the future health consequences of this exposure. To make these projections, one, ideally, would want three types of information:

1. The numbers of workers in each category
2. The distribution of employment periods
3. The times, durations, and intensities of exposure

Selikoff pointed out that, in most cases, such data do not exist. Partial data existed, but, even in these cases, the quality and coverage varied by occupation and industry. The best data available were for primary asbestos manufacturing, shipbuilding, automobile repair, and insulation work. Poorer-quality data were available for secondary asbestos manufacturing, construction, and the maintenance industries.

To make up for the lack of information, Selikoff assumed, first, that the increase in disease due to occupational exposure to asbestos could be described accurately by a linear dose-response function. If this assumption were valid, the projections would not be adversely affected by certain types of error in estimating the size of the exposed population and the duration of exposure.

Second, Selikoff developed a model of the turnover of workers in specific job categories to estimate the total numbers of persons who were newly employed over various periods of time. A model of the workforce was developed using a steady-state assumption.

#### *3.3.2a Task 2a: Workforce turnover model*

Specifically, if $N$ is the number of workers in an industry or occupation, $\alpha$ is the annual fraction of new workers (i.e., $N_{NEW}/N$), and $\beta$ is the annual fraction permanently leaving the occupation or industry, the change with time of the workforce ($dN$) is

$$dN = N(\alpha - \beta)dt.$$

For small changes in $N$ (i.e., $|\Delta N| < 0.05N$, where $\Delta N$ is the net increase or decrease in the number of workers), this may be well approximated by

$$N = N_0 \exp[(\alpha - \beta)t].$$

At steady state, the number of workers entering a job category equals the number leaving (i.e., $\alpha = \beta$) and the average duration of employment is $1/\alpha$ or $1/\beta$.

The equations are slightly different when considering finite changes over a fixed interval (e.g., a year). The change over a year can be expressed as

$$\Delta N = (\alpha - \beta)N.$$

This may be written as

$$\alpha = \beta + (\Delta N/N).$$

Let $T$ be the time necessary to add $N$ new workers to the workforce under steady-state conditions (i.e., $\alpha = \beta$); then,

$$N = \alpha NT,$$

implying $T = 1/\alpha$, as expected from the continuous change model.

Selikoff used this model of labor force turnover to generate estimates of excess mortality from past occupational exposure to asbestos. The excess mortality during a decade, say $M_t$, among a group of persons newly employed in a job category with asbestos exposure is proportional to the number of new hires, $\alpha N$, multiplied by the average duration of their employment $T$, or

$$M_t \propto \alpha NT.$$

Because under steady-state conditions, the average duration of employment is $1/\alpha$, the equation may be written

$$M_t = KN.$$

The proportionality constant, $K$, includes the appropriate risk and exposure variables for the industry. Thus, if the workforce is in a steady state ($\alpha = \beta$), then excess mortality can be simply estimated if the number ($N$) employed in that industry and the associated risk parameter ($K$) are known. Clearly, if the workforce is not in a steady state (i.e., $\alpha \neq \beta$), then information on both new hires and job separation is required.

### 3.3.2b Task 2b: Workforce exposure estimates

To use the model of workforce turnover to estimate excess mortality, it is necessary to assess the size ($N$) and rate of turnover (i.e., $\alpha$ and $\beta$) of the asbestos-exposed workforce. The primary data employed by Selikoff for this task were BLS data on annual employment and earnings in the United States. 1909-1978 in (a) primary manufacturing, (b) select secondary asbestos manufacturing, (c) construction, (d) electric, gas, and utility services, and (e) chemical and oil refining employees. The specific segments of these industries employed, estimates of proportions exposed, and exposure levels were discussed in Task 1. Data on stationary engineers and firemen were also derived from BLS sources. Sources other than the BLS were required for (a) insulation work, (b) shipbuilding and repair, (c) automobile maintenance and repair, (d) railroad steam locomotive repair, and (e) merchant marine engine room work. These included unions, trade associations, the U.S. Navy, and other government sources. The data assembled for the set of 11 industry/occupation categories were judged by Selikoff to be "stable" 1950-1980 and to accurately reflect employment and its change with time (exception: shipbuilding 1948-1952 averaging 189,000 was estimated to be 128,000 in 1950).

Two adjustments had to be made to BLS figures to obtain estimates of the size of the exposed worker population over the period 1940-1979. First, data for some industries do not extend as far back as 1940. In these cases, either (a) BLS series were extrapolated back to 1940 using regressions of the available data on related variables (i.e., numbers of workers in related categories; see Selikoff, 1981, p. 109, Table 2-2) or (b) a straight-line trend was assumed between a census of manufacturing (1939) or census of population (1940) and the earliest year of the relevant BLS series.

Second, there were problems in estimating the rates of turnover (i.e., $\alpha$ and $\beta$) in various industries. For example, for some industries, the available BLS information gives fractions of accessions and separations for given calendar periods subsequent to 1958 for individual establishments. These data may not represent new hires for the industry as a whole (i.e., the numbers of persons who were employed in the industry for the first time). This is especially serious for the construction industry, for which there was a great potential for overestimating the numbers of persons exposed in that industry by treating each accession as a new hire.

To eliminate the problem of duplicate person counts, the numbers of new hires for major industry groups were compared with data from the Continuous Work History Survey (CWHS) of the Social Security Administration (SSA) for the period 1957-1960 (1957 is the first year for which industrial classifications were available in the CWHS). For major industry groups, information was obtained on the number of persons employed in 1960 who were employed in the same industry in 1957. This permitted the calculation of an annual transfer rate from one industry group to another (but not between industries within a group). BLS statistics on permanent retirement or death for each industry were also available. In steady state, the SSA separation rate, plus retirements or deaths, would equal the new-hire rate. The SSA data were compared with the annual rate of new hires from the BLS data for 1958-1960 corrected by the increase/decrease in the workforce size over the 3-year period (such changes, all less than 10%, were attributed to new hires).

In comparing SSA and BLS data, Selikoff found that the fractional gains in the chemical industry and oil refinery operations (0.166 and 0.132, respectively) were close to the number of transfers from the nondurable goods industry (0.111). For these industries, the BLS data on new hires in SIC 28 or 29 were used, but were reduced by 30% to reflect internal transfers. For primary and secondary manufacturing, less transfer was expected between companies, so a figure 80% of that in the BLS data was adopted. For shipyard workers, rehires at another shipyard were expected to be greater, so a rate equal to 50% of the BLS new-hire rate for SIC 3831 was assumed (0.216). For construction trades (except insulation workers), stationary engineers and firemen, and automobile mechanics, the SSA rates were accepted directly. For utilities, the SSA rate was increased by 50%. For insulation workers, the new-hire rate was based on union membership. For nonunion workers, the union hire rate was multiplied by 0.8 for permit workers, by 2.0 for nonunion new hires, and by 1.2 for nonconstruction insulators. Selikoff did not offer a detailed rationale for each of these seemingly arguable adjustments.

Selikoff developed several additional adjustments in response to limitations in BLS data. For example, BLS data on shipbuilding and repair referred only to civilian shipyards. Data on naval shipyards were obtained through the U.S. Navy. It was estimated that 4.3 million men worked in shipyards for a short period of time during World War II, whereas the 1945 estimate of the permanent workforce was 175,000 men. Risks for these WWII workers and for 9000 shipyard insulators were handled separately from other data. Also, because it was not possible to isolate automobile mechanics in SIC 75, SIC 515-2, and SIC 554, census data (and intercensal interpolations) were employed. Finally, employment data reported by the Association of American Railroads were used to estimate exposures to asbestos during railroad steam locomotive repairs. Adjustments to these data included reductions of 45% of men in equipment and stores to exclude carmen, 50% of the residual number to exclude maintenance workers not employed at "back shops," and 11% of the balance to exclude salaried supervisors, coach cleaners, and store laborers.

The estimates of workforce turnover were clearly based on many subjective assumptions. Selikoff cautioned about the lack of precision of these estimates. Selikoff argued, however, that (under a linear dose-response relationship) differences in the estimated numbers of workers exposed would not materially affect his mortality estimates. This was because a misestimate of the new-hire rate could be balanced by a compensatory change in the estimate of employment duration. Having used actual labor force data, Selikoff was not required to deal with artificial "worker equivalents," as was Walker (1982; see Chapter 4).

The estimated annual new-hire rates were applied to annual employment data for each occupation or industry to produce estimates of the numbers of new persons exposed to asbestos each year. Data were then accumulated for each decade since 1940. In industries with workers already included in an occupational group, adjustments were made to employment and new-hire data to eliminate duplication by developing adjustment factors based on the BLS National Industry-Occupation Matrix for asbestos and insulation workers and the 1970 Census of Population for stationary engineers, firemen, and power station operators. No adjustment was necessary for automobile mechanics.

Furthermore, an adjustment was used to eliminate double counting of workers hired since 1940 who previously had asbestos exposure. The adjustment factor was derived from a group of 2544 persons (Selikoff, 1981, p. 79) identified in five studies conducted by the Mt. Sinai Laboratory of operations with identifiable asbestos exposure. The net impact of the adjustment was to reduce the estimate of the exposed population by 10%. This 10% reduction in the number of persons exposed to asbestos was compensated for, under the workforce turnover model, by a 10% increase in duration. Selikoff conceded that the uncertainties in either the estimated size of the exposed population, or in the duration of exposure, probably "greatly" exceed 10%.

Selikoff estimated that 27.5 million persons suffered occupational exposure to asbestos in the 40-year period 1940-1979; 26.0 million were new entrants to the asbestos workforce. Selikoff argued that many of these persons were probably exposed to relatively low levels of asbestos.

From the above data and estimates, using the workforce turnover model, Selikoff (1981, p. 121, Table 2-13) calculated average exposure durations by decade during 1940-1979 from the fractional new-hire rates (i.e., $\alpha$) adjusted by changes in total workforce at different periods in time (i.e., to adjust for $\alpha \neq \beta$). The results for 1950-1969 are used in Chapter 8 in our hybrid projection model (see Table 8.8).

### 3.3.3 Task 3: Estimate Risk Differentials Among Occupations and Industries

To estimate asbestos-related cancer mortality in an industry or occupation using the workforce turnover model, it is necessary to estimate $K$ (i.e., the risk parameter for that job category). Because of the lack of actual exposure data for each industry or occupation group, measures of relative risk were employed. Specifically, the ratio of the mortality risks for each industry or occupation relative to the mortality risks of insulation workers (for equal times of employment) were estimated using one of three types of data:

1. Mortality data
2. Directly measured average asbestos exposure data
3. Prevalence data on X-ray abnormalities after long-term employment

**Table 3.1: Risk of Asbestos-Related Cancer Relative to That of Insulation Workers After Twenty-Five Years of Employment**

| Occupation or Industry | Relative Risk | Source of Data for Estimate |
|---|---|---|
| Primary Manufacturing | 1.00 | Mortality, Exposure |
| Secondary Manufacturing | 0.50 | Exposure |
| Insulation Work | 1.00 | Reference Population |
| Shipbuilding and Repair (excluding insulation) | 0.50 | Mortality, X-Ray |
| Construction Trades (excluding insulation)* | 0.15–0.25** | Mesothelioma |
| Railroad Engine Repair | 0.20 | Mesothelioma |
| Utility Services | 0.30 | Mesothelioma |
| Stationary Engineers and Firemen | 0.15 | X-Ray |
| Chemical Plant and Refinery Maintenance | 0.15 | X-Ray, Mortality |
| Automobile Maintenance | 0.04 | X-Ray, Exposure |
| Marine Engine Room Personnel (excluding U.S. Navy) | 0.10 | X-Ray |

*See text for percentage of construction population considered at risk.
**Risk for years 1958-1972 when the use of sprayed asbestos-fireproofing was common.

Mortality = Group mortality data
Exposure = Exposure measurements
X-Ray = Prevalence of X-ray abnormalities
Mesothelioma = Number of mesothelioma cases in general population

Source: Selikoff (1981, Table 2-16).

To use the third option to develop a relative risk, Selikoff assumed that the percentage of X-ray abnormalities manifest by workers in a particular job category after 20 years of employment was proportional to the total asbestos dose. For industries for which none of these measures was available, relative risks were ascertained from the number of mesothelioma cases identified in McDonald and McDonald's (1980) survey.

From the four sources of data, Selikoff (1981, 1982) calculated a table (Table 3.1) of occupation or industrial relative risks with insulation workers having a relative risk of 1.0.

The relative risks for different job categories varied greatly, from 1.0 for primary manufacturing to 0.04 for automobile maintenance. Selikoff warned that the data used to derive the relative risk estimates were quite limited.

The Occupational Safety and Health Administration (OSHA) has regulated asbestos since 1971. In 1971, the 8-hour time-weighted average permissible exposure limit was set at 5 f/ml. This limit was reduced to 2 f/ml in 1976, to 0.2 f/ml in 1986, and to 0.1 f/ml in 1995. To reflect the initial reductions in exposure, Selikoff (1981, 1982) assumed that the relative risks for manufacturing, insulation work, shipbuilding, and utility employment were reduced to 0.1 for the period 1972-1979. For other groups (except automobile maintenance), the relative risks were assumed to drop to 0.05. Occupational exposures after 1979 were assumed to be negligible.

To assess the reasonableness of his relative risk estimates, Selikoff compared figures in Table 3.1 with estimates for selected occupations made in McDonald and McDonald's (1980) case-control studies. Selikoff argued that the relative risks in Table 3.1 are virtually identical to McDonald and McDonald's estimates. Specifically, McDonald and McDonald found relative risks (compared to unexposed persons) of 46.0 for insulation work, 6.1 for manufacturing, 4.4 for heating trades (utility services), 2.8 for shipyard employment, and 2.6 for construction. For comparison, Selikoff multiplied the relative risks reported in Table 3.1 (assuming an average risk of 0.65 for manufacturing) by the average duration of employment for workers in those categories from 1940 to 1969 (i.e., 13.2, 2.0, 4.7, 1.9, and 6.4 years, respectively). The values obtained for construction workers were divided by 2 because only 50% of construction workers were assumed to be exposed to asbestos. The calculations produced the values, after rounding, 13.2, 1.3, 1.4, 0.95, and 0.5, respectively. By norming these figures to the relative risk of 46.0 obtained for insulators (i.e., multiplying by 46/13.2), Selikoff could compare them with McDonald and McDonald's estimates. The values for insulation workers were not informative because they were forced to be equal. However, the relative risks of 4.6, 4.9, 3.3, and 1.8 could be compared with McDonald and McDonald's estimates of 6.1, 4.4, 2.8, and 2.6, respectively. McDonald and McDonald (1980) did not provide standard errors for these relative risks. However, their relative risks were actually odds ratios, so their standard errors are estimable from Fleiss (1981, p. 63, Eq. 5.19)

$$\mathrm{SE(OR)} = \mathrm{OR} \times \left[\frac{1}{n_{11}} + \frac{1}{n_{12}} + \frac{1}{n_{21}} + \frac{1}{n_{22}}\right]^{1/2},$$

where $n_{11}$, $n_{12}$, $n_{21}$, and $n_{22}$ are the four counts involved in the odds ratio (see Section 2.2.1). For the four values being compared (i.e., 6.1, 4.4, 2.8, and 2.6), the standard errors are 2.7, 1.1, 1.0, and 0.7, respectively. Thus, three of the four values estimated by Selikoff are within one standard error of McDonald and McDonald's estimates. The figures from the two sources, although close, are not "virtually identical."

With relative risk estimates in hand for the 11 occupation/industry groups, Selikoff estimated what fraction of the estimated 27.5 million exposed workers had lower exposure and what fraction had heavier exposure. With the OSHA permissible exposure limit at 2 f/ml in 1976-1986, Selikoff chose a cumulative dose of 2 f-yr/ml as the cutoff between lower and heavier exposure. With the current OSHA permissible exposure limit at 0.1 f/ml, this cutoff corresponds to 20 years at the maximum permissible exposure limit. The same level, however, could have been reached in primary asbestos manufacturing 1940-1970 in as little as 1 month.

To develop these estimates, Selikoff had to account for additional features of employment histories in specific job categories. Specifically, from seniority lists of workforces in selected asbestos-using industries (e.g., a large asbestos manufacturing plant, a major shipyard, a plastics polymer plant, and an asbestos insulation production plant), Selikoff made comparisons with estimates of workforce turnover from BLS and SSA sources. The data from the seniority lists were also used to determine the distribution of employment durations in those industries.

The data from the seniority lists showed that there was a high turnover rate during the first month after hire. Selikoff argued that this was a result of high rates of termination during 1-month probationary periods. Although there was significant variation across industries, Selikoff argued that probationary periods and high early turnover rates are general phenomena in many industries.

Because workers with short periods of employment would likely have low cumulative doses of asbestos, this high rate of early turnover was used to calculate the portion of the total exposed workforce with lower levels of exposure. For example, for persons employed for 2 months in primary asbestos manufacturing or insulation work, the cumulative exposure was only 2-3 f-yr/ml (an annual average exposure of 12-18 f/ml times 1/6 of a year). In secondary asbestos manufacturing or shipbuilding and repair, it would have taken 4 months to attain the same cumulative dose, and in construction trades, 8-12 months. In each case, the multiplier applied to the 2-month interval was inversely related to the relative risk estimate in Table 3.1.

Selikoff assumed that 40% of new hires in primary and secondary manufacturing and 20% of new hires in other industries left employment within 2 months and that the rate of termination ($\beta$) after the 2-month probation was constant. After 2 months, the residual duration of employment was assumed to be exponentially distributed, with survival function $S(t) = e^{-\beta(t-2)}$. Here, $t$ is measured in months, not years.

The analysis of probationary periods suggested the existence of two distinct components of the total workforce – short-term, low exposure populations who worked as little as 2 months and a long-term workforce with a constant rate of separation or dismissal. These assumptions implied that 8.7 million of the total exposed workforce of 27.5 million had lower exposures – less than 2-3 f-yr/ml. Of the 27.5 million workers exposed over the period 1940-1979, Selikoff estimated that 21.0 million survived to 1980. Of the 21.0 million who survived, 14.1 million were estimated to have had exposures greater than 2-3 f-yr/ml and 6.9 million to have had lower exposures.

### 3.3.4 Task 4: Estimate Dose-Response Models for Cancer Risks

To project the number of cancer deaths over the period 1965-2029 from occupational exposure to asbestos, Selikoff determined how cancer risks depend on age and duration of exposure, using the cancer mortality experience of

17,800 insulation workers followed from 1967 to 1976 (see Section 2.3.1; also see Figures 2.2 and 2.4, Selikoff, 1981).

### 3.3.4a Task 4a: Lung and other cancers

An analysis of insulation workers suggested that the relative risks of lung cancer are independent of age and any preexisting risks at the time of initial exposure. The relative risks of lung cancer increased linearly with the duration of exposure up to 40 years, at which time, the increase was more than sixfold. After 40 years, a significant drop in the relative risk was observed.

A linear increase with exposure duration in the relative risk of lung cancer is notable for two reasons. First, under a linear model, as the duration of exposure to asbestos increases, the relative risks continue to increase for 30 or more years, even though the background risk itself exhibits a 10- to 20-fold increase over the same 30-year interval. Second, the backwardly extrapolated line meets the relative risk of 1 at a point at most 10 years after the start of exposure. If the tumor growth started at least 1 or 2 years before becoming clinically manifest and the tumor became clinically manifest 1 or 2 years before death, the elevation in the risk of tumor onset must have begun very shortly (less than 6 years) after the initiation of exposure. Both the short lag time (i.e., 10 years or less) and the independence of the relative risk from the age at which exposure started are consistent with observations made in Seidman et al. (1979).

Selikoff (1981, 1982) integrated the dependency of the relative risk of lung cancer on exposure duration with the industry-specific relative risks reported in Table 3.1. A linear increase in relative risk was assumed to start 7.5 years after exposure initiation and to continue at a fixed rate of 0.23 per year (for insulators) for the period of the average duration of employment for that industry; following that, the relative risk was assumed to remain constant until 40 years after exposure onset and then to decrease linearly to 1.0 over the next 30 years. The magnitude of the increase for other workers was adjusted downward from the 0.23 per year increase for insulators and primary asbestos factory workers according to the relative risks in Table 3.1. The same time course observed for lung cancer was applied to other types of cancer except mesothelioma.

### 3.3.4b Task 4b: Mesothelioma

Mesothelioma was modeled differently because Selikoff argued that there was no effective background rate for these workers in the absence of asbestos exposure. Therefore, a measure of the absolute risk of death should be used. An analysis of mesothelioma deaths for insulation workers provided evidence that the absolute risk of mesothelioma was directly related to the duration of exposure and was independent of age. Selikoff (1981, pp. 90-91) argued that the risk of death from mesothelioma increased as the fourth or fifth power of time from onset of exposure for 40 or 50 years. For the projections, it was argued that the risk of mesothelioma reached a peak of 1.2/100 person-years of exposure after 45 years for persons with 25 or more years of occupational exposure. For insulators with less than 25 years employment, the estimated risk was reduced by the fraction of 25 years employed. For other groups, the estimated risk was jointly scaled by the relative risk of the group (Table 3.1) and the fraction of 25 years employed.

### 3.3.5 Task 5: Project Future Asbestos-Related Cancer Mortality

Using the above-outlined methods, the average annual excess numbers of (1) lung cancer, (2) mesothelioma, (3) gastrointestinal (G.I.) and other asbestos-related cancers, and (4) total excess cancer deaths was projected for each 5-year period, 1965-1969 to 2025-2029, for each cohort of new entrants into each of 11 occupational/industry populations (Tables 3.2 and 3.3). The dominant contributors to asbestos-related cancer deaths were the shipbuilding and construction industries.

Selikoff (1981, 1982) did not provide details on the equations used to conduct the projections. He did indicate that the nonasbestos-related deaths were estimated using U.S. national death rates for total mortality assuming that the rates for 1975-1979 continued to apply through 2030. Recent evidence (SSA, 1992, 1996, 1999) indicates that these nonasbestos mortality rates have, in fact, declined and are expected to continue to decline on the order of 0.5-1.5% per year throughout the projection period. Consequently, Selikoff's projections may underestimate the total number of asbestos-related deaths, with increasing errors in the later years of the projection.

Selikoff argued that although the total number of malignancies is uncertain, the data on the time course of the distribution of malignancies are fairly good. Thus, Selikoff argued that we can be confident that wartime and post-war asbestos effects are about equal (although the wartime experience is more concentrated). Furthermore, Selikoff suggested that excess cancer mortality among construction workers would become important in the future due to the use of sprayed asbestos insulation during 1958-1972. Selikoff projected that of the total mesothelioma incidence during 1940-1999, only a third had occurred by 1980. This implied that the rate of occurrence would be four times higher during 1980-1999 where two-thirds of mesothelioma cases would occur in 20 years as opposed to one-third in 40 years 1940-1979.

### 3.3.6 Task 6: Estimate and Project Deaths Due to Asbestosis

Selikoff (1981, 1982) did not provide detailed projections of the numbers of deaths due to asbestosis because he did not think that future asbestosis mortality would be significant relative to cancer mortality, given the high exposure levels required for asbestosis. By contrast, for Walker (1982), a large portion of the total projected effect of asbestos was linked with asbestosis. Selikoff noted

**Table 3.2: Projected Annual Excess Deaths from All Asbestos-Related Cancer in Selected Occupations and Industries, 1965-2029**

| | Number Deceased in Quinquennium | | | | | | | | | | | | | |
|---|---|---|---|---|---|---|---|---|---|---|---|---|---|---|
| Industry or Occupation | 1965-1969 | 1970-1974 | 1975-1979 | 1980-1984 | 1985-1989 | 1990-1994 | 1995-1999 | 2000-2004 | 2005-2009 | 2010-2014 | 2015-2019 | 2020-2024 | 2025-2029 | Total |
| Primary Asbestos Manufacturing | 1,185 | 1,560 | 1,925 | 2,225 | 2,470 | 2,550 | 2,455 | 2,225 | 1,835 | 1,390 | 935 | 570 | 300 | 21,625 |
| Secondary Manufacturing | 1,180 | 1,520 | 2,015 | 2,535 | 3,050 | 3,295 | 3,370 | 3,245 | 2,920 | 2,445 | 1,835 | 1,260 | 745 | 29,415 |
| Insulation Work | 1,330 | 1,870 | 2,485 | 3,060 | 3,525 | 3,710 | 3,615 | 3,260 | 2,890 | 1,960 | 1,395 | 865 | 470 | 30,435 |
| Shipbuilding and Repair | 7,260 | 9,325 | 11,685 | 12,465 | 13,550 | 12,255 | 10,380 | 8,295 | 6,280 | 4,595 | 3,140 | 2,005 | 1,095 | 102,330 |
| Construction Trades | 3,890 | 5,675 | 8,205 | 10,715 | 12,965 | 15,020 | 16,540 | 16,950 | 15,955 | 13,485 | 9,980 | 6,215 | 3,345 | 138,940 |
| Railroad Engine Repair | 645 | 730 | 810 | 835 | 735 | 650 | 455 | 270 | 140 | 50 | 10 | 0 | 0 | 5,330 |
| Utility Services | 745 | 935 | 1,150 | 1,335 | 1,495 | 1,560 | 1,550 | 1,450 | 1,270 | 1,035 | 760 | 510 | 295 | 14,090 |
| Stationary Engineers and Firemen | 2,170 | 2,635 | 3,155 | 3,605 | 4,080 | 4,325 | 4,375 | 4,095 | 3,640 | 3,010 | 2,245 | 1,520 | 895 | 39,750 |
| Chemical Plant and Refinery Maintenance | 1,025 | 1,345 | 1,685 | 2,020 | 2,285 | 2,410 | 2,360 | 2,185 | 1,875 | 1,505 | 1,085 | 710 | 410 | 20,900 |
| Automobile Maintenance | 880 | 1,180 | 1,520 | 1,920 | 2,350 | 2,620 | 2,890 | 2,930 | 2,880 | 2,690 | 2,290 | 1,730 | 1,110 | 26,990 |
| Marine Engine Room Personnel | 195 | 235 | 280 | 315 | 320 | 300 | 275 | 230 | 190 | 135 | 95 | 60 | 30 | 2,660 |
| Total | 20,505 | 27,010 | 34,915 | 41,030 | 46,825 | 48,695 | 48,265 | 45,135 | 39,875 | 32,300 | 23,770 | 15,445 | 8,695 | 432,465 |
| | Cumulative Total | | | | | | | | | | | | | |
| | 432,465 | 411,960 | 384,950 | 350,035 | 309,005 | 262,180 | 213,485 | 165,220 | 120,085 | 80,210 | 47,910 | 24,140 | 8,695 | |

Source: Selikoff (1981, Table 2-25). Projected quantities converted from annual to quinquennial values.

**Table 3.3: Projected Excess Deaths from Asbestos-Related Cancers Among Exposed Workers, 1965-2029**

| | Number Deceased in Quinquennium | | | | | | | | | | | | | |
|---|---|---|---|---|---|---|---|---|---|---|---|---|---|---|
| Cancer Type | 1965-1969 | 1970-1974 | 1975-1979 | 1980-1984 | 1985-1989 | 1990-1994 | 1995-1999 | 2000-2004 | 2005-2009 | 2010-2014 | 2015-2019 | 2020-2024 | 2025-2029 | Total |
| Mesothelioma | 4,505 | 5,410 | 7,125 | 8,875 | 11,990 | 13,740 | 14,845 | 15,300 | 14,995 | 13,305 | 10,410 | 7,475 | 4,585 | 132,560 |
| Lung cancer | 11,720 | 16,430 | 21,840 | 25,275 | 27,360 | 27,485 | 26,295 | 23,465 | 19,605 | 14,935 | 10,540 | 6,270 | 3,230 | 234,450 |
| GI and other cancer | 4,280 | 5,170 | 5,950 | 6,880 | 7,475 | 7,470 | 7,125 | 6,370 | 5,275 | 4,060 | 2,820 | 1,700 | 880 | 65,455 |
| Total | 20,505 | 27,010 | 34,915 | 41,030 | 46,825 | 48,695 | 48,265 | 45,135 | 39,875 | 32,300 | 23,770 | 15,445 | 8,695 | 432,465 |
| | Cumulative Total | | | | | | | | | | | | | |
| | 432,465 | 411,960 | 384,950 | 350,035 | 309,005 | 262,180 | 213,485 | 165,220 | 120,085 | 80,210 | 47,910 | 24,140 | 8,695 | |

Source: Selikoff (1981, Tables 2-22, 2-23, 2-24, 2-25). Projected quantities converted from annual to quinquennial values.

that, because much higher exposures are required for asbestosis, asbestosis deaths would be concentrated among insulators, manufacturing workers, and long-term shipyard workers.

Selikoff suggested that crude estimates of asbestosis mortality could be made by relating it to cancer mortality forecasts. The mortality risk of asbestosis was estimated to be one-half to three-fourths of the mesothelioma risk among insulators. Because of high labor turnover rates in manufacturing, only one-third as many asbestosis deaths were expected. Long-term but not short-term shipyard workers were thought to be at similar risk. Selikoff estimated that about 200 asbestosis deaths were occurring annually in 1980 – a number which would double during the following two decades and then decline thereafter.

## 3.4 Sensitivity of Selikoff's Projections

Many assumptions were made in Selikoff's projections. One set of assumptions involved the numbers of workers in each of 11 occupational and industrial categories who were exposed to asbestos. Two basic types of assumption were involved in estimating the size of the exposed workforce.

The first involved the level and extent of asbestos exposure of the workforce in each of the 11 categories. The second involved the rate of turnover in the workforce. The impact of these turnover rates, which used BLS and CWHS data, was large and resulted in a doubling of the estimate of the size of the exposed workforce, from the 13.2 million previously reported in Nicholson et al. (1981a) to the 27.5 million reported in Selikoff (1981). Selikoff argued that this doubling of the prior estimate would have little effect on the expected number of excess cancer deaths, assuming a linear dose-response function. Selikoff argued that increasing his estimate of the number of exposed workers from 13.2 million to 27.5 million would be matched by a corresponding reduction in the duration of exposure.

The large proportion of asbestos workers employed for less than 2 months also contributed to the doubling of the estimated size of the exposed workforce. The widespread practice of a 1-month probationary employment period and a rapid turnover of new hires in the first 2 months resulted in the forecasting of a large number of lightly exposed workers (i.e., 8.7 million persons with exposure equivalent to less than 2 months employment as an insulator). Again, the mortality effect of this change was argued to be canceled by the assumption of a linear dose-response relation operating on a relatively constant set of person-years of exposure.

The invariance associated with the linear dose-response assumption applies only to the excess cancer mortality due to asbestos. If compensation is granted for workers in an industry or occupation with significant asbestos exposure, it may have to be granted for all exposed workers with a given disease. In this case, the burden of compensation would be the total number of cancer deaths in the population, which will increase as the estimated number of workers increases and as the estimate of the duration of exposure decreases. Thus, the projection of a large number of lightly exposed persons, although perhaps not contributing much to the excess cancer mortality, would contribute to the total cancer mortality identified for workers in an industry with asbestos exposure. Doubling the estimate of the exposed workforce, from 13.2 to 27.5 million, would double the amount of naturally occurring cancer in the population identified as exposed. Because a significant proportion of total U.S. male cancer mortality would occur among the 21 million survivors of this group, the total number eligible for compensation could be much larger than the excess number of deaths.

Once the estimates of the exposed workforce in each of 11 industrial/ occupational categories were made, the second step in the projection was to apply models which combine (a) the dependence of the level of risk on the duration of exposure and (b) the differential level of risk of the 11 occupational/industrial categories.

The disease-risk models used by Selikoff were different for mesothelioma and lung cancer. For mesothelioma, an "absolute" risk model was developed based on a mathematical form similar to that used by Peto et al. (1982). Another important assumption in Selikoff's projections of mesothelioma was the lack of significant nonoccupational exposure to asbestos among the various industrial/occupational groups. If significant nonoccupational exposure occurred, this would have the effect of increasing the asbestos dose (expressed here as the duration of employment) which could have a nonlinear effect over time.

To assess the overall accuracy of Selikoff's projections of the number of mesothelioma cases, one can compare Selikoff's estimate of 1425 cases for 1977 with an estimate of about 1025 mesothelioma cases generated from the SEER data (Selikoff, 1981, p. 95). The SEER data should be adjusted upward because the program uses only records-based reports, not pathological reports. If we apply Selikoff's estimate of underdiagnosis in death certificates (71 of 175; see Table 2.1) to the SEER estimate, then we obtain a revised estimate of 1725 mesothelioma cases in 1977. This is consistent with Hogan and Hoel (1981), who suggested that the actual number of mesothelioma cases could be as high as 1620, given the statistical variation of the SEER rate estimates. Alternatively, the comparison of Selikoff's 1425 estimate and the SEER 1025 estimate implies an adjustment factor of 1.39 for SEER, not the 1.68 factor implied by Table 2.1. This suggests that the underdiagnosis of mesothelioma in the SEER data may be about 28%, substantially lower than the 41% in death certificate data.

Whereas Selikoff assumed that the absolute risk of mesothelioma increased as the fourth or fifth power of time from onset of exposure, he assumed that the relative risk of lung and other cancers increased linearly with the duration of exposure. This linear increase in the relative risk was multiplied by the age dependence of the tumor type – which, for lung cancer, was approximately the sixth or seventh power of age. This age dependence of tumor risks was again based on data from the insulation workers study. Thus, both mesothelioma and lung cancer risks increased as nonlinear functions of time since onset of exposure; for lung cancer, there were additional nonlinear age effects. The projections of mesothelioma and other cancer deaths depend on data from the insulation workers study. Consequently, their uncertainty depends on the statistical uncertainty of the rates estimated from that study.

Selikoff conceded that the numerical results of his projections were very uncertain but argued that one could have confidence in the timing of the manifestations of the disease (Selikoff, 1981, p. 94). Uncertainty results from such factors as the estimates of the workforce, its exposure, the relative mortality risks, and the data on the time and age dependence of the cancer risks.

Two additional factors could affect the projections in complex ways. First, the strong relationship of cigarette smoking and a number of specific cancer types means that the relative risk model for lung and other cancers may not be as simple as assumed. The projection of the appropriate background rates for these cancers is subject to great uncertainty due to the changing prevalences of cigarette smoking over cohort and time (Manton and Stallard, 1991). Second, the assumed linearity of dose and response for both intensity and duration of exposure applies to the relative risk model but not to the absolute risk model (OSHA, 1983). In the latter case, a more complex adjustment is needed to account for duration and this was not reflected in Selikoff's projections of mesothelioma or asbestosis, which was tied to the mesothelioma projection.

## 3.5 Alternative Projections of Health Implications

In this section, we review alternative projections produced by 1981. This will demonstrate the range of uncertainty associated with such forecasts and will make it easier to understand why we were so concerned about uncertainty in developing our own forecasts for the Court. Table 3.4 summarizes these results.

The first projections were in an unpublished report (Bridbord et al., 1978) jointly produced by investigators at the National Institute for Occupational Safety and Health (NIOSH), the National Cancer Institute (NCI), and the National Institute for Environmental Health Sciences (NIEHS). This report generated considerable debate because it projected, using a strategy logically similar to Selikoff's (1981, 1982), that 13-18% of cancers (58,000-75,000 cases per year) would be produced by occupational exposure to asbestos at the peak of the asbestos "epidemic." These are six to eight times higher than Selikoff's (1981) peak estimates (9739 cases per year) in Table 3.2.

Bridbord et al. (1978) estimated that 8-11 million workers (compared to 27.5 million for Selikoff) had been exposed to asbestos since the start of WWII

**Table 3.4: Alternative Estimates of Excess Cancer Deaths due to Occupational Asbestos Exposure**

| Source | Estimate | Calendar Period or Type of Estimate |
|---|---|---|
| Selikoff (1981) | 432,465 | 1965-2029 |
| | 269,825 | 1980-2009 |
| | 219,730 | 1975-1999 |
| Bridbord et al. (1978) | 2,000,000 | Low 30-year |
| | 2,300,000 | High 30-year |
| Hogan and Hoel (1981) | 236,000 | Low |
| | 497,000 | Best |
| | 726,000 | High |
| Enterline (1981) | 122,520 | 30-year |
| Nicholson et al. (1981) | 221,440 | 1975-1999 |
| McDonald and McDonald (1981) | 30,390 | Low 30-year |
| | 85,110 | Best 30-year |
| | 176,310 | High 30-year |
| Peto et al. (1981) | 150,000 | Best |

Note: See Section 3.5 for details.

with 5.5-7.5 million survivors being subjected to "significant asbestos exposure" and about 4 million of these to "heavy exposure." These estimates were cited in a speech by former DHEW Secretary Joseph Califano on April 26, 1978 but without any detail on how they were produced. Using results from studies by Selikoff (1978a, 1978b) and Selikoff and Hammond (1978), Bridbord et al. (1978) estimated that 35-44% of the heavily exposed group would die from cancer, producing 1.6 million deaths. The remaining group of less heavily exposed workers was projected to experience one-fourth the mortality of the heavily exposed group producing an additional 400,000-700,000 deaths. In total, 2.0-2.3 million deaths were projected to result from asbestos over a 30-35-year period suggesting, an average of 58,000-75,000 asbestos-related cancer deaths per year over this period.

Bridbord's estimates were by far the highest ever produced and were severely criticized on both methodological and substantive grounds (Doll and Peto, 1981). For example, the projections did not take sufficient account of the impact of different intensities and durations of exposure in the affected populations nor were they adjusted to reflect decreases in the exposed population due to death. The sources of the estimates of the exposed population were not identified. The implied numbers of mesothelioma deaths (10,000 per year) were about 10 times the observed level. Peto and Schneiderman (1981) reported that although there was a consensus that the estimates by Bridbord et al. (1978) were inappropriate, there was a wide spectrum of opinions as to what were the most appropriate estimates of the current and future impact of occupational asbestos exposures.

Hogan and Hoel (1981) ostensibly employed the same basic logic as Bridbord et al. (1978), but they provided greater detail on their methods. Hogan and Hoel (1981) estimated that 7.1-8.2 million workers were occupationally exposed to asbestos, of whom more than 50% were still alive in 1980. The largest component of the exposed group was the 4.3-5.4 million shipyard workers estimated to have been exposed in shipyards during World War II. Additional occupationally exposed workers were in a wide range of occupations identified to be "actually or potentially exposed" in the National Occupational Hazard Survey of 1972. The number of workers identified as possibly exposed in this survey was 1.6 million – a number consistent with an estimate of 1.1-1.4 million made in a survey conducted at the Research Triangle Institute. Assuming that 1.4 million exposed workers was a reasonable estimate of the worker population at risk in 1975 based on the two surveys, the total number who have been exposed since 1940 was estimated by assuming the following:

- Levels of exposure had been greatly reduced since 1970 so that the later exposures could be ignored
- The average rate of workforce turnover in these industries was 5% per annum
- The growth of the industries with potential exposure 1940-1970 paralleled the growth of the economy in general

These assumptions yielded an estimate of 2.8 million exposed workers, which, when added to the 4.3-5.4 million estimate of WWII shipyard workers, yielded a total of 7.1-8.2 million workers exposed during 1940-1970.

Although the estimates of the total numbers of exposed workers were similar between Bridbord et al. (1978) and Hogan and Hoel (1981) (i.e., 8-11 million versus 7.1-8.2 million), the number of workers classified as heavily exposed varied drastically between the two studies. Specifically, Hogan and Hoel (1981) estimated that 1.06-1.23 million persons were heavily exposed to asbestos, whereas Bridbord et al. (1978) estimated 4.0 million.

The estimate of the exposed workforce provided by Hogan and Hoel (1981) was more useful than that produced by Bridbord et al. (1978) because the logic and data used by Hogan and Hoel (1981) were described and could be critiqued. For example, Nicholson (1981) argued that Hogan and Hoel (1981) seriously underestimated the size of the exposed workforce. He pointed out that the 1.4 million nonshipyard workers estimated to have asbestos exposure must include 900,000 automobile mechanics leaving 500,000 as an estimate of the working populations of insulation workers, post-World War II shipyard employment, asbestos products manufacturing, chemical and refinery maintenance personnel, power plant workers, and stationary engineers and firemen. In 1980 alone, these categories employed about 1 million persons. Furthermore, Nicholson (1981) concluded that the most serious omission in the estimate of the exposed workforce was the approximately 4 million workers involved in land-based construction – accounting for over two-thirds of all asbestos used in the United States.

Similarly, because Bridbord et al. (1978) did not specify how they estimated the size of their heavily exposed populations, we can only evaluate Hogan and Hoel's (1981) procedure. Hogan and Hoel (1981) recognized the lack of detailed information on the intensity and duration of exposure. The two studies they selected to supply this information (Cochrane et al., 1980; Enterline, 1978) suggested that 3.4-18.5% of the workforce was heavily exposed. The Cochrane study was based on extremely limited data (i.e., 92 workers in a single South African company), so it was thought best to focus on Enterline (1978), who estimated that between 43,000 and 1.7 million workers alive in 1967 (best estimate 250,000) had heavy asbestos exposure. In deriving these estimates, Enterline (1978) noted that "heavy exposure to asbestos causes asbestosis deaths so that one way to estimate the size of the heavily exposed population is to look at deaths due to asbestosis and estimate how large the population must be that produces these [deaths]" (Enterline, 1978). Because Enterline's (1978) procedure may be conservative, a "best estimate of 15%" was assumed by Hogan and Hoel (1981, p. 70). Nicholson (1981, p. 77) commented that although the 15% value was reasonable, it was principally an informed guess based on extremely limited information.

Estimation of excess risk among heavily exposed workers was based on estimates produced by Selikoff and his co-workers and included data on both insulation workers and shipyard workers. Overall, 20-25% of heavily exposed workers were assumed to die from lung cancer, 6-10% from mesothelioma, and 8-9% from gastrointestinal cancer. These were compared to background rates among nonexposed workers of 5%, 0.02%, and 3.5%, respectively.

Estimation of excess risk among less heavily exposed workers was more difficult because of a lack of dose-response information. Analyses of results from Newhouse and Berry (1976) and Blot et al. (1978) suggested that the excess lung cancer risk of less heavily exposed workers was between one-eighth and one-fourth of the excess risk of heavily exposed workers. It was assumed that these ratios applied for cancers other than lung cancer.

Several improvements over the Bridbord et al. (1978) projections were made by Hogan and Hoel (1981). First, they adjusted for excess mortality among workers heavily exposed to asbestos by multiplying the standard U.S. male mortality experience for 1970 by a factor ranging from 1.14 to 1.41; for less heavily exposed workers, the excess risk was reduced proportionately. Second, Hogan and Hoel (1981) made three different sets of projections, illustrating the variation in results attributable to varying assumptions:

1. A lower limit (LL) based on a World War II shipyard workforce of 4.3 million persons, 20% of whom were female, 5% of whom were heavily exposed, and a risk ratio of 1/8 for less heavily exposed persons.
2. A best estimate (BE) assuming a shipyard population of 4.8 million with 10% female workers, 15% heavily exposed workers, and a risk ratio of 1/5 for less heavily exposed workers.
3. An upper limit (UL) based on 5.4 million shipyard workers with 10% female workers, 25% heavily exposed workers, and a risk ratio of 1/4 for less heavily exposed workers.

Hogan and Hoel (1981) projected that 3.1% of cancer deaths after 1980 would be caused by asbestos (ranging from 1.4% to 4.4%) and that the total number of such deaths for all exposures after 1940 would be 497,000 (ranging from 236,000 to 726,000). Hogan and Hoel (1981) evaluated these forecasts in two ways.

First, they calculated that between 1100 and 3500 workers would die annually of mesothelioma – a range that includes the 1150 mesothelioma deaths estimated from SEER in 1976 and its 95% upper bound of 1650. These latter estimates were based on assumptions that the SEER estimate is the weighted average of mesothelioma incidence rates for 10 areas (an assumption used to compute a jackknife estimate of the confidence interval) and that the SEER areas represented an unbiased sample of the total U.S. population occupationally exposed to asbestos.

Hogan and Hoel (1981) also compared their data with a study of the Pearl Harbor shipyard workforce. Kolonel et al. (1980) recorded a relative risk of 1.34 for lung cancer for the Pearl Harbor shipyard workers. Hogan and Hoel (1981) calculated the excess number of cancer deaths assuming the following:

1. Only 63% of those shipyard workers were exposed to asbestos.
2. The risk of dying from cancer in the general population was 20%.
3. The relative risk among nonexposed workers was 0.88.

The calculations based on these assumptions yielded between 470,000 and 540,000 excess cancer deaths due to asbestos. This range bracketed their estimate of 497,000, although it was higher than Selikoff's (1981) estimate of 432,465 deaths.

The approach used by Enterline (1981) was similar to that of Hogan and Hoel (1981) and Selikoff (1981). Enterline first estimated the number of workers in a particular occupation or industry and then applied a relative risk measure to estimate the excess number of deaths.

Enterline (1981) first considered asbestos product factory workers and asbestos insulators. Enterline (1981) used two techniques to estimate the numbers of asbestos-related deaths in these occupational categories. First, he estimated the number of asbestosis deaths occurring in a study of 35 asbestos production facilities ($N = 21{,}755$) (Enterline, 1981) and Selikoff's study of insulation workers ($N = 17{,}800$). Enterline (1981) estimated that these two populations generated about 50% of the U.S. total number of asbestosis deaths. He then estimated the population required to account for all of these deaths at 75,000-85,000 heavily exposed workers in 1981.

Second, Enterline (1981) estimated the total number of exposed workers in these same job categories by examining the census of asbestos workers in the relevant occupations (about 40,000 per year) and, multiplying by 3 to represent turnover, he estimated that 120,000 total workers were heavily exposed over the period 1947-1976. Of the total of 120,000 workers, Enterline (1981) estimated that about 90,000 of this group were alive in 1981.

Enterline's (1981) best estimate of the number of heavily exposed workers alive in 1981 was 80,000. For these 80,000 workers, he projected 20% (16,000) excess lung cancer deaths over the next 30 years (i.e., 530 excess deaths per year).

Enterline (1981) next considered shipyard workers. He divided shipyard workers into WWII and post-WWII employment groups. Using Kolonel et al.'s (1980) estimate that only 60% of the Pearl Harbor shipyard workers were exposed to asbestos, Enterline (1981) estimated that the number of surviving shipyard workers with exposure to asbestos was 1.5 million [compared to Hogan and Hoel's (1981) estimate of 2.5 million surviving shipyard workers].

After reducing the estimate of the number of shipyard workers, however, Enterline (1981) did not make a corresponding adjustment to the estimated risk of the group; that is, if the relative risk for a group is 2.0 but only half of the population has the relevant exposure, then the relative risk for the exposed subgroup could be four times higher (i.e., a higher relative risk among a smaller population can produce the same number of cases). Instead, Enterline (1981) employed an educated guess that shipyard workers had only one-fourth the exposure of primary asbestos workers. This produced an estimate of 900 excess lung cancer deaths annually among World War II shipyard workers.

Enterline (1981) used estimates of exposure in other occupations and straightline extrapolations from exposure levels and years of exposure to produce an annual estimate of 2501 excess lung cancer deaths [compared to Selikoff's (1981) estimate of about 5000]. Adding deaths from other cancers, Enterline (1981) estimated 4084 asbestos-related deaths annually (compared to Selikoff's estimate of about 9000) and, over a 30-year period, a total of 122,520 excess cancer deaths (compared to Selikoff's estimate of 269,825 for 1980-2009). Enterline (1981) estimated that about 7.6 million workers still alive in 1981 were exposed to asbestos, an estimate in line with Bridbord et al. (1978), Hogan and Hoel (1981), and Nicholson et al. (1981a). The primary differences among these estimates are due to differences in the proportion of heavily exposed workers.

Nicholson et al. (1981a) failed to adjust completely for workforce turnover. Nicholson et al.'s (1981a) estimate of 13.2 million persons exposed 1940-1979 (of whom 9.2 million were alive in 1980) was less than half of Selikoff's (1981) estimate. However, because the mortality projections were based on the linear dose-response assumption, the mortality projections did not change significantly (i.e., the increase in the exposed population from 13.2 million to 27.5 million was almost exactly balanced by a corresponding reduction in the average duration of exposure, assuming a linear dose response). The mortality projections for the 25-year period 1975-1999 differed by less than 1% (221,440 vs. 219,730; Table 3.4).

McDonald and McDonald (1981) and Peto et al. (1981) made projections using a logically different projection strategy. They made no effort to ascertain the size of the exposed workforce. Instead, all forecasts were based on epidemiological data reflecting trends in disease incidence and prevalence. McDonald and McDonald (1981) made estimates of mesothelioma incidence from TNCS, SEER, and their own survey of pathologists. From these data, they estimated a mesothelioma rate of 8 per million for males (and increasing) and 2.5 per million for females – suggesting 8 × 105.4 million = 843 mesothelioma cases for U.S. males and 2.5 × 110.6 million = 277 cases for U.S. females, for a total of 1120 cases in 1975 [compared to an estimate of 1254 from Selikoff (1981); see Table 3.3]. McDonald and McDonald (1981) provided numerical estimates for the United States and Canada combined. For comparability, we converted these to estimates for the United States alone.

Based on a review of cohort studies which reported on mesothelioma and other cancers, they estimated a ratio of 2.4 lung cancers and 0.9 digestive cancers for every mesothelioma death. They also assumed that the percentage of mesothelioma cases associated with occupational asbestos exposure (based on their 1980 case-control study) was 75% for men [a figure that can be compared with the value of 54% employed in Walker's (1982) projection – see Chapter 4] and less than 10% for women. With their best estimate implying 843 mesothelioma deaths among U.S. males and 277 among U.S. females in 1975, it followed from the above assumptions that 2719 asbestos-related U.S. male cancer deaths would occur per year – an estimate with a range of 971 to 5632. Although they were reluctant to provide estimates for females, their results suggested an increment of about 4.35% to the male result, which yields a best estimate of 2837 deaths per year – with a range of 1013 to 5877. These estimates depend on the assumed percent of deaths caused by occupational exposure to asbestos and the ratio of the number of exposed worker deaths due to asbestos-caused mesothelioma to the number of deaths due to other asbestos-caused cancers. For comparability, we can convert McDonald and McDonald's (1981) implied annual estimate to an overall estimate for a 30-year period – 85,110 asbestos-related cancer deaths, with a range of 30,390 to 176,310.

Peto et al. (1981) also based their forecasts on epidemiologic data reflecting trends in disease incidence and prevalence. They used estimates of dose-response functions from several cohorts of highly exposed workers (Peto et al., 1982) to infer the size of the exposed cohorts in past years by age and date of first exposure. Because this method is detailed in Chapters 4-5, we summarize their results here.

Peto et al. (1981) utilized an estimate that mesothelioma rises as the 3.5 power of time since first exposure – independent of age, length of exposure, and type of fiber. They estimated that 5000 mesothelioma deaths would arise from WWII shipyard exposure, 3000 from other WWII exposure, and 29,500 from all other exposures prior to 1965 – a total of 37,500. By multiplying these figures by 3 to represent lung cancer deaths and adding 37,500 mesothelioma deaths from all pre-1965 exposures, Peto et al. (1981) obtained a total estimate of 150,000 excess cancer deaths.

# 4

# Forecasts Based on Indirect Estimates of Exposure

## 4.1 Introduction

This chapter reviews the model used by Walker (1982) and collaborators (see Walker et al., 1983) in making indirect estimates of past exposure to asbestos and in projecting the current and future numbers of persons with various asbestos-related diseases such as mesothelioma, lung cancer, and asbestosis. The use of indirect estimates of past exposure distinguishes Walker's model from all but the last two models reviewed in Section 3.5. Other differences between Walker (1982) and Selikoff (1981) are also important: (1) Walker neither required nor provided estimates by occupation; Selikoff did both. (2) Although asbestosis was the most frequent type of claim, it was not included in Selikoff's model; it was included in Walker's. (3) Walker's goal was to estimate the potential number of persons who might file a claim, given that they had a documentable and plausible occupational history of asbestos exposure; Selikoff shared this goal, but he also wished to estimate the total number of persons exposed to asbestos, without regard to their ability to document that exposure to a sufficient degree to file a successful claim should they ultimately develop an asbestos-related disease.

## 4.2 Background

Walker (1982) forecasted the number of asbestos-related diseases for the period 1980-2009. In making these forecasts, he drew heavily on the techniques, data, and parameter estimates reported in Peto et al. (1982) (described in Section 2.3.1d) and Peto et al. (1981) (summarized in Section 3.5). The overall logic of Walker's (1982) mesothelioma model followed the structure given in the Appendix of Peto et al. (1981). This approach is not intuitively natural. We will review the essential elements here.

Peto et al. (1981) made five assumptions, as follows.

*Assumption 1: Independence of dose and time.* Excess mesothelioma incidence increases with increasing dose, and with increasing time, $t$, since start of exposure at age $a$, in the following way:

$$I_a(t) = D(d) \times f(a,t),$$

where

$I_a(t)$ = incidence rate at age $a+t$,
$D(d)$ = response function for dose $d$, independent of age $a$ and time $t$,
$f(a,t)$ = time-dependent function giving incidence at age $a+t$ for unit dose starting at age $a$.

In this formuation, excess mesothelioma incidence is treated as an independent additive risk. This would be appropriate if there were no cases of mesothelioma other than those attributable to asbestos exposure. Alternatively, for applications to occupationally exposed workers where the attributable risk ($AR_H$) is close to unity, this formulation will provide an adequate approximation. Methods for improving the approximation will be considered below.

*Assumption 2: Linearity of dose response.* $D(d)$ can be replaced by a measure of the effective dose, denoted $D$, so that the incidence rate simplifies to

$$I_a(t) = D \times f(a,t).$$

In addition, they assumed that the distribution of $D$ within a cohort indexed by $a$ and $t$ is such that the effects of mortality or morbidity selection will not reduce the mean value significantly over time. Thus, the equation may be applied to a cohort with average dose equal to $D$.

*Assumption 3: Weibull hazard rate.* The time-varying component of the mesothelioma incidence function is a Weibull hazard rate, such that

$$f(a,t) \propto t^k,$$

where $k = 3.5$ (Peto et al., 1982).

*Assumption 4: Cohort model.* The incidence rate $t$ years after initial exposure in a cohort first exposed at age $a$ in year $y$ is

$$I_{ay}(t) = D_{ay} \times t^k,$$

and the expected number of new excess cases (of mesothelioma caused by asbestos exposure) occurring at age $a+t$ in this cohort is

$$E_{ay}(t) = n_{ay} \times D_{ay} \times t^k \times S_{ay}(t),$$

where

$n_{ay}$ = the number of persons initially exposed at age $a$ in year $y$, which is the initial cohort size;
$D_{ay}$ = the average effective dose for the $n_{ay}$ persons;
$S_{ay}(t)$ = the probability that a cohort member alive at age $a$ in year $y$ survives to age $a+t$ in year $y+t$, which can be estimated from national cohort life tables.

*Assumption 5: Effective level of exposure (the calibration step).* The crucial step of Peto et al. (1981) is easy to understand but complex to describe in formal detail, so we give a verbal summary first. In the current year $y_1$, suppose an excess mesothelioma case occurs in a cohort member currently of age $a_1$. His exposure must have begun in some year $y_0$ in the past, when his age was $a_0$. That was $t$ years ago where $t = y_1 - y_0 = a_1 - a_0$. In other words, this person was $a_0 = a_1 - t$ years old, $t$ years ago, in year $y_0 = y_1 - t$, so he is counted as one of the cases enumerated by $E_{a_1-t,\ y_1-t}(t)$. Different men who become excess cases in year $y_1$ at age $a_1$ will have been first exposed at differing numbers of years $t$ in the past, so we count them all by summing over $t$.

Now, we give details. Peto et al. (1981) assumed that the effective level of exposure can be defined using

$$N_{ay} = n_{ay} \times D_{ay}.$$

Because neither $n_{ay}$ nor $D_{ay}$ is known with any degree of certainty, combining them into a single unknown, $N_{ay}$, allows the expected number of excess cases to be rewritten as

$$E_{ay}(t) = N_{ay} \times t^k \times S_{ay}(t).$$

If the expected number of excess cases can be estimated from a sample of current cases, aged $a_1$ in year $y_1$, using information on the age and year of first exposure to distribute the sum $\sum_t E_{a_1-t,\ y_1-t}(t)$ into estimates of each term, $E_{ay}(t)$, then one can solve for $N_{ay}$ using

$$N_{ay} = E_{ay}(t)/[t^k \times S_{ay}(t)].$$

We refer to this calculation as the "calibration" step of the indirect estimation procedure. The indices $a_1$ and $y_1$ will be used to indicate ages and dates of the calibration data. These are distinct from the indices $a_0$ and $y_0$ which will be used in the "projection" step of the indirect estimation procedure to indicate ages and dates of initial exposure to asbestos. When it is not necessary to consider fixed ages and dates, we will simplify the notation and use $a$ and $y$ to index variable ages and calendar years. The nature of the index will be apparent from the specific function being used.

Once the calibration step generates estimates of $N_{ay}$ (or $N_{a_0y_0}$), one can generate projections for any later time using the formula given earlier for $E_{ay}(t)$; that is,

$$E_{a_0y_0}(t) = N_{a_0y_0} \times t^k \times S_{a_0y_0}(t).$$

The indexes $a_0$ and $y_0$ are considered fixed in the projection step, and $t$ denotes the number of years after the initial year $y_0$.

Projections based on $N_{ay}$ are internally consistent. To see why this is so, we rewrite the equations for the expected number of excess cases to refer to time $t+s$, $s \geq 0$:

$$E_{ay}(t+s) = N_{ay} \times (t+s)^k \times S_{ay}(t+s).$$

Substituting for $N_{ay}$ we obtain

$$E_{ay}(t+s) = E_{ay}(t) \times \frac{(t+s)^k}{t^k} \times \frac{S_{ay}(t+s)}{S_{ay}(t)}.$$

For $s = 0$, this is an identity. For $s > 0$, the formula guarantees that the projection will trace a trajectory from the current observed data to all future values consistent with the Weibull model and the selected national life table.

If observed data are available for multiple years, then these may be used to improve the estimates of $N_{ay}$ as follows:

$$N_{ay} = \sum_{s=0} E_{ay}(t+s) / \sum_{s=0} \left[(t+s)^k S_{ay}(t+s)\right].$$

Thus, $N_{ay}$ is estimable. The fact that it does not represent the "real size" of the cohort, however, leads to some important limitations of the method, discussed below.

Peto et al. (1981) assumed that all recorded cases of mesothelioma among occupationally exposed workers were due to asbestos exposure, equivalent to assuming that $AR_H \approx 1$. However, they estimated an attributable risk (AR) of 76% for the general population which was reflected in their model by proportionally reducing the Surveillance, Epidemiology, and End Results Program (SEER) mesothelioma estimates for men for 1974-1978. Walker (1982) followed a similar procedure, except that he used a 54% multiplier in reducing the SEER mesothelioma estimate for 1975-1979, based on his assessment of the percent of cases with documentable occupational exposure to asbestos. In each case, the application of a constant multiplier to control for attributable risk produces an equivalent reduction for each projected value, $E_{ay}(t+s)$. The overall projection is sensitive to the AR-parameter value.

Peto et al. (1981) cautioned that the most severe limitation of their method is the large uncertainty in $N_{ay}$ associated with recent exposures. Because mesothelioma has a long latency period, the counts, say $E_{ay}(t)$, for small values of $t$, are subject to Poisson variation (with variance equal to the mean). Hence, $N_{ay}$ may be highly unstable in these cases. Furthermore, because $N_{ay}$ is proportional to $E_{ay}(t)$ or to $\sum_s E_{ay}(t+s)$, it follows that $N_{ay}$ has the same coefficient of variation as the mesothelioma counts, $E_{ay}(t)$ or $\sum_s E_{ay}(t+s)$, even if uncertainty in the national life table and Weibull hazard is ignored.

The scaling of $N_{ay}$ is arbitrary. Walker (1982) noted that rescaling of $N_{ay}$ using the Weibull scale parameter derived by Peto et al. (1982) from the mesothelioma incidence rates among asbestos insulation workers yielded an interpretation of $N_{ay}$ as the equivalent number of insulation workers required to produce the observed mesothelioma events. Because insulation workers were among the most heavily exposed groups, this means that the insulation-worker-equivalent (IWE) scaling produced a lower bound on the actual number of exposed workers. Walker (1982) argued that it made little difference whether one estimated actual or IWE numbers of exposed workers, as long as the forecasts of the number of mesothelioma and other asbestos-related diseases were accurate.

## 4.3 Walker's Study: General Description

Walker (1982) developed forecasts of the number of cases of asbestos-related diseases (mesothelioma, lung cancer, and asbestosis) for the period 1980-2009 using a modification of the model described by Peto et al. (1981, 1982). Walker (1982) focused on persons with nontrivial exposure to obtain estimates and forecasts of the number of lawsuits which might arise as a result of occupational exposure to asbestos. Whereas prior forecasts had focused almost exclusively on the cancer consequences of asbestos exposure, Walker (1982) included asbestosis in his model because of the potentially large number of claims that could be filed for this common noncancerous result of heavy asbestos exposure.

### 4.3.1 Data

Walker (1982) employed a variety of data. Two important sources were the Third National Cancer Survey (TNCS) and SEER, both sponsored by the National Cancer Institute (NCI). These sources were the basis of estimates of the current numbers of mesothelioma cases and change in mesothelioma incidence rates. Estimates of various other parameters used in Walker's projections were derived from published analyses of the mortality experience of Selikoff's insulation worker cohort (Peto et al., 1982). Data from epidemiological studies of the health risks of occupational exposures to asbestos were used to establish assumptions of the model such as the assumption that the ratio of peritoneal to pleural mesothelioma increased as the level of exposure to asbestos increased. Data from McDonald and McDonald's (1980) case-control study were crucial in determining the proportion of total mesothelioma incidence which was due to heavy exposure to asbestos – or at least to asbestos exposure likely to be recalled by survivors.

In addition to data from TNCS, SEER, and health studies of specific occupationally exposed populations (e.g., insulation workers in Selikoff et al., 1979; factory workers in Seidman et al., 1979; five occupational groups in

McDonald and McDonald, 1980), Walker also made use of data from litigant files maintained by the Johns-Manville Corporation. Whereas data quality could be discussed for the health surveillance and epidemiological data (see Chapter 2), the Johns-Manville data were proprietary and so could not be directly evaluated. It was possible, however, to examine the sensitivity of his projections to this type of data. These assessments are reported in the results of our sensitivity analyses in Chapter 5.

### 4.3.2 Model and Methods

Walker (1982) used the concept of IWE to estimate the number of cases of mesothelioma among cohorts with a common age and date of first exposure to asbestos. Peto et al. (1982) provided a parameterized Weibull hazard rate function giving the incidence rate of mesothelioma among insulation workers as a function of time since first exposure to asbestos. By dividing the observed number of cases by the incidence rate, Walker (1982) estimated the number of exposed insulation workers needed to yield the given number of cases. This latter number is his IWE count of exposed workers. The method of dividing the observed number of cases by incidence rates to estimate implied numbers of exposed workers will be referred to as "indirect estimation of past exposure."

If mesothelioma cases are stratified by occupation and the occupation-specific incidence rates are known, then indirect estimates of past exposure can be obtained with a natural scaling that yields actual numbers of exposed workers. This will be explored further in Chapter 8. Such data were not available to Walker (1982), so he had to select some artificial scaling, and he chose the IWE level.

## 4.4 Walker's Five Tasks

Walker's (1982) projection strategy was complicated. We shall review his methods in the order in which he presented them.

Walker organized his presentation into 5 tasks and 14 subtasks. The *first* task was to apply indirect estimation techniques to current mesothelioma incidence to determine the effective number of past asbestos workers. The *second* task was to project mesothelioma incidence in the exposed population using Peto et al.'s (1982) Weibull hazard rate function. The *third* task was to project lung cancer incidence using Selikoff et al.'s (1980) relative risk factors. The *fourth* task was to estimate current and future asbestosis prevalence. Walker argued that, for asbestosis, the long survival and the frequent lack of awareness of the cause of symptoms made the prevalence of potentially diagnosable asbestosis the key factor for estimating potential lawsuits. The *fifth* task was to estimate the amount of asbestos-related disease likely to occur in women.

### 4.4.1 Task 1: Determine the Effective Number of Past Asbestos Workers

#### *4.4.1a Task 1a: Determine the number of cases of mesothelioma in the United States, 1975-1979*

This was critical to Walker's (1982) report because the forecasts for mesothelioma and asbestosis (and, indirectly, lung cancer) are based on it. Most of this portion of the analysis was newly developed by Walker. Most of the rest of Walker's procedures were derived from Peto et al. (1982) and Peto et al. (1981). We shall see that Walker's conclusions depend on unsupported assumptions.

The primary data sources in this subtask were the 1969-1971 TNCS and 1973-1978 SEER data. From these two datasets, Walker estimated the rate of change in mesothelioma incidence from 1970 to 1975. To estimate the number of mesothelioma cases among U.S. men in 1975-1979, Walker made four assumptions:

1. That the 10% annual increases in mesothelioma incidence between TNCS and SEER were representative of changes in the national incidence of mesothelioma. This assumption supported a 15% upward adjustment of the observed SEER rates, averaged for the period 1973-1978, to account for a 1.5-year shift in time from end of year 1975, to mid-1977.
2. That TNCS and SEER overestimated the level of mesothelioma incidence by 10-15%, with a best guess of about 12%.
3. That SEER diagnostic coding failed to record the occurrence of about 10% of mesotheliomas each year.
4. That the number of mesothelioma cases occurring in the 1975-1979 quinquennium was approximately five times the number occurring in 1977, which accounts for the time shift in Assumption 1.

The TNCS and SEER datasets were the most extensive ones available to assess changes in the incidence of mesothelioma. However, the use of these data to establish changes in cancer incidence over this period became controversial when Pollack and Horm (1980) reported 1.3% and 2.0% annual increases in cancer incidence (as opposed to cancer mortality) for males and females, respectively, based on their analysis of TNCS and SEER data. These estimates, which were accepted by 17 federal agencies and the White House Council on Environmental Quality (CEQ), were subsequently challenged by a number of epidemiologists and industrial groups (Smith, 1980).

Critics of the TNCS and SEER data attacked their representativeness. Specifically, Smith (1980) pointed out the following:

> There is only indirect evidence that the populations surveyed are representative of the total U.S. population. Each survey encompassed 10% of the total population, but underrepresented rural dwellers and

> overrepresented Chinese and Japanese Americans, Indians, and Polynesians. An industry critic complains that it overrepresents shipbuilders who are vulnerable to asbestosis; Harris of CEQ, on the other hand, complains that it underrepresents the industrial Northeast. (p. 1000)
>
> Perhaps more important in the statistical sense is that the survey groups varied considerably from year to year on a nonrandom basis, as cities and regions decided to drop out or were persuaded to join. The survey population in 1976 had only four geographical regions (out of 11) in common with the survey population in 1969 (Atlanta, Detroit, Iowa, and San Francisco). NCI made efforts to ensure continuing regional participation beginning in 1973, but it could not resist the temptation to add new groups until 1976. (p. 1000)

Pollack and Horm (1980) attempted to demonstrate that their results were representative. Smith (1980) reported that their methods were criticized by epidemiologists like Rothman (at Harvard, in a critique prepared for Shell Oil Company) and Morgan (at SRI International, Palo Alto, in a critique prepared for the American Industrial Health Council).

Other critics complained that (Smith, 1980):

1. Case finding techniques had improved enough from 1969 to 1976 to affect the trends (Richard Peto, Oxford University).
2. Migration of cancer victims to NCI-sponsored cancer centers in the region might affect the trend estimates (Abe Lilienfeld, Johns-Hopkins University).

Smith (1980) quoted Harris of the CEQ as concluding that "the data are a long way away from being definitive or permitting conclusions" and that it will be 5 to 10 years, when data are tabulated through at least 1980, before firm conclusions can be reached.

Despite these arguments, Walker first assumed that changes in mesothelioma incidence between TNCS and SEER were representative of changes in the national incidence of mesothelioma. Walker did not assume that the SEER areas were representative of the absolute level of mesothelioma incidence in the U.S. because, he argued, shipbuilding areas were over-represented in SEER.

In evaluating Walker's arguments, it is difficult to understand why SEER and TNCS would produce nationally representative estimates of the rate of change of mesothelioma incidence but not nationally representative estimates of the level of mesothelioma incidence. Although it is theoretically possible to get a representative estimate of the rate of change in incidence but a nonrepresentative estimate of the level of incidence, the evidence on which Walker made his deduction is not clear.

Lacking adequate data, one might conclude that the SEER and TNCS data would produce either representative or nonrepresentative estimates of both the level and rate of change of mesothelioma incidence. If the TNCS and

SEER data were both assumed to be nationally representative, then Walker's first assumption that an appropriate rate of change can be estimated from TNCS and SEER seems reasonable. However, Walker's second assumption, that TNCS and SEER overestimated the level of mesothelioma incidence, seems less reasonable. It is not clear how the estimate of 10-15% reduction in mesothelioma was made nor how the "best guess" of 12% was produced.

Walker assumed, third, that the estimates of mesothelioma incidence were understated by as much as 10% due to the underdiagnosis of mesothelioma. Data might support quite different estimates. In Selikoff et al.'s (1979) study of insulation workers, for example, mesothelioma was under-diagnosed by over 40% (i.e., only 104 of 175 mesothelioma cases were properly identified on the death certificates). Selikoff (1981, p. 95) noted that SEER had not verified pathological diagnoses, so there would be no reason to consider their data more accurate. This was also true for TNCS. Direct comparison of Selikoff's estimates with SEER-based estimates for 1977 in Section 3.4 suggested that mesothelioma cases were underdiagnosed by 28% in SEER data. Walker provided no evidence to justify his lower estimate of the rate of underdiagnosis.

Correcting the second and third assumptions has important implications for the projections. If we assume SEER to be nationally representative (eliminating the 12% reduction), a conclusion reached by Pollack and Horm and Schneiderman's reanalysis (Smith, 1980), and if we estimate under-diagnosis at 39% based on the analysis in Section 3.4, then the estimate of annual mesothelioma incidence in 1977 could be raised to 1365 cases. This is close to the projection of 1425 mesothelioma cases obtained by Selikoff (1981, p. 131). These alternate assumptions would increase Walker's projections of the health effects of asbestos by 40% (i.e., 1365/974 = 1.40).

Although Walker discounted the SEER estimate because he believed the SEER areas overrepresented shipyard workers [as suggested by Smith (1980)], Selikoff (1981) argued that the 10 SEER areas underrepresented industrial areas and metropolitan regions which had significant asbestos activities more than 30 years prior. [Harris of CEQ argued similarly, as reported in Smith (1980).] Consequently, Selikoff (1981, p. 95), in contrast to Walker, concluded that the SEER estimates may be too low.

It is also difficult to rationalize Walker's decision to discount the SEER estimates of mesothelioma cases on the basis that they overrepresent shipbuilding activities because Walker (1982, pp. 18-20) argued that shipyard workers are not a heavily exposed population, based on Kolonel et al.'s (1980) cohort study of Pearl Harbor Naval Shipyard workers. Walker (1982, Table 13) reported the 1.3 multiplier for 10+ years of exposure in Kolonel et al.'s (1980) Table 6 as the maximum relative risk in that study. However, Kolonel et al.'s (1980) Table 5 reported a relative risk of 1.7 for 20-24 years of follow-up, and this was for all exposure durations combined. The result for 10+ years of exposure would presumably be higher than 1.7, so that Walker's use of the 1.3 multiplier to support his conclusion that shipyard workers are not a heavily exposed population does not appear to be appropriate.

This discussion raises two basic questions about the results produced by Walker in Task 1a. First, Walker's contention that "the method presented here is distinguished from other possible techniques in that it is tied to events currently being observed" (1982, p. 1) is not quite accurate. Walker's projections were tied to estimates of the number of mesothelioma cases derived from data that have been the subject of debate (i.e., Smith, 1980). There were different estimates produced by other researchers (i.e., Selikoff, 1981; Hogan and Hoel, 1981) and different estimates obtainable from plausible modifications of Walker's assumptions. Second, given the variability in the estimated number of mesothelioma cases for 1975-1979, the quantity on which the projections were based, it would have been reasonable for Walker to have empirically validated certain of his assumptions (e.g., how representative of the industrial structure of the United States was the industrial composition of the 10 SEER areas?).

#### *4.4.1b Task 1b: Calculate the fraction of mesothelioma cases which have a documented history of asbestos exposure, and estimate what fraction of the exposed are likely to have been heavily exposed*

Peto et al. (1982) (see Section 2.3.1d) estimated the parameters of a model of carcinogenesis from a group of insulation workers who had relatively homogeneous exposure to asbestos. In using those parameter estimates to make projections of mesothelioma for the United States, Walker noted that the parameter estimates taken from Peto et al. (1982) were appropriate only for persons whose asbestos exposure was as high as that of the insulation workers.

Walker assumed that a hypothetical population of "insulation-worker equivalents" (IWEs) could be subclassified according to "light" versus "heavy" exposure. This dichotomization was required to account for the 37% higher overall mortality rate among the heavy exposure subgroups than in the general population (Walker, 1982, p. 14) (see Table 2.1).

*Documentable Asbestos Exposure.* Walker used several types of data to justify his estimate that 54% of mesothelioma cases had a documentable asbestos exposure history because he did not possess documentary evidence of occupational exposure. First, he observed that the lowest incidence rate for SEER areas (Iowa) was only 38% of that for the area with the highest incidence rate (Seattle). Walker suggested that if there were essentially no asbestos exposure in the low-incidence area, the 38% level might indicate the background incidence, with 62% having work-related asbestos exposure.

"Background" incidence refers to the rate at which a disease would occur spontaneously (i.e., without exposure to the risk factor of interest). The rate could be a product of either the disease occurring with no exposure (i.e., as a function of the physiology of the individual) or from the extremely low levels of asbestos exposure normally experienced by persons (e.g., 0.00003-0.003 f/liter in outdoor air in rural areas) (ATSDR, 1990). In the case of mesothelioma, the notion of a background incidence level has been controversial. Selikoff (1981, p. 26) argued that the background incidence level is effectively zero; that is, in the United States, mesothelioma is prima facie evidence of exposure to asbestos. This still leaves open the possibility of significant nonoccupational exposure to asbestos (e.g., household contents, neighborhood sources, and other natural sources of asbestos) (NRC, 1984, EPA, 1986; Camus et al., 1998). In addition, this issue was less critical to Selikoff's (1981) model because he projected the excess numbers of mesothelioma cases attributable to occupational exposure to asbestos, without consideration of the background incidence in the general population. To do this, all Selikoff (1981) really needed was to establish that the attributable risk ($AR_H$) among insulation workers was close to unity. On the other hand, Walker (1982) needed a bona fide estimate of attributable risk (AR) in the general population in order to identify the numbers of "excess" mesothelioma cases required as inputs to Peto et al.'s (1981) model. Although Walker (1982) rejected Peto et al.'s (1981) assumption that AR = 76%, he accepted the rest of their model and significantly extended its application.

Because the SEER data on background levels of incidence were likely to be questioned (e.g., there was likely to be some job-related exposure to asbestos in Iowa), Walker presented results from six studies which reported from 16% to 77% (median 54%) of mesothelioma cases with an occupational history of asbestos exposure. Walker recognized that the range 16-77% was probably a function of such factors as different populations sampled, differences in study design (e.g., source of occupational history), and statistical variation.

Finding the first two types of evidence suggestive of the level of exposure to asbestos but not conclusive, Walker cited the results of the case-control study of McDonald and McDonald (1980) (see Section 2.3.2). Job histories of 344 mesothelioma cases were obtained from interviews with relatives. Cases were matched with the records for 344 controls who died of pulmonary metastases from nonpulmonary primary cancers. Controls were matched for age, sex, hospital, and year of death. The job histories were submitted to four research centers for assessment as to the probability that they entailed exposure to asbestos. These four centers categorized the job histories into definite, probable, possible, and unlikely exposure to asbestos. Walker wished to evaluate how consistently the four centers could rate the job histories as to the likelihood of definite or probable asbestos exposure.

It is difficult to evaluate the results from McDonald and McDonald (1980) because the four categories were not given quantitative definitions (e.g., what probability is associated with a "possible" exposure?). Walker, in assessing the McDonald and McDonald data, noted that there were significant differences between the proportion classified as possible exposures by three centers and the proportion so classified by the Environmental Sciences Laboratory at Mt. Sinai. He judged that the three centers represented a "consensus" and used their average proportion with a definite or probable exposure (54%) as the "best estimate" of the proportion of mesothelioma cases with some occupational exposure to asbestos. This was the same as the median of the six studies cited earlier, but lower than the 62% estimate from SEER data.

Given the imprecision of recall data (especially from secondary sources such as relatives of deceased workers), it would seem equally plausible to group cases with possible, probable, or definite exposure. With this grouping of the categories, there would be a consensus of 73.6% across all four centers. This was the interpretation suggested by McDonald and McDonald in their original report (1980, p. 1652). McDonald and McDonald (1981, p. 78) reiterated their view that the "true proportion" of mesothelioma due to occupational asbestos exposure was about 75% in 1975.

There are several additional questions about Walker's interpretation of these data. For example, the differences between the Mt. Sinai group and the other centers in assigning cases to the probable exposure category could be the result of Mt. Sinai's greater experience in analyzing such data. McDonald and McDonald (1980, p. 1654) recommended using an occupational classification developed by the Mt. Sinai group in preference to the classification of exposure in the four categories (see Section 2.3.2). Walker did not explain why his reinterpretation of the data was preferable to the interpretation and conclusions reached by the original investigators, McDonald and McDonald.

Walker assumed that the proportion of male mesothelioma cases with documentable occupational exposure to asbestos exactly equaled the proportion of mesothelioma cases attributable to occupational exposure to asbestos (54%). Walker's tabulations reported only 54% of the total number of cases projected in the model.

If Walker had accepted McDonald and McDonald's (1981) estimate of 75% of cases with documentable occupational exposure to asbestos, the number of occupationally related cases would have been increased by 38.9%. In adopting the 54% figure, Walker made no adjustment for possible underreporting of occupational exposure to asbestos, misreporting of occupational exposure to asbestos, or misreporting of job histories (leading to errors in interpretation of those histories) in interviews with secondary sources (e.g., family survivors).

*Heavily Exposed Workers.* Walker divided the 54% of mesothelioma cases he attributed to occupational exposure to asbestos into "heavily" and "less heavily" exposed groups. He used the ratio of peritoneal to total mesothelioma as an index of the level of exposure to asbestos. Use of the peritoneal-total mesothelioma ratio to determine the degree of exposure was discussed in other sources (e.g., Lemen et al., 1980, p. 3), but it has never been confirmed and remains highly suspect (see Section 2.5).

Walker attempted to validate the use of the peritoneal-total mesothelioma ratio as an index of the intensity of exposure by examining how that ratio varied over study populations with different levels of exposure.

To do this, he presented data (Walker, 1982, Tables 4-6) from 11 epidemiological studies, with subpopulation results stratified into 3 exposure levels: (1) heavily exposed – 7 subpopulations ranging from 0% to 64% peritoneal (median 44%); (2) less heavily exposed – 3 subpopulations ranging from 13% to 28% peritoneal (median 16%); and (3) general population with no identifiable exposure – 6 subpopulations ranging from 5% to 24% peritoneal (median 12%).

From the evidence presented on the variation of the ratio of peritoneal to total mesothelioma with exposure level, Walker concluded that it was legitimate to use the ratio as an index of exposure. Furthermore, because 189 of 397 cases (48%) in the seven study populations with inferred high exposure had peritoneal mesothelioma, Walker proposed that 50% of workers with heavy exposure who develop mesothelioma would develop peritoneal mesothelioma. Among workers with moderate exposure to asbestos who develop mesothelioma, Walker estimated that 20% would be peritoneal.

Supposing that 50% of mesothelioma among heavily exposed workers was peritoneal, Walker argued that only a small fraction of total U.S. mesothelioma incidence would occur among heavily exposed workers because, he claimed (correctly), only 9-13% of U.S. mesothelioma was peritoneal. Specifically, assuming that 50% of mesothelioma among heavily exposed persons was peritoneal, Walker argued that an upper limit of 26% (actually 18-26%) of U.S. mesothelioma occurred among heavily exposed workers.

Walker argued that the estimate of 26% of mesothelioma cases due to heavy occupational exposure was too high because it did not allow for any peritoneal tumors among persons with little or no known asbestos exposure. Consequently, he suggested that the 20% of the cases McDonald and McDonald (1980) classified as "definite" exposures was a better estimate of the proportion of mesothelioma due to "heavy" exposure levels. Walker chose 34% [i.e., the proportion of workers with "probable" exposure to asbestos in McDonald and McDonald (1980)] as his preferred estimate of the number of cases arising among workers with "less heavy" or "identifiable light" exposure to asbestos.

We will argue that the evidence he presented is not conclusive for several reasons. First, Walker combined data from a range of studies with varying fiber types and sizes, exposure conditions, and study designs. Only one study (McDonald and McDonald, 1980) presented mesothelioma results for all three of Walker's exposure levels, and this study estimated that there were 22.3% peritoneal mesotheliomas for males with moderate and heavy exposure combined (16% and 44% separately) compared to 23.7% peritoneal mesotheliomas for all other males. This study showed no consistent gradient in the peritoneal-pleural ratio over exposure levels. In addition, in the absence of other factors, it would require that Walker's estimate of 9-13% peritoneal mesothelioma in the general population should be increased by a factor of 2. Females had a much lower level of exposure to asbestos (less than 2% with a recorded asbestos work history), yet McDonald and McDonald (1980) indicated that 39% of mesotheliomas among females were peritoneal. To the extent that female mesothelioma cases had lower average exposure than male mesothelioma cases, the finding of 39% female versus 22-24% male peritoneal mesotheliomas appears to refute Walker's assumption that the peritoneal-total mesothelioma ratio indicated the intensity of exposure.

Second, his categorization of the level of exposure for a specific population (i.e., "heavily exposed," "less heavily exposed," and "general population and cases with no identifiable asbestos exposure") was not based on explicit quantitative criteria – so that the exposure level cutpoints assigning a study population to a category cannot be directly evaluated. If we followed McDonald and McDonald's (1980) recommendation to use the Mt. Sinai group's method of classifying occupations at high risk to asbestos exposure, then we could refer to Selikoff's mesothelioma projections (1981, p. 131, Table 2-23) to estimate 17% of mesotheliomas among heavily exposed workers (i.e., asbestos insulation workers, and primary and secondary asbestos factory workers) in 1977 (and 20% in 1997). Taking account of McDonald and McDonald's estimate that 75% of mesothelioma was due to occupational exposure, we estimate that about 13% of combined occupational and nonoccupational mesothelioma in 1977 would have been among heavily exposed workers (15% in 1997). From this perspective, Walker's 20% estimate appears to be too high.

Third, the type of fiber to which the worker was exposed was not controlled in the analyses of the peritoneal-total mesothelioma ratio. Peto et al. (1981, 1982) argued that this ratio was related to fiber type, with amphiboles largely responsible for peritoneal mesotheliomas. McDonald et al. (1980, p. 22) also argued that fiber type was critical in determining the proportion of peritoneal tumors even though studies had major differences in exposure levels.

The EPA (1986) reviewed 41 epidemiologic studies and came to the same conclusion, although they cited greater potency for crocidolite than amosite asbestos within the amphibole mineral group (see Section 2.5). The EPA (1986) also indicated that chrysotile asbestos (a serpentine mineral) rarely produced peritoneal mesothelioma but had roughly comparable potency to the amphiboles in producing pleural mesothelioma. There continues to be some controversy over the role of chrysotile in pleural mesothelioma. For example, Churg (1988) argued that chrysotile rarely produced pleural mesothelioma, whereas Smith and Wright (1996) argued that chrysotile was the main cause of pleural mesothelioma due to its very high market share (90-95% of all commercial asbestos is chrysotile). Either way, the occurrence of peritoneal mesothelioma is indicative only of exposure to amphibole asbestos, not of the intensity of that exposure.

Similarly, the type of industrial processes to which the worker was exposed was not controlled in the analysis of the peritoneal-total mesothelioma ratio. The results in Chapter 2 indicated that the physical properties of the asbestos fibers were critical in the induction of mesothelioma. For example, Lippman (1988) concluded that the critical factor is the size of the fiber: The highest risk for mesothelioma was associated with fibers that are 5-10 $\mu$m in length and less than 0.1 $\mu$m in diameter. For lung cancer, the highest risk was associated with fibers 10-100 $\mu$m in length and 0.3-0.8 $\mu$m in diameter. Thus, the intensity of exposure must take account of the fiber size distribution and this may change according to the type of industrial process involved.

As a consequence, it is not clear how, or even if, the peritoneal-total mesothelioma ratio relates to exposure level because the ratio is confounded with the different mix of fiber types, fiber size distribution, and industrial processes in different study populations over time.

Fourth, the assumption of dose dependence implicitly introduced a nonlinear dose response function that was wholly inconsistent with the rest of Walker's model. Specifically, this assumption contradicts Peto et al.'s (1981) assumption of linearity of dose-response (see Section 4.2), the key assumption of Walker's Task 1d. Peto et al. (1981, p. 52) noted that "the two sites can be amalgamated. Both diseases are quickly fatal, and the incidence of cases caused by asbestos exposure for both appears to be approximately proportional to the 3.5th power of time since first exposure, irrespective of age at first exposure, duration of exposure, or fiber type." By assuming that the peritoneal-total mesothelioma ratio was dose dependent, Walker implicitly introduced a nonlinear dose-response function for one or the other (or both) forms of mesothelioma. Once this was done, the justification for computing effective levels of exposure or insulation-worker equivalents collapsed. This part of Walker's model contradicted the findings of dose independence in McDonald and McDonald (1980) and undermined the logic of the rest of the model.

Fifth, the estimate that 54% of mesothelioma was due to occupational exposure to asbestos, with 20% arising among workers with high exposure and 34% among workers with "less heavy" or "identifiable light" exposure, was based on a single study (McDonald and McDonald, 1980). Furthermore, the estimate was based on interpretations of that study not proposed by the original investigators. Walker argued that assignment of a worker to the "definite" exposure category was equivalent to a high exposure. It is not clear that such an equivalence existed. In addition, Walker assumed a dose-dependent peritoneal-total mesothelioma ratio, whereas McDonald and McDonald's (1980) data suggested dose independence. The increased incidence of peritoneal mesothelioma among insulation and asbestos factory workers was attributed by McDonald and McDonald to crocidolite and amosite exposures, not to a nonlinear dose response.

Our review suggests that the 54% estimate could be as high as 75%, whereas the 20% estimate could be as low as 13%. These differences should be evaluated in the context of the purposes to which the projections will be put. The appropriateness of one or the other set of estimates may be quite different for estimating the number of cases that might deserve compensation as opposed to estimating the number of lawsuits that will eventually be brought to court and are likely to result in an actual award. For example, if family survivors are not aware of prior occupational exposure to asbestos, then they would be unlikely to sue, and Walker's 54% estimate would seem more defensible. Alternatively, it could be argued that the occupations included in the alternative estimates of heavily exposed workers should include heating trades and some fraction of construction workers. For example, Selikoff's (1981) computation of relative risks for heating trades exceeded those of manufacturing workers (4.9 vs. 4.6; see Section 3.3, Task 3). This would make Walker's 20% estimate appear more defensible.

#### *4.4.1c Task 1c: Estimate the timing of exposure in U.S. cases*

Peto et al. (1981) derived the timing of exposure from mesothelioma cases reported for Los Angeles County. Peto et al. (1981) argued that Los Angeles County had mesothelioma incidence rates similar to those of the United States and might be expected to have an exposure history representative of the United States. To estimate the distribution of the timing of exposure, Walker instead used data from 278 litigants of the Johns-Manville Corporation who had mesothelioma. Use of the exposure history of Johns-Manville litigants biases the projections toward the specific experience of Johns-Manville employees and of those who had previously filed suit. Although this made the projections more suitable for Johns-Manville, it also made the projections less applicable to the total U.S. population.

Because the data on the distribution of exposure among the 278 litigants were not presented in Walker's (1982) report, we did not compare the exposure histories of Johns-Manville workers with the Los Angeles County data. However, we did assess the applications made of the data on the exposure histories of the Johns-Manville workers. Section 5.3 discusses use of these data in sensitivity analyses.

Walker used only Johns-Manville litigants with an "analyzable" exposure history. An analyzable exposure history had a plausible age at first exposure (between 15 and 54 years) and a year of first exposure between 1930 and 1954. Walker divided the 278 cases into 5-year age groups and 5-year calendar date of exposure groups. He estimated the age-and-date-of-first-exposure numbers of " insulation-worker equivalents" from the model of mesothelioma described by Peto et al. (1981), using the adjusted age-specific mesothelioma incidence counts estimated from SEER.

This approach was described in Section 4.2. If

$M_{y_1}(a_1)$ = expected number of mesothelioma cases at age $a_1$ in year $y_1$
$E_{ay}(t)$ = expected number of mesothelioma cases at age $a_1$ in year $y_1$ among persons initially exposed in year $y = y_1 - t$ at age $a = a_1 - t$,

then

$$M_{y_1}(a_1) = \sum_{t=0}^{a_1} E_{a_1-t,y_1-t}(t)$$

and

$$F_{ay}(t) = E_{ay}(t)/M_{y+t}(a+t)$$

is the fraction of $M_{y_1}(a_1)$ with initial age $a = a_1 - t$ and date of first exposure $y = y_1 - t$. Hence, if $F_{ay}(t)$ and $M_{y_1}(a_1)$ are obtained from different sources, then one can write

$$E_{ay}(t) = F_{ay}(t) \times M_{y+t}(a+t).$$

Walker obtained $M_{y_1}(a_1)$ from his adjustments to SEER data and $F_{ay}(t)$ from the Johns-Manville litigation data.

A problem with this procedure is that the number of Johns-Manville litigants expected to fall in any age-date quinquennium of exposure cell is small (i.e., 278/40 = 6.95), although not as small as in the Los Angeles County data (69 cases in total). Thus, the estimates of the proportion of the total exposure in each age-date cell will not be statistically stable. Peto et al. (1981) recognized this problem in their efforts to make projections using the Los Angeles County data. They suggested that the experience averaged over a large number of cells would tend toward the "correct" experience (i.e., errors resulting from small expected cell sizes would tend to cancel out because of the constraint to reproduce the current national rates). They evaluated the effect of a substantial change in the assumed distribution of age at first exposure and found the overall projection to be "surprisingly robust." Nonetheless, they cautioned about the impact of small errors in later dates of first exposure – a warning that motivated Walker's Task 1f.

Walker assumed that the age- and date-specific exposure patterns were invariant with respect to the level of exposure. This assumption would seem to be more valid for a specific class of workers (i.e., manufacturing employees of Johns-Manville) than for the United States as a whole. For example, if World War II shipyard workers had low levels of exposure [as suggested by Walker (1982, pp. 18-20)], then the average exposure during WWII would be expected to be lower than before and after the war when proportionately more workers were in manufacturing. In this case, exposure levels would be correlated with the date of exposure.

#### *4.4.1d Task 1d: Calculate the number of workers now alive and exposed at different times in the past which would be required to account for the current observed mesothelioma incidence*

Walker (1982) combined his estimated 54% of mesothelioma cases having an occupational exposure history with data on the incidence of mesothelioma in heavily exposed workers. At the heart of these computations is a theory of carcinogenesis due to Armitage and Doll (1954, 1961) called the multi-hit/multistage theory of carcinogenesis. In this theory, the incidence of a tumor may be described by a simple mathematical function (called the Weibull hazard function; see Section 2.3.1d):

$$I_t = b(t-w)^k,$$

where $t$ is age or time, $k + 1 = m$ is the number of cellular changes ("hits" or "stages" depending on whether they have to occur in a specific order) required before cellular growth control is lost, $b$ is a proportionality constant that represents the product of the probability of each of the $m$ events, $w$ is a waiting or lag time, and $I_t$ is the incidence rate at $t$. If $N_t$ is the size of the population at $t$ that is at risk of incidence at rate $I_t$, then the expected number of cases at time $t$ can be written as

$$E_t = I_t N_t.$$

Walker used two different forms of this function. The first, due to Peto et al. (1982), was

$$I_t = 4.37 \times 10^{-8} t^{3.2}.$$

The second was modified by the inclusion of a "lag" time (Breslow, 1982),

$$I_t = 1.37 \times 10^{-5} (t - 15)^{1.846},$$

to indicate that the likelihood of mesothelioma occurring less than 15 years after the start of exposure was very small. The 15-year adjustment in the second form of the Weibull explains why the two estimates of $b$ (i.e., $4.37 \times 10^{-8}$ and $1.37 \times 10^{-5}$) and $k$ (i.e., 3.2 and 1.846) were very different. This form was attributed by Walker to Newhouse and Berry (1976), although, as Newhouse and Berry indicated, it was employed by Cook et al. (1969).

A number of issues arise in using the Weibull hazard function to project the size of the exposed population. First, the Weibull hazard function was derived from a specific theory of human carcinogenesis. The use of the function for projection must be considered in the context of the validity of that theory. Although the multihit/multistage model is the most generally accepted model of human carcinogenesis for use in risk assessment, alternative theories have been proposed by Armitage and Doll (1957), Burch (1976), Whittemore and Keller (1978), Moolgavkar and Venzon (1979), Portier and Kopp-Schneider (1991), and Tan and Chen (1991), among others. For example, the multistage, multipathway model proposed by Tan and Chen (1991) may allow a more accurate representation of the roles of p53 tumor suppressor genes in asbestos-related cancers (Hemminki et al., 1996).

Second, the estimators of the parameters of any nonlinear function such as the Weibull are frequently highly correlated. In the two forms of the function presented, the introduction of a lag of 15 years reduced the exponent from 3.2 to 1.846.

Third, estimates of individual parameters may be highly uncertain. For example, Peto et al. (1982) found a standard error of estimate of 0.36 for $k$, which suggested that the estimate of 3.2 could be plausibly any value in the range 2.5 to 3.9 (95% confidence interval). Furthermore, because of concerns about biases due to underdiagnosis of mesothelioma in old age and overprediction of the expected numbers at 0-14 years after first exposure, Peto et al. (1982) recommended the use of 3.5 as a point estimate with a plausible range of 3 to 4. Walker did not alert the reader to this uncertainty.

Additional uncertainty in the parameter $b$ was not addressed by Peto et al. (1982) or Walker (1982). The EPA (1986, p. 95) concluded that a reasonable estimate of the 95% confidence limit for mesothelioma incidence would be a factor of 5 (i.e., estimates are divided by 5 and multiplied by 5 to determine the range). These limits are conditional on fixed values of $k$ and $w$, so that the large range is not simply a result of correlated parameter estimates (see Section 2.4 for details).

Fourth, Peto et al. (1982) fitted their function to mortality data, whereas Walker (1982) applied the function to incidence data. Given the parameter correlations and the ability of the Weibull function to accommodate a lag of up to 15 years without a major loss of fit, we can assess the impact of changing from mortality to incidence data by holding $k$ constant and computing the change in $b$ needed to match the incidence and mortality functions at some select time since first exposure $t$, that is,

$$b_I(t-w)^k = b_M t^k$$

so that

$$\frac{b_I}{b_M} = \left(\frac{t}{t-w}\right)^k,$$

where $w$ is the average survival time from diagnosis to mortality. Assuming $t = 35$ and $w = 0.5$, then $b_I$ would be 4.7% larger than $b_M$. Walker (1982) set $b_I = b_M$ without comment.

Because the incidence function was inverted to estimate the size of the exposed population, yielding

$$N_t = E_t/I_t,$$

it will be important to establish the sensitivity of the projections both to the uncertainty in the estimates of its parameters and to other adjustments utilized by Walker. For example, Walker, but not Peto et al. (1982), multiplied the $b$'s (i.e., either $1.37 \times 10^{-5}$ or $4.37 \times 10^{-8}$) by a "correction factor" of 0.8 in the two incidence equations. Walker (1982, p. 13) reported the reason for this adjustment as follows: "Sources reporting to [Johns-Manville's] legal staff indicate that there may be some overstatement in Selikoff's data of the numbers of mesothelioma cases actually occurring. The overstatement certainly appears to be the case for asbestosis, and probably is negligible for lung cancer." Walker did not describe the sources reporting to Johns-Manville's legal staff, nor did he describe the reasons for selecting the value 0.8 as the appropriate adjustment.

Walker correctly cautioned that the number of "exposed workers" produced in the projections was an artificial number. The procedure projected the numbers of "workers" that would have to be exposed at the level of the insulation workers to produce the "observed" number of mesothelioma cases. Clearly, not all workers were exposed to the same level (or fiber type) of asbestos as the insulation workers.

Walker argued that projecting "insulation-worker equivalents," and not the actual number of workers, was not important for forecasting the number of mesothelioma cases. This is reasonable if the dose-response function is linear, because the effective level of exposure can be determined without knowing the absolute number of persons exposed (see Section 4.2).

Based on the subjective impressions of Johns-Manville's legal staff, Walker argued that most lawsuits were coming from heavily exposed workers. He estimated the intensity of exposure using the estimates of the proportions of mesothelioma cases resulting from identifiable occupational exposures to asbestos derived from Task 1c. His tabulations reported only those mesothelioma cases (54%) accepted as due to occupational exposure based on the ability to document occupational exposure to asbestos.

#### *4.4.1e Task 1e: Using actuarial techniques, calculate the size of the originally exposed worker population which would yield the estimated numbers of currently living, previously exposed workers*

The calibration calculations (Section 4.2) were based on assumptions about disease-exposure relations, and the further assumption, not based on epidemiological research, that the estimated age- and date-specific exposure distribution was exactly that of Johns-Manville litigants.

Walker changed the procedure of Peto et al. (1981, 1982) only by multiplying the $b$ coefficient by 0.8 in both forms of the Weibull hazard function and by interpreting each projected population (past and future) $P_{ay}(t)$ aged $a$ in year $y$ after $t$ years from first exposure to asbestos as an insulation-worker equivalent count of exposed persons.

Peto et al. (1981) observed that estimates of cohort survival $S_{ay}(t)$ (Section 4.2) can be obtained from national cohort life tables. They indicated that one should also take account of higher mortality among exposed workers. Walker (1982, p. 13) implemented this recommendation in the following way. First, he cited Selikoff's (1981) observation that insulation workers had a 37% higher mortality rate than the general population (see Table 2.1). He assumed that a relative mortality risk of 1.37 applied to the 20% of surviving IWEs with heavy asbestos exposure. Second, he assumed that the remaining 34% of surviving IWEs had a relative risk of 1.00. He calculated a weighted average relative risk among surviving IWEs as

$$1.14 = \frac{20}{54} \times 1.37 + \frac{34}{54} \times 1.00.$$

He then downward adjusted the national cohort life table survival function, say $S^*_{ay}(t)$, using the approximation

$$S_{ay}(t) = \left[S^*_{ay}(t)\right]^{1.14}.$$

Walker did not identify this adjustment as an approximation to the marginal survival function for a mixed population of high-risk workers and average-risk workers. The exact expression for survival for this mixture is

$$S_{ay}(t) = \frac{20}{54} \times \left[S^*_{ay}(t)\right]^{1.37} + \frac{34}{54} \times S^*_{ay}(t),$$

which includes the same parameters in different roles. Table 4.1 illustrates Walker's approximation. The top panel shows that the approximation is biased downward with an absolute error of 0.63% or less. The relative error increases in size over time, surpassing 2% when 70% of the population has died. These results suggest that the approximation is reasonably accurate.

There are five concerns about Walker's implementation of this procedure. First, the weights should have been adjusted to reflect the relative frequency of moderately exposed actual workers, not their IWE frequency. For example, if each moderately exposed IWE represents 2.6 actual workers (see Task 3a later in this chapter), then the adjusted formula for the weighted average relative risk would be

$$1.07 = \frac{20}{108.4} \times 1.37 + \frac{88.4}{108.4} \times 1.00,$$

using $2.6 \times 34 = 88.4$. In this case, the excess relative risk (0.07) would be half of Walker's estimate (0.14). The bottom panel of Table 4.1 shows the impact of this change on the marginal survival function, using both Walker's approximation and the exact solution. About half of the gap between the cohort survival function and Walker's estimate is removed. Also, with the revised parameter settings, the absolute error in Walker's approximation to the exact solution is reduced by about one-third.

Alternately, if each moderately exposed IWE represents 10 actual workers (see Task 1f), then the adjusted weighted average would be

$$1.02 = \frac{20}{360} \times 1.37 + \frac{340}{360} \times 1.00,$$

with an excess relative risk (0.02) one-seventh that of Walker's estimate.

Second, we noted in Task 1b that the paired estimates of 54% and 20% could plausibly be 75% and 13%, respectively. In this case, Walker's formula for the weighted average relative risk would yield

$$1.06 = \frac{13}{75} \times 1.37 + \frac{62}{75} \times 1.00,$$

with an excess risk of 6%, not Walker's estimate of 14%. However, even this is too high because the weights should reflect assumptions about the actual numbers of workers in each exposure class. For 2.6 workers per IWE, the correct weighted average relative risk would be

**Table 4.1: Evaluation of Errors in Walker's Survival Function Approximation in Task 1e**

| Parameters | | S*(t) | Marginal Survival S(t): Walker's Approximation | Marginal Survival S(t): Exact Solution | Difference: Absolute | Difference: Relative |
|---|---|---|---|---|---|---|
| | | Assuming 1 Worker per IWE | | | | |
| Probability of heavy exposure | = 0.3700 | 0.90 | 0.8871 | 0.8873 | -0.0002 | -0.0002 |
| Probability of moderate exposure | = 0.6300 | 0.80 | 0.7759 | 0.7765 | -0.0006 | -0.0008 |
| Relative risk at heavy exposure | = 1.3700 | 0.70 | 0.6666 | 0.6680 | -0.0013 | -0.0020 |
| Relative risk at moderate exposure | = 1.0000 | 0.60 | 0.5595 | 0.5618 | -0.0023 | -0.0041 |
| Weighted average relative risk | = 1.1369 | 0.50 | 0.4547 | 0.4581 | -0.0034 | -0.0075 |
| | | 0.40 | 0.3528 | 0.3574 | -0.0046 | -0.0129 |
| | | 0.30 | 0.2544 | 0.2601 | -0.0057 | -0.0219 |
| | | 0.20 | 0.1605 | 0.1668 | -0.0063 | -0.0380 |
| | | 0.10 | 0.0730 | 0.0788 | -0.0058 | -0.0739 |
| | | 0.01 | 0.0053 | 0.0070 | -0.0016 | -0.2366 |
| | | Assuming 2.6 Workers per IWE | | | | |
| Probability of heavy exposure | = 0.1845 | 0.90 | 0.8935 | 0.8937 | -0.0001 | -0.0001 |
| Probability of moderate exposure | = 0.8155 | 0.80 | 0.7879 | 0.7883 | -0.0004 | -0.0005 |
| Relative risk at heavy exposure | = 1.3700 | 0.70 | 0.6832 | 0.6840 | -0.0009 | -0.0013 |
| Relative risk at moderate exposure | = 1.0000 | 0.60 | 0.5794 | 0.5809 | -0.0015 | -0.0026 |
| Weighted average relative risk | = 1.0683 | 0.50 | 0.4769 | 0.4791 | -0.0022 | -0.0047 |
| | | 0.40 | 0.3757 | 0.3788 | -0.0030 | -0.0080 |
| | | 0.30 | 0.2763 | 0.2801 | -0.0038 | -0.0135 |
| | | 0.20 | 0.1792 | 0.1834 | -0.0043 | -0.0232 |
| | | 0.10 | 0.0855 | 0.0894 | -0.0040 | -0.0444 |
| | | 0.01 | 0.0073 | 0.0085 | -0.0012 | -0.1399 |

Note: S*(t) is the national cohort life table survival function.

Source: Authors' calculations.

$$1.03 = \frac{13}{174.2} \times 1.37 + \frac{161.2}{174.2} \times 1.00,$$

and for 10 workers per IWE,

$$1.01 = \frac{13}{633} \times 1.37 + \frac{620}{633} \times 1.00.$$

Thus, using these alternative assumptions, the excess risk would be in the range 1-3% – far below Walker's 14% estimate.

Third, it is not clear why Walker assumed that the moderately exposed workers had a relative risk of 1.00. From Selikoff's (1981) analysis (see Table 2.1), it is plausible that this relative risk could be as low as 0.83, in which case the weighted average would be 1.03 using Walker's formula and 0.93 or 0.86 using the corrected formulas with 2.6 or 10 workers per IWE.

Fourth, the use of a constant relative risk assumed that the increase in risk was independent of time since first exposure. This was clearly contradicted by Selikoff (1981) (see Section 3.3.4). In fact, the relative risk for the second decade of follow-up of insulation workers was 1.49 (Selikoff and Seidman, 1990) (see Table 2.1), compared to 1.37 for the first decade. As a consequence, the originally exposed cohort, $N_{ay}(0)$, would be overestimated, but $E_{ay}(u)$, the number of new excess cases at time $u$, would be underestimated for $u > t$.

Fifth, the projection calculations for mesothelioma and lung cancer were stratified according to intensity of exposure (Walker, 1982, p. 13). In this case, it would have been more appropriate to adjust national cohort survival probabilities with different relative risk exponents, 1.00 or 1.37, depending on whether the exposure level was moderate or heavy. Not only would this have yielded an exact solution for the separate survival functions under the constant relative risk assumption, but it could have been implemented without having to specify the number of actual workers per IWE in the weighted average formula used by Walker to adjust the marginal survival function.

In this task, Walker not only chose an inferior method, but he also made an error in formulating the weights used in its implementation. The impact of this error was to bias the exponent in the survival function upward by 7-14%, leading to underpredictions of the number of mesothelioma cases in the later years of the projection period. The impact for the early years of the projection period was minor.

#### *4.4.1f Task 1f: For exposure in the more recent past which would give rise to no current disease, estimate the quantity of exposure, since this could still give rise to future disease*

The Weibull incidence model could not be used to estimate the number of insulation-worker equivalents who were exposed less than 20 years before the estimates of the latest number of mesothelioma cases were made (1975-1979). Due to the long latency time for mesothelioma (i.e., more than 20 years), none of these cases could represent recent exposure. To determine the more recent patterns of exposure, Walker employed data on the "reported age at first entry into an asbestos-related industry" from a survey of workers conducted by Elrick and Lavidge (no published reference given).

In the survey, 214 workers were identified who recalled working in an asbestos-related industry. Of these 214 workers, only 79 recalled exposure to asbestos. Nevertheless, Walker used the age- and date-specific employment experience of all 214 workers. This was done to increase the stability of the age- and date-specific distributions. However, even using data on all 214 workers, the statistical stability of the estimates of the age- and date-specific distributions might be questioned – especially because there were only 77 cases reported for the period after 1955 (and only 31 recalled exposure).

To combine this experience with the distribution of insulation-worker equivalents from the Weibull model, Walker iteratively adjusted the forecasts of the age- and date-specific exposure distribution for the following reasons:

- To maintain the pre-1955 exposure distribution implied by the cases
- To have the post-1955 figure stand in the appropriate proportions to the pre-1955 figures as indicated by the survey (i.e., 36% post-1955)
- To constrain the projected number of mesothelioma cases to equal the estimated numbers of mesothelioma cases in 1975-1979

In making the projection, Walker argued, plausibly, that occupational exposure to asbestos had declined since 1965. To reflect this reduction in exposure, Walker discounted workforce exposure by 10% for 1960-1964, 50% for 1965-1969, 75% for 1970-1974, and 100% for 1975-1979. The data from which these discounts were estimated was not specified.

The results of the above-discussed calculations and assumptions were presented in Table 9 of Walker's (1982, p. 15) report. The numbers in these tables represented only the number of insulation-worker equivalents necessary to produce the 54% of mesothelioma which Walker argued was plausibly due to occupational exposure to asbestos. Walker presented no formal procedure to translate insulation-worker equivalents into the total number of workers. Although Walker suggested that each moderately exposed worker might represent 5 or 10 actual workers, no information was given to justify this translation.

### 4.4.2 Task 2: Project Mesothelioma Incidence

#### *4.4.2a Task 2a: Adjust exposed population and calculate future incidence of mesothelioma*

Walker developed forecasts of new mesothelioma cases for 1980-2009 among men with occupational asbestos exposure histories (Walker's Table 10). The equations used to produce these projections were derived from Peto et al. (1981) (see Section 4.2). The basic equations underlying the projection are identical to the calibration equations used in Task 1e. To review, if the cohort aged $a_1$ in year $y_1$ was first occupationally exposed to asbestos $t_1$ years before $y_1$, then the expected number of excess cases in this cohort $t$ years after year $y_1$ is

$$E_{a_1-t_1,y_1-t_1}(t_1+t) = E_{a_1-t_1,y_1-t_1}(t_1) \times \left(\frac{t_1+t-w}{t_1-w}\right)^k \times S_{a_1y_1}(t).$$

This expression clearly shows that the projection is independent of the estimate of the initial IWE exposure count in prior years. $E_{a_1-t_1,y_1-t_1}(t_1)$ is estimated from mesothelioma mortality surveillance and the survival fraction $S_{a_1y_1}(t)$ from a cohort life table. The mesothelioma mortality count at age $a$ in year $y$ is a sum of counts projected for each possible time since first exposure to asbestos.

These equations assume that the exposure to asbestos has effectively stopped by the calibration date $y$. Otherwise, the projection would be incomplete because the mesothelioma cases due to new exposure after the calibration date would be ignored. This is distinct from the adjustment in Task 1f for exposures shortly before the calibration date.

Once the age- and date-specific exposure distribution of workers was generated, the same function used to generate it was applied to forecast the future numbers of mesothelioma cases. Many of the same issues relevant to the use of the Weibull incidence function to generate the exposure distribution are relevant when the same function is used to forecast future mesothelioma cases. For example, the initial level of incidence was based on adjustments to the SEER data; the number of workers estimated to have entered the workforce at each age and past date was determined from data on Johns-Manville litigants, from the Elrick and Lavidge survey, from the SEER incidence estimates, and from assumptions about the level of exposure and survival. In the Weibull incidence function of the projection equations, both Peto et al.'s (1982) and Breslow's (1982) proportionality constants $b$ were multiplied by 0.8.

Walker further assumed that age-specific mortality rates after 1979 would be unchanged from their late 1970s values. In fact, mortality was declining prior to the calibration period and it has continued to decline since that time (SSA, 1992, 1996, 1999). The effect of this decline is to introduce an implicit increase, on the order of 0.5-1.5% per year, in the relative risk of mortality for all exposed workers over the projection period.

Table 4.2 displays Walker's projections of the future number of mesothelioma cases based on both the Peto et al. (no latency) and Breslow (15-year latency) models. The differences between these two sets of estimates, given the likely impact of other factors and assumptions, were not large. Based on the logic of Task 1b, Walker used an estimate of 2630 mesothelioma cases 1975-1979 instead of the SEER-based estimate of 4870. The difference between 2630 and 4870 reflects Walker's assumption that 46% of all mesothelioma cases would arise in individuals without occupational exposure histories.

**Table 4.2: Projected Numbers of New Mesothelioma Cases, 1980-2009, in Men with Occupational Asbestos Exposure Histories, Using Two Models of Incidence**

| Year | No Latency Period (Peto) | 15-Year Latency Period (Breslow) |
|---|---|---|
| 1975-1979 | 2,630 | 2,630 |
| 1980-1984 | 3,200 | 3,400 |
| 1985-1989 | 3,500 | 3,900 |
| 1990-1994 | 3,600 | 4,200 |
| 1995-1999 | 3,400 | 4,000 |
| 2000-2004 | 2,900 | 3,500 |
| 2005-2009 | 2,100 | 2,500 |
| Total 1980-2009 | 18,700 | 21,500 |

Source: Walker (1982, Table 10).

Walker's projections can be compared with Selikoff's projections, provided we recognize that Selikoff projected all cases of mesothelioma attributable to occupational exposure to asbestos, not just those that can be documented sufficiently to support a tort claim, and that Selikoff (1981, p. 94) was more confident in the timing of his projections than in the actual numerical levels at any given time. The ratio of the projected mesothelioma cases for 2005-2009 versus 1980-1984 averaged 0.7 in Walker's two projections versus 1.7 in Selikoff's projection (see Table 3.3). Walker projected roughly a 30% decline by 2005-2009, whereas Selikoff projected a 70% increase. Walker's peak period was 1990-1994, whereas Selikoff's peak period was 2000-2004. However, Selikoff provided projections for 11 industry/occupation groups and 2 of these (shipbuilding/repair and railroad engine repair) reached their peak mesothelioma incidence in 1985-1989 and had combined ratios for 2005-2009 versus 1980-1984 of 0.8 (and separate ratios of 0.9 and 0.3). Thus, Walker's projection had a different timing of cases than Selikoff's overall projection, but the timing was within the range of some of Selikoff's occupation-specific projections.

#### *4.4.2b Task 2b: Estimate the sensitivity of mesothelioma projections to the assumptions involved*

Walker briefly discussed the results of certain sensitivity analyses he conducted. His presentation lacked sufficient detail for a precise and quantitative evaluation.

The sensitivity of the projections to certain linear factors suggested that they could cause more than a 30% variation in his projections. For example, an increase of 25% in the estimated number of mesothelioma cases for 1975-1979 might arise by using Selikoff's estimate that mesothelioma is underdiagnosed by about 28% in SEER data (see Section 3.4) rather than Walker's estimate of 10%. Important assumptions and uses of data (e.g., those involved in estimating the proportion of workers with a documentable history of occupational exposure to asbestos and the proportion of exposed workers with moderate and heavy exposure) were not explicitly discussed in this section.

Walker claimed that changes in linear factors were not important because they would affect the estimate of the current number of mesothelioma cases, but not the current number of lawsuits. Because the current number of lawsuits was tied to the current number of mesothelioma cases, Walker argued that this would lead to the estimate of the proportion of cases ending up in court being changed in a compensatory fashion; that is, if we know there are 100 court cases and we estimate that these arise from 1000 mesothelioma cases, then our estimate of the proportion of mesothelioma cases who sue is 10%. If our estimates of 1000 mesothelioma cases is wrong and the true number of mesothelioma cases is 2000, then, to produce the 100 court cases, the true likelihood of suing must be 5%. Nonetheless, the fact that we underestimated the number of mesothelioma cases by half could be balanced by overestimating the likelihood of suit by 2. Thus, the error in our estimate of the number of mesothelioma cases can be compensated for by an error in the rate at which lawsuits arise from cases. If this compensating error rate remained fixed over the course of the projections, we would correctly project the future number of lawsuits. Although all of this may be true, its relevance is not clear because Walker did not generate projections of mesothelioma lawsuits.

Walker's discussion of nonlinear parameters was oriented toward a qualitative discussion of their effect on "the shape of the future mesothelioma curve." He argued that the shape of this curve was "relatively robust to large variations in the nonlinear parameters." However, he presented no quantitative evidence to support this argument.

### 4.4.3 Task 3: Project Lung Cancer Incidence

#### *4.4.3a Task 3a: Adjust the size of the exposed population and calculate future incidence of lung cancer*

To project the number of lung cancer cases arising due to asbestos exposure, Walker used estimates of the moderately and heavily exposed population derived from the model of mesothelioma incidence. Consequently, the adequacy of lung cancer projections depends on the adequacy of the assumptions, data, and model used in the analysis of mesothelioma. The mesothelioma-based estimates of the exposed population were used in conjunction with age-specific lung cancer risks from SEER for 1973-1978 and relative risks, specific to time since first exposure, derived by Selikoff et al. (1980). The basic projection equation was the same as for mesothelioma, that is,

$$E_{a_0 y_0}(t) = I_{a_0 y_0}(t) \times N_{a_0 y_0}(t),$$

except that the Weibull form for $I_{a_0 y_0}(t)$ was replaced with a multiplicative relative risk function, that is,

$$I_{a_0 y_0}(t) = L_{a_0+t, y_0+t} \times K_{a_0 y_0}(t),$$

where

$L_{ay}$ = expected lung cancer incidence among the general population aged $a$ in year $y$,

$K_{a_0 y_0}(t)$ = relative risk of lung cancer among cohorts of insulation workers $t$ years after the start of asbestos exposure at age $a_0$ in year $y_0$.

Estimates of $L_{ay}$ were obtained from SEER data for white males for 1973-1978, by 5-year age groups from 20-24 to 75-79. These were assumed to be fixed for all future years at the 1973-1978 level. Estimates of $K_{a_0 y_0}(t)$ were obtained from Selikoff et al. (1980), subject to two adjustments: (1) Beginning with the period 1980-1984, the values of $K_{a-t,y-t}(t)$ were reduced at a rate of 10% per quinquennium and (2) all values were reduced by 50%, with a corresponding doubling of $N_{a_0 y_0}(t)$, to reflect the impact of asbestos-related lung cancer on a worker population "about one-half as intensely exposed to asbestos as insulation workers."

In specifying the first adjustment to $K_{a_0 y_0}(t)$, Walker used a novel interpretation of the insulation worker data analyzed by Selikoff et al. (1980), namely that "Selikoff's multipliers can be interpreted as reflecting a reasonably steady rise through working life, with a decline beginning around the time of retirement, that is, about the time of cessation of asbestos exposure" (Walker, 1982, p. 19). This ignored Selikoff's (1981, Fig. 2-1) demonstration that the relative risk of lung cancer – in particular, the time at which the decline begins – is independent of age at the start of exposure.

Walker admitted that there was no a priori reason to expect such a decline and that other bodies of data did not suggest such a decline. Furthermore, Walker did not discuss the rationale provided by Selikoff for the decline – an explanation based on a phenomenon frequently experienced in occupational cohort studies [i.e., the selective removal of high-risk persons from the cohort by the operation of the disease risk (e.g., Selikoff, 1981, pp. 22, 88, and 180)].

If Selikoff's explanation were true, it would seem appropriate to modify the projection strategy to represent the effects of selection. Under a selection model, the risks for highly exposed individuals would not decline, but the aggregate risk of the population would decrease as the high-risk group died out. To model such an effect, one could use a "distributed parameter" form of the incidence function so that the rate of occurrence of cases, the exhaustion of highly exposed workers, and a possible decline in the aggregate relative risk would be appropriately correlated. Even if Selikoff's argument were not true, it may be appropriate to make other adjustments to Walker's procedures. For example, as Walker suggested, there were other studies in which such a decline was not observed. Consequently, it might have been appropriate to make projections with no decline in relative risk. This was the approach used by OSHA (1983, 1986) and the EPA (1986) after review of the same (and additional) evidence on the time course of relative risks.

In specifying the second adjustment to $K_{a_0y_0}(t)$, Walker observed that the projected number of lung cancer cases was strongly dependent on the average level of asbestos exposure in the population. The lower the average exposure to asbestos, the lower will be the excess risk of lung cancer in the population and the greater the proportion of the total lung cancer incidence that is due to background causes (such as smoking). Among occupationally exposed workers, it may not be possible to distinguish between excess and background cases of lung cancer. The potential pool of litigants for an occupationally exposed population must include all workers who develop lung cancer. To project the number of potential litigants with lung cancer, including background lung cancer cases, one needs an estimate of the number of exposed workers. Because Walker had no formal procedure for determining the actual number of exposed workers, his estimates of lung cancer cases must be considered very uncertain.

To deal with this issue, Walker set his adjustment parameters to represent a worker population that was "about one-half as intensely exposed to asbestos as insulation workers" (Walker, 1982, p. 20). In other words, Walker assumed that, on average, there were two actual workers for each computed IWE. Because the number of actual workers per IWE for heavily exposed workers is 1.00 by assumption, this adjustment allows the implied number of actual workers per IWE for moderately exposed workers to be determined from the identity

$$2.0 = \frac{20}{54} \times 1.0 + \frac{34}{54} \times 2.6,$$

yielding 2.6 moderately exposed worker per IWE – well below the range of 5-10 workers cited by Walker in Task 1f. In this case, moderately exposed workers had risks of asbestos-related lung cancer 38.5% of those of insulation workers. Equivalently, their exposure intensity was assumed to be 38.5% that of insulation workers – about double the mid-range relative risk estimates in Table 3.1 (i.e., for construction trades, railroad engine repair, utility services, stationary engineers and fireman, and chemical plant and refinery maintenance).

In addition to the uncertainty in the two adjustments to the relative risk function, $K_{a_0y_0}(t)$, there was the uncertainty of the function itself. Neither Selikoff et al. (1980) nor Walker addressed this issue. The EPA (1986) reviewed

**Table 4.3: Projected Numbers of New Lung Cancer Cases, 1980-2009, in Men Occupationally Exposed to Asbestos**

| Year | Numbers of New Cases |
|---|---|
| 1975-1979 | 22,248 |
| 1980-1984 | 17,800 |
| 1985-1989 | 13,600 |
| 1990-1994 | 10,200 |
| 1995-1999 | 7,000 |
| 2000-2004 | 4,300 |
| 2005-2009 | 2,220 |
| Total 1980-2009 | 55,120 |

Source: Walker (1982, Table 14).

models of the relative risk of lung cancer among asbestos workers using estimates from 11 separate studies representing a range of fiber types, sizes, and intensities of exposure (see Section 2.4). Their analysis suggested that a 95% confidence limit on the relative risk would be a factor of 2.5 (i.e., estimates are divided by 2.5 and multiplied by 2.5 to determine the range) (EPA, 1986, p. 82). The error would be correlated over age for any given cohort of workers at a given work-site.

In total, Walker projected 55,120 lung cancer deaths for the period 1980-2009 (Table 4.3). Compared with the mesothelioma projections in Table 4.2, this is an increase by a factor of 2.6-2.9. However, the timings of the two diseases are very different. For 1980-1984, the ratio is 5.2-5.6, whereas for 2005-2009, the ratio is 0.9-1.0. Beyond 2005-2009, Walker's projections would yield more mesothelioma than lung cancer cases.

Walker's projected rate of decline in asbestos-related lung cancer deaths was much larger than that projected by Selikoff, both overall and for specific occupations. Overall, the ratio of lung cancer cases projected by Walker for 2005-2009 versus 1980-1984 is 0.12. This ratio is much smaller than the 0.86 ratio projected by Selikoff (1981, Tables 2-22 and 2-27). Selikoff projected both expected and excess lung cancer deaths so that their sum is comparable to Walker's projection. Selikoff (1981, p. 94) was uncertain of the level of his projection; however, he was confident in the timing of the disease.

Selikoff's two highest-risk occupation/industry groups were insulation workers and primary manufacturers, both with a relative risk of 1.0 (Table

3.1). For these groups, Selikoff projected 6695 lung cancer deaths in 1980-1984 and 5790 in 2005-2009, for a ratio of 0.86. Walker's projection actually employed an assumption that the exposed workers were at a relative risk of 0.5 compared to insulation workers. Selikoff reported projections for two occupation/industry groups whose relative risks were 0.5: secondary manufacturers, and shipbuilding and repair. For these groups, Selikoff projected 57,970 lung cancer deaths in 1980-1984 and 29,210 in 2005-2009, for a ratio of 0.50. The ratio for shipbuilding and repair was lower (0.41) due to the depletion of WWII workers. The smallest ratio for the 11 occupation/industry groups analyzed by Selikoff was 0.29 for railroad engine repair, 2625 lung cancer deaths in 1980-1984, and 755 in 2005-2009.

#### *4.4.3c Task 3b: Compare projected and observed lung cancer figures*

In this task, Walker compared the distribution of age and date of onset of exposure for 349 litigants (1975-1981) of Johns-Manville who had lung cancer with the distribution predicted by the model (1975-1979). The distribution of the predicted date of first exposure and the distribution of the observed date of first exposure of litigants (summed over age at diagnosis/lawsuit) correlated reasonably well except for a systematic underprediction by the model of exposures after 1960 (i.e., 10.0% observed and 4.4% predicted).

The predicted distribution of age at diagnosis and the actual distribution of age among litigants showed a much poorer correlation. Whereas 12% of the actual lawsuits came from persons aged 70-79 years (42 of 349), the model predicted that 42% of lung cancer would appear among persons in this age range (9443 of 22,248 cases). Walker argued that this discrepancy represented a decreased propensity to sue at later ages – an observation made by Selikoff (1981) in his analysis of insulation worker data.

The systematic differences in the observed and predicted distributions of age and date of first exposure are significant. Because the model underpredicted cases emerging from recent exposure, Walker may have assumed too rapid rates of decrease of recent exposure – leading to an underprediction of health effects in future years.

### 4.4.4 Task 4: Estimate Current and Future Asbestosis Prevalence

#### *4.4.4a Task 4a: Predict asbestosis using mesothelioma mortality rates in asbestotics*

Walker attempted to estimate the number of persons with asbestosis by combining his estimates of mesothelioma cases with data on the rate of occurrence of mesothelioma and asbestosis simultaneously. He exploited the following relation:

$$A \times I = M \times P,$$

**Table 4.4: Projections of the Numbers of Prevalent Cases of Asbestosis in U.S. Males, 1980-2009**

| Year | Number of Men Alive with Asbestosis | |
|---|---|---|
| | Based on Task 4a | Based on Task 4b |
| 1980-1984 | 65,800 | 64,000 |
| 1985-1989 | 35,400 | 45,300 |
| 1990-1994 | 19,000 | 31,000 |
| 1995-1999 | 9,600 | 19,700 |
| 2000-2004 | 4,400 | 11,400 |
| 2005-2009 | 1,700 | 5,700 |

Source: Walker (1982, Tables 16 and 18).

or

$$A = (M \times P)/I$$

where $A$ is the number of asbestotics, $M$ is the annual number of new mesothelioma cases in the general population (all mesothelioma cases, not just the 54% that Walker assumed had documentable occupational exposure to asbestos), $P$ is the proportion of $M$ with concurrent asbestosis, and $I$ is the annual incidence of mesothelioma in asbestotics.

For the period 1975-1979, Walker estimated $A$ using $M = 974$, $P = 0.28$, and $I = 1/200$, so that

$$A = 974 \times 0.28 \times 200$$
$$= 54{,}544.$$

Estimates of $M$ for later periods require that Walker's projections be converted to annual counts and then be divided by 0.54 to obtain the appropriate $M$ value. For example, for 1980-1984, Table 4.2 indicates 3200 mesothelioma cases under the Peto incidence function. This converts to $M = 1185$, and

$$A = 1185 \times 0.28 \times 200$$
$$= 66{,}360.$$

The value of $A$ for 1980-1984 in Table 4.4 is 65,800, which implies that Walker used $M = 1175$, which is within rounding error of our estimate of $M = 1185$. In his asbestosis projections, Walker estimated $A$ separately for each age group in the mesothelioma projection.

His forecasts (Table 4.4) depend on the estimated values used for $M$, $P$, and $I$. We have already discussed $M$ in detail (e.g., Task 1a), including sources of uncertainty. The estimate of $I$ was based on the average experience in three studies; the estimate of $P$ was based on just one study (Elmes and Simpson, 1976). In the following, we will evaluate the sensitivity of the projections of the number of asbestotics to $I$ and $P$.

The studies used by Walker to provide estimates of $I$ give a broad range of values for $I$: from 1/81 (Finkelstein et al., 1981) to 1/377 (Edge, 1979). If the estimate of $I = 1/377$ were used and if we use the estimate $M = 1175$ cases, then for 1980-1984,

$$A = 1175 \times 0.28 \times 377 = 124{,}033.$$

The estimate of 1/166 from the largest data series (Berry, 1981) gives

$$A = 1175 \times 0.28 \times 166 = 54{,}614.$$

Finkelstein et al.'s (1981) estimate of $I = 1/81$ gives $A = 26{,}649$. The individual studies yield estimates 89% higher or 59% lower than Walker's.

The estimate $P = 0.28$ derives from 70 cases of asbestosis among 247 mesothelioma patients (Elmes and Simpson, 1976, Table 5). However, Walker assumed that only 54% of mesothelioma cases were occupationally exposed to asbestos, whereas asbestosis should be accepted as de facto proof of occupational exposure to asbestos (EPA, 1986, p. 177). Hence, one can improve the estimate of $A$ by making an additional adjustment to obtain $P_a$, the prevalence of asbestosis among mesothelioma cases occupationally exposed to asbestos. To do this, we rely on Elmes and Simpson's Table 2, which indicates that 264 of 277 mesothelioma cases (95.3%) with analyzable occupational histories had definite or probable occupational asbestos exposures. Applying this rate to the 247 mesothelioma patients in their Table 5, we estimate that about 235 were occupationally exposed to asbestos, implying that $P_a = 0.30$ (i.e., 70/235). For consistency, $M$ should be replaced with $M_a$, the number of mesothelioma cases occupationally exposed to asbestos; the 1175 cases of mesothelioma in Walker's formula should be multiplied by 0.54 to yield $M_a = 635$ cases. Thus, we can respecify Walker's model as

$$A = M_a \times P_a / I,$$

which yields

$$A = 635 \times 0.30 \times 200 = 38{,}100,$$

for 1980-1984 – about 58% of Walker's reported estimate of 65,800. If Finkelstein et al.'s (1981) estimate of $I = 1/81$ is used, then $A = 15{,}430$.

The above calculations indicate that the estimate of the number of asbestotics ($A$) for 1980-1984 could plausibly range from 15,430 to 124,033 – a factor of 8.

For the projections of asbestosis in Table 4.3, Walker also assumed that the reduction of asbestos exposure would prevent any new cases of asbestosis from emerging after 1984. There is no reasonable basis for this assumption. It contradicts the assumption in Task 4b that the time course of asbestosis and mesothelioma mortality are very similar (Selikoff et al., 1980), which would imply that significant numbers of asbestosis deaths would occur up to four or five decades after initial exposure to asbestos. In addition, Walker noted that his projections were based on severe asbestosis and that if less severe cases were included, the prevalence of asbestosis could be up to three times larger.

#### *4.4.4b Task 4b: Estimate asbestosis prevalence using the equivalence between asbestosis and mesothelioma mortality*

Walker attempted to verify the prevalence estimates for asbestosis by producing alternate estimates. Walker used two findings. Death rates reported by Selikoff et al. (1980), as a function of time from first exposure, were very similar for mesothelioma and asbestosis. Berry (1981) reported that, among asbestotics, 21.3% of deaths were attributed to asbestosis. Then, assuming equal numbers of deaths from asbestosis and from mesothelioma, we have

$$M_a = 0.213 \times D_A$$
$$D_A = 2.8 \times G \times A.$$

Here, $M_a$ is the annual number of mesothelioma deaths among men exposed to asbestos (available from Walker's mesothelioma projections). $D_A$ is the annual number of deaths among asbestotics. $G$ is the general death rate and is derived on an age-specific basis from vital statistics data. The value of 2.8 represents an estimate of the increase in the risk of mortality among asbestotics derived from Berry (1981) and Finkelstein et al. (1981). Combining these two relations yielded the number of asbestotics as

$$A = M_a/(0.213 \times 2.8 \times G),$$

which yielded the estimate $A$ = 64,000 for 1980-1984 (Table 4.4). In developing the projections in Table 4.4, Walker applied the above equation using age-specific projections of $M_a$ and $G$. The projections of $A$ are sensitive to assumptions and questions arising in the estimate of $M_a$ and the estimate of 21.3%. If the data used to estimate this latter probability were binomially distributed, then it would have a standard deviation of 2.5% and a 95% confidence interval ranging from 16.3% to 26.3%, which could imply a decrease or increase of 23.5% in $A$.

Walker assumed no new incidence of asbestosis after 1984. The different results of Tasks 4a and 4b reflect the much younger age distribution implied by using age-specific $G$ values in the denominator of the formula for $A$. For example, the latter projection for 1995-1999 was 2.05 times larger, and for 2005-2009, it was 3.35 times larger, than the former.

#### *4.4.4c Task 4c: Other methods of projecting asbestosis prevalence*

In this subsection, three "quick and dirty" methods for estimating asbestosis prevalence are presented and critiqued.

1. *U.K. Pneumoconiosis Panels.* In the United Kingdom in 1973-1976, 133 workers were certified annually as asbestotics. If the median survival time of the least disabled workers was 15 years, 2000 workers in the United Kingdom (i.e., 15 × 133) could be projected to have asbestosis. Because the U.S. population was four times larger than the British population, this estimate was multiplied by 4 to get a U.S. estimate of 8000. Because this methodology depends on the completeness of certification of asbestosis in the United Kingdom, this estimate was viewed as a lower bound by Walker.
2. *NCHS.* Walker (1982) cited an unpublished estimate in an unpublished letter by Burnham (1982) to a third party (Breslow) suggested that 427,000 people (with a range of 248,000 to 606,000) had pneumoconiosis in the United States. This letter reported that of 1422 deaths ascribed to pneumoniconiosis, 72 (5%) mentioned asbestos. From these values, an estimate of 21,000 asbestotics was obtained (i.e., 0.05× 427,000), with a range of 12,000 to 30,000.
   Walker noted that the pneumoconiosis estimate was derived from a "questionnaire" (the survey was not identified). Furthermore, reporting of asbestosis on death certificates was very low. For example, Selikoff et al. (1979) (see Table 2.1) reported that 78 cases of asbestosis were correctly diagnosed on 168 death certificates where the best evidence indicated asbestosis. This suggests that the number 72 could plausibly be multiplied by 2.15. This would yield an estimate of 46,500 asbestotics (i.e., 155/1422 × 427,000) with an upper bound of 66,000.
3. *X-ray Abnormalities.* Using his estimates of heavily exposed workers (1982, Table 9) and multiplying by the prevalence of X-ray changes in insulation workers (Walker, 1982, Table 17), Walker estimated 18,000 to 150,000 asbestotics. Again these estimates depend on all of the assumptions in the model of mesothelioma incidence used to forecast the size of the occupationally exposed population. For example, if we use an estimate of 1365 cases, which is 40% higher than the 974 case figure employed (see Task 1a), the upper estimate becomes 210,000 asbestotics.

#### *4.4.4d Task 4d: Derive a general methodology for predicting lawsuits as a function of asbestosis prevalence*

Walker attempted to relate his prevalence estimate of asbestotics to the number of lawsuits generated. Only very general calculations were performed to illustrate that the higher mortality for persons with asbestosis and a 9% annual decrement due to the filing of lawsuits jointly implied that the number

**Table 4.5: Projections of the Numbers of Asbestosis Lawsuits Among U.S. Males, 1980-2009**

| Year | Number of Asbestosis Lawsuits | |
|---|---|---|
| | Based on Task 4a | Based on Task 4b |
| 1980-1984 | 24,800 | 24,100 |
| 1985-1989 | 8,300 | 10,600 |
| 1990-1994 | 2,800 | 4,500 |
| 1995-1999 | 900 | 1,800 |
| 2000-2004 | 200 | 700 |
| 2005-2009 | 100 | 200 |
| Total 1980-2009 | 37,100 | 41,900 |

Source: Walker (1982, Tables 16 and 18).

of lawsuits would decline rapidly over time (again based on the assumption of no new asbestosis cases after 1985; see Table 4.5).

Walker admitted that the rate of filing of lawsuits did not seem to decline as rapidly as these projections indicated, a fact which he attributed not to error in projections but to recent changes in the propensity to sue. With these caveats in mind, but without fully specifying his reasoning, Walker projected that 45,000 lawsuits (cf. 37,100-41,900 in Table 4.5) would be filed in 1980-2009, with a firm lower bound of 30,000 but with an indefinite upper bound of 120,000.

This upper bound of 120,000 was almost twice as large as his estimate of the prevalence of asbestotics (e.g., 64,000 in 1980-1984, and not increasing thereafter). The assumption that 9% of asbestotics would sue per annum implies that, absent death, about 60% of cases would sue over 10 years. According to Selikoff (1981, p. 536), 32% of asbestos-related deaths among insulation workers in 1975-1976 resulted in lawsuits being filed by 1980. Walker's per annum rate of suit takes no account of age variation in the propensity to sue. The likelihood of suit tends to decrease with age (Selikoff, 1981).

### 4.4.5 Task 5: Estimate the Amount of Asbestos-Related Disease Likely to Occur in Women

Walker (1982) estimated the number of asbestos-related disease cases that would occur among women who were occupationally exposed. Despite the absence of data regarding the size of the female workforce and their exposure-

**Table 4.6: Alternative Projections of Mesothelioma and Lung Cancer Cases**

| Source | Mesothelioma | Lung Cancer | Time Interval |
|---|---|---|---|
| Walker (1982) | 18,700–21,500 | 55,120 | 1980–2009 |
| Peto et al. (1981) | 24,375 | 73,125 | 1980–2009 |
| McDonald and McDonald (1981) | 18,970 | 45,520 | 30-Year |
| Enterline (1981) | 10,000 | 75,000 | 30-Year |
| Selikoff (1981) | 79,745 | 149,485 | 1980–2009 |
| Hogan and Hoel (1981) | 65,000 | — | 1980–2009 |

timing distribution, Walker assumed that the number of cases of asbestos-related diseases among women would be roughly in proportion to the observed number of lawsuits filed with the Johns-Manville Corporation among women, relative to those deriving from men. He estimated that women would experience about 5% of the numbers of cases of asbestos-related diseases projected for men and that the absolute number of female cases would diminish to a negligible number by the year 2000. The Johns-Manville lawsuit experience is an inadequate basis for predicting the future occurrence of disease among women. A more reliable projection would require estimation of the size of the exposed female workforce.

## 4.5 Asbestos-Related Disease Projections by Other Authors

Estimates of future numbers of asbestos-related cancer deaths have been made by six authors other than Walker (1982). These are listed in Table 3.4 and were detailed in Section 3.5. Here, we focus only on the five studies that provide direct comparisons with Walker's methods and results (Table 4.6).

Two general approaches have been adopted. The first reconstructs directly the sizes of exposed worker populations in the past, using data compiled by the Department of Labor and other government and private agencies (see Chapter 3). Selikoff (1981), Hogan and Hoel (1981), and Nicholson et al. (1981a) adopted this approach. The second approach indirectly estimates from recent mesothelioma incidence data the numbers of currently living person-exposure equivalents first exposed at various ages and years in the past. McDonald and McDonald (1981), Peto et al. (1981), and Walker (1982) adopted this approach, using a mathematical model that expressed mesothelioma incidence as a function of time from onset of exposure. Enterline's (1981) approach was a hybrid of these two methods. He estimated the heavily exposed population by a back-calculation from recent asbestosis mortality data. His estimates of the presumed lightly exposed workforce were determined from published reports of worker population sizes.

The only projections that can validly be compared are those for mesothelioma and lung cancer, for men, for the 30-year period 1980-2009. For this period, Walker projected 18,700 to 21,500 mesothelioma cases (assuming a 15-year minimum latency for the latter figure) and 55,120 lung cancer cases, resulting from occupational exposure to asbestos. These estimates are in fairly close agreement with those of Peto et al. (1981), who used a procedure very similar to Walker's to project future mesothelioma cases (24,375), and a threefold lung cancer excess relative to mesothelioma to project future lung cancers (73,125).

McDonald and McDonald (1981) estimated 843 mesothelioma deaths among U.S. men in 1975 (see Section 3.5). With their estimate that 75% of male mesothelioma cases were caused by occupational exposure to asbestos and our assumption that this rate represented an average value over the 30-year period, one would expect 18,970 new cases of mesothelioma over 30 years – a projection within Walker's (1982) range of 18,700-21,500. McDonald and McDonald (1981) estimated that there were 2.4 lung cancer deaths for every mesothelioma death attributable to asbestos. Applying this ratio to the above projection yields 45,520 lung cancer deaths for the same 30-year period – almost 10,000 fewer than Walker's projection.

Enterline (1981) projected 10,000 mesotheliomas and 75,000 lung cancers over a 30-year period. Enterline's mesothelioma projections were the lowest of any cited in Table 3.4. His methodology was detailed in Section 3.5. The different results primarily reflected differences in the estimated proportions of heavily exposed workers.

Hogan and Hoel (1981) projected approximately 65,000 mesothelioma cases, and Selikoff (1981) projected 79,745 mesothelioma cases and 149,485 lung cancer cases.

This comparison shows substantial agreement of results when similar methods are used. Walker's (1982), Peto et al.'s (1981), and McDonald and McDonald's (1981) mesothelioma projections are quite similar to each other. General agreement between Walker's and Peto's projections would be anticipated because the same underlying incidence model was used in each approach. Some of the discrepancy between Walker's and Peto's mesothelioma and lung cancer projections may be attributable to the method of determining population-exposure equivalents. Similarly, Hogan and Hoel's (1981) and Selikoff's (1981) mesothelioma projections are in reasonably close agreement. Enterline's (1981) mesothelioma projections were lower than the results from either of the two general approaches, even though his approach was a hybrid of the other two.

The validity of the two general approaches cannot be determined until actual counts of future cases are obtained and compared with predicted numbers.

None of the other authors except Selikoff (1981) projected the future number of asbestosis cases. Selikoff (1981) addressed only the issue of asbestosis deaths. With no elaboration, he projected approximately 10,000 deaths from asbestosis during 1980-2009. In contrast, Walker (1982) projected a prevalence of approximately 65,000 cases in 1980-1984, with roughly a 50% reduction in this number every 5 years to the year 2009.

The differences among the mesothelioma and lung cancer projections can be attributed, at least in part, to discrepancies in estimates of the following factors:

1. The proportion of workers heavily exposed to asbestos, and the relationship between the cumulative exposure of insulation workers and other asbestos-exposed workers
2. The proportion of mesothelioma cases attributable to occupational exposure to asbestos
3. The excess mortality rates in heavily and lightly exposed workers
4. The distribution in the workforce of age and calendar year of first exposure to asbestos
5. The magnitude of the reduction in the 1970s in workforce exposure to asbestos, and the effect of this reduced exposure on the risk of asbestos-related disease over the next 30 years

A numerical assessment of each factor is fraught with uncertainties because the required data were either never directly measured (e.g., individual worker exposure to asbestos) or were not recorded (e.g., the number of workers occupationally exposed to asbestos). Numerical estimates for some factors which are based on appropriate data (such as the proportion of mesothelioma cases attributable to asbestos) vary considerably from study to study.

Overall, Walker's projections were generally low compared to other projections.

## 4.6 Conclusions

1. Walker (1982) adopted a reasonable method of using recent mesothelioma incidence data to estimate the current at-risk population of asbestos-exposed workers.
2. Walker's projections for mesothelioma, although low in comparison with those of other researchers who have attempted to reconstruct directly the actual sizes of historical worker populations, were in reasonably close agreement with those obtained by another investigator who used

a methodology similar to his. Nevertheless, his use of generally conservative projection parameter values led to low projections of new cases of mesothelioma.
3. Walker's future lung cancer projections were lower than those offered by most other authors mainly because he used conservative parameter estimates, most notably recent lung cancer incidence rates for U.S. white males and a 10% per quinquennium diminution of risk. His lung cancer projections were highly dependent on the assumption that the average population exposure was one-half that of a typical insulation worker.
4. Walker's mesothelioma projections did not depend upon the estimated numbers of currently living, previously exposed insulation-worker equivalents, whereas his lung cancer projections increased in direct proportion with the estimates of the number of IWEs. Hence, his mesothelioma projections were inherently more stable than his lung cancer projections.
5. Walker's asbestosis projections were conservative, primarily because he assumed that no new cases would arise after 1984. There is no reasonable basis for this assumption. Asbestosis prevalence would be substantially greater than that predicted by Walker if asbestosis cases continue to occur for just one additional 5-year period beyond 1984.

# 5

# Uncertainty in Forecasts Based on Indirect Estimates

## 5.1 Introduction

Walker's (1982) model built upon Selikoff's (1981; also Selikoff et al., 1979, 1980) estimated incidence functions for asbestos-related diseases derived from epidemiologic analyses of North American insulation workers. This model introduced an artificial construct, insulation-worker equivalent or IWE, the use of which had significant implications for the uncertainty of his projections. Walker's projections were subject to a wide range of variability and uncertainty. To demonstrate the extreme sensitivity of Walker's estimates, we identify the qualitative sources of their uncertainty, summarize two sets of computerized sensitivity analyses, and conclude with some general observations about the ranges of uncertainty in projections of this type.

## 5.2 Qualitative Sources of Uncertainty in Walker's Projections

We now list sources of uncertainty in Walker's (1982) projections, excluding uncertainty about the quantitative values of the parameters used in Walker's projection models. These qualitative sources of uncertainty are of three kinds: (1) those that could make Walker's projections err in either direction, (2) those that could make Walker's projections too high, and (3) those that could make Walker's projections too low. We are concerned with uncertainty that results, for example, from the extrapolation of measurements made on one population to a different population, or from the impossibility of going into the past to make measurements required to project the future reliably, or from the use of epidemiological estimates based on an improper or incomplete analysis of data.

### 5.2.1 Uncertainties in Either Direction

1. Environmental sampling of asbestos concentrations was not done for the populations on which the Peto et al. (1982) curve is based, so the intensity has not been measured at all and the duration only crudely. The time since the onset of asbestos exposure was used as a surrogate for the duration of exposure in the major study of Selikoff et al. (1979).
2. The concentrations of other carcinogens or hazardous substances found in shipyards and other work sites were not measured in the epidemiological studies from which Walker drew his facts, so the risks from asbestos exposure cannot be disentangled from the risks of other exposures.
3. Comparisons of standardized mortality ratios (SMRs) between two or more populations are invalid if there are major differences in the age structures of the populations being compared (Miettinen, 1972). Walker did not verify that the age structure of each population from which an SMR was estimated was approximately the same as the age structure of the population to which he applied the SMR for projection.
4. When the intensity of exposure to asbestos of individuals or groups has not been measured directly, the length of employment is sometimes assumed to be proportional to dose. Yet, a worker may leave an industry sooner because he or she has received an unusually large dose of asbestos. The use of the length of employment as a surrogate measure of dose may obscure the doses received.
5. Risks of asbestos-related disease depend on the types of asbestos fiber and frequency distribution of fiber dimensions. Information about the asbestos fibers is incomplete both for populations studied epidemiologically to estimate dose-response curves and for populations presumed to be at risk in future projections.
6. Asbestosis is difficult to diagnosis in the absence of postmortem examinations. In one study, factory medical officers identified only 65% of true cases (Berry et al., 1979). Consequently, substantial uncertainty surrounds the estimates of asbestosis prevalence in past studies and future projections.
7. Epidemiologic studies reported levels of mortality from asbestosis and mesothelioma that varied from study to study. The variation may result as much from differences in case detection efforts, diagnostic acumen, and length of observation, as from variation in exposures. Case detection, diagnostic acumen, length of observation, and exposures may be different still in the population to which Walker's projections are intended to apply.
8. Most studies of asbestos and lung cancer [with exceptions like Hammond et al. (1979)] had few or no data on the smoking habits of either the asbestos-exposed or the control populations. Because the relative risk of lung cancer resulting from asbestos exposure appears to be sensitive to the prevalence of smoking in a population, estimates of asbestos-related lung cancer risk from populations studied epidemiologically would apply

to projected future populations only if the unknown prevalence of smoking was similar in both.

9. An interaction among mesothelioma, lung cancer, and smoking that Walker appeared to have overlooked is that cigarette smokers are selectively removed from risk of mesothelioma in later years because they suffer excess mortality from lung cancer and cardiovascular diseases earlier in life. If smoking declined in the future, this early selection would also be reduced.
10. The studies of Selikoff et al. (1979, 1980), from which Walker extracted lung cancer multipliers that he used in his projections, did not specify whether retired union members or only active workers were included in the population studied. Retired workers are a special subgroup in that they have survived workplace hazards. The study also did not specify the geographic distribution of other workers studied, so it is unknown if this distribution was comparable to that presumed in Walker's projections.
11. The concept of "insulation-worker equivalent," which was fundamental to the projections of Walker (1982), was poorly defined. In the studies of insulation workers by Selikoff et al. (1979, 1980), the strongest associations between exposure and disease were probably specific to those workers with the highest exposure levels, but the published risk estimates represented averages for the entire cohort.
12. Because the years of asbestos exposure overlapped the years of follow-up for some portion of the cohort studied by Selikoff et al. (1979), the high cumulative dose would be associated with long exposure and long follow-up durations. Consequently, the observed association between dose and death rate would be lower than the true association between dose and death rate. This could explain the increase in relative risk from 1.37 to 1.49 between the first and second decade of follow-up (Selikoff and Seidman, 1991) (see Table 2.1).
13. Walker (1982) assumed that "definite exposure" was equivalent to "heavy exposure" and that "probable exposure" was equivalent to "light exposure." He offered neither evidence nor argument for this assumption, and we know of none.
14. Walker's estimates of the distribution of years of first exposure for workers diagnosed with mesothelioma at a given age were based on 278 cases selected from thousands in the files of Johns-Manville. The cases not chosen lacked data required for Walker's analysis. The selected cases may well not have been representative of the population of all cases in the Johns-Manville files.
15. Even if the selected 278 mesothelioma cases were representative of those in the Johns-Manville files, those cases may not be representative of all U.S. male mesothelioma cases during 1975-1979.
16. Walker's estimates of the distributions by quinquennium of first asbestos exposure of the 311 mesothelioma cases aged 75-79 in 1975-1979 were

based on only 4 cases from the Johns-Manville files and, hence, were statistically unreliable.

17. The mathematical power law of Peto et al. (1982) described the incidence of mesothelioma as a function of time since first exposure to asbestos among heavily exposed workers like insulation workers. It is unknown whether a function of the same form or with the same parameter values described the incidence of mesothelioma in lightly exposed workers, although Walker assumed that the same function applied.
18. In using the results of the Elrick and Lavidge survey (Walker, 1982, p. 14) to estimate the distribution of the year of entry into the asbestos-exposed workforce of workers who entered after 1954, Walker ignored the sampling weights that should have been assigned to each individual according to the design of the survey.

### 5.2.2 Why Walker's Projections May Be Too Low

1. Expected cases of disease among women were underestimated. In textile operations, Peto et al. (1977) found comparable risks among men and women for lung and other cancers and for nonmalignant respiratory diseases, whereas Newhouse and Berry (1979) found higher lung cancer risks among women than among men.
2. When Walker projected the future male population at risk of asbestos-related disease, he assumed that the future age-specific survival probabilities in the population at large would be constant at the values observed in 1975-1979. If, as is more likely, the survival probabilities continued to improve slowly (though not necessarily linearly) as they have over the past half-century, more asbestos-exposed workers would survive and be at risk of future asbestos-related disease than Walker calculated.
3. Walker estimated that the average excess mortality among exposed workers was 14%. He used this figure to further reduce the survival probabilities beyond the reductions implicit in his use of constant probabilities from 1975 to 1979. Our review in Section 4.4.1e suggests that the 14% estimate was too high by at least a factor of 2.
4. Walker omitted all cases of asbestos-related disease among individuals aged 80 and older.
5. Walker assumed that 54% of mesothelioma cases have a documentable history of occupational asbestos exposure. Some published articles suggest that up to 100% of cases may have such histories.
6. Walker's use of the 278 Johns-Manville mesothelioma cases, in conjunction with his assumed restrictions on age at first exposure, eliminated all mesothelioma cases under the age of 40, and therefore underestimated future mesothelioma cases.
7. The Elrick and Lavidge survey considered only men aged 40 and above in 1975-1979, thereby excluding workers aged 15-39 who entered the asbestos-exposed labor force after 1954. Using the figures of Elrick and

Lavidge, Walker underestimated future populations at risk of developing asbestos-related disease.

8. It was reasonable to expect that mortality from lung cancer among U.S. white males aged 40 and older would continue to increase at least until 1990, as it did from 1940 to 1978. In fact, the increases continued to 1990 but reversed afterward (Cole and Rodu, 1996), so that Walker's use of unchanging lung cancer mortality rates may have underestimated the numbers of future cases.
9. Walker discounted both Selikoff's asbestos multipliers and the background lung cancers by 10% per quinquennium, instead of only the asbestos multipliers, thereby understating the likely future number of lung cancers.
10. Walker assumed that no new cases of asbestosis would occur after 1984. However, the first Occupational Health and Safety Administration asbestos standard was enacted in 1971, and environmental data documented substantial levels of asbestos until the early 1980s. If the time course of asbestosis described for insulation workers held for other exposed workers, then new cases would likely continue to develop for 20-30 years after 1971, or until roughly the end of the century, at least.
11. In projecting the prevalence of asbestosis, Walker assumed that the numbers of deaths from asbestosis would be roughly equivalent to the number of deaths from mesothelioma, as was observed among insulation workers by Selikoff et al. (1980). Other studies (Henderson and Enterline, 1979; Seidman et al., 1979) reported far more asbestosis deaths than mesothelioma deaths. Walker's generalizing from insulation workers to all workers may have understated asbestosis deaths and, consequently, asbestosis prevalence.

### 5.2.3 Why Walker's Projections May Be Too High

1. Walker's (1982) lung cancer projections were based on lung cancer "multipliers" taken from Selikoff et al. (1980), which were, in turn, based on a study of 17,800 insulation workers (Selikoff et al., 1979). As Weiss (1983) pointed out in reanalyzing this study, the workers had varying durations of exposure at the start of the study. Workers who entered the workforce long before the start of the study were likely to be older than the others, likely to have had greater average cumulative asbestos exposure, and likely to be at increased risk of subsequent disease. Persons who, at the start of the study, entered the workforce more recently were likely to have had shorter exposure to asbestos and lower cumulative exposures. This cohort structure may exaggerate the excess relative mortality of the workers as follow-up time increases from the onset of exposure.
2. When Walker (1982) projected the future male population at risk of asbestos-related disease, he assumed that lightly exposed "insulation-workers equivalents" would experience no excess mortality. However, if lightly exposed IWEs experienced any excess mortality, as seems likely,

the surviving cohort of asbestos-exposed workers would be smaller than Walker calculated and the resulting incidence would also be smaller.

3. Walker assumed that 54% of mesothelioma cases have a documentable history of occupational asbestos exposure. Some published articles suggest that as few as 10% of cases may have such histories. (This is the opposite extreme from Reason 5 of Section 5.2.2.)
4. Walker used the power law fitted by Peto et al. (1982) to Selikoff's data on insulation workers. Peto et al. (1982) observed that within each time period since first exposure to asbestos, the mesothelioma death rate exhibits little variation in relation to the period of first employment, for workers first exposed between 1922 and 1946. However, workers who were first exposed before 1922 and after 1946 experienced a lower risk. Walker used exclusively the risk estimates based on the period 1922-1946. For projections of future cases of mesothelioma, risk estimates based on the period after 1946 are increasingly relevant, and risk estimates based on 1922-1946 are progressively less relevant, as the earlier-exposed workers die off. Because the later levels of risk appear to be lower, Walker's projections may be too high.
5. Walker used Peto et al.'s (1982) power law for insulation workers to extrapolate future incidence of mesothelioma far beyond the time of cessation of exposure to asbestos. However, Peto et al. (1982, p. 132) cautioned that the rate of increase of the mesothelioma incidence function is lower for briefer exposures and for times far removed from the cessation of exposure. OSHA (1983) presented revised equations to reflect these effects (see Sections 2.4 and 8.4).

## 5.3 Sensitivity Analysis of Walker's Projections

This section summarizes Cohen et al.'s (1984) sensitivity analysis of the numerical projections of Walker (1982). A sensitivity analysis is an attempt to estimate how sensitive the results of a projection are to changes in the assumptions on which the projection is based.

Projections contain two major elements: numerical values for parameters or variables, and a model that describes how the parameters or variables interact. It is possible to analyze the sensitivity of a projection to the numerical values of the variables, taken one at a time, by varying the value of each variable over an estimated range of uncertainty. If there are $n$ variables and two values of each variable are considered in the sensitivity analysis, then $2n$ sets of calculations are required to examine the effect of variation in each variable, taken one at a time. For example, if $n = 17$, $2n = 34$, a number of calculations that is feasible even for a model as complicated as Walker's.

It is much more difficult to analyze the sensitivity of a projection to the numerical values of all variables considered jointly, unless the model happens to be linear, in which case the sensitivities of the variables in the model are added together. In a complicated projection such as Walker's, numerical values of variables are drawn from a variety of separate sources. Usually, no information is available about the joint behavior of the variables. Lacking information about the joint distribution of variables, one might hope to examine all possible combinations of high and low parameter values. However, with as few as $n = 17$ parameters, one would have to generate $2^{17}$, or 131,072, complete model calculations. Even if so many computations were performed, digesting and interpreting them would be difficult.

Still more difficult is a sensitivity analysis of the model itself. There are infinitely many alternatives to any model. Infinitely many of these alternatives cannot be distinguished because they differ only with respect to events that have not happened yet. It is therefore impossible in principle to analyze completely the sensitivity of a projection with respect to alternative models. Yet the assumptions embedded in a model may well reflect greater and more important ignorance than uncertainty about the numerical values or joint distribution of the variables used in the model. The very different models and projections produced by some investigators other than Walker indicate the range of possible models.

The point of emphasizing these difficulties is to make it clear that the sensitivity analysis that follows indicates a minimum uncertainty concerning the projection of asbestos-related diseases. The range of projections that are produced should not be interpreted to mean that whatever happens will certainly or very probably fall within that range. Rather, conditional on the models on which the projections are based, the range of projections shows the uncertainty implied by our recognized ignorance of the true numerical values of the parameters used in the models.

The Johns-Manville court filings included a magnetic tape that contained Walker's computer program for generating his projections. By reading Walker's program carefully, it was possible to write a new computer program to reproduce Walker's calculations. This program was carefully calibrated by establishing that it could reproduce the published numbers of Walker (1982).

Within this program, there were 17 parameters for which it was possible to estimate reasonable ranges of uncertainty. These parameters are listed, grouped according to the submodel (mesothelioma, lung cancer, or asbestosis) in which they first appeared, and alphabetically within each submodel. Mathematically, the parameters were classified as scalar quantities, vectors, and matrices. The scalar quantities are single-element sets. The vectors and matrices are one- and two-dimensional arrays that are manipulated as a set. The dimensionality of the array is indicated in parentheses following the parameter's name in the program. The parameter names that are followed by one number in parentheses are lists of parameter values [e.g., C1(5) is a list of five numbers]. The parameter names that are followed by two numbers in parentheses are tables of parameter values (e.g., M(15, 10) is a table with 15 rows and 10 columns). The parameter descriptions are followed by reference to Walker's (1982) tasks, described in Section 4.4. The total number of numerical values required to specify these 17 parameters for each projection is 204.

Parameters in the mesothelioma projection are as follows:

| | |
|---|---|
| C1(5) | Workforce discount (clean-up) coefficients for 1955-1959 to 1975-1979 (Task 1f) |
| D(12) | U.S. male mesothelioma incidence by age group in 1975-1979 (Task 1a) |
| E2 | Exponent in the Peto et al. (1982) curve for mesothelioma incidence ($k$ in Task 1d) |
| E3 | Constant factor in the Peto et al. (1982) curve for mesothelioma incidence ($b$ in Task 1d) |
| J(10) | Elrick and Lavidge's distribution of year of first exposure (Task 1f) |
| K2 | Proportion of male mesothelioma cases with documentable occupational asbestos exposure (0.54 in Task 1b) |
| K7 | Correction for overestimation of mesothelioma in SEER data (1/1.12 in Task 1a) |
| K8 | Mortality exponent for cohort exposed to asbestos (1.14 in Task 1e) |
| M(15,10) | Mesothelioma cases in 1975-1979 by age at diagnosis and year of first exposure (Task 1c) |
| X3 | Fractional underdiagnosis of mesothelioma in National Cancer Institute's SEER data (0.10 in Task 1a) |

Parameters in the lung cancer projection are as follows:

| | |
|---|---|
| D1 | Multiplicative factor by which the risk of lung cancer declines (0.9 per quinquennium in Task 3a) |
| K9 | Number of real workers per insulation worker equivalent (2.0 in Task 3a) |
| R2(15) | Selikoff asbestos multiplier [$K_{a_0 y_0}(t)$ in Task 3a] |

Parameters in the asbestosis projections are as follows:

| | |
|---|---|
| D0 | Fraction of deaths in asbestotics due to asbestosis (0.213 in Task 4b) |
| D2 | Man-years of asbestosis per mesothelioma case ($1/I$ in Task 4a) |
| K3 | Proportion of mesothelioma cases with asbestosis ($P$ in Task 4a) |
| K4 | Mortality multiplier for asbestotics (2.8 in Task 4b) |

To analyze the sensitivity of projections with respect to variation in a single parameter, it is necessary to specify baseline values for the remaining parameters. The baseline values are identical to those Walker (1982) used in all but one case. For the estimated age-specific U.S. male mesothelioma incidence in 1975-1979 [variable D(12)], more recent data provided by the National Cancer Institute were used in place of the partial data plus extrapolation used by Walker. For all of the remaining parameters, the baseline numerical

**Table 5.1: Projected Totals, Males 1980-2009**

| Type of Estimate | Walker's Calibration | Revised Baseline |
|---|---|---|
| Mesothelioma incidence | 18,700 | 17,600 |
| Lung cancer incidence | 55,120 | 53,600 |
| Asbestosis cumulative prevalance (Task 4a) | 135,900 | 134,700 |
| Asbestosis cumulative prevalance (Task 4b) | 177,100 | 169,500 |

Source: Cohen et al. (1984).

values were the sole or the middle values considered by Walker. The aggregate effect of using the baseline values instead of Walker's calibration values is to project slightly smaller totals, as shown in Table 5.1.

The sensitivity analysis of the projections was carried out in five steps:

1. All of the above-listed parameters were set to their baseline values and a projection was calculated, printed, and summarized by four numbers:
   i. The total predicted incidence of mesothelioma among U.S. men with documentable asbestos exposure, cumulated from 1980 to 2009.
   ii. The total predicted incidence of lung cancer among U.S. men with documentable asbestos exposure, cumulated from 1980 to 2009.
   iii. The cumulated prevalence of asbestosis from 1980 to 2009 among U.S. men with documentable asbestos exposure, calculated from the incidence of mesothelioma in asbestotics [as in Walker (1982) Task 4a] (see Section 4.4.4a).
   iv. The cumulated prevalence of asbestosis from 1980 to 2009 among U.S. men with documentable asbestos exposure, calculated from an equivalence between asbestosis and mesothelioma mortality [as in Walker (1982) Task 4b] (see Section 4.4.4b).
2. The first parameter from the list was set to one extreme of the confidence or plausibility interval proposed for that parameter and a projection was calculated, summarized, and printed.
3. The same parameter was set to the other extreme of the confidence or plausibility interval proposed for that parameter and a projection was calculated, summarized, and printed.
4. All parameters were reset to their baseline values, the next parameter from the list was selected, and steps 2 and 3 were repeated.

5. After the last variable from the list was treated according to steps 2 and 3, the process terminated.

### 5.3.1 Results for Single Parameters

In discussing results, we will round the projected numbers to the nearest thousand, because greater precision has little meaning.

#### *5.3.1a Mesothelioma*

The two variables that caused the largest increases in the mesothelioma projections when varied to one extreme of their confidence or plausibility intervals are X3 and K2.

The most influential variable is X3, the fraction by which the SEER data underdiagnose mesothelioma. Compared to the baseline X3 = 0.10, taking X3 = 0.41 increases the aggregate projected number of mesothelioma cases by more than 52%. The extreme value of X3 of 41% underdiagnosis is an estimate based on a review of insulation worker deaths and death certificates by Selikoff et al. (1980), which is cited by Walker (1982, p. 4). Even 41% could be too small. As physicians' attention is called to mesothelioma, underdiagnosis could be drastically reduced. The 52% increase exceeds the 30% variation indicated in the informal sensitivity analysis of Walker (1982, p. 17) (see Section 4.4.2b).

The variable that causes the next largest increase is K2, the proportion of mesothelioma cases with documentable occupational exposure history. As Walker observed, the projected numbers are directly proportional to the value chosen, or assumed, for this variable. K2 could be substantially higher than the upper value 0.77 assumed here. Because no alternative causes of mesothelioma are known, K2 = 1 is possible and K2 = 0.9 would be a reasonable estimate based on biopsies. Under these alternative upper values, K2 would become the most influential single variable, with increases of 85% and 67%, respectively.

The two variables that cause the largest decrease in total projected male mesothelioma cases when varied to one extreme of their confidence or plausibility intervals are K2 and jointly E3 and E2. When the proportion of mesothelioma cases with documentable occupational exposure history is assumed to be 0.35, the total number of projected mesothelioma cases falls by 35%. When E2 = 2.5, the mesothelioma total is reduced by nearly 24%.

#### *5.3.1b Lung cancer*

The two variables that cause the largest increase in the total number of projected male lung cancer cases with documentable asbestos exposure when varied to one extreme of their confidence or plausibility intervals are K9 and X3. K9 is the number of real workers per insulation worker equivalent. When K9 = 6.7, a value suggested by Walker (1982, p. 15) (equivalent to assuming

10 workers per IWE for the less heavily exposed group), the projected lung cancer cases increase to 106,000, an increase over baseline of more than 97%. X3 is the fraction of underdiagnosis of mesothelioma in the SEER data. When X3 = 0.41, the total number of projected male lung cancer cases increases by more than 52% compared to baseline.

The lowest projected values of total lung cancer incidence result from variations in K2, the proportion of male mesothelioma cases with documentable occupational asbestos exposure, in D1, the quinquennial decline in the risk of lung cancer, and in R2, the Selikoff asbestos multipliers. Variation of K2 to its low value lowers the projected lung cancers by 35% compared to baseline. Variations in the remaining parameters cause a decline of 24% compared to the baseline.

### *5.3.1c Asbestosis*

Walker's use of two methods of projecting asbestosis prevalence represents a rough sensitivity analysis with respect to the models. We shall consider the variation in the results of both methods together.

The highest projected totals of asbestosis prevalence result from setting the fraction of underdiagnosis of mesothelioma in SEER data to X3 = 0.41. Compared to baseline, the increase exceeds 52%. For asbestosis projected by the first method, the next largest value, an increase over baseline of more than 33%, results from assuming that the man-years of asbestosis per mesothelioma case are D2 = 267.

Varying the values of single parameters can reduce the projected asbestosis prevalence substantially. For the first method, the lowest value results from assuming D2 = 140 man-years of asbestosis per mesothelioma case, giving a reduction of 30%. For the second method, assuming that K2, the proportion of male mesothelioma cases with documentable occupational asbestos exposure, is 0.35 lowers the projection by 35% compared to the baseline. When D0, the fraction of deaths in asbestotics due to asbestosis, is 0.290, the projection for the second method is reduced by more than 26% from the baseline.

## 5.3.2 Results for All Variables Jointly

Cohen et al. (1984) considered how to estimate the range of uncertainty about the future quantities of asbestos-related diseases as a function of all 17 variables listed, considered jointly.

To simplify the problem, Cohen et al. (1984) assumed temporarily (1) that all of the formulas in their reconstructed computer program were correct (although Walker may have erred in his treatment of background cases of lung cancer), (2) that the numerical values of all of the variables not on their list of 17 were known precisely, and (3) that they knew exactly the ranges of uncertainty about the 17 variables on their list.

Caution suggested that, for the many variables for which no simple model of sampling variability (e.g., Poisson, binomial) is plausible, one should avoid attempting to fit some "natural" or "standard" parametric family of distributions like the normal, lognormal, or gamma. Also, in the absence of information about the joint variation of, say, the exponent in the Peto et al. (1982) curve for mesothelioma incidence and the number of real workers per insulation-worker equivalent, one should avoid assuming either that those two variables are independent or that they are linked in some way.

Levi (1980, pp. 441-442) advised:

> ... that we should learn to suspend judgment. We should, in the case under consideration, learn to acknowledge that the data justifies and, indeed, obligates us to suspend judgment concerning the objective chance distribution over failure rates within a given range of values. ... Scientific inquiry has furnished us with much knowledge and, in some contexts, with information which justifies appraising risks and expectations with a considerable degree of precision. But although we should prize precision when we can get it, we should never pretend to precision we lack; and we should be ever mindful of our ignorance even when it hurts.

To know how large a problem may be according to current knowledge or best estimates, one could choose for each variable the value within its range of uncertainty that makes the projected incidence or prevalence as large as possible. To know how small a problem may be according to current knowledge or best estimates, one could choose for each variable the value within its range of uncertainty that makes the projected incidence or prevalence as small as possible. If additional data or better models become available, the calculations can be modified. Approximating the shape of an uncertain future is an open-ended process, ever subject to revision.

Using the approach just outlined, Cohen et al. (1984) performed a sensitivity analysis of the projected incidence of mesothelioma and lung cancer and the projected prevalence of asbestosis. The parameter values for the "high" mesothelioma projection were determined by selecting the choice for each parameter that gave the higher mesothelioma total incidence when that parameter was varied by itself; similar selection procedures were used for the "low" mesothelioma projection, as well as the "high" and "low" lung cancer and asbestosis projections.

### *5.3.2a Mesothelioma*

The total number of mesothelioma cases projected for 1980-2009 varied from 7000 to 82,000. Whereas the low projection showed a peak in 1980-1984 and a decline to under 1000 cases for 2000-2004, the high projection showed a steady increase in numbers of mesothelioma cases to over 15,000 cases for 2000-2004, followed by a decline for 2005-2009.

**Table 5.2: Number of Asbestos Workers Employed in Selected Calendar Years (in Thousands)**

| Year | Number |
|---|---|
| 1940 | 1,880 |
| 1945 | 1,773 |
| 1950 | 2,309 |
| 1955 | 2,558 |
| 1960 | 2,766 |
| 1965 | 2,956 |
| 1970 | 3,223 |
| 1975 | 3,095 |

Source: Cohen et al. (1984), based on Selikoff (1981, Table 2-7).

Both low and high projections implied a peak in insulation-worker equivalents exposed during World War II. The high projection implied even greater exposures from 1955 to 1970. This latter trajectory of exposure is consistent with the pattern of annual U.S. consumption of asbestos, which ranged between 600,000 and 800,000 metric tons during this period – close to double the 400,000 metric tons during World War II (NRC, 1984).

Numerical estimates from various sources of the workers at risk of exposure to asbestos from 1940 to 1979 were summarized by Selikoff (1981). In his Table 2-7 (Selikoff, 1981, p. 115), the employed populations potentially exposed to asbestos in selected occupations and industries were estimated as shown in Table 5.2. The estimate for 1945 omitted 4,334,000 temporary wartime insulators and other workers in the shipbuilding industry.

These observations suggested that the large exposures after World War II implied in the high projection of mesothelioma were not a priori implausible.

### *5.3.2b Lung cancer*

The total number of male lung cancer cases projected for 1980-2009 varied from 9000 in the low projection to 928,000 in the high projection. Whereas the low projection showed a steady decline from 1975-1979 onward to fewer than 200 cases in 2005-2009, the high projection showed a peak of 195,000 cases in 1980-1984 and then a decline to 80,000 cases in 2005-2009. The number of lung cancer cases according to the high projection in 2005-2009 alone was nearly nine times the total number of lung cancer cases from 1980 to 2009 according to the low projection.

**Table 5.3: Joint Sensitivities of Walker's Projection Parameters, Males 1980-2009**

| Cumulated Total 1980-2009 | Plausibility Range | | Walker's Calibration | Revised Baseline |
|---|---|---|---|---|
| | Low | High | | |
| Mesothelioma incidence | 7,000 | 82,000 | 19,000 | 18,000 |
| Lung cancer incidence | 9,000 | 928,000 | 55,000 | 54,000 |
| Asbestosis prevalence | 52,000 | 679,000 | 141,000–184,000 | 135,000–169,000 |

Source: Cohen et al. (1984).

As in the mesothelioma projections, both high and low lung cancer projections implied a peak of exposure during World War II. The high projection also implied substantial postwar exposure, with another peak in 1955-1959.

#### *5.3.2c Asbestosis*

The cumulated prevalence of male asbestosis for 1980-2009 ranged from 52,000 in the low projection to 679,000 in the high projection. Both extremes were based on the equivalence of asbestosis and mesothelioma mortality rates. Both the high and low projections, by both methods [Task 4a and Task 4b of Walker (1982)], predicted steady declines in prevalence from 1980-1984 onward. These declines followed from Walker's assumption that there would be no new asbestosis cases after 1984.

### 5.3.3 Summary of Uncertainty Results

This sensitivity analysis indicated a minimum uncertainty concerning the projection of asbestos-related diseases. The range of projections produced does not mean that whatever happens will certainly or even very probably fall within that range. The analysis showed that variation in certain parameters, within the range of uncertainty indicated as plausible by Walker (1982), changed some projections by substantially more than the 30% he expected. For example, increasing the number of real workers per insulation-worker equivalent to the maximum value suggested in Walker (1982) increased the total projected lung cancer cases by 97%.

The uncertainty about the future quantities of asbestos-related diseases as a function of all 17 variables jointly was summarized by the total figures projected for males for the period 1980-2009 (Table 5.3). These numbers

**Table 5.4: Joint Sensitivities of Walker's Projection Parameters, Males 1980-1984 and 1985-1989**

| | Plausibility Range | | | |
|---|---|---|---|---|
| Type of Estimate | Low | High | Low | High |
| | 1980-1984 | | 1985-1989 | |
| Mesothelioma incidence | 1,700 | 8,500 | 1,600 | 11,100 |
| Lung cancer incidence | 3,800 | 194,000 | 2,400 | 194,600 |
| Asbestosis prevalence | 20,100 | 218,100 | 13,700 | 167,100 |

Source: Cohen et al. (1984).

combined near-term and long-term uncertainty into a single aggregated figure for each asbestos-related disease. Experience with population projections suggests that if these projections were reliable at all, it would be in the near-term future. The comparison of the plausibility ranges for males for the first and second quinquennia is illuminating (Table 5.4). For 1980-1984, the ratios of the high projections to the low projections were, in round numbers, 5 for mesothelioma incidence, 51 for lung cancer incidence, and nearly 11 for asbestosis prevalence. For 1985-1989, the ratios of high to low projections for mesothelioma incidence, lung cancer incidence, and asbestosis prevalence were respectively 7, 81, and 12.

For lung cancer and asbestosis, even in forecasts only 5 years ahead of the base period (1975-1979), the variability of the projections was so enormous that there is little reason to have confidence in the particular projections determined by Walker's choice of parameters.

For mesothelioma, the range of uncertainty opened up less explosively, although still very substantially, 5 years ahead of the base period. For 1980-1984, Cohen et al.'s (1984) calculated range of 1700 to 8500 new male mesothelioma cases with documentable occupational asbestos exposure left ample room for Walker's projected figure of 3200 to err.

## 5.4 Further Sensitivity Analysis of Walker's Mesothelioma Projections

This section evaluates plausible alternative scenarios in comparison with Walker's (1982) mesothelioma model. This provides additional insight into the behavior of the model and forms a groundwork for respecifications of the model in Chapters 6-9.

Manton (1983) noted two practical difficulties in conducting a sensitivity analysis of Walker's forecasts, based on his public report alone. One is that much of Walker's data were not publicly available (e.g., data on 278 Johns-Manville litigants; Task 1c). The second is that the detail in Walker's discussion of his projections was not adequate to assure an absolutely faithful reproduction of his projections. Cohen et al. (1984) overcame both of these difficulties through legal access to the computer tapes of Walker's (1982) work.

The portion of Walker's projection methodology which was perhaps best described was that used to project mesothelioma incidence. We focus on this methodology because the projections of mesothelioma cases can be compared directly with Selikoff's results and because uncertainty in the mesothelioma incidence model affects all of Walker's projections.

Walker's matrix of age- and date-specific initial exposures to asbestos from 278 Johns-Manville litigants were crucial only in the attempt to reproduce Walker's forecasts exactly; for that reason, they were obtained for the sensitivity analysis in Section 5.3. However, the Johns-Manville litigant file may not have been representative of national exposure patterns. Any major differences between projections using alternate reasonable data and Walker's data on Johns-Manville litigants would indicate the sensitivity of the projections to questionably reliable data.

The data used in lieu of Walker's data were derived from Table 2 in Peto et al. (1981) – the primary source of the projection methodology utilized by Walker (see Section 4.2). These data have advantages:

1. Peto et al. (1981) suggested that they were nationally representative.
2. They included World War II shipyard exposures.
3. They were not based on a selected sample of litigants through 1982.

Rather than use unpublished estimates of later exposure based on a limited survey (see Section 4.4.1f), we will employ the assumption from Peto et al. (1981, p. 59) that exposure patterns changed little after 1945. It should be stressed that, although these data are different from the data in Walker, their use does not affect our ability to examine the sensitivity of results to other model assumptions because these data are constant in our calculations.

These data also allow us to assess the change in results due to the application of a reasonable alternative to Walker's data and provide an alternative set of projections that may be more representative of the national experience than Walker's estimate.

This sensitivity analysis has three parts. First, we present the projection strategy in sufficient detail to explain how and why certain assumptions were altered. Second, the logic of the alternative projection scenarios is described. Third, we interpret the results of the alternative projections.

### 5.4.1 Projection Methodology

The projection methodology was divided into three logically separate phases: (1) data preparation, (2) model estimation, and (3) model projection. Phases 1 and 2 correspond to Walker's Task 1; Phase 3 corresponds to Walker's Task 2a.

The first phase involved the preparation of basic data. The five steps in this phase were as follows:

1. We obtained age-specific total mortality hazard rates for 5-year intervals 1918-1922, 1923-1927, ..., 1972-1977, for ages 16-20, 21-25, ..., 81-85, from U.S. life tables 1915-1950 and EPA mortality files 1950-1977 for U.S. white males provided by the Health Effects Research Laboratory, Environmental Protection Agency. These data were used to represent the effects of nonmesothelioma mortality. We extended the upper age of our projection to age 85, compared to age 79 for Walker's projection.
2. We obtained mesothelioma incidence data from Walker's Table 1. We converted age groups from 15-19, ..., 75-79 age-coding to 16-20, ..., 76-80 years of age using 0.80-0.20 proportional allocation. We denoted the 1977 counts for age group $a$ to $a+4$ as $M_{1977}(a, a+4)$.
3. We obtained the distribution of initial exposure of mesothelioma cases by age and calendar year from Peto et al.'s (1981) Table 2 (with age and time intervals matching the data in steps 1 and 2). We denoted the counts for 5-year age groups $(a, a+4)$ by 5-year calendar periods $(y, y+4)$ of initial exposure as $E_{a,a+4;y,y+4}(t)$, where $a_1 = a+t$ and $y_1 = y+t = 1975$ are the lower bounds of the age and date-of-diagnosis intervals. In Peto et al.'s Table 2, the diagnosis interval was 1974-1978, but we assumed the same exposure matrix would apply to 1975-1979.
4. We normalized the data in step 3 by cohort so that the cohort sum represented unit probability mass. In doing so, we let $F_{a,a+4;y,y+4}(t)$ denote the individual terms in that sum; that is,

$$F_{a,a+4;y,y+4}(t) = \frac{E_{a,a+4;y,y+4}(t)}{\sum\limits_{s=0,5,10,\ldots} E_{a_1-s,a_1+4-s;y_1-s,y_1+4-s}(s)}.$$

5. We obtained estimates from Walker (1982, p. 6) of the proportion of low, medium, and high levels of exposure among cases of mesothelioma incidence. These fractions are $f_l = 0.46$, 0.34, and 0.20, respectively, where the medium and high levels of exposure are the 54% of workers with documentable occupational exposure to asbestos (Walker's Task 1b).

Each step in the second phase was performed independently for the low-, medium-, and high-exposure subgroups of the mesothelioma incidence counts obtained in step 2. For the high-exposure subgroup, the age-specific hazards for nonmesothelioma mortality were multiplied by 1.37 to conform with the correction to the Walker methodology proposed in Section 4.4.1e.

6. Using the proportions in step 5, we calculated the 1975-1979 incidence level in that exposure group using
$$M^l_{1975-79}(a, a+4) = M_{1977}(a, a+4) \cdot f_l \cdot 5 \cdot (974/853.9),$$
where 974 is Walker's estimate of 1977 mesothelioma incidence and 853.9 is the sum of the counts of mesothelioma incidence for all age groups from step 2.
7. We calculated the exposed population necessary to yield an incidence level of $M^l_{1975-79}(a, a+4)$ if $M^l_{1975-79}(a, a+4)$ was distributed according to the ages and calendar years of initial exposure specified in step 4. For the population $t$ years in the past, we used
$$\begin{aligned} N^l_{a,a+4;y,y+4}(t) &= M^l_{1975-79}(a+t, a+4+t) \times F_{a,a+4;y,y+4}(t) \\ &\div \left[ \exp\left[ -\int_0^t \mu^l(a+2.5-t+x, 1977.5-t+x)dx \right] \right. \\ &\times \left. \left\{ \exp\left[ \frac{-b(t-2.5)^{k+1}}{k+1} \right] - \exp\left[ \frac{-b(t+2.5)^{k+1}}{k+1} \right] \right\} \right]. \end{aligned}$$
The integral involving $\mu^l(a, x)$ was approximated using the nonmesothelioma mortality hazards from step 1. Parameters $b$ and $k$ were $4.37 \times 10^{-8}$ and 3.2, respectively, as given by Walker (1982, p. 10). This was the Weibull distribution function form of adjustment from Walker's Tasks 1d and 1e. The expression in curly brackets {} on the third line of the equation was the difference in the Weibull cumulative distribution function at the two ends of the 5-year interval, of which $t$ is the midpoint.
8. We replaced the exposed population estimates from step 7 for 1955-1959, 1960-1964, 1965-1969, 1970-1974, and 1975-1979 with the age-specific average of exposed population estimates for 1945-1949 and 1950-1954. This is consistent with the comment of Peto et al. (1981, p. 59) that their data "show no evidence of any change in the extent of exposure since 1945."
9. We deflated the later estimates from step 8 using the "workforce discounts" 1960-1964, 10%; 1965-1969, 50%; 1970-1974, 75%; and 1975-1979, 100%. See Walker (1982, p. 15).
10. We renormalized the exposed population estimates from steps 7-9 so that the mesothelioma incidence in 1975-1979 obtained in step 6 was predicted exactly.

The third phase was the projection of future incidence of mesothelioma among persons exposed to asbestos. This phase had just one step.

11. For each age-of-exposure and calendar interval of start of exposure, we projected the number of mesothelioma cases that would occur in later years and at later ages. We accumulated these estimates by 5-year age groups and by quinquennia for display. For the incidence $t$ years in the

future for a population group initially exposed in the calendar period $(y, y+4)$, we reversed the relation in step 7:

$$E^{l}_{a,a+4;y,y+4}(t) = N^{l}_{a,a+4;y,y+4}(t) \times \exp\left[-\int_0^t \mu^l\,(a+2.5+x, y+2.5+x)\,dx\right] \times \left\{\exp\left[-\frac{b(t-2.5)^{k+1}}{k+1}\right] - \exp\left[-\frac{b(t+2.5)^{k+1}}{k+1}\right]\right\}.$$

The total number of projected mesothelioma cases in a given age group and calendar period was obtained by summing over time and level of exposure:

$$M_{y,y+4}(a, a+4) = \sum_{l} \sum_{t=0,5,10,\ldots}^{a} E^{l}_{a-t,a+4-t;y-t,y+4-t}(t).$$

### 5.4.2 Alternative Scenarios

To examine the variation in results due to changes in Walker's assumptions, we conducted sensitivity analyses of the projections in the five areas listed in Table 5.5. This evaluation corresponds to Walker's Task 2b. The changes were as follows:

A1. We changed the "workforce discounts" in step 9 to 1960-1964, 0%; 1965-1969, 10%; 1970-1974, 50%; 1975-1979, 75%.

A2. We changed the distribution of the initial exposure of mesothelioma cases so that there were equal numbers of cases for each age within quinquennia to determine how sensitive the projections are to small changes in the exposure distribution.

A3. We changed $f_l$ in step 5 to 0.25, 0.47, and 0.28, respectively, to reflect an occupational exposure rate of 75% (the sum of medium and high levels) of mesothelioma cases (from McDonald and McDonald, 1980).

A4. We changed the value 974 in step 6 successively to each of five levels: 854 ("observed" numbers of male cases); 974 (Walker's estimate), 1150 [Hogan and Hoel's (1981) estimates of total mesothelioma]; 1425 [Selikoff's (1981) estimate of total cases]; and 1650 [two standard deviation upper bound (Hogan and Hoel, 1981)].

A5. We changed the Weibull parameter estimates ($b$ and $k$) in steps 7 and 11 successively to: $4.37 \times 10^{-8}$ and 3.2; $3.50 \times 10^{-8}$ and 3.2; $3.56 \times 10^{-7}$ and 2.5; $2.85 \times 10^{-7}$ and 2.5; $3.98 \times 10^{-9}$ and 4.0; $3.18 \times 10^{-9}$ and 4.0. The alternate parameter sets reflected Walker's (1982, p. 13) correction factor of 0.8. The high (4.0) and low (2.5) $k$ values deviated by two standard deviations from the Peto et al. (1982) estimate of $k$ (i.e., 3.2). The $b$-values associated with these extreme $k$ values were chosen to make the incidence curve $I_t = bt^k$ coincide at $t = 20$ years.

**Table 5.5: Alternative Projections of Mesothelioma Cases, Males 1980-2009**

| Projection Series | | | | Persons with Low Exposure | Occupationally Exposed i.e., Medium and High Exposures | Total |
|---|---|---|---|---|---|---|
| | | Walker's "Latency" Projections | | 18,315 | 21,500 | 39,815 |
| | | Walker's "No-Latency" Projections | | 15,930 | 18,700 | 34,630 |
| Alternative Projection Scenarios | | | | | | |
| Analysis/Model/Description | | | Comments | | | |
| A0 | 1 | Projections based on Peto et al. (1981) exposure distribution | Baseline model | 32,721 | 37,105 | 69,826 |
| A1 | 1 | Different rate of reduction in occupational exposure levels | | 34,822 | 39,544 | 74,366 |
| A2 | 1 | Projections based on smoothed Peto et al. (1981) exposure distribution | | 31,836 | 36,104 | 67,940 |
| | 2 | Different rate of reduction in occupational exposure levels* | Combines A1 & A2 Model 1 | 33,795 | 38,378 | 72,173 |
| A3 | 1 | "Consensus" estimates of plausibly exposed workers | | 17,783 | 51,520 | 69,303 |
| A4 | | Different estimates of mesothelioma deaths | Estimate = | | | |
| | 1 | "Observed" | 854 | 28,689 | 32,533 | 61,222 |
| | 2 | Walker (1982) | 974 | 32,721 | 37,105 | 69,826 |
| | 3 | Hogan & Hoel (1981) | 1,150 | 38,633 | 43,811 | 82,444 |
| | 4 | Selikoff (1981) | 1,425 | 47,872 | 54,287 | 102,159 |
| | 5 | Upper bound, Hogan & Hoel (1981) | 1,650 | 55,431 | 62,857 | 118,288 |
| A5 | | Different Weibull Exponents | Exponent = | | | |
| | 1 | Peto et al. (1981) estimate | 3.2 | 32,721 | 37,105 | 69,826 |
| | 2 | Peto et al. (1981) estimate | 3.2 ** | 33,162 | 37,590 | 70,752 |
| | 3 | Peto et al. (1981) lower bound | 2.5 | 20,384 | 23,010 | 43,394 |
| | 4 | Peto et al. (1981) lower bound | 2.5 ** | 20,561 | 23,206 | 43,767 |
| | 5 | Peto et al. (1981) upper bound | 4.0 | 63,723 | 72,810 | 136,533 |
| | 6 | Peto et al. (1981) upper bound | 4.0 ** | 64,980 | 74,195 | 139,175 |

*Smoothed matrix of age and time of first exposure.
**Reflects Walker's 0.8 adjustment.
Source: Manton (1983, Table 9).

**Table 5.6: Comparison of Walker's Projection with Our Baseline Projection**

| Period | Walker No-Latency | Baseline Projection of Occupationally Exposed Workers | | | |
|---|---|---|---|---|---|
| | | | Calibration Ages | | |
| | | 21-80 | 21-25 | 26-30 | 31-80 |
| 1975-1979 | 2,630 | 2,630 | 9 | 24 | 2,597 |
| 1980-1984 | 3,200 | 3,595 | 66 | 91 | 3,438 |
| 1985-1989 | 3,500 | 4,516 | 228 | 235 | 4,053 |
| 1990-1994 | 3,600 | 5,492 | 559 | 485 | 4,448 |
| 1995-1999 | 3,400 | 6,617 | 1,113 | 863 | 4,641 |
| 2000-2004 | 2,900 | 7,801 | 1,926 | 1,367 | 4,508 |
| 2005-2009 | 2,100 | 9,083 | 2,986 | 1,959 | 4,138 |
| Total 1980-2009 | 18,700 | 37,105 | 6,878 | 5,000 | 25,226 |

Source: Authors' calculations.

The revised baseline projections can be compared with Walker's estimates to determine the sensitivity of the results to the matrix of occupational exposures. For other comparisons, the revised baseline is the appropriate reference projection.

### 5.4.3 Results

There are three basic outputs of the model. The first is the total number of mesothelioma cases (Table 5.5). Both lung cancer and asbestosis estimates are directly tied to this figure. The second is the estimated number of occupationally exposed workers. The third is the age distribution of projected cases.

#### *5.4.3.a Mesothelioma cases*

For comparison with Walker's two projections of 18,700 and 21,500 mesothelioma cases among occupationally exposed workers, the comparable figures in our baseline projection are the sums of the mesothelioma cases for the medium- and high-level exposure groups. For 1980-2009, we projected 37,105 cases in these groups – about double that of Walker's most comparable value (18,700, based on the "no-latency" model). Table 5.6 compares the projections by quinquiennia, for all cohorts combined, and separately for the cohorts aged 21-25 and 26-30 in 1975-1979.

In describing Task 1f, Walker stated that the projections were "iteratively adjusted until ... the 1975-79 mesothelioma predictions derived from the resulting distribution of exposed workers equalled the numbers actually thought to have occurred" (1982, p. 15). This adjustment is reflected in step 10 of our projection. Walker's 1977 estimates for males aged 20-24 and 25-29 were 2.0 and 6.1 cases, respectively. After adjustments in steps 2 and 6, these were converted to estimates of 9 and 24 cases for ages 21-25 and 26-30, respectively. Despite the extremely small sizes of these estimates, they account for 11,878 cases (32%) of our baseline projection in Table 5.6.

If the two youngest cohorts are removed from our baseline projection, then the projection drops to 25,226 cases for 1980-2009. If all cases occurring above Walker's cutoff at age 75 are also removed, then the projection drops to about 21,840 cases, or about 17% higher than Walker's 18,700 cases and less than 1.6% higher than Walker's 21,500 cases in the "latency" projection.

These comparisons satisfied us that we could explain the differences between Walker's two projections and our baseline. The finding of extreme sensitivity to the mesothelioma counts below age 30 in 1977 alerted us to carefully evaluate this factor in all later projections. Nonetheless, because the younger cohorts were part of the calibration data, we retained their projected values in the baseline and sensitivity analyses.

The "total mesothelioma" projection in Table 5.5 includes persons with "low levels" of exposure. The baseline estimate of the total number of mesothelioma cases (69,826) is, in contrast to the comparison with Walker's results, reasonably close to Selikoff's (1981) projection of 79,745 cases. This latter value is approximately twice the total number of mesothelioma cases implied by Walker's two projections (i.e., 34,630 and 39,815). Our baseline projection is plausible if one considers that the range of Walker's and Selikoff's projections is plausible.

Under scenario A1, extending the period over which occupational exposure to asbestos was reduced slightly increases the projected number of mesothelioma cases (i.e., from 69,826 to 74,366, about 6.5%). The projections under the extension are even closer to Selikoff's (1981) estimates than the baseline projections (74,366 vs. 79,745, 93% of Selikoff's in Table 4.6 of Chapter 4).

Scenario A2 shows how the projections changed if the age and time distribution of exposure was uniform (flat). Changing the exposure distribution reduced the projected values very little from the baseline (Table 5.5). A similar smoothing operation, applied to the exposure distribution matrix, after extending the period of occupational exposure (in effect, combining scenarios A1 and A2) generated projections that were closer to the baseline than the results for A1 alone.

Scenario A3 alters the proportions of occupationally exposed workers (i.e., those with "medium" and "high" levels of exposure). Following Walker's rationale, this would increase the number of potential litigants from about 37,000 to over 51,000.

**Table 5.7: Alternative Estimates of Exposed Population**

| Exposure Level | Baseline | Analysis A2 – Model 1 |
|---|---|---|
| | From Step 7 | |
| Low | 11,965,998 | 6,315,908 |
| Medium | 4,422,217 | 2,334,140 |
| High | 534,349 | 291,185 |
| Total | 16,922,564 | 8,941,233 |
| | From Step 10 | |
| Low | 5,456,271 | 5,396,376 |
| Medium | 2,016,448 | 1,994,313 |
| High | 253,097 | 251,626 |
| Total | 7,725,816 | 7,642,315 |

Source: Manton (1983, Table 10).

Scenario A4 shows that altering the number of mesothelioma cases estimated for 1977 has an approximately linear effect on the projected total estimates. If one accepts the 1425 figure implied by Selikoff (1981), then the projection implies a total of 102,159 mesothelioma cases for 1980-2009.

Scenario A5 changes the exponent of the Weibull hazard function. If one uses an exponent of 2.5, then the sum of medium- and high-exposure cases is very close to Walker's upper estimate (i.e., 23,000 vs. 21,500). Use of the exponent value of 4.0 produces much higher estimates (i.e., about 74,000). Variation of the exponent parameter over its plausible range produces high estimates that are over three times greater than the low estimates.

### *5.4.3b Exposed workers*

The second comparison involves the estimates of the number of exposed workers. As Peto et al. (1981) noted, it is very difficult to estimate the number of exposed workers in this type of model because it is unclear how to translate insulation-worker equivalents obtained in steps 7 and 10 into counts of workers. Nonetheless, the computations may be illustrated using conversion values of 10 to 1 for low exposures and 5 to 1 for moderate exposure, as suggested by Walker (1982, p. 15, Task 1f). These crude calculations are presented in Table 5.7 for the baseline and smoothed (Analysis A2, Model 1) scenarios.

Table 5.7 shows that smoothing, although having a small effect on the number of mesothelioma cases projected, can have a tremendous effect on the estimated number of insulation-worker equivalents. The estimate in step 7 was nearly halved from 16.9 million to 8.9 million because the projected number of exposed workers was greatly decreased in the interval 1965 to 1974. This effect was then almost completely removed by the renormalization in step 10.

#### 5.4.3c Projected age distribution

Walker did not discuss the age distribution of projected cases. As Table 5.6 illustrates, the quinquennial age distributions may reveal cohorts with unstable projections due to model calibration based on small numbers of cases. It is also useful to stratify the projections by age because the propensity to sue varies with age, being highest in middle age (i.e., 24% for persons 45-59) and declining at later ages (i.e., 9% for 65-69 and 6% for 70+; 16% overall) (see Selikoff, 1981, p. 435, Table 5-83). Selikoff's (1981) Table 5-85 showed that the overall tendency to sue increased in later calendar years (i.e., from 3% when the worker died in 1967-1968 to 32% when the worker died in 1975-1976). In addition, the age at death could influence the amount of a settlement if the award were based on a concept such as years-of-life lost. The age distribution of projected mesothelioma cases matters.

Table 5.8 shows that, under the baseline model, 43% of the cases projected to arise from workers with low- and medium-exposure levels and 39% of the projected cases for workers with high levels of exposure occur after age 65. However, 11,878 cases (55%; see Table 5.6) of the total 21,572 cases projected for medium and high levels of exposure for ages 65 or below (Table 5.8) derive from the experience of the two youngest cohorts based on about 8 cases of mesothelioma in 1977. Different parts of the projection have very different levels of uncertainty.

## 5.5 Conclusions

Walker's (1982) projections were based largely on epidemiological studies of asbestos-exposed workers. These studies, although the best then available, suffered limitations of design, measurement, and analysis:

- Design: Periods of exposure to asbestos overlapped periods of follow-up. A survey of asbestos-exposed workers excluded those under age 40.
- Measurement: Concentrations of asbestos fiber in the workplace were not measured. Durations of exposure to asbestos were not measured.
- Analysis: The age structure of exposed populations was not considered. Distributions of year of first asbestos exposure were inferred for large numbers of workers from samples of cases that were not randomly selected or were extremely small in number.

**Table 5.8: Baseline Projections of Mesothelioma Cases by Age and Year, 1980-2009**

| Exposure Level | Age | Quinquennium 1980-1984 | 1985-1989 | 1990-1994 | 1995-1999 | 2000-2004 | 2005-2009 | Total |
|---|---|---|---|---|---|---|---|---|
| | Total | 3,117 | 3,961 | 4,844 | 5,853 | 6,901 | 8,045 | 32,721 |
| Low | ≤65 Yrs. | 1,570 | 1,972 | 2,457 | 2,089 | 4,104 | 5,697 | 18,589 |
| | >65 Yrs. | 1,547 | 1,989 | 2,387 | 2,764 | 2,797 | 2,648 | 14,132 |
| | Total | 2,304 | 2,928 | 3,581 | 4,326 | 5,101 | 5,946 | 24,185 |
| Medium | ≤65 Yrs. | 1,160 | 1,458 | 1,816 | 2,283 | 3,033 | 3,989 | 13,740 |
| | >65 Yrs. | 1,144 | 1,470 | 1,764 | 2,043 | 2,068 | 1,957 | 10,445 |
| | Total | 1,292 | 1,588 | 1,911 | 2,291 | 2,701 | 3,137 | 12,920 |
| High | ≤65 Yrs. | 673 | 839 | 1,043 | 1,308 | 1,728 | 2,241 | 7,832 |
| | >65 Yrs. | 619 | 749 | 869 | 983 | 973 | 896 | 5,088 |
| Medium | Total | 3,596 | 4,516 | 4,772 | 6,617 | 7,802 | 9,083 | 37,105 |
| & | ≤65 Yrs. | 1,833 | 2,297 | 2,859 | 3,591 | 4,761 | 6,230 | 21,572 |
| High | >65 Yrs. | 1,763 | 2,219 | 2,633 | 3,026 | 3,041 | 2,853 | 15,533 |

Source: Manton (1983, Table 11).

Walker's (1982) projections were affected by many sources of qualitative uncertainty, some of which biased his results upward, some downward, and some in either direction.

When we subjected Walker's (1982) projections to a formal sensitivity analysis, using a reconstruction of his computer program, the calculated uncertainty in his projections was much larger than he estimated. For the period 1980-1984, only 5 years ahead of the base period 1975-1979, the ratio of high to low projections from Walker's model was 5 for mesothelioma, 51 for lung cancer, and 11 for asbestosis. This forecast uncertainty resulted only from uncertainty about the numerical values of the parameters used in Walker's model, taking as given many other assumptions Walker made (e.g., that the lower post-World War II mesothelioma incidence rates may be disregarded in favor of the higher 1922-1946 rates, that lung cancer mortality rates would not increase in the future, that no new asbestosis cases would arise after 1984). The uncertainty in the lung cancer and asbestosis projections, only 5 years forward, was so large as to cast doubt on the practical utility of the models for lung cancer and asbestosis. The range of forecasted new cases of mesothelioma among males with documentable asbestos exposure for 1980-1984, from 1700 to 8500, was large enough to leave ample room for Walker's projection of 3200 cases to err.

When Walker's projections were compared with projections of asbestos-related mesothelioma and lung cancer made by others, the range of variation was so large that it was a safe conclusion that some of them would be wrong by a substantial margin. Although Peto et al. (1981) shared the central assumption of Walker's mesothelioma projections and arrived at numerically similar forecasts, others shared neither Walker's approach nor his projections. Ultimately, only when the future happens will it be possible to decide which, if any, of these approaches to projection is sound. This will be explored further in Chapters 6-9.

# 6

# Updated Forecasts Based on Indirect Estimates of Exposure

## 6.1 Introduction

Part of the uncertainty in projections from Walker's (1982) model was due to data limitations. Our modeling for the Rule 706 Panel updated the data used in that model and generated new projections of mesothelioma personal injury claims with substantially narrower uncertainty limits. These mesothelioma projections were done at two levels: Nationally, using the SEER data from NCI, and Manville-specific, using the Trust claim data. Claims of mesothelioma in the Manville Trust data appeared to stabilize at about 40% of the reports of mesothelioma in the SEER data for the periods 1985-1989 and 1990-1994. This relation justified our moving from the SEER level to the Trust level in Chapter 8. Projections of claims from nonmesothelioma diseases were developed using the exposed population estimates from the mesothelioma projections.

Data from the Trust indicated that 97% of all qualified claims were filed by male claimants. Because we conducted a detailed analysis only for the male claim experience, we restricted our projections to the number of future claims for males. Estimates for both sexes may be approximated by multiplying each projected male value by 1.03.

## 6.2 Factors Considered

The projections were based on a computer model with initial calibration conducted in two stages. First, the national incidence of mesothelioma for 1975-1989 was estimated from files prepared by the National Cancer Institute (NCI) through their Surveillance, Epidemiology, and End Results (SEER) Program.

The SEER mesothelioma projections reconstructed the rates and counts of male exposures to asbestos for 1915-1974. From this reconstruction, we generated a forward projection of the number of survivors at each age and calendar year, stratified by the length of time since first exposure (TSFE) to asbestos. For each 5-year period following each age and date of initial exposure, we computed the probability of mesothelioma diagnosis. To generate the projected counts of mesothelioma incidence, the probabilities of mesothelioma diagnosis were multiplied by the counts of survivors, for each combination of age, date, and TSFE. The total number of male mesothelioma diagnoses was obtained by summing over the three dimensions of the projection: age, period, and TSFE.

The counts of survivors were initially scaled to be equivalent to the number of insulation workers who would have been exposed in the past if all current cases of mesothelioma were among insulation workers. The assumption of insulation-worker equivalent (IWE) counts is evaluated in this chapter. The IWE assumption implies that the amount of disease associated with each unit exposure to asbestos does not depend on the actual number of persons exposed. This means that 10 insulation workers, each of whom is exposed to 100 g of airborne asbestos (1 kg total), would exhibit the same disease outcomes as, say, 100 cement workers, each of whom is exposed to 10 g of airborne asbestos.

Second, the Trust claim experience for the period 1990-1994 was tabulated from administrative files maintained by the Trust and extracted for our use on December 2, 1992. Estimates were made for nine categories of alleged disease/injury specified on the Proof of Claim (POC) Form submitted to the Trust by each claimant. The male claim experience from January 1, 1990 to November 30, 1992 (referred to as 1990-1992 for brevity) was extended to the 5 years 1990-1994 by assuming that the monthly claim filing counts were constant within disease and age groups. For each of nine diseases, the claim counts were tabulated by age and TSFE and matched to the corresponding IWE survivor counts (available from the SEER-based projection) by age and TSFE. This allowed computation of disease-specific claim filing rates by age and TSFE for the period 1990-1994. The projection model matched exactly the age, TSFE, and disease-specific male claim counts estimated for the period 1990-1994. This guaranteed that the starting levels of the Trust claim projections were anchored in plausible estimates.

For each 5-year period starting with 1990-1994 and ending with 2045-2049, the disease-specific claim filing rates were multiplied by the IWE surviving population counts, for each combination of age and TSFE, to generate the projected number of future claims against the Trust. The total number of male claims was obtained by summing over the four dimensions of the projection: age, period, TSFE, and disease.

If a claimant claimed multiple diseases, the most severe disease was selected to represent his injury. The hierarchy used by the Trust ranked the injuries by relative severity: (1) mesothelioma; (2) lung cancer; (3) colon/rectal cancer; (4) other cancer; (5) asbestosis; (6) disputed asbestosis; (7) pleural plaques and thickening; (8) nonasbestos-related disease; and (9) unknown disease. These nine categories represented a consolidation by the Trust of listings of hundreds of specific injuries. Thus, in describing category 6 as "disputed asbestosis," the Trust believed it unlikely that the stated injury was actually asbestosis. Describing category 8 as "nonasbestos-related disease" meant that the Trust believed the injury was not caused by asbestos exposure. "Unknown" injuries in category 9 represented incomplete or missing information on the detailed POC form.

The Trust settlement process involved an evaluation of the alleged disease/injury and a reclassification of the claim into a set of "validated" disease/injury categories. Under the original Trust Plan, this occurred in two forms. First, under the "prepetition" settlement process, claims filed before the Johns-Manville bankruptcy petition in 1982 were settled by a process in which groups of claims, organized around law firms, were settled as a group with less intensive scrutiny than for later settlements. Second, under the "postpetition" settlement process, claims filed after the bankruptcy petition were individually evaluated. Unknown disease/injuries were resolved as one of the other eight categories. Almost all of these claims were settled before July 1990.

The Trust was ordered to stop payments under these settlement processes in July 1990 (except for a few cases of extreme hardship), and this stay persisted until the case was finally settled in December 1994. Under the 1994 settlement's Trust Distribution Process (TDP), claims were to be evaluated using a modification of the eight-category classification system that was used under the original postpetition settlement process (Weinstein, 1994). Categories 3 and 4 were combined into a new category 4 (other cancer – specifically: colon/rectum, larynx, esophagus, or pharynx). Category 2 was separated into two new categories distinguishing lung cancer among nonsmokers (category 2) from smokers (category 3). Categories 5 and 6 were redesignated as disabling versus nondisabling bilateral interstitial lung disease.

Table 6.1 displays the distribution of validated disease/injury by most severe alleged disease/injury for claims filed after the Johns-Manville bankruptcy petition in 1982 and settled under the original postpetition settlement process. Overall, 93.4% of alleged cancer claims were validated as cancer and 95.7% of alleged noncancer claims were validated as noncancer. Within these two categories, however, there were substantial shifts. Unknown disease/injury was shifted down to one of the first eight categories, with 80.3% shifted to nonasbestos-related disease. Disputed asbestosis had the lowest validation rate (17.8%), with 57.3% shifted to pleural plaques/thickening. Nonasbestos-related disease had the next lowest validation rate (25.3%), with about equal fractional shifts to categories 5, 6, and 7.

Mesothelioma (93.6%) and lung cancer (91.1%) had the highest validation rates as well as the highest settlement amounts – $150,000-$200,000 for mesothelioma and $60,000-$90,000 for lung cancer, depending on which type of settlement amount (estimated vs. actual) is used.

The scheduled settlement amounts for validated disease/injuries were derived from the final Trust Distribution Process (TDP) (Weinstein, 1994). Under the TDP, the validated disease category corresponding to asbestosis (5) was redefined as disabling bilateral interstitial lung disease, and disputed as-

**Table 6.1: Distribution of Validated Disease/Injury and Average Settlement Amount by Most Severe Alleged Disease/Injury, Post–Bankruptcy-Petition Claims Against the Manville Trust Filed Through November 1992**

| Most Severe Alleged Disease/Injury | Validated Disease/Injury – Post-Petition Settlement Process: 1 | 2 | 3 | 4 | 5 | 6 | 7 | 8 | Number of Claims | Average Settlement Amount ($000s): Estimated[1] | Actual | Ratio of Estimated to Actual Amount |
|---|---|---|---|---|---|---|---|---|---|---|---|---|
| | Percent Distribution | | | | | | | | | | | |
| 1. Mesothelioma | 93.6 | 2.9 | 0.0 | 0.1 | 1.2 | 0.6 | 0.7 | 0.9 | 1,404 | 190.3 | 146.6 | 1.30 |
| 2. Lung cancer | 1.3 | 91.1 | 1.3 | 1.1 | 1.9 | 0.9 | 0.9 | 1.3 | 1,717 | 76.0 | 61.7 | 1.23 |
| 3. Colon/rectal cancer | 0.5 | 0.5 | 65.9 | 15.2 | 4.7 | 4.3 | 2.8 | 6.2 | 211 | 37.5 | 33.4 | 1.12 |
| 4. Other cancer | 0.7 | 35.5 | 4.0 | 42.4 | 7.1 | 2.0 | 2.7 | 5.6 | 448 | 52.0 | 46.8 | 1.11 |
| 5. Asbestosis | 0.6 | 1.9 | 0.5 | 0.8 | 52.2 | 29.9 | 13.8 | 0.4 | 6,179 | 38.4 | 33.6 | 1.14 |
| 6. Disputed asbestosis | 0.7 | 1.7 | 0.3 | 1.8 | 19.4 | 17.8 | 57.3 | 1.1 | 1,678 | 24.6 | 28.5 | 0.86 |
| 7. Pleural plaques/thickening | 0.3 | 1.7 | 0.6 | 1.0 | 19.7 | 20.9 | 55.3 | 0.5 | 1,008 | 24.3 | 22.4 | 1.08 |
| 8. Nonasbestos-related disease | 1.7 | 5.8 | 0.9 | 2.7 | 22.5 | 20.1 | 21.0 | 25.3 | 586 | 28.2 | 28.3 | 0.99 |
| 9. Unknown | 2.4 | 2.4 | 0.5 | 0.5 | 5.3 | 4.3 | 4.3 | 80.3 | 208 | 11.3 | 9.7 | 1.17 |
| Total | 10.5 | 14.6 | 1.7 | 2.6 | 29.6 | 18.8 | 18.9 | 3.3 | 13,439 | 55.9 | 47.4 | 1.18 |
| Cancer (1-4) | 35.5 | 46.7 | 4.8 | 6.4 | 2.4 | 1.1 | 1.2 | 1.9 | 3,780 | 113.5 | 89.9 | 1.26 |
| Noncancer (5-9) | 0.7 | 2.1 | 0.5 | 1.1 | 40.3 | 25.7 | 25.9 | 3.8 | 9,659 | 33.3 | 30.7 | 1.08 |
| Asbestos-related noncancer (5-7) | 0.6 | 1.8 | 0.4 | 1.0 | 42.3 | 26.6 | 26.7 | 0.6 | 8,865 | 34.2 | 31.4 | 1.09 |
| Nonasbestos-related & unknown (8-9) | 1.9 | 4.9 | 0.8 | 2.1 | 18.0 | 16.0 | 16.6 | 39.7 | 794 | 23.8 | 23.4 | 1.01 |

| Item | Validated Disease/Injury – Trust Distribution Process: 1 | 2 | 3 | 4 | 5 | 6 | 7 | 8 | Total |
|---|---|---|---|---|---|---|---|---|---|
| Scheduled settlement amount ($000s)[2] | 200.0 | 78.0 | 40.0 | 40.0 | 50.0 | 25.0 | 12.0 | 0.0 | 55.9 |
| Actual settlement amount ($000s) | 150.1 | 64.4 | 36.0 | 37.3 | 38.2 | 26.4 | 23.0 | 0.0 | 47.4 |
| Ratio of scheduled to actual amount | 1.33 | 1.21 | 1.11 | 1.07 | 1.31 | 0.95 | 0.52 | — | 1.18 |
| Number of claims | 1,409 | 1,967 | 225 | 346 | 3,983 | 2,526 | 2,546 | 437 | 13,439 |

Note 1: Based on scheduled settlement amount for validated disease/injury.

Note 2: The $78,000 amount for lung cancer is based on relative frequencies of the two types of lung cancer claims filed during 1995-1999.

Source: Authors' calculations.

bestosis (6) was redefined as nondisabling bilateral interstitial lung disease. Table 6.1 shows that 52.2% of asbestosis was retained in category 5, and only 17.8% of disputed asbestosis was retained in category 6. The validated lung cancer category had two scheduled payment levels – $60,000 and $90,000, with the higher payment for nonsmokers. In Table 6.1, we assumed an average payment of $78,000.

The scheduled amounts exceeded the actual amounts from the validated claims of the postpetition settlement process by an average of 18%. The greatest excess was for mesothelioma (33%) and the greatest reduction was for pleural plaques/thickening (−48%). The 31% increase for asbestosis reflected the restriction to disabled claimants.

The estimated settlement amounts for alleged disease/injuries were weighted averages of the scheduled settlement amounts, using the percent distribution for alleged diseases as the weights. These estimates agreed well with the actual settlement amounts for alleged disease on the Trust extract file ($R^2 = 99.4\%$), with the largest discrepancy for mesothelioma (30%). The asbestosis excess was 14%, making it just about average (15%). Categories 8 and 9 together had an estimated and actual settlement amount of $23,400, which was comparable with the asbestos-related noncancers in categories 6 and 7. Thus, it was important to retain these categories in claim projections based on alleged disease/injuries.

The Trust claim data played a key role in the first-stage calibration based on the SEER mesothelioma data because the first-stage calibration required estimates of the distribution of TSFE for each age and date of mesothelioma diagnosis. Such estimates could not be obtained from SEER but could be obtained from information in the Trust claim data files on 5482 mesothelioma diagnoses. These estimates were substantially more credible than the estimates in Peto et al. (1981, Table 2) based on 69 male mesothelioma cases or in Walker (1982, Task 1c) based on 278 mesothelioma claims against Johns-Manville.

The main issue in using these estimates in the first-stage calibration was whether they introduced substantial bias into the projections. We argued that they do not for two reasons. First, sensitivity analyses in Section 5.4.2 (scenario A2) indicated that major modifications of Peto et al.'s (1981) exposure distribution had only minor effects on the projections. Given the similar structure of the first-stage calibration in this chapter, we expected a similar lack of sensitivity, so it was not critical to obtain precise estimates of the exposure distribution. Second, Hersch (1992) argued that the Manville Trust's claim experience was representative of the entire asbestos industry. We found that the Trust's mesothelioma diagnosis rates converged to about 40% of the national total, and to perhaps 56-78% of all occurrences related to occupational exposures (see Section 6.5.3). The Trust data represented a large fraction of the national incidence of mesothelioma – up to four times larger than the SEER data, which included only about 10% of mesothelioma incidence – suggesting that biases induced by using the Trust data could be smaller than biases induced by using the SEER data.

Our main interest was in forecasting claims of asbestos-related diseases (or "injuries") against the Trust, so we could tolerate some bias in the national projections if that bias was in the direction of the Trust claim experience. A second-stage calibration was designed to reproduce exactly the Trust claim filing rates in the baseline period 1990-1994, and this could be accomplished with a range of alternative outputs from the first-stage model.

Because of limitations of data in the Trust claim files, detailed projections were done only for males. Simple ratio adjustments were used to make estimates for females. Females accounted for only about 3% of all Trust claims and 6% of mesothelioma claims. In the SEER data, females accounted for 22% of mesothelioma cases in 1985-1989.

## 6.3 Assumptions

The projections were based on the discrete-state stochastic process depicted in Figure 6.1. Under this process, each person is followed from his year of birth $y_b$ to death at any time up to age $a_3$ in year $y_3$. For projections, the number of persons with initial exposures at age $a_0$ in year $y_0$ will be a major focus of interest. The ages are "idealized" in the sense that everyone is potentially at risk to every event. For example, if the initial exposure to asbestos never occurs, we assume that $a_0$ is large (e.g., any value over 125 years).

The next event is the diagnosis of an asbestos-related disease at age $a_1$ in year $y_1$. For this event to occur, a person must not die of some nonasbestos-related disease prior to age $a_1$. Actually, there is a high probability that persons with asbestos exposure will die from nonasbestos-related diseases, especially if the amount of exposure is small.

The next event is the filing of a claim against the Manville Trust at age $a_2$ in year $y_2$. If the claim is filed by the estate of a decedent, then $a_2$ is imputed as $a_2 = y_2 - y_b$.

The final event is the death of a living claimant at age $a_3$ in year $y_3$. Once a claim has been filed, the subsequent death of the claimant will affect the settlement process, but it will not affect projected claim counts. Death from nonasbestos-related disease prior to claim filing reflects a competing risk. This was explicitly taken into account in our projection model.

Our projections were based on a claim hazard rate (CHR) model specific to each of nine diseases. This type of model used data on attained age and TSFE to estimate the rate at which claims were filed in the surviving population with these age and TSFE characteristics during 1990-1992. These claim filing rates were assumed to remain constant for the period 1990-2049, thereby providing a basis for forecasting the future claims for each disease as a fraction of the surviving population in each subsequent period.

Each projection model employed some constancy assumption. In all population forecasting, certain aspects (e.g., parameters) of the forecast are assumed constant (Keyfitz, 1984). It is desirable to minimize the number of

**Figure 6.1: States and Transitions Represented in Projections of Asbestos-Related Disease Claims**

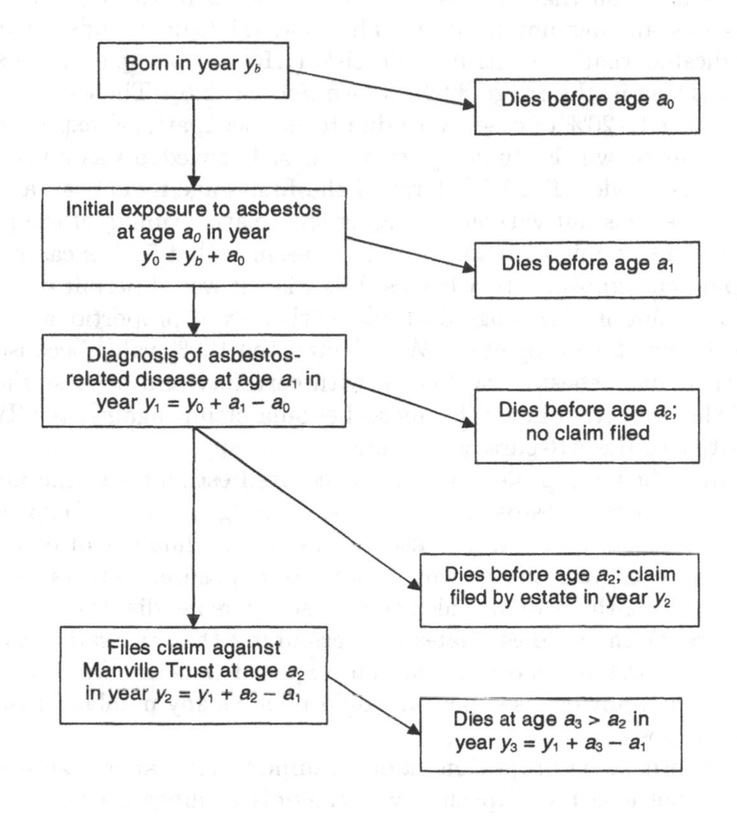


parameters in the projection model while producing a realistic forecast of future events. Comparison of the results of alternative models allows assessment of the effects of various assumptions and ultimately permits one to determine which projections are realistic and which are not.

Our mesothelioma projection model was based on two assumptions. First, it was assumed that all cases of mesothelioma that occur in the United States were due to asbestos. This was the view of the NCI (Reynolds, 1992, p. 561): "Asbestos is the only known risk factor for mesothelioma, a tumor of the membranes lining the chest or abdominal cavities." If this were true, then estimates of the amount of exposure to asbestos in previous years could be made from incident cases of mesothelioma. Second, it was assumed that accurate estimates of the number of incident cases of mesothelioma could be derived from data collected by SEER. If this were true, then both the exposure and incidence counts could be calibrated to reproduce the SEER data in any selected set of baseline years. The incidence predicted for later years would form the projection.

The assumption that all cases of mesothelioma in the U.S. were due to asbestos was an oversimplification. The epidemiological evidence in Chapter 2 indicated that the attributable risk (AR) for occupational exposures to asbestos was in the range 80-90% (see Section 2.6). The extent to which the remaining 10-20% of cases were due to nonoccupational exposures to asbestos was unknown. Fortunately, this lack of knowledge was not critical to our two-stage model. The IWE form of the first-stage model can absorb any multiplicative constant without changing the relative timing of the projected outcomes. This permitted Walker (1982) to assume that 54% of cases were due to occupational exposures to asbestos. Likewise, it would permit us to assume any higher value in the range 54-100% with only a proportional change to the overall first-stage projection. We selected the 100% value because no risk factor other than asbestos was known with certainty and because the second stage of the projection model for mesothelioma claims against the Trust was not sensitive to the AR-parameter value.

Our mesothelioma projection model required estimates of the number of persons exposed to asbestos in prior years, their age at start of exposure, and the date exposure initiated. We also estimated the number of exposed persons alive at each year subsequent to start of exposure. Both tasks required that we estimate the rate at which unexposed persons die, over age and calendar years. With those estimates and assuming that the same rates apply to nonasbestos-related mortality among exposed persons, we estimated the probability that any exposed person would be alive any number of years after his initial exposure.

Our mesothelioma projection model assumed that start of exposure to asbestos, but not length of exposure, was the only significant event in the incidence of mesothelioma. This is only an approximation, however, because there is evidence that very large durations of exposure have incremental effects on the incidence of mesothelioma (OSHA, 1983, 1986; EPA, 1986). Fortunately, the effects are quantifiable: The difference between short-term exposure and long-term exposure at the same constant rate per unit time is manifest in pairs of incidence curves whose ratio is proportional to time since start of asbestos exposure. Furthermore, intermediate durations of exposure yield incidence curves that lie between these two extremes (see Section 8.4). Thus, given that the majority of workforce exposure was for short duration (Selikoff, 1981, Table 2-13), we made the simplifying assumption that time since start of exposure to asbestos was the only (rather than the main) determinant of mesothelioma incidence. This assumption also reflected the limitations of data in the Manville Trust files. In general, the Trust consistently recorded the date of first exposure to Manville asbestos for each claimant. Subsequent exposures and dates of termination of exposure were recorded less consistently.

Throughout this chapter, we use the Weibull function as a parametric model of the claim hazard rate (or claim filing rate), with TSFE as the time variable. This function provided an approximation to the probability that a person who was alive at a given time since first exposure filed a claim within the next time period. Mathematically, this is expressed as

$$\Pr[\text{file claim within one time unit after TSFE, given survival to TSFE}]$$
$$\approx b \times (\text{TSFE})^k \qquad (b, k > 0)\,.$$

The approximation improves as the time units shrink (e.g., from years to months, or weeks). $b$ is a scale parameter that depends on the selected time units (i.e., $b$ is the probability value when TSFE $= 1$). $k$ is a shape parameter that determines the curvature of the function. $k$ may take on fractional values. The $k$ parameter of the Weibull hazard model is generally not restricted to integer values, although it is so restricted in the multihit/multistage model of cancer developed by Armitage and Doll (1954, 1961) (see Section 2.3.1d). However, with variation in duration of exposure, an approximating form of the Weibull model can be developed for the multihit/multistage model that uses noninteger $k$ values (OSHA, 1983, 1986) (see Section 8.4). Thus, we use the best fitting $k$ values without restriction to the set of integers. Five integer $k$ values have important interpretations: (1) $k = 0$ implies that the claim filing rate is constant with respect to TSFE. This is the simplest possible model. (2) $k = 1$ implies that the claim filing rate is proportional to TSFE. This appears to describe the claim filing rates for noncancer claims. (3) $k = 2$ implies that the claim filing rate is a quadratic function of TSFE. This model works well for lung cancer claims. (4) $k = 3$ implies that the claim filing rate is a cubic function of TSFE. This model works well for mesothelioma claims among populations with low-level exposures to asbestos. (5) $k = 4$ implies that the claims filing rate is a quartic function of TSFE. This model works well for asbestosis and mesothelioma mortality among insulation workers.

We used the Weibull function to model incidence, death, and claim filing rates for mesothelioma. Figure 6.1 indicates that a claim must occur after diagnosis, but the model does not restrict the claim to occur only before or only after death. The Trust data for 1958 qualified male claims for mesothelioma injury from the beginning of 1990 to mid-1992 provided useful information on the sequencing of these events. The median lag from diagnosis to death was 0.51 years, and from death to claim filing, it was 0.71 years, for a total lag of 1.22 years from diagnosis to claim filing for the 76.4% of claimants who died before filing. The median lag from diagnosis to claim filing was 0.64 years for the 23.6% of claimants who were alive at filing and 1.06 years for both types of claimants combined. In addition, assuming an exponential distribution of death times following diagnosis, the median survival for the 76.4% implies an overall median survival of 0.73 years – 0.33 years less than the overall median claim lag.

These results were consistent with Selikoff's (1981) observation that most mesotheliomas lead to death within a year of diagnosis. The additional information provided by the Trust was that the claim filings occurred rapidly after diagnosis – with only a moderate delay if death occurred.

Given that (1) the median lags between diagnosis, death, and claim filing were all less than 1 year, (2) the peak death rates for mesothelioma occurred about 45 years after onset of exposure, and (3) there was substantial uncertainty about the $k$ parameter of the Weibull hazard rate, we used a two-parameter Weibull hazard for modeling incidence and claim filing rates for mesothelioma as a function of TSFE, rather than introducing a mathematically more consistent but statistically less stable hazard form for claim filing (or death) based on the convolution of the distributions of the incidence $(a_1 - a_0)$ and continuance $(a_2 - a_1)$ transition times in Figure 6.1. This was consistent with the standard practice of using the Weibull hazard to model both incidence and death rates for cancers that have poor survival. The use of maximum likelihood estimation ensured that the estimated function would be the best fitting function for the selected endpoint of the exposure-occurrence process.

Because the number of mesothelioma deaths in the United States was relatively small compared with the total number of deaths per year (about 0.1% overall; 0.2% for males), we used estimates of total mortality rates for all males to represent nonmesothelioma mortality in the model. We restricted the age ranges to 15-99 years of age because exposures below age 15 were probably not occupational exposures. Exposures associated with survival above 95 years were beyond the range of data from SEER. In effect, we assumed that no claims would arise from incidence of mesothelioma in persons aged 100 and over. As the latest baseline period was 1985-1989, the oldest group reached 95-99 years in 1985-1989. They were mostly born in 1890-1894 and reached age 25 in 1915-1919. We assumed no asbestos exposure prior to 1915, so mortality rate estimates were required for 1915 to 1989, with projections for 1990 to 2049.

Estimates of the number of persons exposed to asbestos were obtained by dividing the number of mesothelioma diagnoses in 1975-1989 by the rate at which mesothelioma occurred among insulation workers. To the extent that this rate was higher than the rate faced by the average worker exposed to asbestos, the estimate of the number exposed was too low. This was not a problem, however, because a low estimate of exposed workers, leading to a low estimate of surviving persons exposed to asbestos, was combined with a high estimate of the mesothelioma incidence rate, yielding the observed number of mesothelioma diagnoses for 1975-1989 and a reasonable extrapolation to future years.

Walker (1982) referred to this canceling effect in describing the concept of "insulation-worker equivalents" (IWEs). Peto et al. (1981) used the term "effective level of exposure" to characterize the same effect. The cancellation effect implied a linear extrapolation from higher doses, where the dose-response effects were easiest to study, to low doses, where the dose-response effects were difficult to study. Such linear extrapolation is standard in environmental risk assessment (EPA, 1986).

One practical issue with the use of IWEs was how to adjust the surviving population when a claim occurred. If the exposure counts were real people, then it would be necessary to adjust the count downward by one person as each claim occurs. On the other hand, if the IWEs were too small by a factor of, say, 40, then the count should only be adjusted downward by 1/40 person as each claim occurs. In this latter case, the depletion due to filed claims would be relatively trivial. The only remaining factor controlling the size of the surviving exposed IWE population was mortality.

We assumed no depletion due to claims in this chapter. This implies that our projections were upwardly biased to a degree that depended on the ratio of the number of IWEs to the true exposure count. We will evaluate the impact of this assumption in Chapter 8.

There were 11 steps in the first-stage baseline projection of mesothelioma incidence at the national level. A 12th step was added to project the surviving number of IWEs by attained age and TSFE. These steps formed the first-stage calibration and they were used to develop several types of projection of the number of claims, by type of disease/injury, for seven classes of alleged disease/injury used by the Trust in their initial classification of claims of asbestos-related injuries. The determination of the claim filing rates for these projections formed the second-stage calibration of the model. Because a person may claim more than one disease/injury, we projected claims according to the most severe disease/injury in order to have a unique classification of claims. We extended the hierarchy of diseases used by the Trust in its settlement negotiation process from seven to nine categories: 1 = mesothelioma; 2 = lung cancer; 3 = colon/rectal cancer; 4 = other cancer; 5 = asbestosis; 6 = disputed asbestosis; 7 = pleural plaques/thickening; 8 = nonasbestos-related disease; and 9 = unknown.

## 6.4 First-Stage Calibration: Overview

The 12 steps in the first-stage baseline model reflected the use of all mesothelioma incidence data for the combined baseline periods 1975-1979, 1980-1984, and 1985-1989. The first 10 steps were updated and modified versions of the first 10 steps described in our sensitivity analysis of Walker's projections in Section 5.4.1. The 11th and 12th steps were formed by separating step 11 of the prior sensitivity analysis into component projections of the at-risk population and the national mesothelioma incidence. Alternative scenarios can be generated by modifying 1 or more of the 12 steps in a manner similar to that in Section 5.4.2. The first-stage calibration may be divided into three logically separate phases: (1) data preparation; (2) model estimation; and (3) model projection.

The first phase focused on data preparation. The steps were as follows:

1. *Nonmesothelioma mortality rates.* We obtained age-specific total mortality rates for 5-year intervals 1915-1919, 1920-1924, ..., 1980-1984, 1985-1989, ..., 2045-2049, for ages 15-19, 20-24, ..., 90-94 from U.S. life tables and vital statistics data from 1910 to 1989 for U.S. males, and projections to 2049. The rates represent the "competing risk" effects of nonmesothelioma mortality. The projections from 1990-1994 to 2045-2049 assumed a constant percentage decline, estimated separately for each 5-year age group.
2. *National estimates of mesothelioma incidence counts.* We obtained national estimates of mesothelioma incidence counts from SEER data for age groups 15-19, ..., 95-99 and the calendar periods 1975-1979, 1980-1984, and 1985-1989. The three periods defined the baseline periods of the projection and, in the baseline projection, were combined into one 15-year period. In Walker's (1982) original methodology, the baseline period was a 5-year period and no method for combining separate 5-year periods was considered. Walker's projections were updated by changing the baseline period from 1975-1979 to 1985-1989, in Chapter 7. Combining baseline periods differed from updating and was a more efficient use of the data.
3. *Distribution of age and date at start of asbestos exposure for mesothelioma incidence among Manville Trust claimants.* We obtained three distributions (one for each baseline period) of the age group and date of start of asbestos exposure claimants from Manville Trust data for mesothelioma diagnoses during each of the three periods 1975-1979, 1980-1984, and 1985-1989. The exposure distributions were smoothed to reduce the variability attributable to stochasticity of small frequencies. We replaced each count by an average of counts from immediately adjacent ages and calendar periods (a type of two-dimensional moving average).
4. *Normalization of exposure.* We normalized each matrix from step 3 by cohort so that the cohort sum represented unit probability mass. Cohort sums were obtained by adding terms along the diagonals of the matrices. More precisely, let $F_{a,a+4;y,y+4}(t)$ denote the individual terms in that sum. Then, for the 1975-1979 diagnoses, we defined

$$F_{a,a+4;y,y+4}(t) = \frac{E_{a,a+4;y,y+4}(t)}{\sum\limits_{s=0,5,10,\ldots} E_{a_1-s,a_1+4-s;y_1-s,y_1+4-s}(s)},$$

where $E_{a,a+4;y,y+t}(t)$ denoted the mesothelioma counts for 5-year groups $(a, a+4)$ by 5-year calendar periods $(y, y+4)$ of initial exposure and where $a_1 = a + t$ and $y_1 = y + t = 1975$ were the lower bounds of the age and date-of-diagnosis interval 1975-1979. Similar expressions were developed for the 1980-1984 and 1985-1989 diagnoses.
5. *Intensity of exposure.* We obtained estimates from Walker (1982, p. 6) of the proportion of low, medium, and high levels of exposure among cases of mesothelioma. These fractions were $f_l = 0.46$, 0.34, and 0.20, respectively,

where the medium and high levels of exposure were the 54% occupationally exposed workers (Walker's Task 1b). The "extra" 46% constituted our estimate of the low-exposure group. Sensitivity analysis of the $f_l$ parameters indicated little sensitivity of the total number of projected claims to variation in these values.

The second phase focused on estimating how many people must have been exposed to asbestos in the past to account for mesothelioma cases in 1975-1989 (see Section 6.6 for details). The calculations were performed separately for low-, medium-, and high-exposure subgroups of the mesothelioma incidence counts obtained from steps 2 and 5. The results of each of these calculations were the numbers of insulation workers (IWEs), not the numbers of persons actually exposed. Insulation workers were generally subject to the highest levels of exposure (see Table 3.1) and one insulation worker could be equivalent to 25 automobile maintenance workers, based on relative risk ratios from Selikoff (1981).

6. *Stratification of national estimates of mesothelioma incidence counts by level of asbestos exposure.* Using the proportions from step 5, we adjusted incidence counts obtained in step 2 for each of three exposure levels. The estimated proportions of low, medium, and high levels of asbestos exposure for cases of mesothelioma were $f_l = 0.46$, 0.34, and 0.20, respectively. The proportions with medium and high levels of exposure (0.34 and 0.20) sum to the 54% of occupationally exposed workers with documentable exposure, defined and discussed by Walker (1982).
7. *Estimation of the IWE population exposed to asbestos prior to 1975 by level of asbestos prior to 1975.* We calculated the population exposed to asbestos, prior to 1975, necessary to yield the national incidence of mesothelioma from step 6 for the periods 1975-1979, 1980-1984, and 1985-1989. These calculations yielded IWE exposure counts by age and date of first exposure and by level of exposure that formed the basis of the projections. In reconstructing these counts using mesothelioma incidence counts from step 6, it was necessary to adjust for both mesothelioma and nonmesothelioma mortality. For the high-exposure subgroup, age-specific hazards for nonmesothelioma mortality (from step 1) were multiplied by 1.37 to conform with Walker's methodology to represent the nonindependence of competing risks (Walker, 1982, p. 14) (see Section 4.4.1e). The medium- and low-exposure subgroups were assumed to have only the standard hazards of nonmesothelioma mortality. We did not distinguish between incidence and mortality rates for mesothelioma because of the short survival time (i.e., less than a year) after diagnosis. For both types of rate, we used the Weibull hazard function as fitted by Peto et al. (1982) to data on mesothelioma mortality among insulation workers over a 50-year exposure period (Selikoff, 1981). Virtually any discrepancy in the use of a common $k$ parameter for both incidence and mortality rates can

be absorbed into the *b* parameter of the incidence function (see Section 4.4.1d).

8. *Adjustments to exposure during 1955-1974, by level of asbestos exposure.* We replaced the IWE exposed populations from step 7 for first exposures in 1955-1959, 1960-1964, 1965-1969, and 1970-1974 with the age-specific average of the IWE exposed population with first exposures in 1945-1949 and 1950-1954. This was consistent with the observation of Peto et al. that their data "show no evidence of any change in extent of exposure since 1945" (Peto et al., 1981, p. 59). The purpose of this step was to avoid instability of estimates of the exposed population due to low numbers of mesothelioma cases among persons exposed more recently.
9. *Adjustments to reflect improvements in the workplace during 1960-1974, by level of asbestos exposure.* We deflated the later data from step 8 using the "workforce discounts" 1960-1964, 10%; 1965-1969, 50%; 1970-1974, 75%; and 1975-1979, 100%. Walker (1982, Task 1f) introduced these discount factors to reflect the effects of improvements in the workplace in the amounts of ambient asbestos fiber. Whereas step 8 implied constant numbers of real workers over time, step 9 implied reduced intensity of exposure per real worker and, hence, in IWEs over time.
10. *Renormalization by level of asbestos exposure.* We renormalized the IWE exposed populations from step 9 so that the mesothelioma incidence by cohort for the combined periods 1975-1979, 1980-1984, and 1985-1989 obtained in step 6 was predicted exactly. The renormalized values were critical parameters for the projection model because they represented the inferred IWE population exposed to asbestos by age and by date of first exposure to asbestos. They were central, not only to mesothelioma projections but also to projections of the other eight diseases. On the assumption that mesothelioma never occurs without prior asbestos exposure, mesothelioma is an unambiguous marker of asbestos exposure (as opposed to chronic respiratory problems or lung cancer which can be caused by other agents) and the mesothelioma-based IWE estimate can be used validly to project the other asbestos-related diseases. This will also be true if a small constant fraction of mesothelioma is due to some cause other than asbestos, although the IWE estimates will be too large by a corresponding amount.

The third phase comprised the projections of the survival of the at-risk IWE population and the future incidence of mesothelioma at the national level. This phase completed the first-stage baseline projection model.

11. *Forward projection of the at-risk IWE population by level of asbestos exposure.* For each age and date of start of exposure, we projected the number of IWEs that would survive to later years and later ages. These estimates were accumulated by 5-year age groups and 5-year calendar periods. The detailed IWE results were used in the national mesothelioma projections

in step 12 and in the Manville Trust claim projections in the second-stage model.

12. *Forward projection of mesothelioma incidence by level of asbestos exposure.* For each age and date of start of exposure, we projected the number of mesothelioma cases that would occur nationally in later years and at later ages. These estimates were accumulated by 5-year age groups and by 5-year calendar periods.

Once the baseline model had been developed, it was possible to forecast claims against the Trust in a second stage, using a claim hazard rate (CHR) model based on the assumption that a select, detailed set of claim filing rates observed in 1990-1992 would be constant over the entire period 1990-2049. The historical exposure estimates from step 10 were combined with the Trust claim data for 1990-1992 to calibrate a separate nonparametric hazard function for each of the alleged nine diseases. The nonparametric hazard rates for mesothelioma and for each of the other eight alleged diseases were then used to replace the parametric hazard rates for mesothelioma used in step 12. This yielded direct and independent projections of the claim counts for each of the nine alleged diseases.

## 6.5 Data Preparation

Steps 1-5 involved acquisition and manipulation of the various data inputs for the mesothelioma projection model. In this section we describe the data sources and discuss their limitations.

### 6.5.1 Step 1: Nonmesothelioma Mortality Rates

Construction of nonmesothelioma mortality rates required observed mortality and population data for the period 1915-1989 and projected mortality rates for the period 1990-2049. The data used in this analysis came from many sources and were adjusted using estimates of census underenumeration (Siegel, 1974; Keyfitz, 1979). In general, adjustments for census underenumeration were greater for blacks than whites, especially for young adult males.

For the white population for decennial years 1910 to 1950, life table estimates were made by Coale and Zelnik (1963). For the black population, for decennial years 1910 to 1950, official life tables were adjusted for enumeration error using the estimates of net census undercounts by Coale and Rives (1973). Mortality rates for single years of ages 0, 1, ..., 89 for whites and 0, 1, ..., 84 for blacks were obtained by nonlinear interpolation of the 5-year mortality rates in each set of life tables. The interpolation preserved the 5-year total hazards while keeping the endpoints of adjacent 5-year intervals as close as possible. Both sets of mortality rates were extrapolated to age 99 using the procedure described in this section.

Mortality rates for the total sex-specific populations were estimated from adjusted data assuming that the white rates applied to 87.5% of the population and the black rates to 12.5% (i.e., to all nonwhites). Rates for other than decennial years were obtained by linear interpolation between adjacent decennial rate estimates.

For 1950-1989, separate estimates of the population of whites and nonwhites were obtained by linear interpolation of race- and single-year age-specific counts (ages 0, 1, ..., 94) from the 1950, 1960, 1970, and 1980 U.S. Censuses and intercensal estimates for 1981-1989; these data were adjusted for underenumeration, race misclassification, and age misreporting errors (Coale and Zelnik, 1963; Passel et al., 1982; Siegel, 1974). Siegel's adjustments employed the Coale-Zelnik and Coale-Rives methodology for estimating underenumeration.

We obtained counts of deaths for whites and nonwhites for the period 1950-1989 from the National Center for Health Statistics. These were combined with the corresponding population data, collapsed on race (after adjustment), to produce gender- and age-specific annual mortality rates.

For all years, we extrapolated mortality rates up to age 99 by assuming that the rate of increase of known mortality rates continued to ages where mortality rates were unknown. Specifically, the mortality rate at age $a$ (in year $y$) was estimated from the rate at ages $a - 5$ and $a - 10$ using

$$m_{a,y} = m_{a-5,y} \times (m_{a-5,y}/m_{a-10,y}) \qquad (a > a^*).$$

For 1910-1949, the youngest age $a^*$ was 90 for whites and 85 for blacks; for 1950 to 1989, the youngest age $a^*$ was 95 for both whites and nonwhites. Combining the two sets of mortality rates, we obtained annual mortality rates for the period 1910-1989, by sex and age for ages 0 to 99 years.

Walker's (1982) projections used quinquennial age and date intervals in defining fundamental quantities because further age or time detail would be difficult to justify as credible, given the small size of the mesothelioma counts in the baseline period in SEER data – under 200 cases per year. We accepted the argument that quinquennial age and date periods were most appropriate for these data and we developed the estimates and projection equations using quinquennial time units. Although this is not difficult, it does require careful consideration of a number of technical issues.

We converted the single year of age and date mortality rates into 5-year mortality hazards for quinquennial age groups 15-19, 20-24, ..., 90-94 and quinquennial calendar periods 1915-1919, 1920-1924, ..., 1980-1984. For the population at age $a$ in year $y$, we defined the 5-year mortality hazard as

$${}_5h_{a,y} = \frac{1}{2}m_{a,y} + \sum_{t=1}^{4} m_{a+t,y+t} + \frac{1}{2}m_{a+5,y+5}.$$

This definition reflected the hazard faced by a population of size $N_{a,y}$ defined at the midpoint of age $a$ and year $y$, followed to the midpoint of age $a+5$ and year $y+5$. Thus, the number of survivors $N_{a+5,y+5}$ was

$$N_{a+5,y+5} = N_{a,y} \times {}_5S_{a,y},$$

where

$${}_5S_{a,y} = \exp\{-{}_5h_{a,y}\},$$

$\exp\{\cdot\}$ is the exponential function, and ${}_5S_{a,y}$ is the 5-year survival function.

In defining ${}_5h_{a,y}$ and ${}_5S_{a,y}$, the left subscripted 5 denotes the number of time units over which the indicated function (cumulative hazard or survival probability) applies. If this subscript is deleted, then the function applies to one time unit, by convention.

We introduced simplified notation that allows us to move between annual time units and the alternative quinquennial time units: (1) lowercase age, year, and time indexes and subscripts referred to annual time units; and (2) uppercase age, year, and time indexes and subscripts referred to quinquennial time units. For example, to generate quinquennial projections, we defined the average 5-year hazard for ages $A = \{a, a+1, a+2, a+3, a+4\}$ to $A+1 = \{a+5, a+6, a+7, a+8, a+9\}$ and calendar years $Y = \{y, y+1, y+2, y+3, y+4\}$ to $Y+1 = \{y+5, y+6, y+7, y+8, y+9\}$ as the unweighted average of the ${}_5h_{a,y}$ functions. Thus,

$$h_{A,Y} = \frac{1}{25}\sum_{b=0}^{4}\sum_{z=0}^{4} {}_5h_{a+b,y+z}.$$

Furthermore, since 1915-1919 was the first quinquennial calendar period, we used the designation $Y$ = 1915-1919 and $Y = 1$, interchangeably, with similar extensions such as $Y$ = 1985-1989 equivalent to $Y = 15$. Quinquennial age groups were treated likewise; for example, $A$ = 15-19 was equivalent to $A = 1$, because 15-19 was the youngest age group considered. Thus the above expression yielded values of $h_{A,Y}$ for $A$ = 0-4, 5-9, ..., 90-94 and $Y$ = 1915-1919, 1920-1924, ..., 1980-1984.

A population of size $N_{A,Y}$ can be defined for the quinquennium $(A, Y)$. We adopted the convention that $N_{A,Y}$ is the average number of persons at risk at each point of calendar period $Y$. This is consistent with the usual definition of $N_{a,y}$ as the average number of persons at risk at each point of year $y$. Hence, we can express $N_{A,Y}$ as a function of $N_{a,y}$ in one of two ways:

$$N_{A,Y} \approx \sum_{b=0}^{4} N_{a+b,y+2}$$

or

$$N_{A,Y} \approx \frac{1}{5}\sum_{b=0}^{4}\sum_{z=0}^{4} N_{a+b,y+z}.$$

In working with the quinquennial counts $N_{A,Y}$, we do not need to make an explicit choice between these two approximations. However, we do need to ensure that we appropriately account for the events in the 5-year interval between observations of $N_{A,Y}$ and $N_{A+1,Y+1}$.

For a general population not subjected to asbestos exposure, the quinquennial population $N_{A,Y}$ projects to $N_{A+1,Y+1}$ using

$$N_{A+1,Y+1} = N_{A,Y} \times S_{A,Y},$$

where $S_{A,Y}$ is the 5-year survival function defined by

$$S_{A,Y} = \exp\{-h_{A,Y}\}.$$

To project these values to future years corresponding to $Y$ = 1985-1989, 1990-1994, ..., 2045-2049, we assumed the hazard rates obey

$$h_{A,Y} = h_{A,\text{1980-84}} \times p_A^T,$$

where $T$ is the number of quinquennial periods from 1980-1984 to $Y$, $p_A$ is the empirical "hazard-projection rate" for quinquennial age $A$, which was set equal to the average rate of change in $h_{A,Y}$ over the period 1950-1954 to 1980-1984, and $p_A^T$ is the $T$th power of $p_A$. To help stabilize the results, the changes for age groups 15-19 through 85-89 were computed with data for 15-year age intervals using

$$p_A^6 = \frac{h_{A-1,\text{1980-84}} + 2h_{A,\text{1980-84}} + h_{A+1,\text{1980-84}}}{h_{A-1,\text{1950-54}} + 2h_{A,\text{1950-54}} + h_{A+1,\text{1950-54}}},$$

which yields $p_A$ as the sixth root of $p_A^6$.

This procedure was used to project $h_{A,Y}$ for $A$ = 15-19, 20-24, ..., 90-94 and $Y$ = 1985-1989, 1990-1994, ..., 2045-1949 (Table 6.2).

This model produced an exponential decrease (constant percentage decline) in mortality of about 1% per year, depending on age, which was consistent with the analysis of Lee and Carter (1992) using time series methods. This level of decrease was also consistent with recent recommendations of the Social Security Advisory Board (SSA, 1999).

### 6.5.2 Step 2: National Estimates of Mesothelioma Incidence Counts

The projection model required estimates of the number of cases of mesothelioma occurring in the United States by quinquennial age groups (15-19, 20-24, ..., 90-94, 95-99) and quinquennial calendar periods (1975-1979, 1980-1984, and 1985-1989) by gender. Such estimates were generated using the computerized data from the National Cancer Institute's (NCI) Surveillance, Epidemiology, and End Results (SEER) program for 1973-1989. SEER sampled cancer incidence in approximately 10% of the population and represented the best source of data for estimating the national incidence of mesothelioma. The

**Table 6.2: Projected Mortality Hazard Rates ($h_{A,Y}$; in Percent) for U.S. Males by Age and Quinquennium**

| | Quinquennium | | | | | | | | | | | | | |
|---|---|---|---|---|---|---|---|---|---|---|---|---|---|---|
| Age | 1985-1989 | 1990-1994 | 1995-1999 | 2000-2004 | 2005-2009 | 2010-2014 | 2015-2019 | 2020-2024 | 2025-2029 | 2030-2034 | 2035-2039 | 2040-2044 | 2045-2049 | Projection Rate |
| 15-19 | 0.80 | 0.80 | 0.80 | 0.70 | 0.70 | 0.70 | 0.70 | 0.70 | 0.70 | 0.70 | 0.70 | 0.70 | 0.70 | 0.9957 |
| 20-24 | 0.80 | 0.80 | 0.80 | 0.80 | 0.80 | 0.80 | 0.80 | 0.80 | 0.80 | 0.80 | 0.80 | 0.80 | 0.80 | 0.9964 |
| 25-29 | 0.90 | 0.90 | 0.90 | 0.80 | 0.80 | 0.80 | 0.80 | 0.80 | 0.80 | 0.80 | 0.80 | 0.80 | 0.80 | 0.9873 |
| 30-34 | 1.00 | 1.00 | 0.90 | 0.90 | 0.90 | 0.90 | 0.80 | 0.80 | 0.80 | 0.80 | 0.70 | 0.70 | 0.70 | 0.9689 |
| 35-39 | 1.30 | 1.20 | 1.10 | 1.10 | 1.00 | 1.00 | 0.90 | 0.90 | 0.80 | 0.80 | 0.70 | 0.70 | 0.70 | 0.9461 |
| 40-44 | 1.90 | 1.70 | 1.60 | 1.50 | 1.40 | 1.30 | 1.20 | 1.20 | 1.10 | 1.00 | 1.00 | 0.90 | 0.80 | 0.9357 |
| 45-49 | 3.00 | 2.80 | 2.70 | 2.50 | 2.30 | 2.20 | 2.10 | 1.90 | 1.80 | 1.70 | 1.60 | 1.50 | 1.40 | 0.9381 |
| 50-54 | 4.90 | 4.60 | 4.30 | 4.10 | 3.90 | 3.60 | 3.40 | 3.20 | 3.10 | 2.90 | 2.70 | 2.60 | 2.40 | 0.9432 |
| 55-59 | 7.80 | 7.40 | 7.00 | 6.60 | 6.30 | 5.90 | 5.60 | 5.30 | 5.00 | 4.80 | 4.50 | 4.30 | 4.10 | 0.9473 |
| 60-64 | 12.10 | 11.50 | 11.00 | 10.50 | 10.00 | 9.50 | 9.10 | 8.70 | 8.30 | 7.90 | 7.50 | 7.20 | 6.90 | 0.9540 |
| 65-69 | 18.80 | 18.00 | 17.30 | 16.70 | 16.00 | 15.40 | 14.80 | 14.20 | 13.70 | 13.10 | 12.60 | 12.10 | 11.60 | 0.9609 |
| 70-74 | 28.60 | 27.50 | 26.50 | 25.50 | 24.50 | 23.50 | 22.60 | 21.80 | 20.90 | 20.10 | 19.40 | 18.60 | 17.90 | 0.9616 |
| 75-79 | 42.80 | 41.00 | 39.20 | 37.40 | 35.80 | 34.20 | 32.70 | 31.30 | 29.90 | 28.60 | 27.40 | 26.20 | 25.00 | 0.9562 |
| 80-84 | 62.90 | 59.90 | 57.10 | 54.40 | 51.90 | 49.40 | 47.10 | 44.90 | 42.80 | 40.70 | 38.80 | 37.00 | 35.20 | 0.9529 |
| 85-89 | 92.00 | 87.50 | 83.20 | 79.10 | 75.20 | 71.50 | 68.00 | 64.70 | 61.50 | 58.50 | 55.60 | 52.90 | 50.30 | 0.9510 |
| 90-94 | 127.40 | 121.00 | 114.90 | 109.20 | 103.70 | 98.50 | 93.50 | 88.80 | 84.40 | 80.10 | 76.10 | 72.30 | 68.70 | 0.9498 |

Source: Stallard and Manton (1993, Table 2).

nine SEER sites were San Francisco-Oakland Standard Metropolitan Statistical Area, Connecticut, Metropolitan Detroit, Hawaii, Iowa, New Mexico, Seattle (Puget Sound), Utah, and Metropolitan Atlanta. The records in the SEER database included information from death certificates, from hospital medical records, and from various other sources, including private laboratories, nursing homes, and other health care providers.

The SEER file for 1973-1989 contained records on 1,454,079 primary tumors occurring among 1,307,892 individuals. When more than one primary site was recorded, all sites were searched so that no mesotheliomas were missed in tabulating the file.

Associated with the SEER files were population counts, for each geographic site, by quinquennial age groups (0-4, 5-9, ..., 80-84, and 85+) and single calendar years, and corresponding population counts for the total United States, but with age 85+ broken out as 85-89, 90-94, 95-99, and 100+. The SEER populations at age 85+ were pro rata distributed into the age groups 85-89, 90-94, and 95-99 using U.S. national data, by sex and year, to establish the relative weights for each subgroup. These modified data were used to generate estimates of age-specific mesothelioma incidence counts for the total U.S. population aged 0-99. Because no mesothelioma cases occurred above age 94, we terminated the calculations using 95-99 as the oldest age group.

These counts were summed within 5-year calendar periods to produce the quinquennial estimates in Table 6.3. The counts for 1973 and 1974 were summed and multiplied by 2.5 to represent 1970-1974. Nearly 24% of mesothelioma incidence occurred among females; but only 5.4% of mesothelioma claims against the Manville Trust were filed by females (see Table 6.4). This suggested that forecasts of the total societal impact of mesothelioma may be more difficult to develop than forecasts of the number of claims of mesothelioma disease/injury against the Manville Trust where documented exposure history may be required. In our analysis, where the goal was to forecast claims, we focused only on the experience of males. The incidence estimates for males for 1975-1979, 1980-1984, and 1985-1989 were input into the projection model.

Walker (1982, Task 1a) defined upward and downward adjustments to the SEER-based estimates, with a net downward adjustment of 2%. We did not use these adjustments because the claim projection methodology is insensitive to the adjustments and because the rationale for the net result of −2% was not compelling (see Section 4.4).

### 6.5.3 Step 3: Distribution of Age and Date at Start of Asbestos Exposure for Mesothelioma Incidence Among Manville Trust Claimants

The Manville Trust maintained data files on the claim settlement process, in both hard-copy and computer-readable form. An extract from the claims database was prepared on December 2, 1992 that contained data for all claims filed from November 28, 1988 through November 30, 1992. The extract included

Table 6.3: Estimated U.S. Mesothelioma Incidence Counts by Age, Quinquennium, and Gender, 1970-1989

| | Males | | | | | Females | | | | |
|---|---|---|---|---|---|---|---|---|---|---|
| | Quinquennium | | | | | Quinquennium | | | | |
| Age | 1970-1974 | 1975-1979 | 1980-1984 | 1985-1989 | Total | 1970-1974 | 1975-1979 | 1980-1984 | 1985-1989 | Total |
| 0-4 | 0 | 0 | 0 | 0 | 0 | 0 | 0 | 10 | 0 | 10 |
| 5-9 | 0 | 0 | 0 | 0 | 0 | 0 | 0 | 0 | 0 | 0 |
| 10-14 | 0 | 0 | 0 | 0 | 0 | 0 | 0 | 0 | 0 | 0 |
| 15-19 | 0 | 11 | 11 | 0 | 22 | 0 | 22 | 0 | 0 | 22 |
| 20-24 | 33 | 11 | 21 | 11 | 75 | 0 | 21 | 11 | 11 | 42 |
| 25-29 | 27 | 30 | 0 | 10 | 67 | 0 | 30 | 20 | 10 | 61 |
| 30-34 | 85 | 59 | 10 | 20 | 175 | 0 | 40 | 20 | 20 | 80 |
| 35-39 | 91 | 50 | 60 | 91 | 293 | 28 | 52 | 0 | 51 | 130 |
| 40-44 | 140 | 83 | 103 | 193 | 519 | 57 | 42 | 94 | 92 | 284 |
| 45-49 | 270 | 342 | 283 | 195 | 1,090 | 60 | 95 | 96 | 126 | 376 |
| 50-54 | 347 | 454 | 494 | 307 | 1,602 | 151 | 137 | 118 | 151 | 557 |
| 55-59 | 559 | 581 | 812 | 797 | 2,749 | 131 | 225 | 227 | 241 | 824 |
| 60-64 | 544 | 745 | 1,158 | 996 | 3,443 | 348 | 248 | 312 | 258 | 1,165 |
| 65-69 | 488 | 899 | 1,331 | 1,857 | 4,575 | 208 | 284 | 384 | 343 | 1,218 |
| 70-74 | 633 | 799 | 1,411 | 1,780 | 4,622 | 231 | 155 | 285 | 428 | 1,099 |
| 75-79 | 411 | 603 | 1,243 | 1,246 | 3,504 | 129 | 185 | 294 | 386 | 993 |
| 80-84 | 91 | 295 | 551 | 780 | 1,717 | 149 | 100 | 262 | 220 | 731 |
| 85-89 | 0 | 151 | 208 | 276 | 635 | 0 | 11 | 87 | 110 | 207 |
| 90-94 | 0 | 0 | 44 | 77 | 121 | 29 | 11 | 0 | 55 | 94 |
| 95-99 | 0 | 0 | 0 | 0 | 0 | 0 | 0 | 0 | 0 | 0 |
| Total | 3,718 | 5,113 | 7,740 | 8,637 | 25,207 | 1,520 | 1,656 | 2,220 | 2,500 | 7,896 |

Source: Stallard and Manton (1993, Table 5); based on SEER data, 1973-1989.

**Table 6.4: Incidence of Most Severe Disease/Injury by Claim Filing Year – Qualified Claims Against the Manville Trust, 1988-1992**

| Most Severe Alleged Disease/Injury | Year of Claim | | | | | Total |
|---|---|---|---|---|---|---|
| | 1988 | 1989 | 1990 | 1991 | 1992 | |
| | | | Both Sexes | | | |
| 1. Mesothelioma | 1,437 | 4,555 | 777 | 798 | 683 | 8,250 |
| 2. Lung cancer | 2,081 | 8,013 | 1,079 | 1,147 | 1,038 | 13,358 |
| 3. Colon/rectal cancer | 273 | 978 | 103 | 153 | 145 | 1,652 |
| 4. Other cancer | 966 | 1,837 | 212 | 189 | 129 | 3,333 |
| 5. Asbestosis | 17,480 | 43,846 | 6,449 | 8,010 | 9,366 | 85,151 |
| 6. Disputed asbestosis | 4,611 | 13,425 | 2,095 | 2,425 | 2,686 | 25,242 |
| 7. Pleural plaques/thickening | 3,123 | 19,605 | 2,717 | 2,556 | 2,625 | 30,626 |
| 8. Non-asbestos related disease | 2,327 | 14,142 | 1,604 | 3,053 | 337 | 21,463 |
| 9. Unknown | 96 | 1,382 | 1,644 | 254 | 140 | 3,516 |
| Total | 32,394 | 107,783 | 16,680 | 18,585 | 17,149 | 192,591 |
| | | | Males | | | |
| 1. Mesothelioma | 1,373 | 4,300 | 736 | 739 | 639 | 7,787 |
| 2. Lung cancer | 2,036 | 7,812 | 1,058 | 1,110 | 1,013 | 13,029 |
| 3. Colon/rectal cancer | 269 | 944 | 102 | 152 | 143 | 1,610 |
| 4. Other cancer | 938 | 1,789 | 203 | 184 | 124 | 3,238 |
| 5. Asbestosis | 16,937 | 42,473 | 6,184 | 7,734 | 9,086 | 82,414 |
| 6. Disputed asbestosis | 4,478 | 13,038 | 2,042 | 2,387 | 2,662 | 24,607 |
| 7. Pleural plaques/thickening | 3,077 | 19,162 | 2,659 | 2,511 | 2,588 | 29,997 |
| 8. Non-asbestos related disease | 2,254 | 13,800 | 1,557 | 3,042 | 329 | 20,982 |
| 9. Unknown | 85 | 1,175 | 1,573 | 237 | 136 | 3,206 |
| Total | 31,447 | 104,493 | 16,114 | 18,096 | 16,720 | 186,870 |
| | | | Females | | | |
| 1. Mesothelioma | 60 | 253 | 41 | 59 | 44 | 457 |
| 2. Lung cancer | 41 | 193 | 21 | 37 | 25 | 317 |
| 3. Colon/rectal cancer | 4 | 34 | 1 | 1 | 2 | 42 |
| 4. Other cancer | 22 | 45 | 9 | 5 | 5 | 86 |
| 5. Asbestosis | 435 | 1,237 | 259 | 271 | 276 | 2,478 |
| 6. Disputed asbestosis | 100 | 337 | 50 | 37 | 24 | 548 |
| 7. Pleural plaques/thickening | 39 | 358 | 51 | 43 | 36 | 527 |
| 8. Non-asbestos related disease | 61 | 268 | 40 | 11 | 8 | 388 |
| 9. Unknown | 9 | 191 | 70 | 15 | 4 | 289 |
| Total | 771 | 2,916 | 542 | 479 | 424 | 5,132 |
| | | | Unknown Sex | | | |
| 1. Mesothelioma | 4 | 2 | 0 | 0 | 0 | 6 |
| 2. Lung cancer | 4 | 8 | 0 | 0 | 0 | 12 |
| 3. Colon/rectal cancer | 0 | 0 | 0 | 0 | 0 | 0 |
| 4. Other cancer | 6 | 3 | 0 | 0 | 0 | 9 |
| 5. Asbestosis | 108 | 136 | 6 | 5 | 4 | 259 |
| 6. Disputed asbestosis | 33 | 50 | 3 | 1 | 0 | 87 |
| 7. Pleural plaques/thickening | 7 | 85 | 7 | 2 | 1 | 102 |
| 8. Non-asbestos related disease | 12 | 74 | 7 | 0 | 0 | 93 |
| 9. Unknown | 2 | 16 | 1 | 2 | 0 | 21 |
| Total | 176 | 374 | 24 | 10 | 5 | 589 |

Source: Stallard and Manton (1993, Table 8).

information from the Proof of Claim (POC) Forms used by the Manville Trust as well as summary data from the claim settlement process.

The extract consisted of three files. The first contained 206,810 records, with 1 record per claimant. The second contained 316,110 records, with 1 record for each alleged asbestos-related disease/injury diagnosis due to exposure to asbestos or products containing asbestos manufactured, sold, or distributed by the Johns-Manville Corporation or related companies. The third contained 252,100 records with 1 record for each employment-related instance where the claimant was exposed to Johns-Manville-produced asbestos or asbestos products. Each of the three files assigned a unique claim identifier that allowed all data for a claimant to be linked.

To process these files, we created a workfile in which all data for each claimant were linked and summarized. Certain decisions were encoded in this file that affected the analysis.

First, each alleged injury-diagnosis was classified into one of nine categories:

1. Mesothelioma
2. Lung cancer
3. Colon/rectal cancer
4. Other cancer
5. Asbestosis
6. Disputed asbestosis
7. Pleural plaques and pleural thickening
8. Nonasbestos-related disease
9. Unknown or missing disease/injury

We defined the most severe diagnosis using this hierarchy. Thus, mesothelioma outranked all other injuries and the use of the most severe diagnosis misses no cases of mesothelioma. Occurrences of other diseases were not counted when higher ranked diseases were present for a claimant. All of our disease-specific estimates were based on the application of this hierarchy to uniquely assign to each claimant the most severe disease alleged on the POC Forms.

The most severe disease, as just defined, differs from the definition of a related data field, "evaluated disease type," used by the Trust for settled claims. This field was subject to change based on updated information provided just prior to or during settlement negotiations and was set to indicate nonasbestos-related disease for 167,648 claimants in the extract. To evaluate the consistency of the two disease definitions, we tabulated them from the workfile for settled claims with nonzero settlement amounts (Table 6.1). There was 93.6% concordance (i.e., 1314 of 1404 claims alleging mesothelioma were validated by the Trust as mesothelioma) between the two classifications for mesothelioma.

Certain claims were excluded from further consideration by the Trust. These were based on the claim status codes for "dead," deactivated, disqualified, and void claims. The claim statuses retained were settled, unsettled, and work-in-progress, representing a total of 192,591 "qualified" claims. The distribution of the most severe disease/injury by claim year for these persons is displayed in Table 6.4.

Of the 192,591 personal injury claims analyzed, 186,870 (97.03%) were for males, 5132 (2.66%) for females, and 589 (0.31%) for unknown gender. This latter group of 589 was excluded from gender-specific tabulations but was included in tabulations for "both sexes."

Figure 6.2 displays claims filings by month from November 28, 1988 to November 30, 1992. This is the longest period that could be used in calibrating the projection model. The Trust could not accept claims prior to November 28, 1988. During the period November 1988 to November 1989, a substantial backlog of claims was filed (i.e., 138,583 claims, of which approximately 17,000 were filed prior to May 1982). Thus, the accumulation of unfiled claims during the period May 1982 to November 1989 averaged 16,033 claims per year – a rate that is within 11% of the 17,991 per year average observed for 1990-1992.

From December 1, 1989 to November 30, 1992, although some months had very high filing rates, there was no temporal trend (Figure 6.3). The jump in August 1992 and subsequent decline in September to November 1992 reflected effects of a change in the POC Form in that period. Figure 6.3 shows no observable impact of the judicial stay issued in July and August 1990 that prevented the Trust from making payments except in cases of extreme hardship. The claim filing process for this period was relatively stable and at a level consistent with the emergence of new cases of disease.

Figure 6.4 presents corresponding statistics for mesothelioma claims. A slight downward trend was clearly attributable to the period September to November 1992. When mesothelioma was expressed as a percent of total claims, the downward trend was clearer (Figure 6.5), but when the period September to November 1992 was deleted, the trend disappeared. We assumed that the claims filing rates in the period 1990-1992 were constant and applied to the entire period 1990-1994. The lack of trend in Figures 6.2-6.5 was important to this aspect of the projection model.

Figure 6.6 displays trends in claims filing percentages for three classes of disease groups: (1) asbestos-related cancers (mesothelioma, lung cancer, colon/rectal cancer, and other cancer); (2) asbestos-related noncancers (asbestosis, disputed asbestosis, and pleural plaques/thickening); and (3) nonasbestos-related and unknown diseases. As for mesothelioma, class 1 (asbestos-related cancers) showed no temporal trend. On the other hand, there was a diverging and compensating trend between class 2 (asbestos-related noncancers), and class 3 (nonasbestos-related and unknown diseases). From Table 6.4, it was plausible that a significant number of claims had shifted from the latter category to asbestosis, the main component of asbestos-related noncancers. This might reflect increasing awareness among claimants and their representatives that claims alleging nonasbestos-related or unknown diseases were less likely to result in compensation than claims alleging an asbestos-

**Figure 6.2: Total Qualified Claims Filed Against the Manville Trust from November 1988 Through November 1992 (Source: Authors' Calculations)**

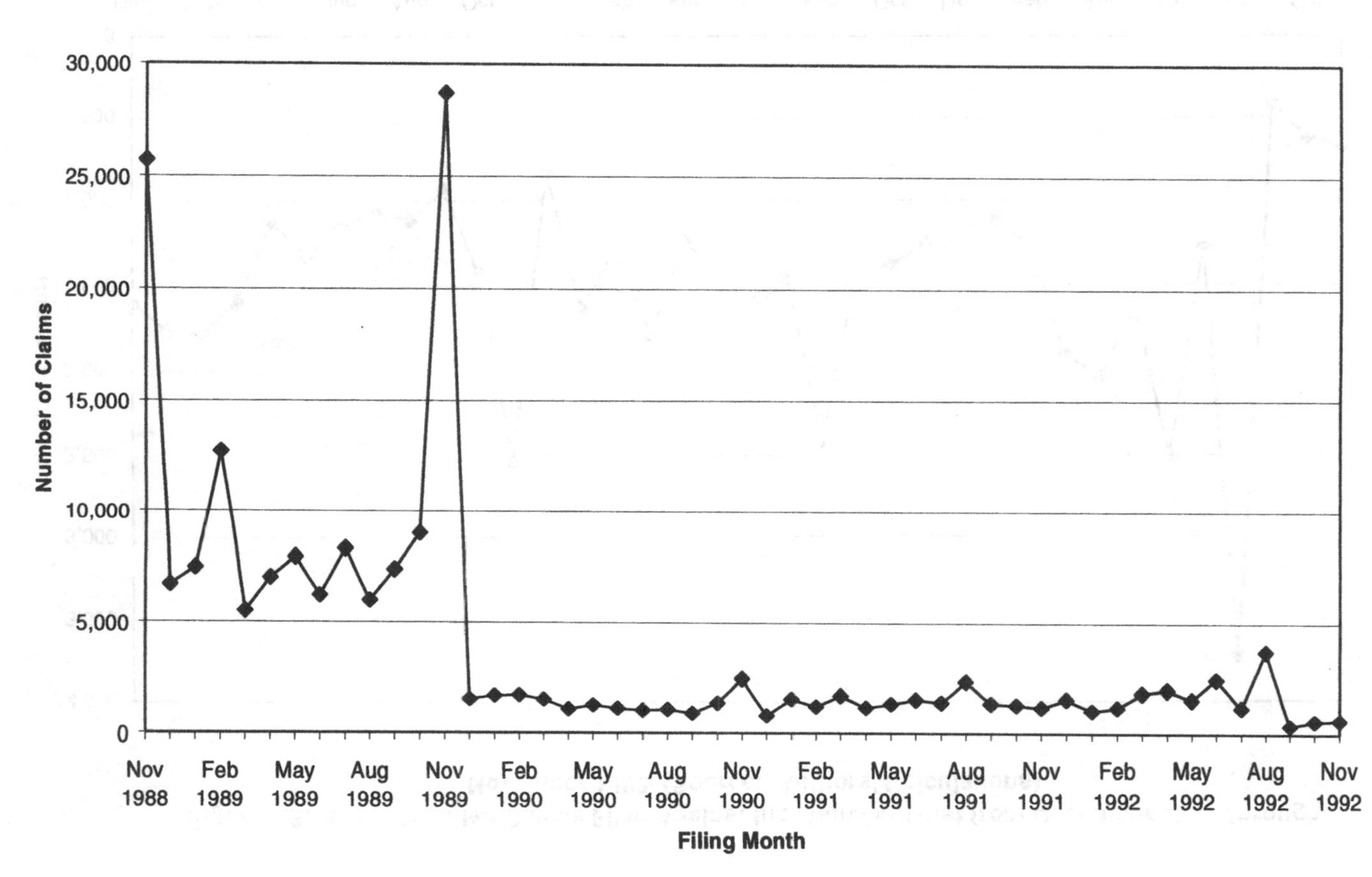

**Figure 6.3: Total Qualified Claims Filed Against the Manville Trust from December 1989 Through November 1992 (Source: Authors' Calculations)**

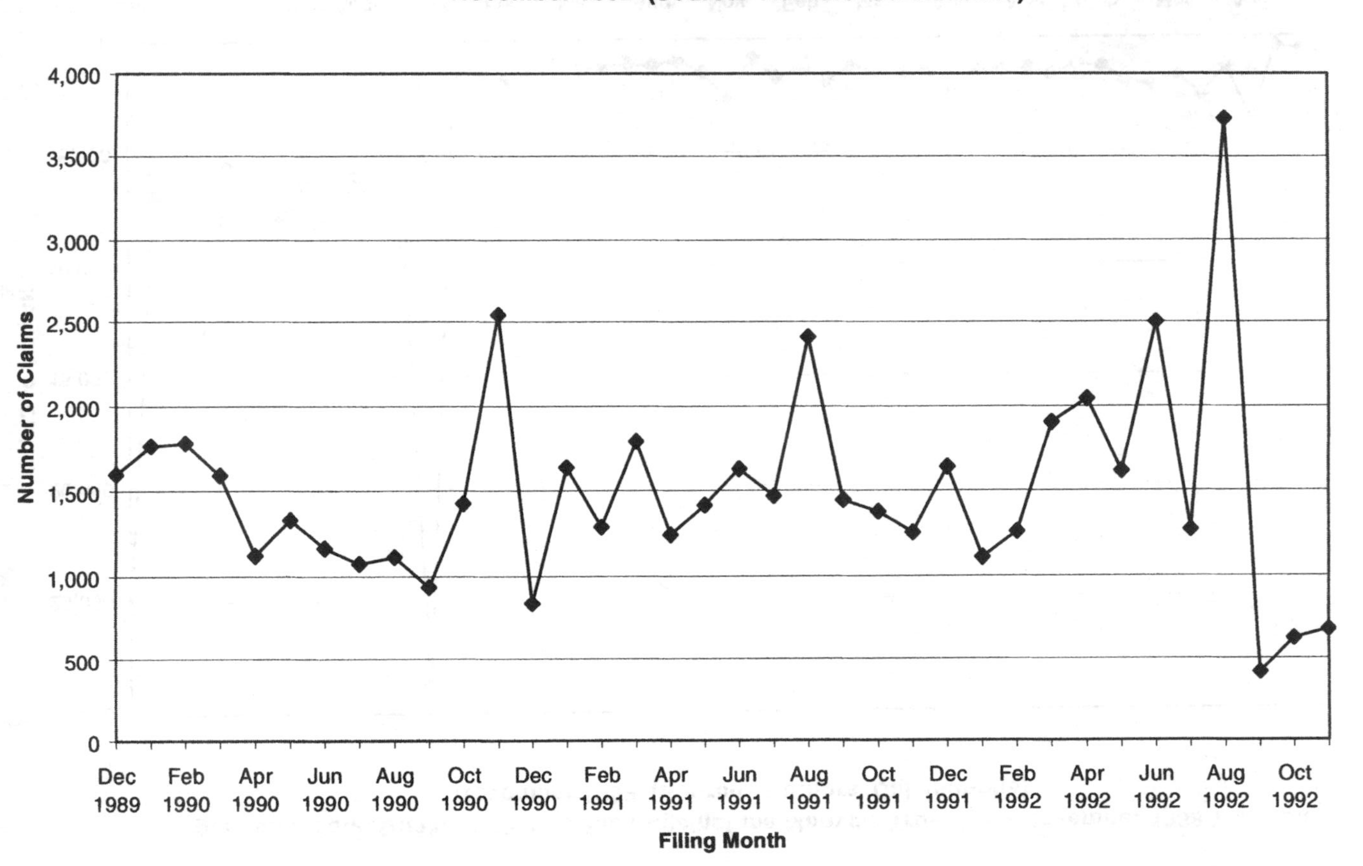

**Figure 6.4: Total Qualified Mesothelioma Claims Filed Against the Manville Trust from December 1989 Through November 1992 (Source: Authors' Calculations)**

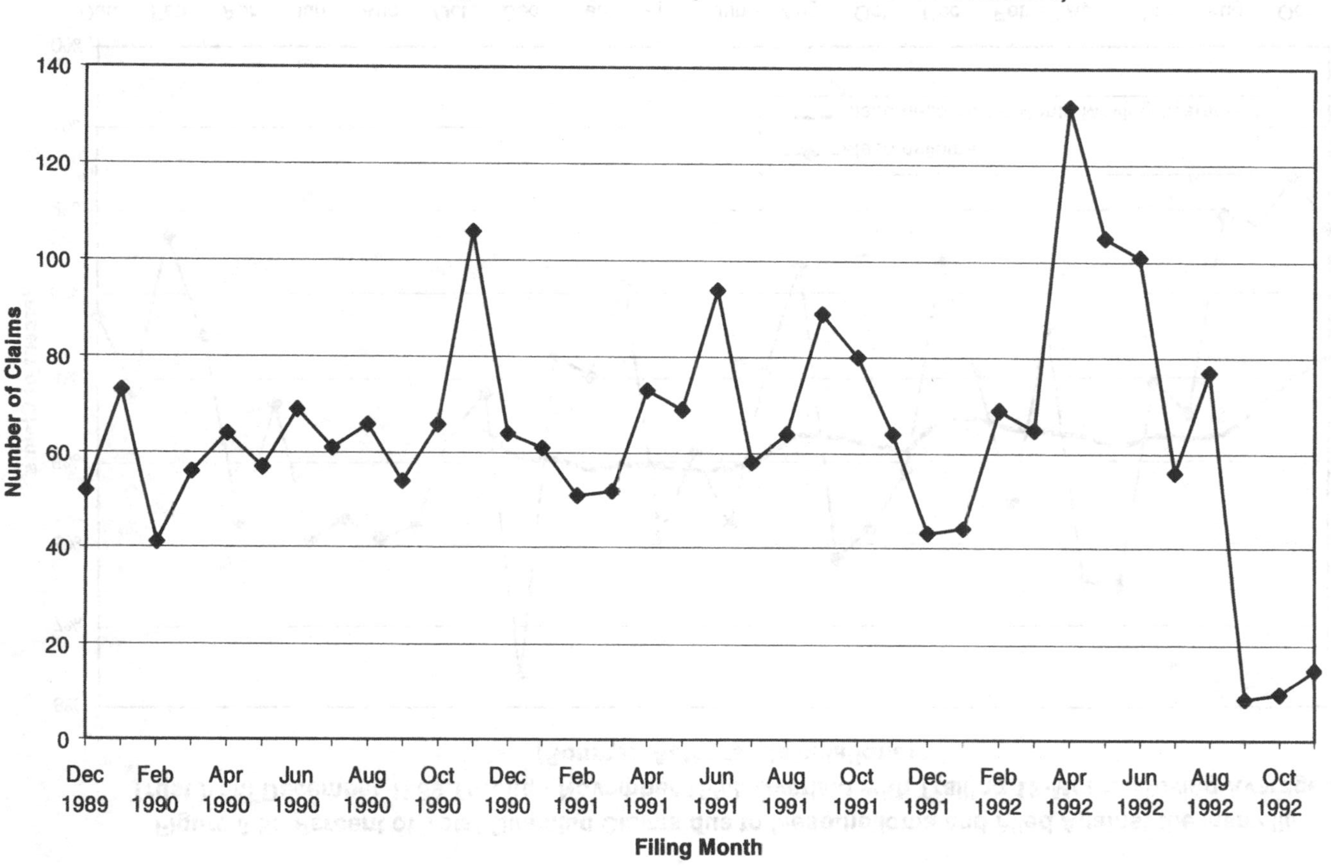

**Figure 6.5: Percent of Total Qualified Claims due to Mesothelioma and Filed Against the Manville Trust from December 1989 Through November 1992, Overlaid with Trailing 12-Month Moving Average (Source: Authors' Calculations)**

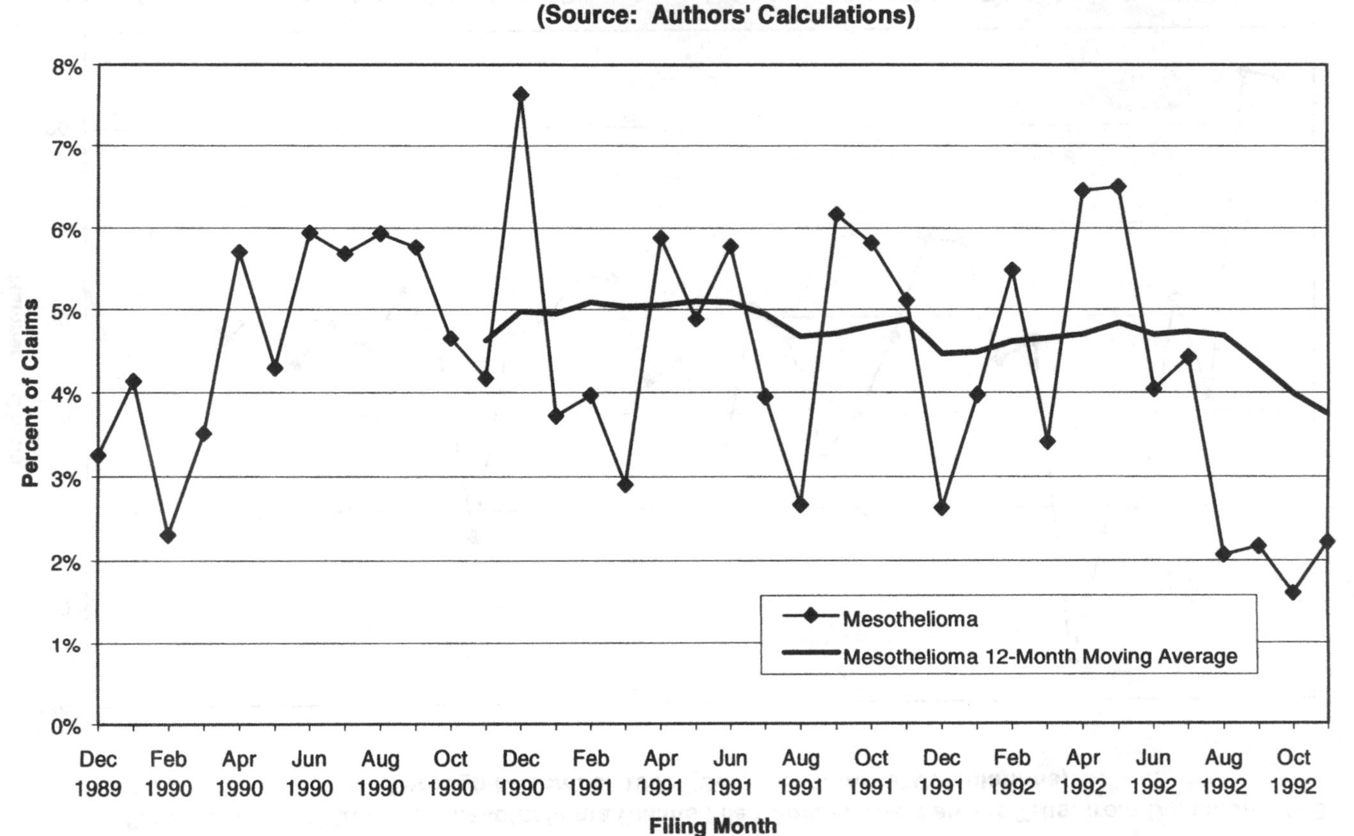

related disease. In our projections, we assumed that the average claim rates for 1990-1992 would continue for 1990-1994. To the extent that a shift had occurred from class 3 to 2, our projections of class 3 should be combined with class 2. Figures 6.3 and 6.6 together imply that a shift in alleged disease classes occurred during 1990-1992, but not an increase in the total number of asbestos claims. This is consistent with Table 6.1, which showed that only 40% of alleged class 3 diseases were validated as class 3 injuries in the settlement process.

Table 6.5 presents the distribution of age and date of diagnosis, by gender, for 8244 claimants with mesothelioma. Age was defined as the difference between date of diagnosis and date of birth. If date of diagnosis was unknown, then age was the difference between date of claim filing and date of birth. If date of birth was unknown, then age was unknown. The date of diagnosis was defined as the earliest alleged diagnosis date for mesothelioma for the 25% of mesothelioma claims with more than one mesothelioma diagnosis record.

Of the 7787 males in Table 6.5, 99.0% provided information on employment-related instances where the claimant was exposed to Johns-Manville-produced asbestos or asbestos products. Of the 457 females in Table 6.5, 98.2% provided similar information on employment-related exposure. Information on nonoccupational exposure to Johns-Manville asbestos or asbestos products was asked for on the POC Form, but was not released to us. Hence, all references to asbestos exposure in this analysis are to employment-related exposures. Nonoccupational exposures may be more important for females and may explain why females suffer 23.9% of mesothelioma incidence nationally in the SEER estimates (Table 6.3), but file only 5.5% of mesothelioma claims against the Manville Trust (Table 6.5).

The Trust data for 611 qualified male claims for mesothelioma filed in 1992 indicated that 93.0% of diagnoses occurred in 1989-1992 and 97.7% occurred in 1984-1992. If the same distribution applied to claims filed in 1993, then the male diagnosis counts for 1985-1989 would increase by 30 cases (0.9%) and for 1980-1984 by 9 cases (0.5%). In view of (1) the rapid lethality of mesothelioma, (2) the fact that these are only alleged cases of disease, not validated cases, and (3) the statutes of limitation that require timely filing of claims, it is likely that the diagnosis counts in Table 6.5 are relatively stable for 1980-1984 and earlier, subject to modest upward revisions for 1985-1989 (less than 1% increase with each new claim year, trending to 0% within, say, 10 years – about a 5% ultimate increase) and subject to substantial revisions for 1990-1992 (a 62.9% increase is needed just to match the diagnosis rates for 1985-1989).

The mesothelioma diagnoses for Manville Trust claimants in Table 6.5 are comparable to the national estimates in Table 6.3 and indicate the relative frequency of claims among affected persons: for males, 9.5% (487/5113) in 1975-1979, 24.9% in 1980-1984, and 40.0% in 1985-1989; for females, 1.51% in 1975-1979, 4.60% in 1980-1984, and 9.28% in 1985-1989.

**Figure 6.6: Percent of Total Qualified Claims, by Alleged Disease Group, Filed Against the Manville Trust from December 1989 Through November 1992, Overlaid with Trailing 12-Month Moving Averages (Source: Authors' Calculations)**

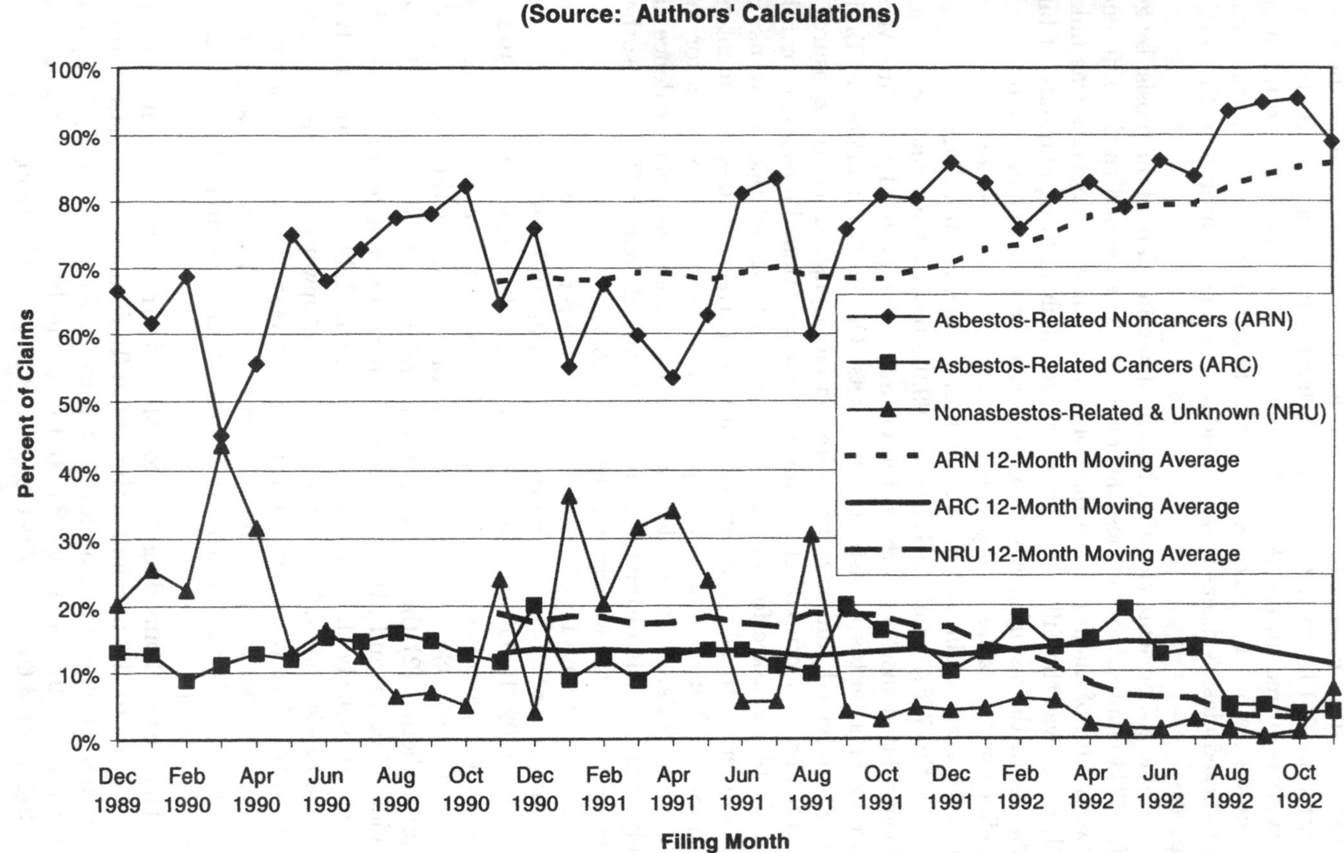

**Table 6.5: Incidence of Mesothelioma by Age, Date of Diagnosis, and Gender – Qualified Claims Against the Manville Trust, 1988-1992**

| | Males | | | | | | | Females | | | | | | |
|---|---|---|---|---|---|---|---|---|---|---|---|---|---|---|
| | Year of Diagnosis | | | | | | | Year of Diagnosis | | | | | | |
| Age | <1975 | 1975-1979 | 1980-1984 | 1985-1989 | 1990-1992 | Unknown | Total | <1975 | 1975-1979 | 1980-1984 | 1985-1989 | 1990-1992 | Unknown | Total |
| 0-4 | 0 | 0 | 0 | 0 | 0 | 0 | 0 | 0 | 0 | 0 | 0 | 0 | 0 | 0 |
| 5-9 | 0 | 0 | 0 | 0 | 0 | 0 | 0 | 0 | 0 | 0 | 0 | 0 | 0 | 0 |
| 10-14 | 0 | 0 | 1 | 0 | 0 | 0 | 1 | 0 | 0 | 0 | 0 | 0 | 0 | 0 |
| 15-19 | 0 | 0 | 0 | 0 | 0 | 0 | 0 | 0 | 0 | 0 | 0 | 0 | 0 | 0 |
| 20-24 | 0 | 1 | 0 | 1 | 0 | 0 | 2 | 0 | 0 | 1 | 2 | 0 | 0 | 3 |
| 25-29 | 0 | 2 | 4 | 3 | 1 | 1 | 11 | 0 | 1 | 1 | 0 | 0 | 0 | 2 |
| 30-34 | 3 | 7 | 8 | 12 | 4 | 0 | 34 | 0 | 1 | 3 | 3 | 1 | 0 | 8 |
| 35-39 | 1 | 6 | 14 | 33 | 8 | 1 | 63 | 0 | 0 | 5 | 4 | 1 | 0 | 10 |
| 40-44 | 8 | 14 | 41 | 66 | 13 | 4 | 146 | 0 | 2 | 2 | 9 | 2 | 0 | 15 |
| 45-49 | 17 | 48 | 90 | 126 | 47 | 6 | 334 | 0 | 4 | 6 | 12 | 7 | 0 | 29 |
| 50-54 | 22 | 55 | 183 | 220 | 82 | 17 | 579 | 2 | 3 | 9 | 23 | 3 | 2 | 42 |
| 55-59 | 26 | 105 | 320 | 403 | 162 | 35 | 1,051 | 0 | 6 | 22 | 28 | 7 | 1 | 64 |
| 60-64 | 18 | 109 | 392 | 637 | 209 | 79 | 1,444 | 1 | 3 | 16 | 46 | 15 | 1 | 82 |
| 65-69 | 10 | 78 | 381 | 756 | 277 | 87 | 1,589 | 0 | 1 | 20 | 48 | 17 | 4 | 90 |
| 70-74 | 4 | 47 | 307 | 629 | 238 | 102 | 1,327 | 1 | 2 | 9 | 25 | 12 | 1 | 50 |
| 75-79 | 3 | 11 | 132 | 378 | 152 | 88 | 764 | 0 | 2 | 5 | 21 | 7 | 1 | 36 |
| 80-84 | 0 | 4 | 39 | 137 | 64 | 53 | 297 | 0 | 0 | 2 | 9 | 5 | 3 | 19 |
| 85-89 | 0 | 0 | 9 | 39 | 17 | 37 | 102 | 0 | 0 | 1 | 1 | 1 | 1 | 4 |
| 90-94 | 0 | 0 | 1 | 4 | 3 | 8 | 16 | 0 | 0 | 0 | 0 | 0 | 0 | 0 |
| 95-99 | 0 | 0 | 0 | 0 | 0 | 6 | 6 | 0 | 0 | 0 | 0 | 0 | 1 | 1 |
| 100+ | 0 | 0 | 0 | 0 | 0 | 0 | 0 | 0 | 0 | 0 | 0 | 0 | 0 | 0 |
| Unknown | 0 | 0 | 5 | 8 | 2 | 6 | 21 | 0 | 0 | 0 | 1 | 0 | 1 | 2 |
| Total | 112 | 487 | 1,927 | 3,452 | 1,279 | 530 | 7,787 | 4 | 25 | 102 | 232 | 78 | 16 | 457 |

Source: Stallard and Manton (1993, Table 9).

Because a claim must be filed before a diagnosis is included in Table 6.5, it is plausible that the relative claim frequency for males ultimately would reach 42.0% for 1985-1989 and 25.5% for 1980-1984. The upper limit of the relative claim frequency could range from 54% (Walker, 1982) to 75% (McDonald and McDonald, 1981), based on estimates of the fraction of mesothelioma cases due to occupational exposure to asbestos. If Walker's (1982) estimate were correct, then about 78% of such cases would file a claim against the Manville Trust. McDonald and McDonald's (1981) estimate would reduce this estimate to 56%. These estimates can be compared with Johns-Manville asbestos market share estimates in the range 25-40%, averaging near 30% (Hersch, 1992). The claim filing rates are two to three times the market share estimates, and this supports Hersch's argument that the Manville Trust's experience serves "as an effective proxy for industry-wide personal injury liability" (Hersch, 1992, p. 6).

The female claim filing rate of 9.28% in 1985-1989 implies that about 17.4% of mesothelioma cases among females were due to occupational exposure to asbestos, using a 5% ultimate increase in claims diagnosed in 1985-1989 and assuming this to be 56% of all mesotheliomas due to occupational exposure. In other words, of the 2500 female mesothelioma cases diagnosed in 1985-1989, about 2065 cases were not associated with occupational exposures. Because of the lack of an occupational association and the lack of a linkage with tobacco usage, it is reasonable to expect a corresponding number of nonoccupational mesothelioma cases among males. With an estimated total of 8637 male mesothelioma diagnoses in 1985-1989, these assumptions would imply that about 6572 (76.1%) cases would be due to occupational exposure to asbestos. Similar computations indicate that occupational exposures could have been responsible for 68.5% of male mesothelioma diagnoses in 1975-1979 and 73.7% in 1980-1984. These results are broadly consistent with McDonald's and McDonald's (1981) estimate of 75% and substantially higher than Walker's (1982) estimate of 54%.

The projection model required the distribution of age at start of asbestos exposure and date of start of asbestos exposure by gender for persons with incidence (diagnosis) of mesothelioma in 1975-1979, 1980-1984, or 1985-1989. These persons corresponded to the claimants in Table 6.5 with known age and date of diagnosis. Only claimants for whom the date of first exposure was known were counted. If the date of first exposure fell in the same 5-year period as the date of diagnosis of mesothelioma, then the claim was deleted from the exposure tabulation. The diagnosis information was retained, but the initiation date for start of asbestos exposure was regarded as unreliable. In the exposure distribution (Table 6.6), age at start of exposure was defined as the difference between date of first exposure and date of birth. For 1975-1979, 92.2% of males in Table 6.5 were included in the exposure matrix in Table 6.6; for 1980-1984, 93.7% were included; and for 1985-1989, 93.5% were included.

The exposure matrices for males in Table 6.6 were input into the projection model under two assumptions. First, we assumed that the relative

**Table 6.6: Observed Distribution of Start of Asbestos Exposure by Age at Start of Exposure, Date of Start of Exposure, and Date of Diagnosis of Mesothelioma – Qualified Male Mesothelioma Claims Against the Manville Trust, 1988-1992**

| | Date of First Exposure to Asbestos | | | | | | | | | | | | | | | |
|---|---|---|---|---|---|---|---|---|---|---|---|---|---|---|---|---|
| Age | 1915-1919 | 1920-1924 | 1925-1929 | 1930-1934 | 1935-1939 | 1940-1944 | 1945-1949 | 1950-1954 | 1955-1959 | 1960-1964 | 1965-1969 | 1970-1974 | 1975-1979 | 1980-1984 | 1985-1989 | Total |
| | Diagnosis in 1975-1979 | | | | | | | | | | | | | | | |
| 15-19 | 2 | 2 | 10 | 9 | 16 | 34 | 14 | 5 | 2 | 1 | 2 | 1 | 0 | 0 | 0 | 98 |
| 20-24 | 0 | 0 | 9 | 6 | 19 | 45 | 19 | 20 | 4 | 2 | 5 | 0 | 0 | 0 | 0 | 129 |
| 25-29 | 1 | 0 | 0 | 3 | 10 | 44 | 15 | 13 | 10 | 2 | 0 | 0 | 0 | 0 | 0 | 98 |
| 30-34 | 0 | 0 | 0 | 0 | 4 | 26 | 9 | 10 | 4 | 1 | 0 | 0 | 0 | 0 | 0 | 54 |
| 35-39 | 0 | 0 | 0 | 0 | 1 | 13 | 9 | 5 | 4 | 1 | 1 | 1 | 0 | 0 | 0 | 35 |
| 40-44 | 0 | 0 | 0 | 0 | 1 | 7 | 1 | 4 | 3 | 1 | 2 | 0 | 0 | 0 | 0 | 19 |
| 45-49 | 0 | 0 | 0 | 0 | 0 | 1 | 0 | 0 | 1 | 5 | 2 | 1 | 0 | 0 | 0 | 10 |
| 50+ | 0 | 0 | 0 | 0 | 0 | 0 | 0 | 1 | 0 | 2 | 2 | 1 | 0 | 0 | 0 | 6 |
| Total | 3 | 2 | 19 | 18 | 51 | 170 | 67 | 58 | 28 | 15 | 14 | 4 | 0 | 0 | 0 | 449 |
| | Diagnosis in 1980-1984 | | | | | | | | | | | | | | | |
| 15-19 | 6 | 13 | 38 | 22 | 57 | 163 | 61 | 50 | 21 | 9 | 2 | 1 | 0 | 0 | 0 | 443 |
| 20-24 | 1 | 8 | 13 | 24 | 79 | 161 | 85 | 57 | 25 | 9 | 5 | 3 | 1 | 0 | 0 | 471 |
| 25-29 | 0 | 1 | 4 | 11 | 51 | 183 | 74 | 58 | 18 | 10 | 10 | 1 | 0 | 0 | 0 | 421 |
| 30-34 | 0 | 0 | 0 | 2 | 23 | 116 | 42 | 32 | 21 | 7 | 3 | 0 | 0 | 0 | 0 | 246 |
| 35-39 | 0 | 0 | 0 | 1 | 7 | 45 | 26 | 30 | 16 | 8 | 0 | 1 | 0 | 0 | 0 | 134 |
| 40-44 | 0 | 0 | 0 | 0 | 1 | 6 | 9 | 17 | 7 | 7 | 8 | 0 | 0 | 0 | 0 | 55 |
| 45-49 | 0 | 0 | 0 | 0 | 0 | 4 | 0 | 3 | 6 | 5 | 5 | 2 | 0 | 0 | 0 | 25 |
| 50+ | 0 | 0 | 0 | 0 | 0 | 1 | 0 | 0 | 2 | 0 | 3 | 2 | 2 | 0 | 0 | 10 |
| Total | 7 | 22 | 55 | 60 | 218 | 679 | 297 | 247 | 116 | 55 | 36 | 10 | 3 | 0 | 0 | 1,805 |
| | Diagnosis in 1985-1989 | | | | | | | | | | | | | | | |
| 15-19 | 5 | 15 | 49 | 39 | 108 | 316 | 120 | 89 | 58 | 22 | 20 | 4 | 0 | 0 | 0 | 845 |
| 20-24 | 0 | 9 | 18 | 37 | 134 | 345 | 175 | 120 | 74 | 38 | 21 | 4 | 3 | 1 | 0 | 979 |
| 25-29 | 0 | 1 | 1 | 11 | 72 | 273 | 143 | 99 | 66 | 22 | 20 | 5 | 5 | 0 | 0 | 718 |
| 30-34 | 0 | 0 | 0 | 1 | 19 | 146 | 65 | 67 | 28 | 18 | 19 | 1 | 1 | 1 | 0 | 366 |
| 35-39 | 0 | 0 | 0 | 1 | 7 | 47 | 27 | 44 | 20 | 16 | 14 | 1 | 1 | 0 | 0 | 178 |
| 40-44 | 0 | 0 | 0 | 0 | 0 | 15 | 11 | 20 | 13 | 17 | 10 | 2 | 2 | 0 | 0 | 90 |
| 45-49 | 0 | 0 | 0 | 0 | 0 | 0 | 3 | 4 | 2 | 8 | 7 | 3 | 1 | 0 | 0 | 28 |
| 50+ | 0 | 0 | 0 | 0 | 0 | 0 | 0 | 0 | 6 | 1 | 11 | 2 | 2 | 2 | 0 | 24 |
| Total | 5 | 25 | 68 | 89 | 340 | 1,142 | 544 | 443 | 267 | 142 | 122 | 22 | 15 | 4 | 0 | 3,228 |

Source: Stallard and Manton (1993, Table 10).

**Table 6.7: Credibility and Event Counts**

| Maximum Acceptable Departure from the Expected Count | Probability of Observed Count Falling Within the Acceptable Range | | |
|---|---|---|---|
| | 90% | 95% | 99% |
| | Minimum Required Expected Count | | |
| +/–2.5% | 4,329 | 6,146 | 10,616 |
| +/–5.0% | 1,082 | 1,537 | 2,654 |
| +/–7.5% | 481 | 683 | 1,180 |
| +/–10% | 271 | 384 | 663 |
| +/–20% | 68 | 96 | 166 |
| +/–30% | 30 | 43 | 74 |
| +/–40% | 17 | 24 | 41 |
| +/–50% | 11 | 15 | 27 |

Source: Based on Longley-Cook (1962).

distribution of age, date of start of asbestos exposure, and date of diagnosis of mesothelioma for those claimants missing one or more dates was the same as for those with all three dates known. Second, we assumed that the relative distribution of age, date of start of asbestos exposure, and date of diagnosis of mesothelioma for the national incidence represented in Table 6.3 was the same as for the subset of the mesothelioma cases who also were claimants against the Manville Trust.

Because the exposure matrices in Table 6.6 were used to estimate historical numbers of workers exposed to asbestos, we smoothed these estimates to remove variability due to the stochasticity of small frequencies. As a first approximation, the standard errors of the numbers in Table 6.6 are approximately equal to the square roots of the expected counts in each cell. More precise calculations of probable errors are presented in Table 6.7 (Longley-Cook, 1962). An expected count of 1082 cases implies that 10% of observed counts would deviate by more than 5% from the expected count. With an expected count of 271 cases, 10% of observed counts would deviate by more than 10% from the expected count. The deviation increases markedly as the expected count decreases. Because most cell entries in Table 6.6 are below 271, they have low credibility and would benefit from smoothing (Longley-Cook, 1962).

In smoothing, however, we did not want to move too far from the observed counts at each age and date. We adopted the following rule: If $E_{A,Y}^{Y_1}$ denotes the entry in Table 6.6 for age group $A$ and calendar period $Y$ for diagnosis in calendar period $Y_1$, then the "smoothed" estimator was the average of the five adjacent cells centered at $(A, Y)$, with double weighting given to cell $(A, Y)$.

Hence,

$$\hat{E}^{Y_1}_{A,Y} = \frac{1}{6}\left( \sum_{B=-1}^{1} E^{Y_1}_{A+B,Y} + \sum_{Z=-1}^{1} E^{Y_1}_{A,Y+Z} \right).$$

For cells on the edges of Table 6.6, we used

$$\hat{E}^{Y_1}_{A,Y} = \begin{cases} \frac{1}{4}\left( E^{Y_1}_{A,Y} + \sum_{B=-1}^{1} E^{Y_1}_{A+B,Y} \right) & \text{for sides} \\ \frac{1}{4}\left( E^{Y_1}_{A,Y} + \sum_{Z=-1}^{1} E^{Y_1}_{A,Y+Z} \right) & \text{for top or bottom.} \end{cases}$$

For corner cells, no smoothing was performed. Hence,

$$\hat{E}^{Y_1}_{A,Y} = E^{Y_1}_{A,Y} \qquad \text{for corners.}$$

Under this smoothing procedure, the total number of exposed claimants was closely matched; most of the embedded zero cells (i.e., above the main diagonal) were eliminated, and only modest shifts of cases counts over age or time (no more than $\pm 5$ years) occurred.

Once smoothing of Table 6.6 was completed, all exposures initiating in 1975-1979 or later were zeroed out for consistency with step 9.

### 6.5.4 Step 4: Normalization of Exposure

The projection model assumed that the relative distribution of age, date of start of asbestos exposure, and date of diagnosis of mesothelioma for the national incidence data in Table 6.3 was the same as the smoothed distribution of mesothelioma cases who were claimants against the Manville Trust. The relative distribution is defined as the fraction of mesothelioma diagnoses in a baseline period (i.e., 1975-1979, 1980-1984, or 1985-1989) whose exposure to asbestos initiated at each past age and date. These fractions were obtained from the smoothed exposure counts based on Table 6.6 by "normalization."

Specifically, let $A_1$ be any age group in Table 6.3 and $Y_1$ be any baseline period 1975-1979, 1980-1984, or 1985-1989. Define the sums, $M_{A_1,Y_1}$, from the smoothed version of Table 6.6 by adding entries from right to left, moving backward and upward along each diagonal; that is,

$$M_{A_1,Y_1} = \sum_{T=1} \hat{E}^{Y_1}_{A_1-T,Y_1-T},$$

where the upper limit of $T$ was such that subscripts $A_1 - T$ and $Y_1 - T$ are both within Table 6.6. The normalized distribution of exposure was then obtained by dividing each entry in the smoothed version of Table 6.6 by the corresponding sum for the diagonal that contains the entry. For example, for 1975-1979, $Y_1 = 13$ and $Y = 1, \ldots, 12$. Hence,

$$F^{1975-79}_{A,Y} = \hat{E}^{1975-79}_{A,Y} / M_{A+(13-Y),13}.$$

Similarly,

$$F_{A,Y}^{1980-84} = \hat{E}_{A,Y}^{1980-84}/M_{A+(14-Y),14}$$

and

$$F_{A,Y}^{1985-89} = \hat{E}_{A,Y}^{1985-89}/M_{A+(15-Y),15}.$$

These definitions ensure that the equality

$$1 = \sum_{T=1} F_{A_1-T,Y_1-T}^{Y_1}$$

is satisfied, where the range of $T$ is the same as above. These computations are essentially the same as in Walker's (1982) Task 1e.

### 6.5.5 Step 5: Intensity of Exposure

The next step was to stratify the national estimates of mesothelioma incidence counts in Table 6.3 by level of exposure to asbestos. Walker (1982, Task 1b) concluded that the "best" estimate of the proportion of male mesothelioma cases with a "definite" or "probable" history of occupational exposure to asbestos was 54%, with 20% having "definite" exposure. We defined a three-level exposure variable with 46% low, 34% medium, and 20% high. The terms low, medium, and high describe confidence that these persons had a documentable history of occupational exposure.

It is difficult to evaluate these estimates, or even to know precisely what is a "documentable" history of asbestos exposure. Walker's (1982) rationale for assuming an equivalence between the 20% with definite exposure to asbestos and the 20% assumed to be heavily exposed is not convincing (see Section 4.4.1b).

It may be that Walker (1982) was distinguishing subgroups who could provide documentable evidence of injury by products of the Johns-Manville Corporation or other asbestos companies. In this case, the 54% might be interpreted as the upper limit of the fraction of male mesothelioma cases that could provide evidence of the type required in the Manville Trust's POC Forms. When we compared Tables 6.3 and 6.5, we saw that the implied fraction of incident cases filing such claims was 9.5% in 1975-1979, 24.9% in 1980-1984, and 40.0% in 1985-1989 – all of which were below 54%. Thus, this interpretation is consistent with the available data, although, as indicated in step 3, the 75% estimate of McDonald and McDonald (1981) appears more plausible as an upper limit and is consistent with the lack of occupational exposure for the majority of female mesothelioma cases.

The issue of which upper-limit estimate, 54% or 75%, is better was not critical to our model development because we retained all three levels of exposure, as defined by Walker (1982). By contrast, Walker (1982) ignored the 46% with "low exposure" on the basis that any asbestos exposure they might have had was not documentable.

## 6.6 Model Estimation

Steps 6-10 involved manipulation of the input data to yield estimates of the number of workers exposed to asbestos by age and date of start of exposure, using IWEs to count these workers. In this section, we describe the manipulations and discuss our assumptions.

### 6.6.1 Step 6: Stratification of National Estimates of Mesothelioma Incidence Counts, by Level of Asbestos Exposure

The mesothelioma incidence counts for males in Table 6.3 were multiplicatively allocated to the low-, medium-, and high-exposure groups using Walker's (1982) exposure fractions 0.46, 0.34, and 0.20, respectively.

Stratification of the mesothelioma projections by exposure level had no effect on the final projected values because they were simply summed and then rescaled in the claim forecasts to reflect the Trust's experience. However, Walker (1982, p. 14) cited evidence presented by Selikoff et al. (1979) (see Table 2.1) that the most heavily exposed workers (i.e., 17,800 asbestos insulation workers in North America) had 37% excess mortality compared with the general population. More recent data cited by Selikoff and Seidman (1991) revised this estimate to 43% for the 20-year period January 1, 1967 to December 31, 1986 (the 37% excess was for the first 10 years), and Table 2.1 shows the excess to have been 49% for the second 10-year period January 1, 1977 to December 31, 1986. The "excess" included deaths due to mesothelioma. The determination that death was due to mesothelioma was performed twice using two separate criteria: one based solely on the death certificate, and the other based on "best evidence." Under Selikoff's best-evidence criterion, the determination was based on review of autopsy, surgical, and clinical material, including histopathology review and confirmation. Under the death-certificate criterion, the determination was based only on information recorded on the death certificate. Selikoff's best-evidence criterion yielded approximately 2.5 times as many mesothelioma deaths as the death-certificate criterion in this cohort of insulation workers. When mesothelioma was excluded, the 20-year excess was 30.1% or 38.1%, depending on the criterion used to assign mesothelioma. The nonmesothelioma excess for the first 10 years was 26.3% or 30.6% depending on the criterion; for the second 10 years, the corresponding excess was 33.6% or 45.0%, respectively. The nonmesothelioma excess appeared to increase with each decade of follow-up.

In view of this, we retained the estimate of 37% excess mortality cited by Walker (1982), but applied it only to nonmesothelioma causes of death for the high-exposure group. Nonmesothelioma mortality rates in Table 6.2 were multiplied by 1.37 to account for this excess, and the high-exposure population was projected separately from the low- and medium-exposure populations.

Walker (1982) projected the medium and high exposures (54% of cases) as separate groups that were subject to the same mortality rates. To do this, he multiplied his rates for total mortality by 1.14, based on the assumption that the average excess in the two groups was 14% (Walker 1982, Task 1e); that is, the excess in the medium group was 0%, the excess in the high group was 37%, and, therefore,

$$0.14 = \left(\frac{34}{54}\right) \times 0.0 + \left(\frac{20}{54}\right) \times 0.37.$$

This is mathematically incorrect because it assumes that there is one actual worker for each IWE in the medium-exposure group. However, Walker (1982) indicated that there may be 2.6, 5, or 10 workers per IWE in the medium-exposure group, and these would require additional adjustments to the weights in the above formula (see Section 4.4.1e). To avoid these problems, we projected each exposure group separately.

### 6.6.2 Step 7: Estimation of the IWE Population Exposed to Asbestos Prior to 1975 by Level of Asbestos Exposure

The next step was to estimate the insulation-worker equivalent number of persons in the past exposed to asbestos, stratified by level of exposure as defined in step 6. This was the critical step in model calibration. This exposure underlies the national incidence of mesothelioma in Table 6.3. Here, we present the structural relationships between the (unknown) original exposure counts and the known incidence counts in Table 6.3.

Let $A_1$ be an age group in Table 6.3 and let $Y_1$ be one of the baseline periods 1975-1979, 1980-1984, or 1985-1989. Let $M_{A_1,Y_1}$ be an entry in Table 6.3 denoting the number of cases of mesothelioma diagnosed at age $A_1$ in year $Y_1$. Let $l$ be the index of exposure level defined in step 6 and let $f_l = 0.46$, 0.34, or 0.20, depending on $l$. Following step 6, let $M^l_{A_1,Y_1} = f_l \times M_{A_1,Y_1}$ be the number of cases of mesothelioma diagnosed at age $A_1$ in calendar period $Y_1$, among persons with level $l$ = low, medium, or high exposure. Let $E^{Y_1,l}_{A,Y} = E^{Y_1,l}_{A_1-T,Y_1-T}$ be the number of persons included in $M^l_{A_1,Y_1}$ whose exposure to asbestos started at age $A = A_1 - T$ in calendar period $Y = Y_1 - T$ at level $l$ = low, medium, or high exposure. This number was estimated as the product of the normalized factors from step 4, the exposure fractions from step 5, and the incidence counts from step 2, or

$$E^{Y_1,l}_{A,Y} = F^{Y_1}_{A,Y} \times f_l \times M_{A_1,Y_1}, \text{ where } A_1 = A + Y_1 - Y.$$

The sum of $E^{Y_1,l}_{A,Y}$ along diagonals of the matrix yields $M^l_{A_1,Y_1}$, so

$$M_{A_1,Y_1} = \sum_l \sum_{T=1} E^{Y_1,l}_{A_1-T,Y_1-T}.$$

As earlier, the upper limit of $T$ is such that subscripts $A_1 - T$ and $Y_1 - T$ were both within Table 6.6. It follows that all persons in $E^{Y_1,l}_{A,Y}$ were alive in calendar period $Y_1$. The normalized exposure fractions, $F^{Y_1}_{A,Y}$, were assumed constant over level of exposure. Walker's method does not allow these fractions to be estimated specific to exposure level. This restriction is relaxed in Chapter 8 where exposure fractions were estimated separately for occupational groups with different average exposure levels.

We now consider estimation of the original number of persons exposed in calendar period $Y$, not just the subset who were still alive in $Y_1$.

Let $N^l_{A,Y}$ be the IWE estimate of the number of persons whose exposure to asbestos started at age $A$ in calendar period $Y$ at level $l$. Let $Q^{T,l}_{A,Y}$ be the probability that a person included in $N^l_{A,Y}$ was diagnosed with mesothelioma during the $T$th quinquennial period following period $Y$. These assumptions imply that $E^{Y_1,l}_{A,Y}$ is obtained from $N^l_{A,Y}$ as

$$N^l_{A,Y} \times Q^{(Y_1-Y),l}_{A,Y} = E^{Y_1,l}_{A,Y}$$

or

$$N^l_{A,Y} \times Q^{T,l}_{A,Y} = E^{Y+T,l}_{A,Y}.$$

For $T = Y_1 - Y$, the two equations are the same. We refer to the first equation as the backward projection equation and to the second equation as the forward projection equation, or the forecasting equation.

For a single baseline period $Y_1$, the backward projection equation can be solved for the original IWE exposure counts, yielding

$$N^l_{A,Y} = E^{Y_1,l}_{A,Y} / Q^{(Y_1-Y),l}_{A,Y},$$

provided that $Q^{(Y_1-Y),l}_{A,Y}$ is known. This is the projection calibration equation proposed by Peto et al. (1981) (see Section 4.2) and used by Walker (1982). Updating the Walker projection could recognize that $Y_1$ is changed from 1975-1979 to 1980-1984 or 1985-1989.

For multiple baseline periods 1975-1979, 1980-1984, and 1985-1989, the following identity is obtained from the backward projection equation:

$$N^l_{A,Y} \times \sum_{Y_1=13}^{15} Q^{(Y_1-Y),l}_{A,Y} = \sum_{Y_1=13}^{15} E^{Y_1,l}_{A,Y},$$

where the summation on $Y_1$ is for the three baseline periods $13 \equiv$ 1975-1979, $14 \equiv$ 1980-1984, and $15 \equiv$ 1985-1989. Solving this identity for $N^l_{A,Y}$ yields

$$N^l_{A,Y} = \frac{\sum_{Y_1=13}^{15} E^{Y_1,l}_{A,Y}}{\sum_{Y_1=13}^{15} Q^{(Y_1-Y),l}_{A,Y}}.$$

This exemplifies the general form of the exposure calibration equation for multiple baseline periods.

Application of the exposure calibration equation requires that we estimate $Q^{T,l}_{A,Y}$. This function has two components. First, a person whose exposure starts at age group $A$ in calendar period $Y$, at exposure level $l$, must survive both mesothelioma incidence and nonmesothelioma causes of death for $T$ quinquennial periods. We write the $T$-period survival as

$$N^{T,l}_{A,Y} = N^{l}_{A,Y} \times S^{T,l}_{A,Y},$$

where

$$S^{T,l}_{A,Y} = \exp\left\{-\left(\sum_{N=0}^{T-1} h_{A+N,Y+N}\right) \times r_l - \int_0^T I_N\, dN\right\}$$

where $r_l = 1$, 1, or 1.37 depending on whether $l$ is low, medium, or high exposure (see step 6), $t_T = 5T$ is the time since first exposure (in years), and $I_T$ is the mesothelioma incidence rate at time $T$ in quinquennial time units.

Second, Peto et al. (1982) analyzed data described by Selikoff et al. (1979) on mortality in insulation workers as a function of elapsed time since first exposure to asbestos. The determination of mesothelioma deaths was based on Selikoff's best-evidence criterion. The model fitted by Peto et al. (1982) was used by Walker (1982) to express the hazard rate for mesothelioma incidence in the form of a power, $k$, of time since first exposure in years; that is

$$I_t = \left(4.37 \times 10^{-8}\right) \times t^{3.2} = bt^k,$$

where $I_t$ is the mesothelioma incidence rate at $t$. The constants $b = 4.37 \times 10^{-8}$ and $k = 3.2$ are parameters of the Weibull function. Converting from annual to quinquennial time units, we obtain

$$I_T = \left(3.768 \times 10^{-5}\right) \times T^{3.2},$$

with parameter $B = b \times 5^{4.2} = 3.768 \times 10^{-5}$, and for the integrated hazard rate

$$\int_0^T I_N\, dN = B\, T^{k+1}/(k+1)$$
$$= 8.97 \times 10^{-6}\, T^{4.2}.$$

Combining the survival and incidence functions, under the assumption that the instantaneous diagnosis rate approximates the quinquennial diagnosis rate, we obtain

$$Q^{T,l}_{A,Y} = S^{T,l}_{A,Y} \times I_T.$$

The original IWE number of persons exposed is the sum over levels of exposure of $N^{l}_{A,Y}$; that is,

$$N_{A,Y} = \sum_l N^l_{A,Y}$$
$$= \sum_l \left\{ \frac{\sum_{Y_1=13}^{15} E^{Y_1,l}_{A,Y}}{\sum_{Y_1=13}^{15} S^{(Y_1-Y),l}_{A,Y} \times I_{Y_1-Y}} \right\}.$$

Walker (1982) described $N_{A,Y}$ as a measure of IWEs because the estimates were calibrated using mesothelioma incidence among insulation workers. Walker (1982) noted that the actual size of $N_{AY}$ was not critical to the projections because each estimate, $N^l_{A,Y}$, is an intermediate quantity that can be eliminated; for example, the forward projection equation can be revised to

$$E^{Y+T,l}_{A,Y} = E^{Y_1,l}_{A,Y} \times Q^{T,l}_{A,Y} / Q^{(Y_1-Y),l}_{A,Y},$$

which shows that only the ratios of the $Q$'s matter in the projections.

When exposure calibration is based on multiple baseline periods such as 1975-1979, 1980-1984, and 1985-1989, the forward projection equation can be written as

$$E^{Y+T,l}_{A,Y} = Q^{T,l}_{A,Y} \times \frac{\sum_{Y_1=13}^{15} E^{Y_1,l}_{A,Y}}{\sum_{Y_1=13}^{15} Q^{(Y_1-Y),l}_{A,Y}},$$

which shows again that only the ratios of the $Q$'s matter in the projections. The IWE exposure estimates cancel out of the forward projection equation.

To gain further insight into the model, we considered the forward projection equation for a single-period calibration, with the ratio of the $Q$'s simplified as follows:

$$E^{Y+T,l}_{A,Y} = E^{Y_1,l}_{A,Y} \times Q^{T,l}_{A,Y} / Q^{T_1,l}_{A,Y}$$
$$= E^{Y_1,l}_{A,Y} \exp\left\{ -\sum_{N=T_1}^{T-1} h_{A+N,Y+N} \times r_l - \int_{T_1}^{T} I_N \, dN \right\},$$

where $T_1 = Y_1 - Y$. We were interested in evaluating the relative change in the projection of $E^{Y+T,l}_{A,Y}$ induced by a large relative change in $B$. This allowed us to assess the potential impact of the use of IWEs as our basic measure of exposure. Because insulators had relatively high exposures, we were interested in the maximum impact of reducing $B$. A 100% reduction in $B$ (yielding $B = 0$) results in a relative increase in $E^{Y+T,l}_{A,Y}$ by the multiplicative factor,

$$\exp\left\{ \frac{B}{k+1} \left[ T^{k+1} - T_1^{k+1} \right] \right\}.$$

To evaluate this function, we assumed $T_1 = 4$ (or $t$=20) so that the earliest diagnoses would be at least 20 years after onset of exposure. For $T = 6$ (or $t$=30), the factor is 1.014; for $T = 8$, the factor is 1.054; for $T = 10$, the factor is 1.149. In other words, 30 years into the projection, the maximum increase is less than 15%. The following 30 years exhibit large increases, however, with $T = 16$ (or $t$=80) having a multiplier of 2.775. The doubling point occurs at $t = 73$ years, or 53 years into the projection. These calculations suggested that the IWE assumption may induce significant bias in long-range asbestos projections if a substantial number of claims occur more than 30 years beyond the baseline period.

The estimates of $N_{A,Y}$ are presented in Table 6.8, for the combined baseline periods 1975-1979, 1980-1984, and 1985-1989. Peto et al. noted that their data showed "no evidence of any change in extent of exposure since 1945" (Peto et al., 1981, p. 59). However, Table 6.8 shows total IWE exposures of about 75,300 in 1955-1959, 81,100 in 1960-1964, 124,000 in 1965-1969, and 115,800 in 1970-1974. These compare with 104,600 in 1945-1949 and 89,900 in 1950-1954. The period 1945-1974 can be characterized by declining exposure initiations through 1955-1959, followed by increases through 1970-1974, contrary to Peto et al. (1981).

Three types of evidence needed to be considered in evaluating the results in Table 6.8. First, Selikoff (1981, Table 2-12) estimated the numbers of new entrants to industries and occupations with asbestos exposures as 4.35 million in 1950-1959, 4.91 million in 1960-1969, and 5.48 million in 1970-1979 – an overall increase of 26%. The corresponding IWE estimates from Table 6.8 were about 165,000, 205,000, and 231,000 – an overall increase of 40%. Although both sets of estimates implied substantial increases for this period, Selikoff's increases were substantially smaller than the IWE increases.

Second, more detailed analysis of Selikoff's (1981, Table 2-12) estimates showed that the increases were primarily attributable to two large occupational groups with relatively low levels of exposure: automobile maintenance and construction trades. When these two groups were removed from the totals, the estimates were reduced to 1.80 million, 1.76 million, and 1.73 million for the three decades – reversing the strong upward trend to a modest 4% decrease for this period. Because these two occupational groups had substantially reduced exposures, the IWE increases implied by Selikoff's results must fall somewhere in the range 4-26%, exaggerating the discrepancy noted earlier.

Third, Walker (1982, p. 14) argued that the calibration of exposure in step 7 breaks down when applied to first exposures less than 20 years prior to the baseline. In his projection, special consideration had to be given to first exposures in the period 1955-1974. Walker (1982) used data on the distribution of age at first employment in an asbestos-related occupation or industry from a survey by Elrick and Lavidge. Although the results of using these data were illustrated in Walker's (1982) report, the details were not provided.

The sensitivity analysis of Walker's projection in Section 5.4.3 indicated that the IWE exposure estimates for the youngest cohorts were subject to significant fluctuations due to stochastic variability of the SEER mesothelioma counts. Our results in Table 6.8 were based on approximately 15 times more data (i.e., 1975-1989 vs. 1977), which improved the calibration for the younger cohorts, but still did not completely eliminate the fluctuations. For example,

**Table 6.8: Estimated Number of Insulation Worker Equivalents Exposed to Asbestos by Age at Start of Exposure and Date of Start of Exposure**

| | Date of Start of Asbestos Exposure | | | | | | | | | | | | |
|---|---|---|---|---|---|---|---|---|---|---|---|---|---|
| Age | 1915-1919 | 1920-1924 | 1925-1929 | 1930-1934 | 1935-1939 | 1940-1944 | 1945-1949 | 1950-1954 | 1955-1959 | 1960-1964 | 1965-1969 | 1970-1974 | Total |
| 15-19 | 3,482 | 4,657 | 4,462 | 4,633 | 9,058 | 12,714 | 14,417 | 9,319 | 8,755 | 15,032 | 20,250 | 18,310 | 125,088 |
| 20-24 | 4,209 | 4,539 | 6,067 | 6,203 | 13,044 | 23,124 | 15,426 | 16,026 | 11,451 | 10,292 | 26,207 | 20,748 | 157,334 |
| 25-29 | 8,848 | 3,351 | 4,571 | 6,755 | 16,796 | 28,563 | 19,670 | 13,787 | 13,814 | 9,222 | 11,674 | 18,172 | 155,223 |
| 30-34 | 0 | 389 | 1,219 | 4,940 | 18,362 | 29,722 | 17,580 | 12,926 | 9,189 | 8,428 | 7,989 | 7,122 | 117,864 |
| 35-39 | 0 | 0 | 464 | 4,142 | 15,782 | 30,482 | 14,826 | 12,378 | 9,090 | 6,856 | 10,180 | 5,189 | 109,389 |
| 40-44 | 0 | 0 | 0 | 5,870 | 17,741 | 24,793 | 13,101 | 10,526 | 8,403 | 9,153 | 9,875 | 10,640 | 110,101 |
| 45-49 | 0 | 0 | 0 | 0 | 16,015 | 21,811 | 5,329 | 7,464 | 6,921 | 9,493 | 13,124 | 8,907 | 89,063 |
| 50-54 | 0 | 0 | 0 | 0 | 0 | 13,380 | 2,783 | 4,235 | 5,691 | 8,270 | 10,016 | 10,468 | 54,843 |
| 55-59 | 0 | 0 | 0 | 0 | 0 | 0 | 1,439 | 3,268 | 1,954 | 3,988 | 9,106 | 10,110 | 29,865 |
| 60-64 | 0 | 0 | 0 | 0 | 0 | 0 | 0 | 0 | 0 | 343 | 4,845 | 6,180 | 11,368 |
| 65-69 | 0 | 0 | 0 | 0 | 0 | 0 | 0 | 0 | 0 | 0 | 697 | 0 | 697 |
| Total | 16,540 | 12,936 | 16,784 | 32,543 | 106,797 | 184,588 | 104,570 | 89,929 | 75,268 | 81,075 | 123,961 | 115,844 | 960,834 |
| | Cumulative Total | | | | | | | | | | | | |
| | 960,835 | 944,295 | 931,359 | 914,576 | 882,033 | 775,235 | 590,648 | 486,077 | 396,149 | 320,881 | 239,806 | 115,844 | |

Source: Stallard and Manton (1993, Table 13).

the youngest cohort, aged 15-19 in 1970-1974, yielded 18,310 IWEs, based on just 3 mesothelioma diagnoses in the SEER data (30.79 cases in Table 6.3). The third cohort, aged 15-19 in 1960-1964, yielded 59,411 IWEs over the 3 periods 1960-1964, 1965-1969, and 1970-1974, based on 31 mesothelioma diagnoses in the SEER data (312.46 cases in Table 6.3). The intermediate cohort, aged 15-19 in 1965-1969, yielded 40,998 IWEs over the periods 1965-1969 and 1970-1974, based on 13 mesothelioma diagnoses in the SEER data (130.58 cases in Table 6.3). In evaluating the age patterns of IWEs in Table 6.8, we saw that the IWEs for younger cohorts had low credibility because of the small number of SEER cases underlying each computation.

Taken together, the above considerations supported some modifications to Table 6.8 in the period 1955-1974, but did not indicate the precise form that these modifications should take. To deal with this, we developed a three-step approach (steps 8-10) that continued to reproduce the aggregate mesothelioma counts in the baseline period 1975-1989 while sacrificing some of the age detail.

### 6.6.3 Step 8: Adjustments to Exposure During 1955-1974, by Level of Asbestos Exposure

In step 8, we introduced the assumption that the absolute number and the age distribution of new exposures of IWEs over the period 1955-1974 were unchanged from their values in 1945-1954. To do this, we replaced the IWE exposed population counts from step 7 for the periods 1955-1959, 1960-1964, 1965-1969, and 1970-1974 with the age-specific average counts for 1945-1949 and 1950-1954. This was done separately by level of asbestos exposure. This adjustment had only a minor effect on the estimated total number of exposed IWEs (−1.8%, 1955-1974). However, for the later periods, it provided a much smoother progression of counts over adjacent age groups and calendar periods – especially for the youngest three cohorts whose IWE estimates appeared to be substantially higher than for older cohorts at the same ages and, at the same time, were subject to concerns about their credibility.

### 6.6.4 Step 9: Adjustments to Reflect Improvements in the Workplace During 1960-1974, by Level of Asbestos Exposure

Walker (1982, p. 15) indicated that significant reductions in the amounts of ambient asbestos faced by workers in asbestos-using industries probably had begun by 1965, and by 1975 the exposures had been essentially eliminated. To reflect these trends, we used Walker's (1982) discount factors of 10% for 1960-1964; 50% for 1965-1969; 75% for 1970-1974; and 100% for 1975 and later.

To account for these reductions in the projection model, we could have assumed that the hazard rate for mesothelioma incidence was reduced proportionally, so that $b$ or $B$ was 10% smaller for exposures that initiated in 1960-1964, 50% smaller for 1965-1969, 75% smaller for 1970-1974, and 100%

smaller thereafter (yielding $B = 0$). Alternatively, Walker (1982) assumed that one could keep $b$ or $B$ at fixed levels but have the number of workers exposed reduced proportionally, so that the estimated IWE number of workers with first exposure to asbestos in 1965-1969 was 50% smaller than estimated in step 8. Walker (1982) argued, that because the estimates were not actual counts of people but were insulation-worker equivalents, reducing their value by, say, 50% in the second method was equivalent to reducing $b$ or $B$ by 50% in the first method, but keeping the count of workers fixed.

Walker's (1982) argument that the two methods give the same result is not correct. The first method implies a reduction in $B$ by an amount, say, $\Delta B$ and, as discussed in step 7, this will increase the projection of $E_{A,Y}^{Y+T,l}$ by a time-dependent multiplicative factor

$$\exp\left\{\frac{\Delta B}{k+1}\left[T^{k+1} - T_1^{K+1}\right]\right\},$$

which for $T_1 = 4$ and $\Delta B = 0.5B$ could be as large as 7.2% after 50 years, and 66.6% after 80 years. In contrast, the second method reduces the projection of $E_{A,Y}^{Y+T,l}$ by a fixed time-independent multiplicative factor. The projected mesothelioma incidence counts under the two methods would agree for the initial calibration period but would gradually diverge during the later years of the projection period.

Some narrowing of the divergence would occur under a hybrid model that recognized reductions in both the numbers of exposed workers and the levels of exposure. Given the uncertainties involved in specifying such a model, in step 9, we implemented the second method, multiplying each of the period-specific estimates from step 8 by the fractions 0.9 for 1960-1964, 0.5 for 1965-1969, and 0.25 for 1970-1974. This maintained comparability with Walker's (1982) method while allowing additional adjustments in step 10 to ensure that the aggregate mesothelioma counts in the calibration period were reproduced.

### 6.6.5 Step 10: Renormalization by Level of Asbestos Exposure

Step 9 yielded the number of IWEs exposed in the past to asbestos by age at first exposure and calendar period of first exposure. Estimates for 1915-1954 were obtained from step 7; estimates for 1955-1974 reflected adjustments in steps 8 and 9. An unwanted effect of these two adjustments was that the forward projection equation for 1975-1979, 1980-1984, and 1985-1989 no longer reproduced the incidence of mesothelioma cases in Table 6.3. An additional adjustment was required to accomplish this.

Let $\widehat{N}_{A,Y}^{l}$ be the step 9 estimate of the number of IWEs whose exposure to asbestos started at age group $A$ in calendar period $Y$ at exposure level $l$. Prior to 1955, this is just $N_{A,Y}^{l}$ from step 7. For 1955-1974, $\widehat{N}_{A,Y}^{l}$ reflects computations in steps 8 and 9. Let $\widehat{E}_{A,Y}^{Y_1,l}$ be the number of mesothelioma diagnoses in calendar period $Y_1$ arising from $\widehat{N}_{A,Y}^{l}$. Under the forward projection equation,

$$\widehat{E}_{A,Y}^{Y_1,l} = \widehat{N}_{A,Y}^{l} \times Q_{A,Y}^{(Y_1-Y),l}.$$

These incidence counts were summed as in step 7 to define $\widehat{M}_{A_1,Y_1}^{l}$:

$$\widehat{M}_{A_1,Y_1}^{l} = \sum_{T=1} \widehat{E}_{A_1-T,Y_1-T}^{Y_1,l}.$$

Thus, $\widehat{M}_{A_1,Y_1}^{l}$ was the total predicted number of mesothelioma diagnoses at age group $A_1$ in baseline period $Y_1$, based on the exposure counts at step 9, stratified by level of asbestos exposure. As noted, we anticipated that $\widehat{M}_{A_1,Y_1}^{l}$ would underestimate $M_{A_1,Y_1}^{l}$, the incident counts obtained from Table 6.3 by allocating the level of asbestos exposure. Hence, we defined a renormalization

$$R_{A_1}^{l} = \frac{\sum_{T=0}^{2} M_{A_1+T,13+T}^{l}}{\sum_{T=0}^{2} \widehat{M}_{A_1+T,13+T}^{l}},$$

where $Y_1 = 13$ corresponds to 1975-1979. To apply this factor, we defined our final estimate of $N_{A,Y}^{l}$ as

$$\widetilde{N}_{A,Y}^{l} = \widehat{N}_{A,Y}^{l} \times R_{A+13-Y}^{l}.$$

The results are in Table 6.9, summed over the level of asbestos exposure.

Comparison of Tables 6.9 and 6.8 shows the overall impact of steps 8-10. As expected, the main changes occurred in 1955-1974, with increases of 20.6% and 13.0% in 1955-1959 and 1960-1964, respectively, and decreases of 44.1% and 54.1% in 1965-1969 and 1970-1974, respectively. The IWE estimates in Table 6.9 at ages 15-19 and 20-24 during 1960-1974 appeared high in comparison with older ages and earlier calendar periods. Neither age group appeared to benefit from the workplace improvements implemented in step 9. This suggested that step 10 effectively nullified the impact of step 9 for these age groups. This is an issue of obvious concern that will be addressed in Chapter 7.

## 6.7 Model Projection

Steps 11 and 12 were the model projection steps. Step 11 used the $T$-period survival function defined in step 7 to project the at-risk population from the age and date of initial exposure to the projection periods 1990-1994, 1995-1999, ..., 2045-2049. This step constituted the first-stage calibration of the two-stage projection model.

Step 12 used the forward projection equation defined in step 7 to project mesothelioma diagnoses throughout the period 1990-2049. This section presents both projection methods and compares the results with Walker's and Selikoff's projections.

**Table 6.9: Revised Estimates of Number of Insulation Worker Equivalents Exposed to Asbestos by Age at Start of Exposure and Date of Start of Exposure**

| | Date of Start of Asbestos Exposure | | | | | | | | | | | | |
|---|---|---|---|---|---|---|---|---|---|---|---|---|---|
| Age | 1915-1919 | 1920-1924 | 1925-1929 | 1930-1934 | 1935-1939 | 1940-1944 | 1945-1949 | 1950-1954 | 1955-1959 | 1960-1964 | 1965-1969 | 1970-1974 | Total |
| 15-19 | 3,482 | 4,707 | 4,535 | 4,654 | 8,977 | 12,040 | 13,750 | 7,731 | 9,618 | 21,301 | 22,358 | 18,310 | 131,463 |
| 20-24 | 4,209 | 4,538 | 6,132 | 6,305 | 13,105 | 22,916 | 14,608 | 15,285 | 13,045 | 11,469 | 15,680 | 14,812 | 142,103 |
| 25-29 | 8,848 | 3,351 | 4,571 | 6,828 | 17,072 | 28,697 | 19,494 | 13,056 | 15,952 | 12,486 | 6,777 | 8,338 | 145,469 |
| 30-34 | 0 | 389 | 1,219 | 4,940 | 18,560 | 30,209 | 17,662 | 12,810 | 14,440 | 13,088 | 6,323 | 3,088 | 122,729 |
| 35-39 | 0 | 0 | 464 | 4,142 | 15,781 | 30,811 | 15,069 | 12,437 | 13,476 | 11,587 | 6,482 | 2,818 | 113,069 |
| 40-44 | 0 | 0 | 0 | 5,870 | 17,741 | 24,791 | 13,243 | 10,699 | 11,866 | 10,531 | 5,589 | 2,814 | 103,143 |
| 45-49 | 0 | 0 | 0 | 0 | 16,015 | 21,811 | 5,329 | 7,545 | 6,500 | 5,781 | 3,167 | 1,512 | 67,660 |
| 50-54 | 0 | 0 | 0 | 0 | 0 | 13,380 | 2,783 | 4,235 | 3,546 | 3,209 | 1,761 | 868 | 29,782 |
| 55-59 | 0 | 0 | 0 | 0 | 0 | 0 | 1,439 | 3,268 | 2,353 | 2,140 | 1,195 | 590 | 10,986 |
| 60-64 | 0 | 0 | 0 | 0 | 0 | 0 | 0 | 0 | 0 | 0 | 0 | 0 | 0 |
| 65-69 | 0 | 0 | 0 | 0 | 0 | 0 | 0 | 0 | 0 | 0 | 0 | 0 | 0 |
| Total | 16,540 | 12,986 | 16,922 | 32,739 | 107,250 | 184,655 | 103,376 | 87,064 | 90,797 | 91,592 | 69,332 | 53,151 | 866,403 |
| | Cumulative Total | | | | | | | | | | | | |
| | 866,403 | 849,863 | 836,878 | 819,956 | 787,217 | 679,967 | 495,312 | 391,936 | 304,871 | 214,075 | 122,483 | 53,151 | |

Source: Stallard and Manton (1993, Table 15).

### 6.7.1 Step 11: Forward Projection of the At-Risk IWE Population by Level of Asbestos Exposure

This step used the $T$-period survival equation defined in step 7:

$$N_{A,Y}^{T,l} = N_{A,Y}^{l} \times S_{A,Y}^{T,l} \,,$$

where the estimates of $N_{A,Y}^{l}$ were obtained from step 10. For a fixed period $Y$, we rewrote the age indexes to refer to attained age in period $Y$:

$$N_{A-T,Y-T}^{T,l} = N_{A-T,Y-T}^{l} \times S_{A-T,Y-T}^{T,l} \,.$$

Summing over levels of exposure and introducing new notation, we obtained the total surviving number of IWEs:

$$P_{A,Y}^{T} = N_{A-T,Y-T}^{T} = \sum_{l=1} N_{A-T,Y-T}^{T,l} \,.$$

The results of this equation (Table 6.10) estimated that there were 361,505 IWEs alive in 1990-1994 and at risk of mesothelioma. The estimated total number of IWEs ever exposed to asbestos was 866,403 (Table 6.9), so that Table 6.10 implies that 41.7% were still alive in 1990-1994.

The IWEs were stratified by TSFE and attained age. The TSFE with the largest number of IWEs was 30 years ($T = 6$) with 67,138 IWEs, which represented 73.3% of the 91,592 IWEs initially exposed in 1960-1964 (Table 6.9). The attained age with the largest number of IWEs is 60-64 with 50,386.

Selikoff (1981, Table 2-18) estimated that during 1940-1979, 18.8 million workers had exposures to asbestos equivalent to 2+ months of insulation work, with about 17.7 million initiating exposure during 1940-1979. This contrasts with our estimate of 679,967 IWEs (3.84%) for the same period (Table 6.9), with 344,240 (50.6%) surviving to 1990-1994. Selikoff's estimates of actual workers are about 26 times larger than our estimates of IWEs. This ratio drops to about 13 to 1 if we exclude automobile maintenance and construction trades from Selikoff's totals. This is above the upper estimate of 10 to 1 suggested by Walker (1982).

### 6.7.2 Step 12: Forward Projection of Mesothelioma Incidence by Level of Asbestos Exposure

Let $N_{A,Y}^{l}$ be the estimate from step 10 of the number of IWEs whose exposure to asbestos started at age group $A$ in calendar period $Y$ at exposure level $l$. Let $E_{A,Y}^{Y+T,l}$ be the expected number of mesothelioma diagnoses in calendar period $Y + T$ arising from $N_{A,Y}^{l}$. Under the forward projection equation defined in step 7,

$$E_{A,Y}^{Y+T,l} = N_{A,Y}^{l} \times Q_{A,Y}^{T,l} .$$

**Table 6.10: Estimated Number of Insulation Worker Equivalents Surviving to 1990-1994 by Attained Age and Time Since First Exposure**

| | Time Since First Exposure | | | | | | | | | | | | |
|---|---|---|---|---|---|---|---|---|---|---|---|---|---|
| Age | 20 | 25 | 30 | 35 | 40 | 45 | 50 | 55 | 60 | 65 | 70 | 75 | Total |
| 35-39 | 17,529 | 0 | 0 | 0 | 0 | 0 | 0 | 0 | 0 | 0 | 0 | 0 | 17,529 |
| 40-44 | 14,114 | 20,983 | 0 | 0 | 0 | 0 | 0 | 0 | 0 | 0 | 0 | 0 | 35,097 |
| 45-49 | 7,862 | 14,564 | 19,436 | 0 | 0 | 0 | 0 | 0 | 0 | 0 | 0 | 0 | 41,862 |
| 50-54 | 2,837 | 6,133 | 10,199 | 8,358 | 0 | 0 | 0 | 0 | 0 | 0 | 0 | 0 | 27,527 |
| 55-59 | 2,467 | 5,440 | 10,554 | 10,770 | 6,184 | 0 | 0 | 0 | 0 | 0 | 0 | 0 | 35,415 |
| 60-64 | 2,275 | 5,121 | 10,127 | 12,054 | 11,183 | 9,626 | 0 | 0 | 0 | 0 | 0 | 0 | 50,386 |
| 65-69 | 1,081 | 3,863 | 7,800 | 9,470 | 8,290 | 8,857 | 6,870 | 0 | 0 | 0 | 0 | 0 | 46,231 |
| 70-74 | 515 | 1,788 | 5,734 | 7,117 | 6,545 | 9,502 | 10,483 | 3,783 | 0 | 0 | 0 | 0 | 45,467 |
| 75-79 | 261 | 724 | 2,258 | 4,452 | 4,505 | 6,082 | 9,259 | 3,882 | 1,237 | 0 | 0 | 0 | 32,660 |
| 80-84 | 0 | 310 | 775 | 1,491 | 2,357 | 3,130 | 5,854 | 3,031 | 1,000 | 625 | 0 | 0 | 18,573 |
| 85-89 | 0 | 0 | 255 | 396 | 805 | 1,316 | 2,832 | 1,556 | 510 | 396 | 255 | 0 | 8,321 |
| 90-94 | 0 | 0 | 0 | 95 | 162 | 186 | 791 | 455 | 126 | 101 | 83 | 51 | 2,050 |
| 95-99 | 0 | 0 | 0 | 0 | 30 | 23 | 160 | 116 | 24 | 6 | 14 | 14 | 387 |
| Total | 48,941 | 58,926 | 67,138 | 54,203 | 40,061 | 38,722 | 36,249 | 12,823 | 2,897 | 1,128 | 352 | 65 | 361,505 |
| | Cumulative Total | | | | | | | | | | | | |
| | 361,505 | 312,564 | 253,638 | 186,500 | 132,297 | 92,236 | 53,514 | 17,265 | 4,442 | 1,545 | 417 | 65 | |

Source: Stallard and Manton (1993, Table 15).

These cases were projected to occur in calendar period $Y+T$ at attained age $A+T$. They were summed to obtain $M_{A,Y}$:

$$M_{A,Y} = \sum_{l} \sum_{T=1} E^{Y,l}_{A-T,Y-T},$$

where $A$ is attained age in calendar period $Y$, where $Y$ indexes all quinquennia in the range 1970-1974, ..., 2045-2049.

The projected national incidence counts of mesothelioma diagnoses among males in the United States, by age and calendar period, are in Table 6.11 for 1990-2049. Also presented are model-based estimates for 1970-1989. These may be compared with estimates in Table 6.3 for 1970-1989 from SEER data. The results are summarized in Table 6.12.

For 1975-1979, the SEER estimate was 5112; the baseline model estimate was 5941 (+16.2%). For 1980-1984, the SEER estimate was 7740 cases; the baseline model estimate was 7218 (−6.7%). For 1985-1989, the SEER estimate was 8637; the baseline model estimate was 8267 (−4.3%). The SEER mesothelioma counts rose more rapidly than predicted, suggesting that there are factors other than age and TSFE that are not fully accounted for by the baseline model. Given that the projected mesothelioma counts in Table 6.11 peaked at 9206 in 1995-1999 (11.1% higher than predicted in 1985-1989), we expected that the SEER counts would also peak at or near the 1995-1999 period, but at a higher level than predicted by the baseline model. This discrepancy would have been an issue of concern if our primary goal were to forecast the SEER data. It was of less concern in our task of forecasting claims against the Trust because any single multiplicative adjustment developed to raise the estimates of the at-risk population in step 11 to better match the SEER mesothelioma counts in 1985-1989 or later would cancel out in step 2 of the second-stage calibration (Section 6.8).

The baseline model did not fully capture the rising trend in diagnosed mesothelioma incidence. Such a trend might have been a result of (a) increasing awareness of mesothelioma among physicians, (b) a real rise in incidence, or (c) statistical fluctuation. Nonetheless, the baseline model predicted 5-year counts within about ±10%. For the 15-year period 1975-1989, the SEER estimate was 21,489; the baseline model estimate was 21,426 (−0.3%) – with the discrepancy due to 63 cases reaching ages 15-19 after 1970-1974 in Table 6.3, an age group not in the projections. In this case, the baseline model was calibrated to reproduce almost exactly the occurrence of over 21,400 cases of mesothelioma among males aged 20-94 years in the 15-year period 1975-1989.

Also included in Table 6.12 are Walker's (1982) and Selikoff's (1981) mesothelioma projections. Walker presented additional estimates assuming a 15-year latency period, during which no incidence of mesothelioma could occur. For 1985-1989, this model projected 3900 cases (+11.43%) compared to 3500 under his no-latency assumption. Both models substantially underpredicted the SEER 1985-1989 estimate of 8637 cases. Selikoff's 1985-1989 estimate of 11,990 cases substantially overpredicted the SEER estimate of 8637 cases. The 4.3%

**Table 6.11: Estimated and Projected Incidence of Mesothelioma Among U.S. Men by Age and Date of Diagnosis, 1970-2049**

| | Date of Diagnosis | | | | | | | | | | | | | | | | |
|---|---|---|---|---|---|---|---|---|---|---|---|---|---|---|---|---|---|
| Age | 1970-1974 | 1975-1979 | 1980-1984 | 1985-1989 | 1990-1994 | 1995-1999 | 2000-2004 | 2005-2009 | 2010-2014 | 2015-2019 | 2020-2024 | 2025-2029 | 2030-2034 | 2035-2039 | 2040-2044 | 2045-2049 | Total |
| 15-19 | 0 | 0 | 0 | 0 | 0 | 0 | 0 | 0 | 0 | 0 | 0 | 0 | 0 | 0 | 0 | 0 | 0 |
| 20-24 | 1 | 1 | 0 | 0 | 0 | 0 | 0 | 0 | 0 | 0 | 0 | 0 | 0 | 0 | 0 | 0 | 2 |
| 25-29 | 9 | 9 | 7 | 0 | 0 | 0 | 0 | 0 | 0 | 0 | 0 | 0 | 0 | 0 | 0 | 0 | 24 |
| 30-34 | 17 | 33 | 34 | 23 | 0 | 0 | 0 | 0 | 0 | 0 | 0 | 0 | 0 | 0 | 0 | 0 | 107 |
| 35-39 | 45 | 47 | 89 | 88 | 57 | 0 | 0 | 0 | 0 | 0 | 0 | 0 | 0 | 0 | 0 | 0 | 326 |
| 40-44 | 156 | 106 | 104 | 190 | 184 | 113 | 0 | 0 | 0 | 0 | 0 | 0 | 0 | 0 | 0 | 0 | 854 |
| 45-49 | 275 | 304 | 208 | 197 | 349 | 330 | 197 | 0 | 0 | 0 | 0 | 0 | 0 | 0 | 0 | 0 | 1,861 |
| 50-54 | 543 | 476 | 522 | 359 | 329 | 570 | 530 | 308 | 0 | 0 | 0 | 0 | 0 | 0 | 0 | 0 | 3,638 |
| 55-59 | 706 | 850 | 741 | 807 | 555 | 497 | 844 | 776 | 442 | 0 | 0 | 0 | 0 | 0 | 0 | 0 | 6,219 |
| 60-64 | 822 | 1,018 | 1,204 | 1,046 | 1,130 | 778 | 682 | 1,142 | 1,039 | 583 | 0 | 0 | 0 | 0 | 0 | 0 | 9,443 |
| 65-69 | 872 | 1,073 | 1,318 | 1,541 | 1,334 | 1,431 | 986 | 850 | 1,407 | 1,271 | 703 | 0 | 0 | 0 | 0 | 0 | 12,786 |
| 70-74 | 616 | 1,000 | 1,241 | 1,519 | 1,748 | 1,512 | 1,614 | 1,115 | 947 | 1,554 | 1,396 | 764 | 0 | 0 | 0 | 0 | 15,026 |
| 75-79 | 353 | 606 | 997 | 1,242 | 1,509 | 1,719 | 1,493 | 1,589 | 1,103 | 928 | 1,513 | 1,354 | 735 | 0 | 0 | 0 | 15,141 |
| 80-84 | 193 | 296 | 502 | 825 | 1,031 | 1,256 | 1,428 | 1,253 | 1,340 | 940 | 786 | 1,281 | 1,148 | 620 | 0 | 0 | 12,899 |
| 85-89 | 0 | 122 | 194 | 322 | 540 | 685 | 844 | 968 | 865 | 936 | 668 | 559 | 916 | 827 | 447 | 0 | 8,892 |
| 90-94 | 0 | 0 | 56 | 90 | 152 | 263 | 343 | 433 | 507 | 467 | 515 | 377 | 319 | 529 | 484 | 263 | 4,798 |
| 95-99 | 0 | 0 | 0 | 17 | 30 | 51 | 93 | 127 | 166 | 200 | 193 | 219 | 166 | 143 | 242 | 226 | 1,873 |
| Total | 4,608 | 5,941 | 7,218 | 8,267 | 8,948 | 9,206 | 9,054 | 8,560 | 7,816 | 6,878 | 5,774 | 4,555 | 3,284 | 2,119 | 1,172 | 489 | 93,889 |
| | Cumulative Total | | | | | | | | | | | | | | | | |
| | 93,889 | 89,281 | 83,341 | 76,123 | 67,856 | 58,908 | 49,702 | 40,648 | 32,088 | 24,271 | 17,394 | 11,620 | 7,065 | 3,780 | 1,662 | 489 | |

Source: Stallard and Manton (1993, Table 16).

**Table 6.12: Alternative Estimates of U.S. Mesothelioma Incidence Counts, Males 1970-1989**

| Estimate/Projection | Type | Date of Diagnosis/Death 1970-1974 | 1975-1979 | 1980-1984 | 1985-1989 |
|---|---|---|---|---|---|
| SEER | Table 6.3 | 3,718 | 5,112 | 7,740 | 8,637 |
| First-stage baseline | Table 6.11 | 4,608 | 5,941 | 7,218 | 8,267 |
| Walker (1982) | No latency | — | 2,630 | 3,200 | 3,500 |
| | 15-Year latency | — | 2,630 | 3,400 | 3,900 |
| Selikoff (1981)[1] | No latency | 5,410 | 7,125 | 8,875 | 11,990 |

Note 1: Selikoff's (1981) estimates include a small but unspecified number of females.

Source: Stallard and Manton (1993, Table 17).

underprediction of the baseline model compares favorably with the 59.5% underprediction of Walker's no-latency model and the 38.8% overprediction of Selikoff's model.

Table 6.11 presents the projected number of incident cases of mesothelioma for each quinquennium. Table 6.13 compares these values with projections by Walker (1982, Table 10) and Selikoff (1981, Table 2-23).

Two projections were presented by Walker (1982). For 1990-1994, Walker projected 3600 and 4200 cases of mesothelioma. Both substantially underpredicted (−59.8% and −53.1%) the baseline model's 8948 cases. For 1990-2009, Walker projected 12,000 and 14,200 cases of mesothelioma. Again, both substantially underpredicted (−66.5% and −60.3%) the first-stage baseline model's 35,769 cases. Two adjustments can be made to Walker's projections to resolve these discrepancies. First, both of Walker's projections can be increased by the factor $100/54 = 1.852$, because Walker's results were only for the 54% documentably exposed workers, not for all incidence of mesothelioma (see step 6). With this adjustment, Walker's no-latency projection more closely matches that of our baseline model. For 1990-1994 and 1990-2009, both of Walker's adjusted projected totals are still below our baseline projection. However, Walker excluded persons aged 80 and over from his projections. To be comparable, we present results for the baseline model with such persons excluded. In addition, Cohen et al. (1984) attempted to replicate Walker's projections using the same methods and data sources (see Section 5.3). Cohen's analysis resulted in upward revisions of 18.2% in the no-latency model and 9.8% in the 15-year latency model. The revised no-latency model is virtually identical to our baseline model for age 15-79 for the period 1990-2009 (i.e.,

**Table 6.13: Alternative Estimates of U.S. Mesothelioma Incidence Counts, Males 1990-2049**

| | | Date of Diagnosis/Death | | | |
|---|---|---|---|---|---|
| Estimate/Projection | Type | 1990-1994 | 1990-2009 | 1990-2029 | 1990-2049 |
| First-stage baseline | Table 6.11 | 8,948 | 35,769 | 60,791 | 67,856 |
| | Table 6.11, Age 15-79 | 7,196 | 26,272 | 41,276 | 42,011 |
| Walker (1982) | No latency | 3,600 | 12,000 | — | — |
| | 185.2% | 6,667 | 22,223 | — | — |
| | Revised[2] | — | 26,264 | — | — |
| Walker (1982) | 15-Year latency | 4,200 | 14,200 | — | — |
| | 185.2% | 7,778 | 26,296 | — | — |
| | Revised[2] | — | 28,864 | — | — |
| Selikoff (1981)[1] | No latency | 13,740 | 58,880 | 94,655 | — |
| | 94.5% | 12,984 | 55,642 | 89,449 | — |
| | 74.7% | 10,264 | 43,983 | 70,707 | — |
| | 65.1% | 8,948 | 38,345 | 61,643 | — |

Note 1: Selikoff's (1981) estimates include a small but unspecified number of females.
Note 2: Based on Cohen et al. (1984, p.19); reflects revision of Walker's (1982) 1980-2009 projections from 18,700 to 22,100 (no latency) and from 21,500 to 23,600 (15-year latency).

Source: Stallard and Manton (1993, Table 18).

26,264 vs. 26,272 cases). Thus, our model is fully consistent with the method proposed by Walker. The results show that a sizable number of mesothelioma cases will occur above age 80 (i.e., 38.1% of all cases 1990-2049).

Selikoff's (1981) projection is presented as reported and as modified to reflect three adjustments. The original projection for 1990-1994 was 13,740 cases, 53.6% higher than our baseline projection. Accepting at face value the SEER estimates for 1980-1984 and 1985-1989 in Table 6.12, this estimate appears highly implausible. There could be several reasons for this result. Selikoff's projection included women; the number is "small" but unknown. Therefore, we made two adjustments to his projections: one assuming that 5.5% of his total were women (94.5% male) and the second that 25.3% were women (74.7% male). These percentages correspond to (1) the number of female claimants against the Manville Trust with mesothelioma diagnoses (Table 6.5) and (2) the number of female cases of mesothelioma in the 1985-1989 national estimates from SEER (Table 6.3). Neither adjustment brought Selikoff's projection near the baseline model.

A third adjustment introduced a scale factor (0.651) to Selikoff's 1990-1994 projection to exactly match our baseline model. This almost completely eliminated the overall difference for the period 1990-2029: Selikoff's adjusted projection was 1.4% higher than the baseline projection. However, some differences were manifest early in the period: Selikoff's adjusted projection for 1990-2009 was 7.2% higher. The adjustment suggested that the relative changes in both sets of projections over the period 1990-2029 were in general agreement, even though absolute levels differed.

The conclusion to be reached from these comparisons is that the general time course of the incidence of mesothelioma in both Walker's and Selikoff's projections was similar to that of our first-stage baseline model. The differences in the absolute levels of projected cases were large and only the first-stage baseline model was calibrated to match the levels observed in any version of the SEER data. The relatively good agreement on the time course of the various projections gave us confidence that the modest discrepancies between the baseline model and the SEER data over the period 1975-1989 were not a source of significant bias in our claim projections. However, the use of Walker's (1982) or Selikoff's (1981) unadjusted projections to estimate absolute levels of mesothelioma incidence could not be recommended.

## 6.8 Nonparametric Hazard Modeling of Claim Filing Rates: CHR Model

The second-stage model used the Trust's claim experience for 1990-1994 to project the number and timing of claims for 1990-2049 in three steps. Step 1 was the data preparation step in which claims against the Trust were tabulated by attained age, time since first exposure, and type of disease/injury. Step 2 was the estimation of the claim filing rates by attained age and TSFE. The numerators of these rates were the estimated numbers of claims for each of nine disease/injuries during 1990-1994 from step 1. The denominators of these rates were the IWE survival estimates developed in the first-stage calibration (Table 6.10). Step 3 was the claim projection step in which the claim hazard rates were multiplied by the IWE survival distributions based on Table 6.10.

### 6.8.1 Step 1: Distribution of 1990-1994 Claims by Attained Age, TSFE, and Disease/Injury

The characteristics of the claim data maintained by the Manville Trust were described in Sections 6.2 and 6.5 (step 3). Each qualified claim was assigned to one of nine categories based on the selection of the most severe disease/injury alleged on the Proof of Claim (POC) Form submitted to the Trust. For each category, the claims were tabulated by attained age and TSFE as of the date each claim was filed with the Trust.

We had to decide which months of data to include in the tabulations for this step. Figure 6.2 showed that the period from November 1988 to November 1989 was distinctively different from the following periods. This was the first year of operation of the Trust and there were requirements that backlogged claims be filed within the first 12 months. The only other detectable shift in Figure 6.2 was the dropoff in September 1992 when the POC Form was revised. To retain the maximum number of months in our analysis, we tabulated all claims in the period January 1990 to November 1992 and used these tabulations to impute the number of claims in the period December 1992 to December 1994. The result was an estimate of the total number of claims for the period 1990-1994 by age and TSFE, for each of the nine disease/injuries. The estimates for the 60-month period were obtained by multiplying the 35-month count for each combination of age and TSFE by the ratio 60/35. The marginal distribution of claims by age and disease/injury is shown in Table 6.14.

### 6.8.2 Step 2: Estimation of Claim Hazard Rates by Attained Age, TSFE, and Disease/Injury

This step employed a nonparametric hazard model to calibrate a separate nonparametric hazard for each of the nine categories of disease used by the Manville Trust in classifying claims. The age-specific claim data in Table 6.14 were stratified by TSFE and matched with the corresponding projected IWE number of survivors in Table 6.10 to generate estimates of the claim hazard rates (CHRs) for 1990-1994. The hazard rates in future years were assumed to be constant for each combination of age and TSFE. Because the distribution of TSFE will shift upward in future years (i.e., assuming no new exposures after 1974), this assumption specified one potential mechanism through which cohort effects could operate. By stratifying on both age and TSFE, we avoided assuming either equality or proportionality of the underlying claim hazard rates over age.

Let, $P_{A,Y}^{T}$ represent the number of IWEs alive at age group $A$ in period $Y$ whose TSFE is $T$. The $P_{A,1990-94}^{T}$ were presented in Table 6.10.

Let the corresponding claim counts be denoted by $C_{A,Y}^{T,d}$ for age group $A, Y =$ 1990-1994, $T =$ TSFE, and disease $d$. Two problems were that (1) a large number of cases had unknown TSFE (3-4% for asbestos-related diseases; 17% for disputed asbestosis; 28% for nonasbestos-related diseases; and 80% for unknown disease) and (2) some known TSFEs fell in cells for which there was zero exposure in Table 6.10. Our solution to this was to (1) include the second group with the first (i.e., regarding their TSFE as unknown) and (2) allocate the unknown TSFE pro rata among the known TSFE. This preserved the age-specific claim totals which are presented in Table 6.14. Less serious was that about 0.32% of Trust cases had unknown age; these were deleted. In addition, about 0.95% of cases had a reported age below 35 years and were not included in the at-risk population in Table 6.10; these also were deleted.

The CHRs for each disease were obtained as ratios of the claim count to the population at risk:

**Table 6.14: Estimated Number of Qualified Male Claims Against the Manville Trust by Age and Alleged Disease/Injury 1990-1994 – Based on Observed Claims, 1990-1992**

| | Most Severe Alleged Disease/Injury | | | | | | | | | |
|---|---|---|---|---|---|---|---|---|---|---|
| Age | Mesothelioma | Lung Cancer | Colon/Rectal Cancer | Other Cancer | Asbestosis | Disputed Asbestosis | Pleural Plaques/ Thickening | Nonasbestos Related Disease | Unknown Disease | Total |
| 15-19 | 0 | 0 | 0 | 0 | 0 | 0 | 0 | 0 | 0 | 0 |
| 20-24 | 0 | 0 | 0 | 0 | 0 | 0 | 2 | 0 | 2 | 3 |
| 25-29 | 0 | 0 | 0 | 2 | 9 | 7 | 3 | 26 | 41 | 88 |
| 30-34 | 12 | 5 | 0 | 0 | 142 | 98 | 45 | 202 | 225 | 729 |
| 35-39 | 19 | 15 | 0 | 3 | 964 | 266 | 233 | 343 | 245 | 2,089 |
| 40-44 | 45 | 39 | 3 | 9 | 2,295 | 520 | 631 | 443 | 273 | 4,257 |
| 45-49 | 106 | 160 | 15 | 22 | 3,115 | 883 | 1,048 | 460 | 225 | 6,034 |
| 50-54 | 214 | 257 | 21 | 33 | 3,859 | 1,295 | 1,480 | 695 | 216 | 8,070 |
| 55-59 | 357 | 542 | 65 | 101 | 4,972 | 1,643 | 1,875 | 830 | 295 | 10,680 |
| 60-64 | 578 | 1,053 | 112 | 185 | 6,898 | 2,273 | 2,612 | 1,561 | 427 | 15,698 |
| 65-69 | 722 | 1,242 | 187 | 199 | 7,287 | 2,355 | 2,544 | 1,751 | 444 | 16,731 |
| 70-74 | 741 | 1,024 | 137 | 156 | 5,137 | 1,566 | 1,599 | 1,086 | 408 | 11,853 |
| 75-79 | 499 | 746 | 98 | 110 | 3,098 | 852 | 825 | 628 | 288 | 7,144 |
| 80-84 | 237 | 271 | 33 | 36 | 1,158 | 274 | 283 | 292 | 161 | 2,744 |
| 85-89 | 79 | 79 | 7 | 10 | 359 | 77 | 74 | 67 | 52 | 803 |
| 90-94 | 10 | 12 | 3 | 2 | 52 | 9 | 14 | 17 | 10 | 129 |
| 95-99 | 0 | 3 | 0 | 2 | 9 | 2 | 0 | 2 | 7 | 24 |
| Total | 3,619 | 5,449 | 681 | 870 | 39,351 | 12,119 | 13,266 | 8,401 | 3,319 | 87,073 |

Source: Stallard and Manton (1993, Table 21a).

$$H_{A,Y_2}^{T,d} = C_{A,Y_2}^{T,d} / P_{A,Y_2}^{T} \text{ where } Y_2 = 1990\text{-}1994.$$

These were quinquennial rates based on 35 months of observed claim filings. The variances can be approximated by a Poisson process with expected number of claims proportional to $P_{A,Y_2}^T$:

$$\text{var}\left[H_{A,Y_2}^{T,d}\right] \approx \frac{H_{A,Y_2}^{T,d}}{P_{A,Y_2}^T} \times \left(\frac{60}{35}\right),$$

conditional on $P_{A,Y_2}^T$, which is the number of exposed persons alive $T$ quinquennial periods after the start of exposure. The final multiplier (60/35) has an exponent of 1, not 2, because a second appearance of that multiplier was absorbed into the numerator of the hazard rate, $H_{A,Y_2}^{T,d}$, in inflating the claim counts from 35 to 60 months.

The filing of a claim may occur before or after the time of death, so a more appropriate definition of the claim hazard rate would take full account of the endpoints in Figure 6.1. We used the surviving exposed population, $P_{A,Y_2}^T$, to approximate the at-risk population because stratification on both age and TSFE allowed any systematic errors to be absorbed into $H_{A,Y_2}^{T,d}$.

The CHRs are additive over disease/injuries. Table 6.15 presents the CHRs for all claims (i.e., summed over the index $d$). The row and column totals refer to the marginal CHRs by TSFE and age, and the overall total refers to the overall average claim filing rate. The overall rate was 23.9%. The age-specific rates were unimodal with a peak of 36.2% at age 65-69. The TSFE-specific rates were bimodal with peaks of 40.2% at 45 years and 37.3% at 65 years.

When the CHRs were stratified according to cancer and noncancer claims, both types of claim rates peaked at age 65-69. The overall cancer claim rate was 2.9% and the rate rose fairly consistently over TSFE, with a peak at 70 years. The overall noncancer claim rate was 20.9%, and this represented 87.7% of all claims.

The cancer CHRs fell into four disease categories. The overall mesothelioma claim rate was 1.0% and the rates peaked at age 70-74. The rates increased monotonically with TSFE up to 50 years, with a secondary increase beginning at 65 years. The overall lung cancer claim rate was 1.5%, the age peak occurred at 65-69, and the TSFE pattern was similar to mesothelioma up to 60 years.

The CHRs for colon/rectal and other cancers jointly accounted for 0.4% of all claims and followed the cancer pattern with an age peak at 65-69 years, increasing monotonically with TSFE through 50 years.

The noncancer CHRs included two asbestosis categories. The first category (undisputed asbestosis) accounted for about 10.8% of claims; the second (disputed asbestosis) accounted for 3.3%. The age peak occurred at 65-69 years. The TSFE rates peaked at 45 years and 40 years, respectively, but there was not any clear trend thereafter.

The overall rate for pleural plaques and thickening was 3.7%; the peak age was 65-69 years; and the peak TSFE was 40 years.

**Table 6.15: Estimated Claim Hazard Rates (per 1000 Insulation Worker Equivalents) for Total Claims, 1990-1994, by Attained Age and Time Since First Exposure**

| | Time Since First Exposure | | | | | | | | | | | | |
|---|---|---|---|---|---|---|---|---|---|---|---|---|---|
| Age | 20 | 25 | 30 | 35 | 40 | 45 | 50 | 55 | 60 | 65 | 70 | 75 | Total |
| 35-39 | 118.97 | 0.00 | 0.00 | 0.00 | 0.00 | 0.00 | 0.00 | 0.00 | 0.00 | 0.00 | 0.00 | 0.00 | 118.97 |
| 40-44 | 162.52 | 93.56 | 0.00 | 0.00 | 0.00 | 0.00 | 0.00 | 0.00 | 0.00 | 0.00 | 0.00 | 0.00 | 121.29 |
| 45-49 | 148.81 | 203.58 | 97.68 | 0.00 | 0.00 | 0.00 | 0.00 | 0.00 | 0.00 | 0.00 | 0.00 | 0.00 | 144.13 |
| 50-54 | 256.26 | 336.31 | 291.02 | 276.59 | 0.00 | 0.00 | 0.00 | 0.00 | 0.00 | 0.00 | 0.00 | 0.00 | 293.15 |
| 55-59 | 212.96 | 231.95 | 170.82 | 352.57 | 532.38 | 0.00 | 0.00 | 0.00 | 0.00 | 0.00 | 0.00 | 0.00 | 301.56 |
| 60-64 | 191.25 | 197.65 | 121.29 | 233.47 | 482.63 | 499.91 | 0.00 | 0.00 | 0.00 | 0.00 | 0.00 | 0.00 | 311.56 |
| 65-69 | 252.60 | 168.51 | 93.72 | 154.49 | 429.89 | 623.72 | 658.73 | 0.00 | 0.00 | 0.00 | 0.00 | 0.00 | 361.90 |
| 70-74 | 180.40 | 119.64 | 56.84 | 98.65 | 267.86 | 323.18 | 412.79 | 361.44 | 0.00 | 0.00 | 0.00 | 0.00 | 260.69 |
| 75-79 | 165.72 | 112.77 | 61.06 | 69.30 | 163.35 | 251.74 | 271.50 | 356.88 | 328.27 | 0.00 | 0.00 | 0.00 | 218.72 |
| 80-84 | 0.00 | 122.38 | 66.90 | 52.63 | 92.60 | 155.22 | 157.19 | 131.56 | 273.86 | 445.70 | 0.00 | 0.00 | 147.74 |
| 85-89 | 0.00 | 0.00 | 57.72 | 13.72 | 73.15 | 94.25 | 84.80 | 58.49 | 119.53 | 311.39 | 329.53 | 0.00 | 96.46 |
| 90-94 | 0.00 | 0.00 | 0.00 | 18.73 | 10.84 | 72.48 | 35.10 | 34.08 | 63.51 | 188.31 | 405.51 | 138.00 | 62.71 |
| 95-99 | 0.00 | 0.00 | 0.00 | 0.00 | 0.00 | 173.08 | 34.95 | 58.82 | 89.29 | 0.00 | 281.25 | 0.00 | 57.78 |
| Total | 156.23 | 173.93 | 136.43 | 211.83 | 374.95 | 401.97 | 346.47 | 254.66 | 259.21 | 373.42 | 347.41 | 108.95 | 238.58 |

Source: Authors' calculations.

**Table 6.16: Projected Number of Qualified Male Claims Against the Manville Trust by Most Severe Alleged Disease/Injury and Summarized Dates of Claim, 1990-2049 – Baseline CHR Model**

| | Date of Claim | | | |
|---|---|---|---|---|
| Most Severe Alleged Disease/Injury | 1990-1994 | 1990-2009 | 1990-2029 | 1990-2049 |
| 1. Mesothelioma | 3,607 | 14,139 | 22,871 | 24,343 |
| 2. Lung cancer | 5,444 | 20,301 | 31,869 | 33,545 |
| 3. Colon/rectal cancer | 681 | 2,638 | 4,174 | 4,282 |
| 4. Other cancer | 864 | 3,276 | 5,409 | 5,794 |
| 5. Asbestosis | 39,200 | 133,280 | 190,339 | 195,782 |
| 6. Disputed asbestosis | 12,014 | 42,293 | 59,527 | 60,987 |
| 7. Pleural plaques/thickening | 13,217 | 46,098 | 64,631 | 66,392 |
| 8. Nonasbestos-related disease | 8,171 | 26,633 | 38,587 | 39,830 |
| 9. Unknown | 3,051 | 9,836 | 14,219 | 14,797 |
| Total | 86,248 | 298,495 | 431,626 | 445,751 |
| Cancer (1-4) | 10,596 | 40,355 | 64,323 | 67,964 |
| Noncancer (5-9) | 75,653 | 258,140 | 367,303 | 377,787 |
| Asbestos-related noncancer (5-7) | 64,431 | 221,671 | 314,497 | 323,161 |
| Nonasbestos-related & unknown (8-9) | 11,222 | 36,469 | 52,806 | 54,626 |

Source: Authors' calculations.

Nonasbestos-related and unknown disease categories jointly account for 3.1% of claims. Nonasbestos-related disease rates peaked at age 65-69; unknown disease rates peaked at age 95-99 but the age curve was basically flat, as was the TSFE curve. The TSFE rates for nonasbestos-related disease peaked at 45 years.

### 6.8.3 Step 3: Claim Projections

The projected claim counts for 1990-1994, 1990-2009, 1990-2029, and 1990-2049 based on the disease-specific CHRs are summarized in Table 6.16. The detailed projections by quinquennia for 1990-2049 are provided in Table 6.17. The claim total for 1990-1994 was 99.05% of the total in Table 6.14. The difference of 0.95% was due to deletion of claims below age 35 in the CHR calibration. Accounting for the 0.32% unknown ages would make the total in Table 6.16 equal to 98.73% of the true total for males.

For mesothelioma, the total projected claim count for males was 24,343 for 1990-2049. For all diseases, the total projected claim count for males was 445,751 for 1990-2049.

With Cohen et al.'s (1984) adjustments, Walker's (1982) estimates of mesothelioma incidence were close to our estimates for 1990-2009 (see Ta-

**Table 6.17: Projected Number of Qualified Male Claims Against the Manville Trust by Most Severe Alleged Disease/Injury and Quinquennial Dates of Claim, 1990-2049 – Baseline CHR Model**

| | Date of Claim | | | | | | | | | | | | | |
|---|---|---|---|---|---|---|---|---|---|---|---|---|---|---|
| Most Severe Alleged Disease/Injury | 1990-1994 | 1995-1999 | 2000-2004 | 2005-2009 | 2010-2014 | 2015-2019 | 2020-2024 | 2025-2029 | 2030-2034 | 2035-2039 | 2040-2044 | 2045-2049 | Total | % |
| 1. Mesothelioma | 3,607 | 3,558 | 3,519 | 3,456 | 3,188 | 2,520 | 1,895 | 1,128 | 798 | 503 | 171 | 0 | 24,343 | 5.5 |
| 2. Lung cancer | 5,444 | 5,183 | 4,943 | 4,732 | 4,304 | 3,291 | 2,413 | 1,560 | 1,013 | 484 | 139 | 40 | 33,545 | 7.5 |
| 3. Colon/rectal cancer | 681 | 674 | 655 | 628 | 596 | 465 | 308 | 167 | 85 | 24 | 0 | 0 | 4,282 | 1.0 |
| 4. Other cancer | 864 | 815 | 794 | 804 | 792 | 624 | 439 | 278 | 179 | 123 | 84 | 0 | 5,794 | 1.3 |
| 5. Asbestosis | 39,200 | 34,894 | 31,098 | 28,088 | 24,080 | 16,803 | 10,559 | 5,617 | 3,223 | 1,688 | 533 | 0 | 195,782 | 43.9 |
| 6. Disputed asbestosis | 12,014 | 11,092 | 10,073 | 9,114 | 7,510 | 5,077 | 3,097 | 1,552 | 927 | 447 | 86 | 0 | 60,987 | 13.7 |
| 7. Pleural plaques/thickening | 13,217 | 12,080 | 10,938 | 9,863 | 8,361 | 5,409 | 3,199 | 1,564 | 900 | 604 | 211 | 46 | 66,392 | 14.9 |
| 8. Nonasbestos-related disease | 8,171 | 6,950 | 5,893 | 5,619 | 5,158 | 3,904 | 1,900 | 991 | 551 | 427 | 184 | 81 | 39,830 | 8.9 |
| 9. Unknown | 3,051 | 2,490 | 2,319 | 1,976 | 1,738 | 1,258 | 852 | 535 | 287 | 149 | 141 | 0 | 14,797 | 3.3 |
| Total | 86,248 | 77,736 | 70,231 | 64,280 | 55,727 | 39,351 | 24,661 | 13,393 | 7,961 | 4,449 | 1,548 | 167 | 445,751 | 100.0 |
| Cancer (1-4) | 10,596 | 10,229 | 9,910 | 9,620 | 8,880 | 6,901 | 5,055 | 3,133 | 2,073 | 1,134 | 393 | 40 | 67,964 | 15.2 |
| Noncancer (5-9) | 75,653 | 67,507 | 60,321 | 54,660 | 46,847 | 32,451 | 19,606 | 10,260 | 5,888 | 3,315 | 1,155 | 127 | 377,787 | 84.8 |
| Asbestos-related noncancer (5-7) | 64,431 | 58,066 | 52,109 | 47,065 | 39,951 | 27,288 | 16,854 | 8,733 | 5,050 | 2,739 | 830 | 46 | 323,161 | 72.5 |
| Nonasbestos-related & unknown (8-9) | 11,222 | 9,441 | 8,212 | 7,595 | 6,896 | 5,162 | 2,752 | 1,527 | 838 | 576 | 326 | 81 | 54,626 | 12.3 |

Source: Authors' calculations.

**Table 6.18: Projected Number of Qualified Male Claims Against the Manville Trust by Attained Age or Age in 1990-1994 and Quinquennial Dates of Claim, 1990-2049 – Baseline CHR Model**

| Item | Date of Claim | | | | | | | | | | | | | |
|---|---|---|---|---|---|---|---|---|---|---|---|---|---|---|
| | 1990-94 | 1995-99 | 2000-04 | 2005-09 | 2010-14 | 2015-19 | 2020-24 | 2025-29 | 2030-34 | 2035-39 | 2040-44 | 2045-49 | Total | % |
| Attained Age | | | | | | | | | | | | | | |
| 35-39 | 2,086 | | | | | | | | | | | | 2,086 | 0.5 |
| 40-44 | 4,257 | 1,611 | | | | | | | | | | | 5,868 | 1.3 |
| 45-49 | 6,034 | 4,801 | 1,639 | | | | | | | | | | 12,473 | 2.8 |
| 50-54 | 8,069 | 11,765 | 9,270 | 4,449 | | | | | | | | | 33,554 | 7.5 |
| 55-59 | 10,680 | 9,118 | 15,084 | 14,136 | 8,024 | | | | | | | | 57,042 | 12.8 |
| 60-64 | 15,698 | 10,740 | 9,191 | 15,095 | 13,752 | 6,821 | | | | | | | 71,299 | 16.0 |
| 65-69 | 16,731 | 17,870 | 12,715 | 10,873 | 17,714 | 15,527 | 7,751 | | | | | | 99,182 | 22.3 |
| 70-74 | 11,853 | 9,858 | 10,442 | 7,345 | 5,842 | 8,787 | 7,408 | 3,408 | | | | | 64,945 | 14.6 |
| 75-79 | 7,143 | 7,719 | 6,578 | 6,838 | 4,747 | 3,724 | 5,620 | 4,805 | 2,256 | | | | 49,429 | 11.1 |
| 80-84 | 2,744 | 3,091 | 3,704 | 3,365 | 3,494 | 2,355 | 2,070 | 3,545 | 3,427 | 2,005 | | | 29,800 | 6.7 |
| 85-89 | 803 | 904 | 1,229 | 1,591 | 1,371 | 1,483 | 1,090 | 1,000 | 1,732 | 1,678 | 844 | | 13,726 | 3.1 |
| 90-94 | 129 | 217 | 320 | 508 | 656 | 537 | 629 | 527 | 458 | 715 | 632 | 167 | 5,496 | 1.2 |
| 95-99 | 22 | 39 | 59 | 79 | 127 | 116 | 93 | 107 | 88 | 51 | 72 | 0 | 853 | 0.2 |
| Total | 86,248 | 77,736 | 70,231 | 64,280 | 55,727 | 39,351 | 24,661 | 13,392 | 7,961 | 4,449 | 1,548 | 167 | 445,751 | 100.0 |
| Age in 1990-1994 | | | | | | | | | | | | | | |
| 35-39 | 2,086 | 1,611 | 1,639 | 4,449 | 8,024 | 6,821 | 7,751 | 3,408 | 2,256 | 2,005 | 844 | 167 | 41,061 | 9.2 |
| 40-44 | 4,257 | 4,801 | 9,270 | 14,136 | 13,752 | 15,527 | 7,408 | 4,805 | 3,427 | 1,678 | 632 | 0 | 79,695 | 17.9 |
| 45-49 | 6,034 | 11,765 | 15,084 | 15,095 | 17,714 | 8,787 | 5,620 | 3,545 | 1,732 | 715 | 72 | | 86,163 | 19.3 |
| 50-54 | 8,069 | 9,118 | 9,191 | 10,873 | 5,842 | 3,724 | 2,070 | 1,000 | 458 | 51 | | | 50,397 | 11.3 |
| 55-59 | 10,680 | 10,740 | 12,715 | 7,345 | 4,747 | 2,355 | 1,090 | 527 | 88 | | | | 50,287 | 11.3 |
| 60-64 | 15,698 | 17,870 | 10,442 | 6,838 | 3,494 | 1,483 | 629 | 107 | | | | | 56,561 | 12.7 |
| 65-69 | 16,731 | 9,858 | 6,578 | 3,365 | 1,371 | 537 | 93 | | | | | | 38,532 | 8.6 |
| 70-74 | 11,853 | 7,719 | 3,704 | 1,591 | 656 | 116 | | | | | | | 25,639 | 5.8 |
| 75-79 | 7,143 | 3,091 | 1,229 | 508 | 127 | | | | | | | | 12,099 | 2.7 |
| 80-84 | 2,744 | 904 | 320 | 79 | | | | | | | | | 4,048 | 0.9 |
| 85-89 | 803 | 217 | 59 | | | | | | | | | | 1,079 | 0.2 |
| 90-94 | 129 | 39 | | | | | | | | | | | 168 | 0.0 |
| 95-99 | 22 | | | | | | | | | | | | 22 | 0.0 |
| Total | 86,248 | 77,736 | 70,231 | 64,280 | 55,727 | 39,351 | 24,661 | 13,392 | 7,961 | 4,449 | 1,548 | 167 | 445,751 | 100.0 |

Source: Authors' calculations.

ble 6.13). For lung cancer, Walker (1982) developed a disease-specific projection model that can be compared with our claims estimates in Table 6.16. For 1990-1994, we projected 5444 claims; he projected 10,200 cases (+87.4%; Table 4.3). For 1990-2009, we projected 20,301 claims; he projected 23,720 (+16.8%). Comparison of Tables 4.3 and 6.17 shows that the timing of cases was dramatically different.

Selikoff (1981) projected excess deaths due to asbestos-related lung cancer (Table 3.3). For 1990-1994, he projected 27,485 deaths (+404.9%). Neither Walker nor Selikoff was close to the 5444 lung cancer claims that we estimated were filed over 1990-1994.

Walker (1982, Task 4d) developed a disease-specific claim projection model for asbestosis. For 1990-1994, we projected 39,200 claims; he projected from 2800 (−92.9%) to 4500 (−88.5%) lawsuits from asbestosis injury (Table 4.5). For 1990-2009, we projected 133,280 asbestosis claims; he projected 4000-7200 lawsuits for the same period. Given the observed number of 22,944 claims from January 1, 1990 to November 30, 1992 in the Trust extract file, it was obvious that Walker's asbestosis projections were at least one order of magnitude too low.

Comparison of Tables 6.13 and 6.16 shows that the projected mesothelioma claims represented 40.3% of mesothelioma diagnoses in 1990-1994, 39.5% in 1990-2009, 34.9% in 2010-2029, and 20.8% in 2030-2049. The decline in later years reflects different patterns of increase for the Weibull versus the nonparametric claim hazard rates as functions of TSFE. The Weibull increased as the 3.2 power of TSFE, whereas the nonparametric CHRs peaked at 50 years and then declined temporarily before starting a second increase at 65 years.

The top panel of Table 6.18 displays the projected claims 1990-2049 by attained age and date of filing; the bottom panel displays the same data by cohort, based on each claimant's age in 1990-1994. The age peak occurred at 65-69 years with 22.3% of all projected claims. The age range 60-74 years contained 52.9% of all projected claims. The bottom panel shows that the cohort with the highest number of projected claims was aged 45-49 years in 1990-1994, followed by the cohort aged 40-44. These patterns are consistent with the exposure estimates for these two cohorts in Table 6.9 (i.e., workers aged 15-19 in 1960-1964 and 1965-1969).

Chapter 7 presents sensitivity analyses for these projections. Section 7.13 presents our conclusions.

# 7

# Uncertainty in Updated Forecasts

## 7.1 Introduction

The Rule 706 Panel conducted sensitivity analyses of the updated claim hazard rate (CHR) model (Stallard and Manton, 1993). Selected results will be presented, discussed, and used to evaluate the range of uncertainty of the model.

We selected for sensitivity analyses the characteristics of the model that we felt were most likely to be influential on the projections. We refer to the projection in Tables 6.16-6.18 as the baseline projection and we address the issue of how different that projection could be under some plausible modifications of the model. The sensitivity analyses are grouped into nine areas and involve the calculation of from one to five alternative projections. Table 7.1 summarizes the characteristics of the baseline and 27 alternative projections. Table 7.2 contains the corresponding claim summaries for 1990-2049.

The nine areas in which sensitivity analyses were conducted are as follows:

S1. *Constant age-specific claim runoff.* The CHR projections in Section 6.8 established the major role of age as a factor in claim filing. In this analysis, we assumed that the age-specific number of claims in each future quinquennium would be the same as in 1990-1994. The incidence of new exposure was assumed to end in 1975-1979, implying that all claims will have ceased in 2050-2054, when persons aged 15-19 in 1970-1974 are age 95-99. For consistency with other projections, we assumed no claims after 2045-2049.

S2. *Ratio estimation of nine asbestos-related diseases: Propensity to sue (PTS) model.* The national projections of mesothelioma diagnoses in Section 6.7 were weighted to produce projections of mesothelioma claims among Trust claimants, which were, in turn, weighted to produce projections of claims for the other eight diseases.

S3. *Parametric claim hazard rate model.* The CHR model in Section 6.8 was nonparametric. In this analysis, we fitted the claim data using a para-

**Table 7.1: Sensitivity Analyses Conducted to Assess the Uncertainty in Projections of Qualified Male Claims Against the Manville Trust, 1990-2049**

| Analysis | Model | # | Description |
|---|---|---|---|
| S0 | | 0 | Baseline CHR model in Section 6.8 |
| S1 | 1 | 1 | Constant age-specific claim runoff |
| S2 | 1 | 2 | Constant age-specific propensity to sue for each disease set proportional to national mesothelioma projection |
| S3 | | | Parametric models of claim hazard rates |
| | 0 | 3 | $k$ = 3.2 (and $b$ = 3.58 $10^{-8}$) |
| | 1 | 4 | $k$ varies |
| | 2 | 5 | $k$ varies and $c > 0$ |
| | 3 | 6 | $k$ varies and $b$ depends on age |
| | 4 | 7 | $k$ varies, $c > 0$, and $b$ depends on age |
| S4 | | | Alternative mesothelioma mortality models |
| | 0 | 8 | $k$ = 3.2 |
| | 1 | 9 | $k$ = 4.2 |
| | 2 | 10 | $k$ = 5.3 and $c > 0$ |
| | 3 | 11 | $b$ = 1 $10^{-8}$ |
| | 4 | 12 | $b$ = 1 $10^{-9}$ |
| S5 | | | Alternative assumptions about the calibration of the youngest workers in the 1960s and 1970s |
| | 1 | 13 | Delete step 10 |
| | 2 | 14 | Delete steps 8, 9, and 10 |
| | 3 | 15 | Delete step 10 only for young cohorts |
| S6 | | | Alternative to national SEER mesothelioma data for exposure calibration |
| | 1 | 16 | Use only SEER costal sites |
| | 2 | 17 | Use only SEER inland sites |
| S7 | | | Alternative projected mortality rates |
| | 1 | 18 | Low mortality assumption |
| | 2 | 19 | High mortality assumption |
| S8 | | | Alternative size and excess mortality of the heavy-exposure group |
| | 0 | 20 | No excess mortality |
| | 1 | 21 | 30.1% excess mortality |
| | 2 | 22 | 45.0% excess mortality |
| | 3 | 23 | 16.0% of claimants had heavy exposure |
| | 4 | 24 | 25.9% of claimants had heavy exposure |
| S9 | | | Alternative rates of decline in future claim filing rates |
| | 1 | 25 | 1% per year decline |
| | 2 | 26 | 3% per year decline |
| | 3 | 27 | 7% per year decline |
| S1–S9 | | 28 | Median of 27 models in analyses S1–S9 |
| S10 | | | Bridge model linking baseline model (Ch. 6) with hybrid model (Ch. 8) |
| | 1 | 29 | Exposure calibrated using Trust mesothelioma diagnoses 1975-1989 |
| | 2 | 30 | Exposure calibrated using Trust mesothelioma diagnoses 1985-1989 |
| | 3 | 31 | Exposure calibrated using Trust mesothelioma claims 1990-1994 |

Source: Stallard and Manton (1993).

metric Weibull and generalized Weibull to smooth out irregularities in the nonparametric estimates. This allowed evaluation of the Weibull exponents for all nine diseases and provided insight into the assumption of the constant ratio PTS models in alternative S2.

S4. *Mesothelioma incidence function.* The Weibull hazard function used in Section 6.6 was generalized and refitted to more recent and comprehensive data on mortality among insulation workers (Selikoff and Seidman, 1991). These results suggested that the use of the 3.2 power of age is incorrect for insulation workers, although not necessarily for all asbestos workers.

S5. *Adjustments to the insulation-worker equivalent (IWE) exposed population.* Table 6.9 provided estimates of the IWE exposed population. We evaluated the sensitivity of the projections to modifications of these estimates. In contrast to other sensitivity analyses, this analysis identified a major source of sensitivity.

S6. *National mesothelioma incidence counts.* The counts of mesothelioma incidence based on the Surveillance, Epidemiology, and End Results Program (SEER) data in Table 6.3 were replaced by SEER data stratified by coastal versus inland sites to evaluate effects of shipyard and other traditional sources of exposure. This analysis addressed concerns about the representativeness of the SEER data and its potential impact on model calibration in Section 6.6.

S7. *Nonmesothelioma mortality rates.* The projected mortality rates in Table 6.2 were modified to reflect the range of future mortality as projected by Lee and Carter (1992).

S8. *Excess mortality among insulation workers.* Step 7 of the first-stage calibration in Section 6.6.2 assumed that 20% of mesothelioma incidence occurred among heavily exposed workers who suffered 37% excess mortality. This step evaluated the effects of changing the 20% heavy exposure to 0%, 16.0%, and 25.9%, and the 37% excess mortality to 0%, 30.1%, and 45.0%. The results showed only minor sensitivity to these variations. The lack of convincing evidence for the assumptions in step 7 need not be a major concern.

S9. *Decline in claim filing rates.* The CHR model in Section 6.8 assumed that age and time since first exposure (TSFE) were the only factors in filing claims. In this analysis, we evaluated the effects of declines of 1%, 3%, and 7% per calendar year in these claim filing rates. Although the analysis did not detect trends in claim filing rates after 1989 (see Figures 6.2-6.5), such declines could have occurred after the Manville Trust litigation was completed. Except for this simple illustration, the interactions between litigation events and claims filing rates were beyond the scope of our analysis.

Following the individual sensitivity analyses, we consider the overall sensitivity of the model to the various assumptions, assess the plausible range of

**Table 7.2: Alternative Projections of Qualified Male Claims Against the Manville Trust, 1990-2049, by Most Severe Alleged Disease/Injury**

| Analysis/Model/# | | | Mesothelioma Diagnoses | Number of Claims: Most Severe Alleged Disease/Injury[1] | | | | | | | | | | | | | Total |
|---|---|---|---|---|---|---|---|---|---|---|---|---|---|---|---|---|---|
| | | | | 1 | 2 | 3 | 4 | 5 | 6 | 7 | 8 | 9 | 1-4 | 5-9 | 5-7 | 8-9 | 1-9 |
| Baseline S0 | | 0 | 67,856 | 24,343 | 33,545 | 4,282 | 5,794 | 195,782 | 60,987 | 66,392 | 39,830 | 14,797 | 67,964 | 377,787 | 323,161 | 54,626 | 445,751 |
| S1 | 1 | 1 | — | 25,368 | 37,991 | 4,859 | 5,979 | 232,805 | 71,384 | 77,428 | 49,371 | 17,719 | 74,196 | 448,705 | 381,616 | 67,089 | 522,901 |
| S2 | 1 | 2 | 67,856 | 23,077 | 34,139 | 4,335 | 5,405 | 212,861 | 65,010 | 70,655 | 44,775 | 16,758 | 66,956 | 410,059 | 348,526 | 61,533 | 477,014 |
| S3 | 0 | 3 | 67,856 | 27,410 | 41,368 | 5,174 | 6,569 | 297,895 | 91,300 | 100,436 | 62,092 | 23,187 | 80,521 | 574,909 | 489,631 | 85,278 | 655,431 |
| | 1 | 4 | 67,856 | 26,239 | 35,447 | 4,962 | 5,741 | 190,191 | 60,340 | 64,607 | 39,613 | 13,574 | 72,388 | 368,325 | 315,138 | 53,187 | 440,714 |
| | 2 | 5 | 67,856 | 23,294 | 32,376 | 4,203 | 5,254 | 188,636 | 57,869 | 62,215 | 39,614 | 13,574 | 65,127 | 361,908 | 308,720 | 53,188 | 427,035 |
| | 3 | 6 | 67,856 | 25,612 | 33,442 | 4,606 | 5,455 | 188,175 | 59,344 | 63,665 | 38,570 | 13,713 | 69,115 | 363,467 | 311,184 | 52,283 | 432,582 |
| | 4 | 7 | 67,856 | 25,136 | 33,442 | 4,423 | 5,455 | 188,173 | 59,345 | 63,665 | 38,569 | 13,713 | 68,456 | 363,465 | 311,183 | 52,283 | 431,921 |
| S4 | 0 | 8 | 70,471 | 24,605 | 33,869 | 4,316 | 5,862 | 197,122 | 61,369 | 66,815 | 40,115 | 14,918 | 68,653 | 380,338 | 325,306 | 55,032 | 448,991 |
| | 1 | 9 | 133,238 | 39,550 | 53,318 | 6,593 | 9,412 | 291,568 | 90,481 | 98,504 | 59,544 | 22,038 | 108,873 | 562,134 | 480,553 | 81,581 | 671,008 |
| | 2 | 10 | 97,094 | 43,268 | 58,918 | 7,553 | 10,236 | 339,258 | 107,154 | 116,978 | 69,348 | 24,325 | 119,975 | 657,063 | 563,389 | 93,673 | 777,038 |
| | 3 | 11 | 80,429 | 25,519 | 34,998 | 4,433 | 6,100 | 201,740 | 62,681 | 68,278 | 41,105 | 15,341 | 71,050 | 389,145 | 332,699 | 56,446 | 460,195 |
| | 4 | 12 | 84,516 | 25,860 | 35,419 | 4,475 | 6,190 | 203,445 | 63,163 | 68,819 | 41,475 | 15,499 | 71,944 | 392,401 | 335,427 | 56,974 | 464,346 |
| S5 | 1 | 13 | 54,839 | 18,172 | 25,558 | 3,330 | 4,363 | 145,440 | 44,432 | 47,852 | 30,073 | 11,512 | 51,423 | 279,309 | 237,724 | 41,585 | 330,731 |
| | 2 | 14 | 71,962 | 25,322 | 35,248 | 4,499 | 6,072 | 209,686 | 65,192 | 70,965 | 42,248 | 15,935 | 71,140 | 404,025 | 345,842 | 58,183 | 475,165 |
| | 3 | 15 | 51,277 | 17,391 | 24,436 | 3,184 | 4,175 | 140,898 | 43,152 | 46,554 | 29,046 | 11,105 | 49,186 | 270,755 | 230,604 | 40,151 | 319,941 |
| S6 | 1 | 16 | 74,157 | 23,390 | 32,107 | 4,083 | 5,503 | 189,215 | 58,943 | 64,143 | 38,181 | 14,339 | 65,083 | 364,821 | 312,301 | 52,521 | 429,904 |
| | 2 | 17 | 54,808 | 28,331 | 39,492 | 5,101 | 6,994 | 227,140 | 70,880 | 77,381 | 47,262 | 16,849 | 79,918 | 439,512 | 375,401 | 64,111 | 519,431 |
| S7 | 1 | 18 | 73,804 | 25,358 | 34,765 | 4,419 | 6,122 | 201,009 | 62,367 | 67,958 | 40,964 | 15,339 | 70,665 | 387,637 | 331,333 | 56,304 | 458,301 |
| | 2 | 19 | 62,356 | 23,305 | 32,287 | 4,141 | 5,488 | 190,432 | 59,543 | 64,780 | 38,684 | 14,256 | 65,221 | 367,695 | 314,754 | 52,940 | 432,916 |
| S8 | 0 | 20 | 70,999 | 24,443 | 33,661 | 4,295 | 5,818 | 196,188 | 61,092 | 66,496 | 39,910 | 14,837 | 68,217 | 378,523 | 323,776 | 54,747 | 446,740 |
| | 1 | 21 | 68,340 | 24,356 | 33,560 | 4,284 | 5,797 | 195,832 | 61,000 | 66,405 | 39,839 | 14,801 | 67,997 | 377,877 | 323,237 | 54,641 | 445,875 |
| | 2 | 22 | 67,340 | 24,330 | 33,530 | 4,281 | 5,792 | 195,735 | 60,975 | 66,381 | 39,820 | 14,792 | 67,932 | 377,703 | 323,090 | 54,612 | 445,635 |
| | 3 | 23 | 68,485 | 24,364 | 33,569 | 4,285 | 5,799 | 195,866 | 61,009 | 66,414 | 39,846 | 14,805 | 68,017 | 377,940 | 323,289 | 54,651 | 445,957 |
| | 4 | 24 | 66,929 | 24,312 | 33,508 | 4,278 | 5,787 | 195,653 | 60,954 | 66,359 | 39,804 | 14,783 | 67,885 | 377,553 | 322,966 | 54,588 | 445,438 |
| S9 | 1 | 25 | 67,856 | 20,896 | 29,008 | 3,726 | 4,957 | 172,949 | 54,003 | 58,801 | 35,141 | 13,011 | 58,586 | 333,905 | 285,753 | 48,152 | 392,491 |
| | 2 | 26 | 67,856 | 16,080 | 22,611 | 2,923 | 3,796 | 139,728 | 43,759 | 47,686 | 28,353 | 10,457 | 45,410 | 269,983 | 231,173 | 38,810 | 315,392 |
| | 3 | 27 | 67,856 | 10,911 | 15,645 | 2,025 | 2,564 | 101,614 | 31,860 | 34,786 | 20,619 | 7,595 | 31,144 | 196,474 | 168,260 | 28,214 | 227,618 |
| Median S1-9 | | 28 | 67,856 | 24,443 | 33,569 | 4,316 | 5,792 | 195,832 | 61,000 | 66,405 | 39,839 | 14,801 | 68,217 | 377,877 | 323,237 | 54,641 | 445,875 |
| S10 | 1 | 29 | 27,084 | 31,716 | 41,097 | 5,229 | 9,428 | 222,585 | 65,247 | 72,423 | 46,835 | 18,827 | 87,469 | 425,918 | 360,256 | 65,662 | 513,387 |
| | 2 | 30 | 33,934 | 26,670 | 35,287 | 4,500 | 7,598 | 194,810 | 58,096 | 63,666 | 40,472 | 16,007 | 74,055 | 373,052 | 316,573 | 56,479 | 447,107 |
| | 3 | 31 | 27,734 | 25,693 | 34,444 | 4,428 | 5,617 | 184,109 | 54,893 | 59,627 | 38,811 | 15,578 | 70,181 | 353,018 | 298,629 | 54,390 | 423,200 |

Note 1: See Table 6.16 for descriptive labels for disease categories and subtotal groupings.

(Continued)

**Table 7.2 (Continued)**

| Analysis | Model | # | Mesothelioma Diagnosis | Difference (%) Between Alternative Model and Baseline Model (S0): Most Severe Alleged Disease/Injury[1] 1 | 2 | 3 | 4 | 5 | 6 | 7 | 8 | 9 | 1-4 | 5-9 | 5-7 | 8-9 | Total 1-9 |
|---|---|---|---|---|---|---|---|---|---|---|---|---|---|---|---|---|---|
| S1 | 1 | 1 | — | 4.2 | 13.3 | 13.5 | 3.2 | 18.9 | 17.0 | 16.6 | 24.0 | 19.7 | 9.2 | 18.8 | 18.1 | 22.8 | 17.3 |
| S2 | 1 | 2 | 0.0 | -5.2 | 1.8 | 1.2 | -6.7 | 8.7 | 6.6 | 6.4 | 12.4 | 13.3 | -1.5 | 8.5 | 7.8 | 12.6 | 7.0 |
| S3 | 0 | 3 | 0.0 | 12.6 | 23.3 | 20.8 | 13.4 | 52.2 | 49.7 | 51.3 | 55.9 | 56.7 | 18.5 | 52.2 | 51.5 | 56.1 | 47.0 |
| | 1 | 4 | 0.0 | 7.8 | 5.7 | 15.9 | -0.9 | -2.9 | -1.1 | -2.7 | -0.5 | -8.3 | 6.5 | -2.5 | -2.5 | -2.6 | -1.1 |
| | 2 | 5 | 0.0 | -4.3 | -3.5 | -1.9 | -9.3 | -3.6 | -5.1 | -6.3 | -0.5 | -8.3 | -4.2 | -4.2 | -4.5 | -2.6 | -4.2 |
| | 3 | 6 | 0.0 | 5.2 | -0.3 | 7.6 | -5.9 | -3.9 | -2.7 | -4.1 | -3.2 | -7.3 | 1.7 | -3.8 | -3.7 | -4.3 | -3.0 |
| | 4 | 7 | 0.0 | 3.3 | -0.3 | 3.3 | -5.9 | -3.9 | -2.7 | -4.1 | -3.2 | -7.3 | 0.7 | -3.8 | -3.7 | -4.3 | -3.1 |
| S4 | 0 | 8 | 3.9 | 1.1 | 1.0 | 0.8 | 1.2 | 0.7 | 0.6 | 0.6 | 0.7 | 0.8 | 1.0 | 0.7 | 0.7 | 0.7 | 0.7 |
| | 1 | 9 | 96.4 | 62.5 | 58.9 | 54.0 | 62.4 | 48.9 | 48.4 | 48.4 | 49.5 | 48.9 | 60.2 | 48.8 | 48.7 | 49.3 | 50.5 |
| | 2 | 10 | 43.1 | 77.7 | 75.6 | 76.4 | 76.7 | 73.3 | 75.7 | 76.2 | 74.1 | 64.4 | 76.5 | 73.9 | 74.3 | 71.5 | 74.3 |
| | 3 | 11 | 18.5 | 4.8 | 4.3 | 3.5 | 5.3 | 3.0 | 2.8 | 2.8 | 3.2 | 3.7 | 4.5 | 3.0 | 3.0 | 3.3 | 3.2 |
| | 4 | 12 | 24.6 | 6.2 | 5.6 | 4.5 | 6.8 | 3.9 | 3.6 | 3.7 | 4.1 | 4.7 | 5.9 | 3.9 | 3.8 | 4.3 | 4.2 |
| S5 | 1 | 13 | -19.2 | -25.4 | -23.8 | -22.2 | -24.7 | -25.7 | -27.1 | -27.9 | -24.5 | -22.2 | -24.3 | -26.1 | -26.4 | -23.9 | -25.8 |
| | 2 | 14 | 6.1 | 4.0 | 5.1 | 5.1 | 4.8 | 7.1 | 6.9 | 6.9 | 6.1 | 7.7 | 4.7 | 6.9 | 7.0 | 6.5 | 6.6 |
| | 3 | 15 | -24.4 | -28.6 | -27.2 | -25.6 | -27.9 | -28.0 | -29.2 | -29.9 | -27.1 | -24.9 | -27.6 | -28.3 | -28.6 | -26.5 | -28.2 |
| S6 | 1 | 16 | 9.3 | -3.9 | -4.3 | -4.6 | -5.0 | -3.4 | -3.4 | -3.4 | -4.1 | -3.1 | -4.2 | -3.4 | -3.4 | -3.9 | -3.6 |
| | 2 | 17 | -19.2 | 16.4 | 17.7 | 19.1 | 20.7 | 16.0 | 16.2 | 16.6 | 18.7 | 13.9 | 17.6 | 16.3 | 16.2 | 17.4 | 16.5 |
| S7 | 1 | 18 | 8.8 | 4.2 | 3.6 | 3.2 | 5.7 | 2.7 | 2.3 | 2.4 | 2.8 | 3.7 | 4.0 | 2.6 | 2.5 | 3.1 | 2.8 |
| | 2 | 19 | -8.1 | -4.3 | -3.8 | -3.3 | -5.3 | -2.7 | -2.4 | -2.4 | -2.9 | -3.7 | -4.0 | -2.7 | -2.6 | -3.1 | -2.9 |
| S8 | 0 | 20 | 4.6 | 0.4 | 0.3 | 0.3 | 0.4 | 0.2 | 0.2 | 0.2 | 0.2 | 0.3 | 0.4 | 0.2 | 0.2 | 0.2 | 0.2 |
| | 1 | 21 | 0.7 | 0.1 | 0.0 | 0.0 | 0.0 | 0.0 | 0.0 | 0.0 | 0.0 | 0.0 | 0.0 | 0.0 | 0.0 | 0.0 | 0.0 |
| | 2 | 22 | -0.8 | -0.1 | 0.0 | 0.0 | 0.0 | 0.0 | 0.0 | 0.0 | 0.0 | 0.0 | 0.0 | 0.0 | 0.0 | 0.0 | 0.0 |
| | 3 | 23 | 0.9 | 0.1 | 0.1 | 0.1 | 0.1 | 0.0 | 0.0 | 0.0 | 0.0 | 0.1 | 0.1 | 0.0 | 0.0 | 0.0 | 0.0 |
| | 4 | 24 | -1.4 | -0.1 | -0.1 | -0.1 | -0.1 | -0.1 | -0.1 | -0.1 | -0.1 | -0.1 | -0.1 | -0.1 | -0.1 | -0.1 | -0.1 |
| S9 | 1 | 25 | 0.0 | -14.2 | -13.5 | -13.0 | -14.4 | -11.7 | -11.5 | -11.4 | -11.8 | -12.1 | -13.8 | -11.6 | -11.6 | -11.9 | -11.9 |
| | 2 | 26 | 0.0 | -33.9 | -32.6 | -31.7 | -34.5 | -28.6 | -28.2 | -28.2 | -28.8 | -29.3 | -33.2 | -28.5 | -28.5 | -29.0 | -29.2 |
| | 3 | 27 | 0.0 | -55.2 | -53.4 | -52.7 | -55.8 | -48.1 | -47.8 | -47.6 | -48.2 | -48.7 | -54.2 | -48.0 | -47.9 | -48.4 | -48.9 |
| Median S1-9 | | 28 | 0.0 | 0.4 | 0.1 | 0.8 | 0.0 | 0.0 | 0.0 | 0.0 | 0.0 | 0.0 | 0.4 | 0.0 | 0.0 | 0.0 | 0.0 |
| S10 | 1 | 29 | -60.1 | 30.3 | 22.5 | 22.1 | 62.7 | 13.7 | 7.0 | 9.1 | 17.6 | 27.2 | 28.7 | 12.7 | 11.5 | 20.2 | 15.2 |
| | 2 | 30 | -50.0 | 9.6 | 5.2 | 5.1 | 31.1 | -0.5 | -4.7 | -4.1 | 1.6 | 8.2 | 9.0 | -1.3 | -2.0 | 3.4 | 0.3 |
| | 3 | 31 | -59.1 | 5.5 | 2.7 | 3.4 | -3.1 | -6.0 | -10.0 | -10.2 | -2.6 | 5.3 | 3.3 | -6.6 | -7.6 | -0.4 | -5.1 |

Note 1: See Table 6.16 for descriptive labels for disease categories and subtotal groupings.

Source: Authors' calculations based on Stallard and Manton (1993).

uncertainty in these projections, and identify additional issues to be resolved in developing a more complete analysis of uncertainty in this class of models.

Finally, we present a set of analyses (S10) that forms a "bridge" between the updated form of Walker's (1982) model in Chapters 6 and 7 and the hybrid form of Walker's (1982) and Selikoff's (1981) model in Chapters 8 and 9. These analyses dropped the SEER data from step 2 of the first-stage calibration (see Section 6.5.2) and replaced them with claim data from the Manville Trust. This was necessary because the hybrid model was stratified by occupation/industry. The claim data provided the required information on occupation/industry, but the SEER data did not. The bridge model was needed to assess the amount of distortion, if any, induced by replacing the SEER data in step 2 with claim data.

## 7.2 Analysis S1: Constant Age-Specific Claim Runoff

In Section 6.5, we assumed that the claim counts for 1990-1994 were simple multiples of the counts from January 1, 1990 through November 30, 1992. This is reflected in Table 6.14 and is consistent with Figures 6.2-6.6. Figure 6.6 suggests that there was a shift in alleged disease/injury from nonasbestos-related and unknown diseases to asbestos-related noncancers, primarily asbestosis. Because the total claims in Figure 6.2 were constant through November 1992, as were the percentages due to asbestos-related cancer in Figure 6.6, it is likely that this shift reflected better claims preparation on the part of claimants and their legal representatives, rather than a real change in the underlying distribution of disease. Under this assumption, the average claim filing rates for each disease for 1990-1992 were applied to 1990-1994. It was instructive to determine the consequences of extending this assumption throughout the period 1990-2049, with deletion of the youngest age group for each successive quinquennium, starting with age 15-19 in 1975-1979. This yielded a claim projection with a minimum number of assumptions and provided a benchmark against which the first- and second-stage baseline calibrations can be evaluated. The results are summarized in Table 7.2.

For mesothelioma, the total projected claims count of 25,368 males for 1990-2049 was 4.2% higher than baseline. For all diseases, the total projected claim count of 522,901 males for 1990-2049 was 17.3% higher than baseline.

For lung cancer, the projected increase over baseline for 1990-2049 was 13.3%; for asbestosis, 18.9%; for asbestos-related noncancer, 18.1%; and for nonasbestos-related and unknown diseases, 22.8%. These results indicate that important aspects of the dynamics of mesothelioma and other diseases, especially noncancers, cannot be represented by a constant claim runoff model.

## 7.3 Analysis S2: Ratio Estimation of Nine Asbestos-Related Diseases – PTS Model

In this alternative, we scaled the mesothelioma incidence counts in Table 6.11 to the levels corresponding to each of the nine categories of disease used by the Manville Trust in classifying alleged disease/injuries. These categories were discussed in Sections 6.2 and 6.5.

We estimated (Table 6.14) 3607 male mesothelioma claims filed with the Trust for 1990-1994 at ages 35 and above. The corresponding national estimate of incident cases of mesothelioma among males from Table 6.11 was 8948. Thus, each entry in Table 6.11 could be multiplied by the ratio 0.4031 = 3607/8948, representing propensity to sue (PTS), to scale estimates to the level filed against the Manville Trust. However, because the mesothelioma CHRs underlying Table 6.16 exhibited unimodal age-specific patterns, it was more realistic to assume that the PTS rates also vary by attained age at the claim filing date. This ensured that the model reproduced the age-specific distribution of claims for 1990-1994 for each of the nine disease/injuries.

In general, let $n^7_{A,1990-94}$ be the number of mesothelioma claims at age group $A$ in Table 6.14. Let $n^d_{A,1990-94}$ be the corresponding number in Table 6.14 for any other disease, $d$. Then, the age-specific ratio estimates for each disease $d$ are

$$\begin{aligned} n^d_{A,Y} &= M_{A,Y} \times \frac{n^7_{A,1990-94}}{M_{A,1990-94}} \times \frac{n^d_{A,1990-94}}{n^7_{A,1990-94}} \\ &= M_{A,Y} \times \frac{n^d_{A,1990-94}}{M_{A,1990-94}}, \end{aligned}$$

where $M_{A,Y}$ is the projected mesothelioma incidence count at age group $A$ in calendar period $Y$ under the first-stage calibration model (Table 6.11; see Section 6.7). This calculation replaced the second-stage calibration based on the CHR model with an alternative second-stage calibration based on the PTS model with constant age-specific PTS ratios over disease/injuries. The results are summarized in Table 7.2.

For 1990-1994, analysis S2 reproduced the age-specific baseline claim counts. The total claims for 1990-2049 were 7.0% higher than baseline (i.e., 477,014 vs. 445,751). For mesothelioma, the claims for 1990-2049 were 5.2% lower than baseline (23,077 vs. 24,343). The projected average PTS was 34.0% (vs. 40.3% for 1990-1994) for mesothelioma for 1990-2049, based on comparison of Table 7.2 and Table 6.13 (i.e., 23,077 vs. 67,856). The use of age-specific PTSs implied lower projected claim counts for mesothelioma and other cancers, but higher claim counts for lung cancer and colon/rectal cancer. The projected claim counts were higher for all noncancers. Overall, the changes were modest and supported the use of the CHR model in the second-stage baseline calibration.

## 7.4 Analysis S3: Parametric Claim Hazard Rate Model

In Section 6.8, we developed nonparametric hazard rate estimates for the claim filing rates for nine asbestos-related diseases. In this section, we develop model-based hazard rates estimated for the same data using Weibull and gamma-mixed Weibull models. Such models smooth out irregularities in the nonparametric estimates underlying Table 6.15. They also allow comparison of the parameters across different disease/injuries. In particular, we were interested in the extent to which the Weibull $k$-parameter estimates agree. If they were similar for all diseases, it would justify the assumption in alternative S2 that the nonmesothelioma claim rates were proportional to the mesothelioma claim rates. To the extent that the $k$ parameters were dissimilar, it lends further credibility to the baseline CHR model.

We fitted five models to the claim data used in the second-stage calibration of the CHR model (see Section 6.8). The hazard rates were modeled as follows with parameters that maximize the Poisson likelihood:

$$L_d = \prod_A \prod_T \left(\widetilde{P}^T_{A,Y_2} \mu^{T,d}_A\right)^{\widetilde{n}^{T,d}_{A,Y_2}} \exp\left(-\widetilde{P}^T_{A,Y_2} \mu^{T,d}_A\right) / \widetilde{n}^{T,d}_{A,Y_2}!,$$

where $d$ indexes disease, $T$ indexes TSFE (in quinquennia), $A$ indexes age group, $Y_2 = 1990\text{-}1994$, and $\mu^{T,d}_A$ is the claim hazard rate for disease $d$, at age $A$ and TSFE $T$. The observed claim counts for 1990-1992 were used to get unbiased test statistics. The exposure indicator $P^T_{A,Y_2}$ was adjusted to reflect this change; that is, each entry in Table 6.10 was divided by 1.715 to generate $\widetilde{P}^T_{A,Y_2}$, the person-years of exposure underlying the claims in Table 6.14. Similarly, the claim estimates in Table 6.14 were divided by 1.715 to generate $\widetilde{n}^T_{A,Y_2}$, the corresponding values of the TSFE claim distribution. These adjustments imply that the model based (quinquennial) estimator of $H^{T,d}_{A,Y_2}$ is

$$\widehat{H}^{T,d}_{A,Y_2} = \mu^{T,d}_A.$$

With $T = (t - 2.5, t + 2.5)$, and $t$ measured in years, we defined Model 0 (the null model for this section):

$$\mu^{T,d}_A = b_d t^k \qquad (k \equiv 3.2),$$

with a disease-specific proportionality constant $b_d$.

In Section 6.8, we observed that the age-specific marginal CHRs were unimodal with a peak generally at age 65-69. This implied that, at younger and older ages, the CHRs were lower than at the peak age. We developed formal tests of this hypothesis by specifying appropriate alternative models to the one earlier.

We used two age specifications for the proportionality constants: no age variation (Models 1 and 2) versus four age bands (Models 3 and 4 – ages 15-49, 50-64, 65-74, and 75-99). In addition, we specified two functional forms:

Weibull (Models 1 and 3) versus gamma-mixed Weibull (Model 2 and 4). The Weibull implies a monotonic increase in CHR with increasing TSFE. The gamma-mixed Weibull starts out similar to the Weibull, but it can reach a peak value and then plateau or decrease with increasing TSFE. The four models were as follows:

Model 1 (releases constraint on $k$ in Model 0):

$$\mu_A^{T,d} = b_d t^{k_d},$$

Model 2 (distributed risk):

$$\mu_A^{T,d} = b_d t^{k_d} / \left[1 + b_d c_d t^{k_d+1}/(k_d+1)\right],$$

Model 3 (age-specific constant):

$$\mu_A^{T,d} = b_{A,d} t^{k_d},$$

Model 4 (age-specific distribution of risk):

$$\mu_A^{T,d} = b_{A,d} t^{k_d} / \left[1 + b_{A,d} c_d t^{k_d+1}/(k_d+1)\right].$$

Parameter restrictions were $b > 0$ and $c \geq 0$. If $c_d = 0$, then Models 2 and 4 simplify to Models 1 and 3, respectively.

The parameter $c_d$ in Model 2 is the squared coefficient of variation of a risk distribution representing individual differences in the parameter $b_d$ in Model 1. Manton and Stallard (1979) showed that if individual susceptibilities to risk were gamma distributed with mean equal to $b_d$ and variance equal to $b_d^2 \times c_d$, then the Weibull marginal hazard in Model 1 would be transformed to the gamma-mixed form in Model 2.

In Models 3 and 4, $b_{A,d}$ is constant over quinquennia within four broad age groups: 15-49, 50-64, 65-74, and 75-99 years. The parameter estimates and goodness-of-fit chi-squared statistics are in Table 7.3.

All five models were estimated using 80 observations (see Table 6.15). Hence, the degrees of freedom (d.f.) for evaluating chi-squared statistics are 79, 78, 77, 75, and 74 d.f., respectively. At 78 d.f., $\chi^2_{.05} = 99.3$ and $\chi^2_{.01} = 109.2$: no model achieved an acceptable fit for any disease. Model 4 provided improvement (i.e., $c_d > 0$) only for mesothelioma and colon/rectal cancer. Model 3 is the most detailed model that is fitted for all nine diseases. Compared to Model 1, Model 3 was highly significantly improved for every disease ($\chi^2$ difference, 3 d.f.). The absolute $\chi^2$ values for Model 3 indicated substantial variation in quality of fit. The rank order of fit dichotomized the nine diseases into cancers versus noncancers:

1. Cancers
   - Colon/rectal cancer ($\chi^2 = 120.6$; $k = 3.14 \pm 0.29$)
   - Other cancer ($\chi^2 = 146.4$; $k = 2.78 \pm 0.25$)

**Table 7.3: Parameter Estimates and Standard Errors for Five Models of Claim Hazard Rates – Qualified Male Claims Against the Manville Trust, 1990-1994**

| | | | | Age Groups for Models 3-4 | | | | | | | | | | |
|---|---|---|---|---|---|---|---|---|---|---|---|---|---|---|
| | | | | All ages/15-49 | | 50-64 | | 65-74 | | 75-99 | | | | |
| Disease | Model | $k$ | Std. Error $k$ | $b$ | Std. Error $b$ | $b$ | Std. Error $b$ | $b$ | Std. Error $b$ | $b$ | Std. Error $b$ | $c$ | Std. Error $c$ | Chi-Squared |
| 1. Mesothelioma | 0 | 3.20 | — | 1.77E-08 | 3.9E-10 | | | | | | | 0.00 | — | 546.6 |
| | 1 | 2.96 | 0.09 | 4.38E-08 | 1.4E-08 | | | | | | | 0.00 | — | 538.7 |
| | 2 | 4.29 | 0.22 | 4.72E-10 | 3.7E-10 | | | | | | | 11.03 | 1.76 | 484.7 |
| | 3 | 3.61 | 0.13 | 3.39E-09 | 1.5E-09 | 5.79E-09 | 2.8E-09 | 4.42E-09 | 2.2E-09 | 2.23E-09 | 1.2E-09 | 0.00 | — | 306.8 |
| | 4 | 3.92 | 0.26 | 1.24E-09 | 1.1E-09 | 2.01E-09 | 1.8E-09 | 1.52E-09 | 1.4E-09 | 7.35E-10 | 7.1E-10 | 2.34 | 1.71 | 304.8 |
| 2. Lung cancer | 0 | 3.20 | — | 2.67E-08 | 4.7E-10 | | | | | | | 0.00 | — | 1,102.1 |
| | 1 | 2.36 | 0.07 | 6.16E-07 | 1.5E-07 | | | | | | | 0.00 | — | 946.5 |
| | 2 | 3.57 | 0.18 | 1.10E-08 | 6.7E-09 | | | | | | | 7.59 | 1.07 | 882.7 |
| | 3 | 2.34 | 0.10 | 2.62E-07 | 8.6E-08 | 8.86E-07 | 3.1E-07 | 8.42E-07 | 3.1E-07 | 4.36E-07 | 1.7E-07 | 0.00 | — | 513.5 |
| | 4 | 2.34 | 0.21 | 2.62E-07 | 1.8E-07 | 8.86E-07 | 6.2E-07 | 8.42E-07 | 5.9E-07 | 4.36E-07 | 3.2E-07 | 0.00 | 1.09 | 513.5 |
| 3. Colon/rectal cancer | 0 | 3.20 | — | 3.33E-09 | 1.7E-10 | | | | | | | 0.00 | — | 188.7 |
| | 1 | 2.97 | 0.20 | 7.98E-09 | 5.9E-09 | | | | | | | 0.00 | — | 187.3 |
| | 2 | 6.08 | 0.64 | 1.93E-13 | 4.3E-13 | | | | | | | 127.94 | 25.57 | 148.4 |
| | 3 | 3.14 | 0.29 | 1.71E-09 | 1.7E-09 | 5.31E-09 | 5.6E-09 | 5.76E-09 | 6.4E-09 | 2.37E-09 | 2.8E-09 | 0.00 | — | 120.6 |
| | 4 | 4.57 | 0.66 | 1.65E-11 | 3.6E-11 | 3.79E-11 | 8.7E-11 | 4.56E-11 | 1.1E-10 | 1.36E-11 | 3.3E-11 | 54.38 | 22.45 | 113.7 |
| 4. Other cancer | 0 | 3.20 | — | 4.23E-09 | 1.9E-10 | | | | | | | 0.00 | — | 257.5 |
| | 1 | 2.46 | 0.17 | 6.62E-08 | 4.1E-08 | | | | | | | 0.00 | — | 238.8 |
| | 2 | 3.53 | 0.45 | 1.88E-09 | 2.9E-09 | | | | | | | 41.52 | 16.59 | 231.1 |
| | 3 | 2.78 | 0.25 | 9.13E-09 | 7.8E-09 | 3.18E-08 | 2.9E-08 | 2.51E-08 | 2.4E-08 | 1.13E-08 | 1.1E-08 | 0.00 | — | 146.4 |
| | 4 | 2.78 | 0.54 | 9.13E-09 | 1.6E-08 | 3.18E-08 | 5.8E-08 | 2.51E-08 | 4.6E-08 | 1.13E-08 | 2.2E-08 | 0.00 | 18.77 | 146.4 |
| 5. Asbestosis | 0 | 3.20 | — | 1.92E-07 | 1.3E-09 | | | | | | | 0.00 | — | 16,254.4 |
| | 1 | 0.87 | 0.02 | 1.00E-03 | 7.9E-05 | | | | | | | 0.00 | — | 6,113.3 |
| | 2 | 1.03 | 0.07 | 6.15E-04 | 1.3E-04 | | | | | | | 0.19 | 0.08 | 6,106.8 |
| | 3 | 1.41 | 0.03 | 1.53E-04 | 1.7E-05 | 2.01E-04 | 2.5E-05 | 1.49E-04 | 1.9E-05 | 6.71E-05 | 9.1E-06 | 0.00 | — | 3,690.2 |
| | 4 | 1.41 | 0.09 | 1.53E-04 | 4.1E-05 | 2.01E-04 | 5.4E-05 | 1.49E-04 | 4.2E-05 | 6.71E-05 | 2.1E-05 | 0.00 | 0.09 | 3,690.2 |

(Continued)

**Table 7.3 (Continued)**

| | | | | Age Groups for Models 3-4 | | | | | | | | | | |
|---|---|---|---|---|---|---|---|---|---|---|---|---|---|---|
| | | | | All ages/15-49 | | 50-64 | | 65-74 | | 75-99 | | | | |
| Disease | Model | $k$ | Std. Error $k$ | $b$ | Std. Error $b$ | $b$ | Std. Error $b$ | $b$ | Std. Error $b$ | $b$ | Std. Error $b$ | $c$ | Std. Error $c$ | Chi-Squared |
| 6. Disputed asbestosis | 0 | 3.20 | — | 5.88E-08 | 7.0E-10 | | | | | | | 0.00 | — | 5,073.8 |
| | 1 | 1.04 | 0.04 | 1.69E-04 | 2.4E-05 | | | | | | | 0.00 | — | 2,437.8 |
| | 2 | 2.09 | 0.13 | 6.34E-06 | 2.7E-06 | | | | | | | 3.89 | 0.47 | 2,361.8 |
| | 3 | 1.73 | 0.06 | 1.44E-05 | 3.0E-06 | 2.16E-05 | 4.9E-06 | 1.43E-05 | 3.4E-06 | 5.09E-06 | 1.3E-06 | 0.00 | — | 1,178.0 |
| | 4 | 1.73 | 0.15 | 1.44E-05 | 6.9E-06 | 2.16E-05 | 1.1E-05 | 1.43E-05 | 7.4E-06 | 5.09E-06 | 2.9E-06 | 0.00 | 0.46 | 1,178.0 |
| 7. Pleural plaques/ thickening | 0 | 3.20 | — | 6.47E-08 | 7.4E-10 | | | | | | | 0.00 | — | 6,210.9 |
| | 1 | 0.91 | 0.04 | 2.97E-04 | 4.1E-05 | | | | | | | 0.00 | — | 2,908.0 |
| | 2 | 1.93 | 0.13 | 1.27E-05 | 5.3E-06 | | | | | | | 3.52 | 0.44 | 2,835.8 |
| | 3 | 1.63 | 0.06 | 2.28E-05 | 4.5E-06 | 3.54E-05 | 7.5E-06 | 2.20E-05 | 5.0E-06 | 7.40E-06 | 1.8E-06 | 0.00 | — | 1,352.3 |
| | 4 | 1.63 | 0.14 | 2.28E-05 | 1.0E-05 | 3.54E-05 | 1.6E-05 | 2.20E-05 | 1.1E-05 | 7.40E-06 | 3.9E-06 | 0.00 | 0.41 | 1,352.3 |
| 8. Nonasbestos-related disease | 0 | 3.20 | — | 4.00E-08 | 5.8E-10 | | | | | | | 0.00 | — | 3,913.6 |
| | 1 | 0.87 | 0.05 | 2.12E-04 | 3.7E-05 | | | | | | | 0.00 | — | 1,792.7 |
| | 2 | 0.87 | 0.14 | 2.12E-04 | 8.9E-05 | | | | | | | 0.00 | 0.74 | 1,792.7 |
| | 3 | 1.12 | 0.07 | 7.57E-05 | 1.8E-05 | 1.10E-04 | 2.9E-05 | 1.00E-04 | 2.8E-05 | 4.44E-05 | 1.3E-05 | 0.00 | — | 1,347.8 |
| | 4 | 1.12 | 0.18 | 7.57E-05 | 4.1E-05 | 1.10E-04 | 6.1E-05 | 1.00E-04 | 5.7E-05 | 4.44E-05 | 2.8E-05 | 0.00 | 0.93 | 1,347.8 |
| 9. Unknown | 0 | 3.20 | — | 1.49E-08 | 3.5E-10 | | | | | | | 0.00 | — | 1,896.5 |
| | 1 | 0.46 | 0.08 | 3.40E-04 | 9.4E-05 | | | | | | | 0.00 | — | 764.4 |
| | 2 | 0.46 | 0.27 | 3.40E-04 | 2.6E-04 | | | | | | | 0.00 | 4.76 | 764.4 |
| | 3 | 0.83 | 0.12 | 1.13E-04 | 4.4E-05 | 9.19E-05 | 3.9E-05 | 8.78E-05 | 4.0E-05 | 6.96E-05 | 3.3E-05 | 0.00 | — | 744.1 |
| | 4 | 0.83 | 0.56 | 1.13E-04 | 1.8E-04 | 9.19E-05 | 1.6E-04 | 8.78E-05 | 1.5E-04 | 6.96E-05 | 1.2E-04 | 0.00 | 9.10 | 744.1 |

Source: Authors' calculations based on Stallard and Manton (1993).

- Mesothelioma ($\chi^2 = 306.8$; $k = 3.61 \pm 0.13$)
- Lung cancer ($\chi^2 = 513.5$; $k = 2.34 \pm 0.10$)

with all $k$'s in the range 2.3 to 3.6.

2. Noncancers
   - Unknown disease ($\chi^2 = 744.1$; $k = 0.83 \pm 0.12$)
   - Disputed asbestosis ($\chi^2 = 1{,}178.0$; $k = 1.73 \pm 0.06$)
   - Nonasbestos-related disease ($\chi^2 = 1{,}347.8$; $k = 1.12 \pm 0.07$)
   - Pleural plaques/thickening ($\chi^2 = 1{,}352.3$; $k = 1.63 \pm 0.06$)
   - Asbestosis ($\chi^2 = 3{,}690.2$; $k = 1.41 \pm 0.03$)

   with all $k$'s in the range 0.8 to 1.7.

The $k$-parameter estimates under Model 1 fell into three distinct groups:

1. Values near 0.9, for noncancers and nonasbestos-related diseases
2. Values near 2.4, for lung cancer and other cancer
3. Values near 3.0, for mesothelioma and colon/rectal cancer

A $k$ value of 1 implies that the claim filing rate is proportional to TSFE.

These results indicate that the null hypothesis in Model 0, that $k = 3.2$ for all nine diseases, was wrong. The hypothesis was plausible for the four cancers, but not for the noncancers where $k < 2$. The $k$-parameter estimates were higher under Model 3 (except for lung cancer). In addition, the $b$ parameters in Model 3 peaked at ages 50-64 or 65-74, with lower values at both younger and older ages. These patterns provide insight into the behavior of the age-specific PTS model in analysis S2. The assumption of the PTS model that nonmesothelioma claims increased with TSFE at the same rate as mesothelioma claims was reasonable for the other three cancers, but yielded overpredictions for asbestos-related noncancers and nonasbestos-related diseases. The overpredictions were ameliorated by the use of age-specific PTS ratios.

The projection results for the parametric CHR models are in Table 7.2. The most serious discrepancies with the baseline CHR model occurred with Model 0, where the total claims 1990-2049 were 47.0% higher. The total claims 1990-2049 in Model 1 were only 1.1% lower than in the baseline model. Model 1 assumed no age effects, only TSFE effects. Comparison of Model 3 with the baseline model showed a 3.0% decrease in total claims 1990-2049. Although neither Model 1 nor 3 achieved a statistical fit, their predictions were close to those of the nonparametric CHR model. The two models reinforced each other and provided consistent sets of outcomes.

Comparison of Models 1 and 3 showed the effects of age specific $b$'s on the CHR rates. The total claims 1990-2049 declined 1.8% in Model 3. Once the effects of disease-specific $k$'s had been represented, the additional effects of age on the CHRs were less important.

The findings of significant $c$-parameter values for mesothelioma and colon/rectal cancer in Model 4 meant that the upward rate of increase of the CHR model was slower than predicted by the Weibull model. This reduced the Model 3 projection for 1990-2049 for mesothelioma by 1.9% and colon/rectal cancer by 4.0%. The resulting projections were each 3.3% higher than baseline.

The $k$-parameter estimate for mesothelioma under Model 3 was 3.61 – 12.8% higher than the value 3.20 assumed for mesothelioma claims in the first-stage projection. The corresponding estimate under Model 1 was 2.96 – 7.5% lower than the assumed first-stage value of 3.20. The average of the Model 1 and Model 3 estimates was 3.29 – within 2.8% of the assumed value of 3.20. These comparisons suggest that the Manville mesothelioma claims may provide an appropriate alternative source of data for the first-stage calibration (see Section 7.12).

## 7.5 Analysis S4: Mesothelioma Incidence Function

The Weibull function expresses the incidence rate for mesothelioma as a function of TSFE. This function was derived by Peto et al. (1982), who showed that the incidence was independent of age at first exposure in a cohort of 17,800 North American insulation workers. The parameters $b$ and $k$ were estimated by Peto et al. (1982) based on 180 mesothelioma deaths occurring in 48,812 man-years of exposure (i.e., counting each year of survival after first exposure as 1 man-year). Their estimate of $k = 3.20$ had a standard error of $\pm 0.36$, so a 95% confidence interval was 2.49 to 3.91. Their preferred estimate was 3.50. In addition, their estimate was based on man-years of exposure below age 80, for which the year of first exposure was between 1922 and 1946. The period of observation was 1967-1979.

More recent data were provided by Selikoff and Seidman (1991) for all 17,800 insulation workers for 1967-1986 (see Table 2.1). During this period, 458 mesothelioma deaths (173 pleural; 285 peritoneal) were observed in 301,593 man-years of exposure (239,937 at 15+ years from onset of asbestos exposure; 6151 at 50+ years). This represents a fivefold increase over the data available to Peto et al. (1982). The data are displayed in Table 7.4 by TSFE categories.

Also in Table 7.4 are data on 427 asbestosis deaths and 1168 deaths due to lung cancer. The standardized mortality ratio (SMR – the ratio of observed to expected deaths) for lung cancer was 4.35 (Selikoff and Seidman, 1991), implying 899 excess deaths. The SMR for all deaths was 1.43, implying 1498 excess deaths. In addition, there was a deficit for noncancer, nonrespiratory causes of 398 deaths (a healthy worker effect), for a "true" excess of 1896 deaths. Hence, mesothelioma (458), asbestosis (427), and lung cancer (899) account for 94.1% (1784 of 1896) of the true excess.

Table 7.5 presents the parameter estimates for three forms of Weibull models fitted to the data in Table 7.4, both excluding and including data for 50+ years of exposure. The average TSFE above 50 years was not provided. For 15-49 years, we used the midpoint of each quinquennium to represent TSFE. Exposures below 15 years were deleted.

**Table 7.4: Asbestos-Related Deaths by Cause and Time Since First Exposure to Asbestos Among 17,800 North American Insulation Workers**

| Time Since First Exposure (TSFE) | | Person-Years | Cause of Death | | | | |
|---|---|---|---|---|---|---|---|
| | | | Mesothelioma | | | Lung | Asbest- |
| Interval | Midpoint | Exposure | Total | Pleural | Peritoneal | Cancer | osis |
| 0-14 | 7.5 | 61,655 | 0 | 0 | 0 | 9 | 1 |
| 15-19 | 17.5 | 52,710 | 5 | 2 | 3 | 37 | 14 |
| 20-24 | 22.5 | 57,595 | 18 | 10 | 8 | 95 | 31 |
| 25-29 | 27.5 | 50,519 | 73 | 33 | 40 | 183 | 52 |
| 30-34 | 32.5 | 37,166 | 105 | 40 | 65 | 281 | 59 |
| 35-39 | 37.5 | 20,340 | 91 | 33 | 58 | 239 | 84 |
| 40-44 | 42.5 | 10,201 | 59 | 17 | 42 | 155 | 80 |
| 45-49 | 47.5 | 5,257 | 58 | 27 | 31 | 75 | 33 |
| 50+ | 52.5 | 6,151 | 49 | 11 | 38 | 94 | 73 |
| Total | | 301,593 | 458 | 173 | 285 | 1,168 | 427 |

Source: Selikoff and Seidman (1991, Tables 4, 5, 6, and 7).

Model 0 (the null model for this section) is the original Weibull in Section 2.3.1d with $k = 3.2$, except that $b$ was allowed to vary to maximize the Poisson likelihood,

$$L_d = \prod_T (P_T \mu_{T,d})^{n_{T,d}} \exp(-P_T \mu_{T,d}) / n_{T,d}!,$$

where $d$ indexes disease, $T$ indexes quinquennia, $P_T$ is the man-years of exposure at $T \equiv (t - 2.5, t + 2.5)$, and $n_{T,d}$ is the corresponding number of deaths. With $t$ measured in years, at the midpoint of each exposure interval,

$$\mu_{T,d} \equiv \mu_{t,d} = b_d t^{k_d}.$$

With one free parameter, the chi-squared statistics (based on the likelihood ratio transformation), have either 6 or 7 d.f., depending on whether TSFE 50+ years is included (using $t = 52.5$ for the final observation). Our discussion focuses on the data with TSFE $< 50$ years.

For mesothelioma, Model 0 produced $\chi^2 = 40.43$ (6 d.f., $p < 0.001$) and $b = 3.58 \pm 0.18 \times 10^{-8}$. Model 0 failed to fit either or both forms of mesothelioma. It also failed to fit asbestosis and lung cancer. The $b$ estimate of $3.58 \times 10^{-8}$ for mesothelioma was significantly lower than the $4.37 \times 10^{-8}$ in Section 2.3.1d.

Model 1 freed $k$ to be the maximum likelihood estimate. For mesothelioma, $\chi^2 = 17.08$ (5 d.f., $0.01 > p > 0.001$), $k = 4.18 \pm 0.21$. For peritoneal mesothelioma, $\chi^2 = 12.32$ (5 d.f., $0.05 > p > 0.01$), $k = 4.35 \pm 0.27$; for pleural mesothelioma, $\chi^2 = 11.22$ (5 d.f., $0.05 > p > 0.01$), $k = 3.93 \pm 0.32$.

**Table 7.5: Parameter Estimates and Standard Errors for Asbestos-Related Deaths Among 17,800 North American Insulation Workers**

| Disease | Model | $k$ | Std. Error $k$ | $b$ | Std. Error $b$ | $c$ | Std. Error $c$ | Chi-Squared |
|---|---|---|---|---|---|---|---|---|
| | | | Estimates based on TSFEs 15-49 | | | | | |
| 1. Mesothelioma | 0 | 3.20 | — | 3.58E-08 | 1.8E-09 | 0.00 | — | 40.43 |
| | 1 | 4.18 | 0.21 | 1.14E-09 | 8.5E-10 | 0.00 | — | 17.08 |
| | 2 | 5.28 | 0.52 | 3.20E-11 | 5.5E-11 | 8.91 | 3.94 | 10.66 |
| 1.1 Pleural | 0 | 3.20 | — | 1.42E-08 | 1.1E-09 | 0.00 | — | 16.45 |
| | 1 | 3.93 | 0.32 | 1.11E-09 | 1.3E-09 | 0.00 | — | 11.22 |
| | 2 | 4.20 | 0.76 | 4.52E-10 | 1.1E-09 | 6.31 | 15.57 | 11.05 |
| 1.2 Peritoneal | 0 | 3.20 | — | 2.16E-08 | 1.4E-09 | 0.00 | — | 31.47 |
| | 1 | 4.35 | 0.27 | 3.73E-10 | 3.6E-10 | 0.00 | — | 12.32 |
| | 2 | 6.02 | 0.71 | 1.56E-12 | 3.7E-12 | 20.91 | 8.43 | 3.81 |
| 2. Lung cancer | 0 | 3.20 | — | 9.33E-08 | 2.9E-09 | 0.00 | — | 32.76 |
| | 1 | 3.26 | 0.12 | 7.54E-08 | 3.2E-08 | 0.00 | — | 32.50 |
| | 2 | 4.52 | 0.29 | 1.38E-09 | 1.3E-09 | 4.84 | 1.06 | 5.90 |
| 3. Asbestosis | 0 | 3.20 | — | 3.09E-08 | 1.6E-09 | 0.00 | — | 26.78 |
| | 1 | 3.80 | 0.22 | 3.73E-09 | 2.9E-09 | 0.00 | — | 18.82 |
| | 2 | 3.92 | 0.41 | 2.54E-09 | 3.4E-09 | 1.19 | 3.45 | 18.70 |
| | | | Estimates based on TSFEs 15-49 and 50+ | | | | | |
| 1. Mesothelioma | 0 | 3.20 | — | 3.42E-08 | 1.6E-09 | 0.00 | — | 46.77 |
| | 1 | 3.63 | 0.17 | 7.42E-09 | 4.5E-09 | 0.00 | — | 40.03 |
| | 2 | 5.54 | 0.45 | 1.36E-11 | 2.1E-11 | 11.80 | 2.74 | 11.52 |
| 1.1 Pleural | 0 | 3.20 | — | 1.29E-08 | 9.8E-10 | 0.00 | — | 28.22 |
| | 1 | 3.24 | 0.26 | 1.12E-08 | 1.1E-08 | 0.00 | — | 28.19 |
| | 2 | 5.24 | 0.74 | 1.69E-11 | 4.2E-11 | 37.28 | 13.99 | 16.23 |
| 1.2 Peritoneal | 0 | 3.20 | — | 2.13E-08 | 1.3E-09 | 0.00 | — | 31.90 |
| | 1 | 3.87 | 0.22 | 1.90E-09 | 1.5E-09 | 0.00 | — | 21.73 |
| | 2 | 5.85 | 0.59 | 2.76E-12 | 5.5E-12 | 17.95 | 5.22 | 4.03 |
| 2. Lung cancer | 0 | 3.20 | — | 8.66E-08 | 2.5E-09 | 0.00 | — | 79.35 |
| | 1 | 2.88 | 0.10 | 2.68E-07 | 9.5E-08 | 0.00 | — | 69.27 |
| | 2 | 4.62 | 0.27 | 1.02E-09 | 8.9E-10 | 5.40 | 0.82 | 6.54 |
| 3. Asbestosis | 0 | 3.20 | — | 3.18E-08 | 1.5E-09 | 0.00 | — | 28.73 |
| | 1 | 3.73 | 0.17 | 4.75E-09 | 3.0E-09 | 0.00 | — | 19.13 |
| | 2 | 3.93 | 0.35 | 2.48E-09 | 2.9E-09 | 1.29 | 2.00 | 18.70 |

Source: Stallard and Manton (1993; Table 37a and 37b).

Thus, Model 1 was substantially improved although the fit was still not good. For asbestosis, Model 1 yielded $\chi^2 = 18.82$ (5 d.f., $0.01 > p > 0.001$), $k = 3.80 \pm 0.22$, a substantial improvement. For lung cancer, Model 1 yielded $\chi^2 = 32.50$ (5 d.f., $p < 0.001$), $k = 3.26 \pm 0.12$, no improvement.

Model 2 introduced an additional parameter, $c_d$, representing the squared coefficient of variation of a gamma-mixed Weibull hazard function:

$$\mu_{t,d} = b_d t^{k_d} / \left[1 + b_d c_d t^{k_d+1} / (k_d + 1)\right] \qquad (c_d \geq 0)\,.$$

Model 2 provided an acceptable fit ($p > 0.05$) for both lung cancer and peritoneal mesothelioma, a marginally acceptable fit ($0.05 > p > 0.01$) for pleural mesothelioma and undifferentiated mesothelioma, and an unacceptable fit ($p < 0.01$) for asbestosis.

The same general conclusion was reached when data for 50+ years of exposure were included. However, the probability level for pleural mesothelioma fell to an unacceptable level (to $0.01 > p > 0.001$). Closer inspection of Table 7.4 reveals the reason for this: The rate dropped from 514/100,000 at TSFE 45-49 to 179/100,000 at TSFE 50+. This is inconsistent with the hazard rate reaching a plateau at this duration. Instead, it may signal differences in exposure levels between insulators with TSFE < 50 versus > 50 years.

Independent of the level of fit, the results for 15-49 years TSFE indicated that for all five disease models, including asbestosis and lung cancer, the average $k$ for insulation workers was at least 3.2. The $k$ values for Model 1 were significantly lower than for Model 2, except for pleural mesothelioma and asbestosis. The $k$ values for Model 1 were significantly higher than in analysis S3, Model 3 (see Table 7.3), with the largest difference for asbestosis (3.80 vs. 1.41), the smallest difference for mesothelioma (4.18 vs. 3.61), and lung cancer falling in between (3.26 vs. 2.34). These results indicate that there is substantial variation in susceptibility to asbestos-related disease mortality among insulators, and that the rate of increase with TSFE of asbestos-related disease claims is substantially lower than for asbestos-related disease mortality. The fact that 62% of mesothelioma deaths in Table 7.4 were peritoneal suggests that there may be further differences between insulators and other workers in their disease-specific hazard rates as a function of TSFE. We will consider these issues in Chapter 8.

In the remainder of this section, we evaluate the sensitivity of the baseline projection model to changes in both the $b$ and $k$ parameters of the Weibull model.

### 7.5.1 Sensitivity to the $b$ Parameter

Peto et al. (1982) estimated a $b$ value equal to $4.37 \times 10^{-8}$, whereas we estimated $3.58 \times 10^{-8}$ (Model 0 – 18.1% lower). Table 7.2 shows that our revised estimate resulted in 0.7% more total claims and 1.1% more mesothelioma claims, 1990-2049. The mesothelioma diagnoses were projected to increase

3.9% over the same period. The two-stage calibration was relatively insensitive to the $b$ value, even though the first-stage projection of mesothelioma diagnoses was somewhat more sensitive. The first-stage sensitivity was expected based on the discussion in Section 6.6.2.

This raised the issue of the sensitivity of the baseline projection to the IWE assumption. As indicated, Walker (1982) suggested that the number of actual workers per IWE may be 2, 5, or 10. In Section 6.7.1, we used Selikoff's (1981) estimates to determine that the rates could also be 13 or 26 actual workers per IWE. The ratio may vary by an order of magnitude.

We evaluated the extremes of these ranges by defining two additional models (3 and 4) that elaborate on the null model in Table 7.5. Table 7.2 shows the impact of setting $b$ equal to $1.0 \times 10^{-8}$ (Model 3 – corresponding to a ratio near 4 to 1) and $1.0 \times 10^{-9}$ (Model 4 – corresponding to a ratio near 40 to 1). The latter ratio was high enough to include the lowest levels of exposure considered by Selikoff (1981).

Under the 4:1 ratio, total claims increased by 3.2% and mesothelioma claims increased by 4.8% during 1990-2049. The corresponding increases under the 40:1 ratio were 4.2% and 6.2%, respectively. As expected, the projected numbers of mesothelioma diagnoses increased more – 18.5% and 24.6%. Although the claim sensitivity was modest and the diagnosis sensitivity was larger, both were capped at maximum values respectively about 6% and 25% higher than projected.

### 7.5.2 Sensitivity to the $k$ Parameter

Peto et al. (1982, p. 132) commented that brief exposure yields a $k$ value 1 unit less than continuous exposure at the same dose (see Section 2.4). The types of exposure leading to claims against the Manville Trust include all 11 occupation/industry categories identified by Selikoff (1981) (see Table 3.1), so the "true" $k$ value for the first-stage calibration is between $k$ and $k-1$, where $k$ is determined from Table 7.5.

In Table 7.2, we show the impact of replacing the mesothelioma incidence function in the baseline model with Models 1 and 2, as estimated in Table 7.5 for TSFE in the range 15-49 years. Under Model 1, the total number of claims increased 50.5% for 1990-2049; under Model 2, the corresponding increase was 74.3%. Comparable increases occurred for the nine specific disease/injuries.

We now consider why these increases occurred. Model 1 increased $k$ to 4.18. The projected mesothelioma diagnoses for 1990-2049 increased 96.4%. Model 2 increased $k$ to 5.28. Mesothelioma diagnoses increased 43.1%. Both models yielded large increases in total exposure initiated in 1970-1974: 126.2% (Model 1) and 247.2% (Model 2). These increases were primarily for the age groups 15-19, 20-24, and 25-29, with the largest increases for 15-19: 186.1% (Model 1) and 431.3% (Model 2). However, the total number of male mesothelioma diagnoses in SEER 1975-1989 at or below age 29 was 10. Projections based on such small numbers have low credibility.

**Table 7.6: Alternative Projections of the Total Number of Qualified Male Claims Against the Manville Trust, 1990-2049, by Age in 1990-1994**

| # | Age in 1990-1994 | Birth Cohorts Included | Total Claims: Analysis S0 Baseline Model | Total Claims: Analysis S4 Model 1 | Total Claims: Analysis S4 Model 2 | Ratio of Total Claims from S4 to Total Claims from S0: Model 1 | Ratio of Total Claims from S4 to Total Claims from S0: Model 2 |
|---|---|---|---|---|---|---|---|
| 1 | 35-39 | 1951-1959 | 41,061 | 96,425 | 156,110 | 2.348 | 3.802 |
| 2 | 40-44 | 1946-1954 | 79,695 | 151,515 | 200,765 | 1.901 | 2.519 |
| 3 | 45-49 | 1941-1949 | 86,163 | 136,219 | 151,596 | 1.581 | 1.759 |
| 4 | 50-54 | 1936-1944 | 50,397 | 69,340 | 69,326 | 1.376 | 1.376 |
| 5 | 55-59 | 1931-1939 | 50,287 | 62,697 | 59,362 | 1.247 | 1.180 |
| 6 | 60-64 | 1926-1934 | 56,561 | 65,904 | 59,578 | 1.165 | 1.053 |
| 7 | 65-69 | 1921-1929 | 38,532 | 42,780 | 38,592 | 1.110 | 1.002 |
| 8 | 70-74 | 1916-1924 | 25,639 | 27,851 | 24,836 | 1.086 | 0.969 |
| 9 | 75-79 | 1911-1919 | 12,099 | 12,781 | 11,702 | 1.056 | 0.967 |
| 10 | 80-84 | 1906-1914 | 4,048 | 4,197 | 3,929 | 1.037 | 0.971 |
| 11 | 85-89 | 1901-1909 | 1,079 | 1,106 | 1,055 | 1.025 | 0.978 |
| 12 | 90-94 | 1896-1904 | 168 | 172 | 167 | 1.021 | 0.993 |
| 13 | 95-99 | 1891-1899 | 22 | 22 | 22 | 1.000 | 1.000 |
| | | | Cumulative Totals | | | | |
| # 1-13 | 35-99 | 1891-1959 | 445,751 | 671,008 | 777,038 | 1.505 | 1.743 |
| # 2-13 | 40-99 | 1891-1954 | 404,690 | 574,583 | 620,928 | 1.420 | 1.534 |
| # 3-13 | 45-99 | 1891-1949 | 324,995 | 423,068 | 420,163 | 1.302 | 1.293 |
| # 4-13 | 50-99 | 1891-1944 | 238,832 | 286,849 | 268,567 | 1.201 | 1.125 |

Source: Authors' calculations.

Table 7.6 presents the total claim projections for 1990-2049, stratified to isolate the effects of each age group (or cohort). The total claims for Models 1 and 2 exhibited the largest relative increases for the youngest age group (134.8% and 280.2%, respectively). These increases declined substantially over the next two older age groups. For males in the 10 oldest age groups, the increases were even smaller (averaging 20.1% and 12.5%, respectively). Model 1 generally projected more claims than Model 2 for these groups.

With all but the youngest age group included in the projections, the average relative increases fell to 42.0% and 53.4% respectively for Models 1 and 2. Given the lack of credibility for the youngest age group, we would argue that these latter values better capture the potential impact of Models 1 and 2 on the baseline projection.

Given the sensitivity observed, it is worth asking now whether there is any theoretical evidence to support the use of $k = 3.2$ in the baseline model. As noted earlier, Peto et al. (1982) observed that the $k$ value is 1 unit less for brief exposure than for continuous exposure. Selikoff (1981; his Table 2-

13) estimated that the average duration of employment for insulation workers (15.9 years in 1960-1969) was much longer than in primary asbestos manufacturing (3.8 years in 1960-1969) and shipbuilding/repair (4.2 years in 1960-1969). Other occupations had average employment durations of 4.0 to 8.7 years in 1960-1969, but these were all at lower risk of mesothelioma. Selikoff and Seidman's (1991) data were for insulation workers, all of whom were in the insulators' union in 1967. Their average duration of exposure must have been at least as long as, if not substantially longer than, that of the average insulation worker. In the Manville Trust data, only 8.8% of the 7787 qualified male mesothelioma claims were insulation workers. It is reasonable to conclude that the actual number of exposed workers was a large multiple of (up to 40 times) the IWE number displayed in Table 6.10; that average daily exposure levels were substantially below that of insulation workers, and that the average duration of exposure was substantially shorter than for insulation workers. With Model 1 yielding $k = 4.18 \pm 0.21$, a value of $k$ 1 unit lower (i.e., $k = 3.20$) appears plausible for use with the general population of workers exposed to asbestos.

This argument was used by neither Walker (1982) nor Selikoff (1981). Walker adopted Peto et al.'s (1982) original estimate of $k$ equal to 3.2 for insulation workers, without suggesting a downward adjustment, although, by not using Peto's preferred estimate of 3.5, Walker implicitly used a small downward adjustment. Selikoff (1981, Task 4b) argued that the risk of death from mesothelioma increased as the fourth or fifth power of TSFE for 40-50 years. His projections were consistent with the $k$ values in Models 1 and 2. We would argue, however, that Selikoff (1981) should have lowered $k$ in his projections for occupations other than insulation workers. We will consider how to do this in Chapter 8.

## 7.6 Analysis S5: Adjustments to the IWE Exposed Population

Steps 8-10 of the first-stage calibration of the baseline model introduced smoothing, cleanup (of asbestos), and recalibration of the IWE estimates for exposures initiating in 1955-1974. This section evaluates the impact of these adjustments using three alternative models. The results are summarized in Table 7.2.

Model 1 deleted step 10 from the first-stage calibration of the baseline model. This had the effect of replacing the exposed population estimates in Table 6.9 with the "discounted" estimates from step 9. For example, the estimate of 304,871 male IWEs with first exposure during the period 1955-1974 was reduced to 257,711 male IWEs. In addition, some of the "bumpiness" in the age distribution at or below age 25-29 was removed. The net effect was to decrease the 1990-2049 mesothelioma diagnoses projection by 19.2%, mesothelioma claims by 25.4%, and total claims by 25.8%.

Model 2 deleted steps 8, 9, and 10 from the first-stage calibration of the baseline model. This had the effect of replacing the IWE exposed population estimates in Table 6.9 with the initial estimates from step 7 in Table 6.8. The exposure estimates for 1955-1974 were comparable to those for 1945-1954, but with some "bumpiness" in the age distribution for later periods. The net effect was to increase the 1990-2049 mesothelioma diagnoses projection by 6.1%, mesothelioma claims by 4.0%, and total claims by 6.6%. The adjustments in steps 8-10, considered as a set, had only a modest impact on the projections. However, as shown by Model 1, implementation of steps 8 and 9 alone led to a sizable reduction in projected claims because the first-stage calibration no longer was constrained to reproduce the observed mesothelioma diagnosis counts for 1975-1989. This is undesirable if the observed mesothelioma counts are credible. An exception was considered in Model 3 for the three youngest cohorts, for which the credibility of the observed mesothelioma counts was suspect.

Model 3 modified the baseline model so that the renormalization in step 10 of the first-stage calibration was not applied to the three youngest cohorts, aged 15-19, 20-24, and 25-29 in 1970-1974. These are the same three cohorts identified in analysis S4. Compared to the baseline, Model 3 showed a decrease of 24.4% in mesothelioma diagnoses 1990-2049, a decrease of 28.6% in mesothelioma claims, and a decrease of 28.2% in total claims. Without the resolution in step 10, the exposed population at age 15-19 in 1970-1974 was 13.0% of the total, and at age 20-24, it was 17.3% of the total. For the period 1965-1969, the corresponding values were 13.2% and 17.4%, respectively. These were substantially lower than Selikoff's (1981; his Table 2-21) estimates for 1965 of 15.1% for ages 18-19 and 27.3% for ages 20-24. The projection in Model 3 could be viewed as a lower bound. On the other hand, the baseline model (Table 6.9) produced estimates of 32.2% for ages 15-19 and 22.6% for ages 20-24 in 1965-1969, the first of which was substantially higher than but the second lower than Selikoff's estimates. For the period 1970-1974, the corresponding estimates were 34.4% and 27.9%, respectively. Again the first was substantially higher than but the second was close to Selikoff's estimates. The baseline projection could be viewed as an upper bound using these criteria.

The conclusion from this section is that there was substantial sensitivity to the exposure calibration at younger ages in the 1960s and 1970s.

## 7.7 Analysis S6: National Mesothelioma Incidence Counts

The counts of diagnoses of mesothelioma for 1975-1989 for the United States in Table 6.3 were based on data from the SEER program sponsored by NCI. These data reflected mesothelioma incidence in about 10% of the U.S. population. There are concerns about how representative the SEER data are of the entire United States.

The projections were calibrated both to the SEER data 1975-1989 and to the Trust claims 1990-1992. In this section, we evaluate two alternative models. Model 1 replaced the estimates in Table 6.3 with corresponding estimates based on the five coastal sites in SEER: San Francisco-Oakland SMSA, Connecticut, Metropolitan Detroit (Lake Michigan), Hawaii, and Seattle(Puget Sound). Model 2 replaced the estimates in Table 6.3 with corresponding estimates based on the four inland sites in SEER: Metropolitan Atlanta, Iowa, New Mexico, and Utah. The replacements for Table 6.3 are in Table 7.7.

Table 7.7 shows several differences between the estimates based on coastal versus inland SEER sites: (1) the coastal sites yield higher overall estimates (+51.8% for 1970-1989); (2) the coastal sites have a steeper rate of secular increase (6.6% vs. 3.3% per year from 1970-1974 to 1985-1989); and (3) there is a crossover in the age-specific estimates between ages 40-44 and 45-49, with the inland sites yielding higher estimates at younger ages and the coastal sites yielding higher estimates at older ages. This suggests that the exposures at the inland sites were relatively more recent than at the coastal sites.

The projections are summarized in Table 7.2. Compared with the baseline model, the coastal sites (Model 1) yielded 9.3% more mesothelioma diagnoses 1990-2049, 3.9% fewer mesothelioma claims 1990-2049, and 3.6% fewer total claims 1990-2049. The increase in mesothelioma diagnoses is consistent with the higher overall estimate for coastal sites in Table 7.7. The decrease in mesothelioma and total claims is consistent with the relatively older age of mesothelioma cases in the coastal sites, combined with the constraints imposed by the second-stage calibration.

The reverse pattern is seen for the inland sites (Model 2). Compared with the baseline model, the inland sites yielded 19.2% fewer mesothelioma diagnosis, but 16.4% and 16.5% more mesothelioma and total claims for 1990-2049. These reversals are also consistent with the differences seen in Table 7.7.

The average number of claims for 1990-2049 from Models 1 and 2 was 6.5% higher than in the baseline model (474,667 vs. 445,751). Given the substantial differences between the coastal and inland sites, these results indicated that the SEER data, taken as a set, provided reasonable estimates of mesothelioma diagnoses for use in the first-stage calibration of the baseline model. We consider alternative data sources in Section 7.12 in preparation for replacing the SEER data with Manville Trust data in Chapters 8 and 9. However, this replacement was due to the lack of occupational information in SEER data, not to concerns about its representativeness of the entire United States.

## 7.8 Analysis S7: Nonmesothelioma Mortality Rates

The projected mortality rates in Table 6.2 were based on an assumed exponential decrease of about 1% per year. The exact factor applied to each quinquennium is given in the last column of Table 6.2. These estimates, $\hat{p}_A$,

**Table 7.7: Estimated U.S. Male Mesothelioma Incidence Counts by Age and Quinquennium, Coastal vs. Inland Sites in SEER Data, 1970-1989**

| | Quinquennium | | | | |
|---|---|---|---|---|---|
| Age | 1970-1974 | 1975-1979 | 1980-1984 | 1985-1989 | Total |
| | | | Costal | | |
| 0-14 | 0 | 0 | 0 | 0 | 0 |
| 15-19 | 0 | 0 | 17 | 0 | 17 |
| 20-24 | 0 | 16 | 16 | 16 | 49 |
| 25-29 | 0 | 45 | 0 | 15 | 60 |
| 30-34 | 117 | 29 | 15 | 15 | 176 |
| 35-39 | 45 | 60 | 45 | 106 | 256 |
| 40-44 | 151 | 46 | 77 | 197 | 471 |
| 45-49 | 253 | 453 | 341 | 197 | 1,244 |
| 50-54 | 345 | 472 | 520 | 346 | 1,682 |
| 55-59 | 442 | 606 | 914 | 954 | 2,915 |
| 60-64 | 550 | 825 | 1,338 | 1,238 | 3,950 |
| 65-69 | 509 | 988 | 1,478 | 2,157 | 5,132 |
| 70-74 | 779 | 968 | 1,700 | 2,074 | 5,521 |
| 75-79 | 568 | 721 | 1,177 | 1,408 | 3,874 |
| 80-84 | 137 | 364 | 708 | 950 | 2,159 |
| 85-89 | 0 | 200 | 272 | 341 | 812 |
| 90-94 | 0 | 0 | 17 | 85 | 102 |
| 95-99 | 0 | 0 | 0 | 0 | 0 |
| Total | 3,896 | 5,792 | 8,635 | 10,099 | 28,423 |
| | | | Inland | | |
| 0-14 | 0 | 0 | 0 | 0 | 0 |
| 15-19 | 0 | 31 | 0 | 0 | 31 |
| 20-24 | 103 | 0 | 30 | 0 | 134 |
| 25-29 | 104 | 0 | 0 | 0 | 104 |
| 30-34 | 0 | 123 | 0 | 31 | 153 |
| 35-39 | 214 | 31 | 92 | 61 | 398 |
| 40-44 | 109 | 159 | 156 | 185 | 608 |
| 45-49 | 334 | 99 | 160 | 190 | 784 |
| 50-54 | 338 | 414 | 432 | 226 | 1,411 |
| 55-59 | 877 | 521 | 584 | 468 | 2,451 |
| 60-64 | 546 | 565 | 765 | 481 | 2,357 |
| 65-69 | 432 | 715 | 1,020 | 1,223 | 3,390 |
| 70-74 | 306 | 458 | 836 | 1,178 | 2,778 |
| 75-79 | 97 | 375 | 1,372 | 924 | 2,768 |
| 80-84 | 0 | 163 | 262 | 462 | 887 |
| 85-89 | 0 | 61 | 93 | 156 | 310 |
| 90-94 | 0 | 0 | 93 | 63 | 155 |
| 95-99 | 0 | 0 | 0 | 0 | 0 |
| Total | 3,460 | 3,715 | 5,897 | 5,646 | 18,719 |

Source: Authors' tabulation of SEER data, 1973-1989.

were based on the historical rate of change in hazard rates, $h_{A,Y}$, from 1950-1954 to 1980-1984.

The hazard rate projections were described in Section 6.5.1. As noted in that discussion, this projection model was consistent with the analysis of Lee and Carter (1992, p. 664) which reported that the choice of starting date (1930 to 1960) made "little difference to either the point forecasts or the confidence intervals." The relative standard error of the forecasted rate of decline was about 23% for the forecast from 1990 to 2050. Applying this to our forecasted age-specific declines of about 1% per year, we approximated a 95% confidence interval by introducing a ±0.5% per year (±2.5% per quinquennium) adjustment. Let $N$ be the number of quinquennia from 1980-1984 to $Y$. In Model 1, we modified the hazard rate projection to represent low mortality,

$$h_{A,Y} = h_{A,1980-84} \times p_A^N \times 1.025^{-N},$$

and in Model 2, we introduced a complementary modification to represent high mortality,

$$h_{A,Y} = h_{A,1980-84} \times p_A^N \times 1.025^N.$$

The results from the low- and high-mortality models are compared with the baseline model in Table 7.2.

Under low mortality (Model 1), the projected mesothelioma diagnoses for 1990-2049 increased 8.8%, mesothelioma claims increased 4.2%, and total claims increased 2.8%. Under high mortality, mesothelioma diagnoses decreased 8.1%, mesothelioma claims decreased 4.3%, and total claims decreased 2.9%. In both cases, the effects were small.

## 7.9 Analysis S8: Excess Mortality Among Insulation Workers

Walker's (1982) assumptions about excess mortality among heavily exposed persons were discussed in Sections 4.4.1b, 5.4.2, and 6.5.5. The cases of mesothelioma incidence were divided into three groups: 46% without documented history of occupational exposure (our low-dose exposure group), 34% with probable occupational exposure (our medium-dose exposure group), and 20% with definite occupational exposure (our high-dose exposure group). The 20% with definite exposure was equated by Walker (1982, p. 6) to heavy exposure, which is why these workers formed our high-dose exposure group. Walker deleted the low-dose exposure group from his projections. We retained them so that our first-stage calibration would exactly match the SEER data (Table 6.3). The remaining uncertainty involved the assumption that there was an excess of 37% nonmesothelioma mortality among a subgroup of the population that yielded 20% of mesothelioma cases.

In this section, we evaluate changes in both statistics using excess mortality percentages of 0.0% (Model 0), 30.1% (Model 1), and 45.0% (Model 2). These values were discussed in Section 6.6.1. In addition, we evaluate high-dose exposure percentages of 0.0% (Model 0), 16.0% (Model 3), and 25.9% (Model 4). These percentages were derived from the Trust's claim data files, where of 7787 qualified male mesothelioma claims, 686 were insulators and 557 were asbestos factory workers (16.0%; relative risk = 1), and 1547 were shipyard workers (19.9%; relative risk = 0.5), using relative risks from Selikoff (1981) (see Table 3.1). The weighted sum of these claims, using the relative risks as weights, was 2016.5, which was 25.9% of the total, 7787. The 25.9% was almost identical to Walker's upper limit estimate of 26% (Walker, 1982, p. 6).

If the fraction of mesothelioma diagnoses with occupational exposure were indeed substantially below 100% [e.g., 75% as suggested by McDonald and McDonald (1981)], then the high-dose fractions for Models 4 and 5 could be adjusted down by, say, 25%. We retained the higher values to evaluate the sensitivity of the baseline projection over a broader range of alternatives. The results are in Table 7.2.

Model 0 eliminated both the high-exposure subgroup and their excess mortality. Compared to the baseline model, the mesothelioma diagnoses 1990-2049 increased 4.6%, the mesothelioma claims increased 0.4%, and the total claims increased 0.2%. This was the most extreme of the five models in Table 7.2. The baseline projection model was relatively insensitive to these assumptions.

## 7.10 Analysis S9: Decline in Claim Filing Rates

The baseline CHR model assumed that age and TSFE were the only factors in filing claims. Figures 6.2-6.5 displayed the rates at which filings occurred from December 1989 to November 1992. There was no clear trend discernible from those data. The trends in Figure 6.6 showed an increasing percentage of claims due to asbestos-related noncancers. We argued that this was merely a shift away from claims based on nonasbestos-related and unknown diseases, not a true increase in the rate of asbestosis claims. On the other hand, some lawyers were concerned that the claim filing rates may decline as the assets of the various defendants are depleted. For example, in the National Gypsum bankruptcy case, a decline of near 7% per year was assumed in the bench ruling (Felsenthal, 1993). In view of this, we evaluated the effects of declines of 1%, 3%, and 7% per calendar year in the claims filing rates of the CHR model. The results are compared with the baseline model in Table 7.2.

At 1% (Model 1), mesothelioma claims 1990-2049 declined 14.2%; total claims declined 11.9%. At 3% (Model 2), mesothelioma claims 1990-2049 declined 33.9%; total claims declined 29.2%. These declines were less than three times those in Model 1, implying nonlinear effects. At 7% (Model 3), mesothelioma claims 1990-2049 declined 55.2%; total claims declined 48.9%. The declines for other diseases were of similar magnitude. For example, at 7%, the decline in asbestosis claims 1990-2049 was 48.1%.

The declines in claim filing rates also had an alternate interpretation, in the case of noncancer diseases. This alternative was based on the clear distinction between the effects of age and TSFE on cancer versus noncancer diseases found in Section 7.4 (see Table 7.3). For the cancers, the scientific evidence was much stronger that age and TSFE were the major determinants of claims. Some additional effects were also attributable to duration of exposure, but these were approximated by reducing the $k$ parameter up to 1 unit in value. For asbestosis (the main noncancer disease), the evidence was less clear that age and TSFE were the major determinants of claims. To the extent that intensity of exposure had nonlinear effects, the assumption that the exposed population can be calibrated in IWEs may have overestimated the future occurrence of asbestosis claims. The fact that the $k$ value for asbestosis deaths in Table 7.5 was substantially higher than for asbestosis claims in Table 7.3 (3.8 vs. 1.4) supported this interpretation.

Walker (1982) assumed the virtual cessation of new diagnoses of asbestosis after 1984. Unfortunately, that model underpredicted asbestosis claims for 1990-1994 by 88.5-92.9% (see Section 6.8.3).

## 7.11 Overall Sensitivity: Analyses S1-S9

The results of the nine sensitivity analyses indicate that the projections conducted under the baseline model in Section 6.8 can be accepted with a reasonable degree of confidence. The mesothelioma incidence projections in Section 6.7 compared favorably with those of Walker (1982) and Selikoff (1981), differing in absolute levels, but agreeing on a relative basis. The comparisons with Walker and Selikoff highlighted the sources of the discrepancy between these two projections, a discrepancy that was the focus of two lengthy reviews (Manton, 1983; Cohen et al., 1984). The updated model in Chapter 6 had the advantage of using a much broader database, and, with its CHR submodel in Section 6.8, was directly calibrated to the number of claims against the Manville Trust.

These projections cannot be treated as exact predictions of the future number of claims against the Trust. Indeed, no population projection can be treated as an exact prediction. They are the numerical consequence of the data and assumptions built into the model. Thus, an understanding of the likely stability of the projections is based on understanding the steps involved in the projections. This is the reason we discussed the detailed steps of the two stages of calibration of the updated projection model. This is also the reason we conducted 28 projections. A listing of the 28 projections with a brief description of their characteristics is in Table 7.1. More detailed information and discussion was given in Sections 7.2-7.10.

The 27 projections selected as alternatives to the baseline CHR model fell into 9 classes that reflected our judgment of the major areas of sensitivity of the model. These classes were not an exhaustive listing of areas of sensitivity.

Instead, they were intended to be the starting point for further consideration and analysis. Their purpose was to alert the user of these methods and projections to the potential range of variation in outcomes.

Table 7.8 displays the results for total claims 1990-2049, sorted in ascending order within each of the nine sensitivity analyses, making it easy to identify the models with the largest impact. The two largest increases (+74.3% and +50.5%) were for analysis S4, which evaluated the impact of alternative mesothelioma incidence/mortality models. This analysis was sensitive to the IWE estimates for the youngest cohort: dropping the results for this cohort reduced the 74.3% increase to 53.4% (see Table 7.6). The next largest increase (+47.0%) was for the null model in analysis S3, which evaluated the parametric form of the CHRs. The fourth largest increase (+17.3%) was for analysis S1, which evaluated the assumption of a constant age-specific claim runoff.

The two largest decreases (−48.9% and −29.2%) were for analysis S9 which evaluated the impact of secular declines in CHRs. The next largest decrease (−28.2%) was for analysis S5, which evaluated the impact of the IWE adjustments in steps 8-10 of the first-stage calibration. The decline was associated with the deletion of the renormalization in step 10, which primarily affected the IWE estimates for the youngest cohorts.

An alternative approach is to array the projection outcomes, selecting the median (i.e., the 14th of 27 in rank order) as the best single estimate, identifying the lowest and highest as the likely bounds of uncertainty.

Table 7.9 displays the results for total and disease-specific claims 1990-2049, sorted in ascending order across the nine sensivity analyses based on the percentage differences in Table 7.2. The bottom row of Table 7.9 (same as row 14) compares the medians of the 27 projections with the corresponding results from the baseline projection. The differences are minor. This suggests that, taken as a whole, the 27 projections provide a balanced representation of the uncertainty in the future claims faced by the Manville Trust.

Figure 7.1 graphs the total claims by quinquennia for 1990-2049 for each of the 27 projections, and Figure 7.2 presents the same data cumulated by quinquennia. The alternative projections cluster near the baseline in both cases. Figure 7.2 shows that the uncertainty increases with increasing duration of the projection.

Any projection can be viewed as a function of various data inputs, all of which are subject to random variation. The component of variation in each projection attributable to variation in data is not reflected in our calculations of projection limits. In addition, the timing of projected claims is subject to random variation and this is also not reflected in our calculation of projection limits. These sources of variation could be modeled by computer simulation methods, but we have not done so. Such variation is "within-model" variation. Our uncertainty limits, in contrast, refer to "between-model" variation. We focus on the between-model component of variation because it is the dominant source of uncertainty. Our knowledge of future claims of asbestos-related

**Table 7.8: Comparisons of Alternative Projections of the Total Number of Qualified Male Claims Against the Manville Trust, 1990-2049, with Models Sorted by Increasing Impact Within Sensitivity Analysis Groupings**

| Analysis | Model | # | Total Claims 1990-2049 | Difference (%) Between Alternative Model and Baseline Model (S0) |
|---|---|---|---|---|
| Baseline S0 | | 0 | 445,751 | 0.0 |
| S1 | 1 | 1 | 522,901 | 17.3 |
| S2 | 1 | 2 | 477,014 | 7.0 |
| S3 | 2 | 5 | 427,035 | -4.2 |
| | 4 | 7 | 431,921 | -3.1 |
| | 3 | 6 | 432,582 | -3.0 |
| | 1 | 4 | 440,714 | -1.1 |
| | 0 | 3 | 655,431 | 47.0 |
| S4 | 0 | 8 | 448,991 | 0.7 |
| | 3 | 11 | 460,195 | 3.2 |
| | 4 | 12 | 464,346 | 4.2 |
| | 1 | 9 | 671,008 | 50.5 |
| | 2 | 10 | 777,038 | 74.3 |
| S5 | 3 | 15 | 319,941 | -28.2 |
| | 1 | 13 | 330,731 | -25.8 |
| | 2 | 14 | 475,165 | 6.6 |
| S6 | 1 | 16 | 429,904 | -3.6 |
| | 2 | 17 | 519,431 | 16.5 |
| S7 | 2 | 19 | 432,916 | -2.9 |
| | 1 | 18 | 458,301 | 2.8 |
| S8 | 4 | 24 | 445,438 | -0.1 |
| | 2 | 22 | 445,635 | 0.0 |
| | 1 | 21 | 445,875 | 0.0 |
| | 3 | 23 | 445,957 | 0.0 |
| | 0 | 20 | 446,740 | 0.2 |
| S9 | 3 | 27 | 227,618 | -48.9 |
| | 2 | 26 | 315,392 | -29.2 |
| | 1 | 25 | 392,491 | -11.9 |
| Median S1–S9 | | 28 | 445,875 | 0.0 |
| S10 | 3 | 31 | 423,200 | -5.1 |
| | 2 | 30 | 447,107 | 0.3 |
| | 1 | 29 | 513,387 | 15.2 |

Source: Authors' calculations.

**Table 7.9: Difference (%) Between Alternative and Baseline Projections of the Number of Qualified Male Claims Against the Manville Trust, 1990-2049, by Most Severe Alleged Disease/Injury – Separately Sorted by Increasing Impact Within Each Category of Most Severe Alleged Disease/Injury**

| | | Mesothelioma | Most Severe Alleged Disease/Injury[1] | | | | | | | | | | | | | Total |
|---|---|---|---|---|---|---|---|---|---|---|---|---|---|---|---|---|
| Analysis/Model/# | | Diagnosis | 1 | 2 | 3 | 4 | 5 | 6 | 7 | 8 | 9 | 1-4 | 5-9 | 5-7 | 8-9 | 1-9 |
| Sorted S1–S9 | 1 | -24.4 | -55.2 | -53.4 | -52.7 | -55.8 | -48.1 | -47.8 | -47.6 | -48.2 | -48.7 | -54.2 | -48.0 | -47.9 | -48.4 | -48.9 |
| | 2 | -19.2 | -33.9 | -32.6 | -31.7 | -34.5 | -28.6 | -29.2 | -29.9 | -28.8 | -29.3 | -33.2 | -28.7 | -28.6 | -29.0 | -29.2 |
| | 3 | -19.2 | -28.6 | -27.2 | -25.6 | -27.9 | -28.0 | -28.2 | -28.2 | -27.1 | -24.9 | -27.6 | -28.2 | -28.5 | -26.5 | -28.2 |
| | 4 | -8.1 | -25.4 | -23.8 | -22.2 | -24.7 | -25.7 | -27.1 | -27.9 | -24.5 | -22.2 | -24.3 | -26.1 | -26.4 | -23.9 | -25.8 |
| | 5 | -1.4 | -14.2 | -13.5 | -13.0 | -14.4 | -11.7 | -11.5 | -11.4 | -11.8 | -12.1 | -13.8 | -11.6 | -11.6 | -11.9 | -11.9 |
| | 6 | -0.8 | -5.2 | -4.3 | -4.6 | -9.3 | -3.9 | -5.1 | -6.3 | -4.1 | -8.3 | -4.2 | -4.4 | -4.5 | -4.3 | -4.2 |
| | 7 | 0.0 | -4.3 | -3.8 | -3.3 | -6.7 | -3.9 | -3.4 | -4.1 | -3.2 | -8.3 | -4.2 | -3.8 | -3.7 | -4.3 | -3.6 |
| | 8 | 0.0 | -4.3 | -3.5 | -1.9 | -5.9 | -3.6 | -2.7 | -4.1 | -3.2 | -7.3 | -4.0 | -3.7 | -3.7 | -3.9 | -3.1 |
| | 9 | 0.0 | -3.9 | -0.3 | -0.1 | -5.9 | -3.4 | -2.7 | -3.4 | -2.9 | -7.3 | -1.5 | -3.3 | -3.4 | -3.1 | -3.0 |
| | 10 | 0.0 | -0.1 | -0.3 | 0.0 | -5.3 | -2.9 | -2.4 | -2.7 | -0.5 | -3.7 | -0.1 | -2.6 | -2.6 | -2.6 | -2.9 |
| | 11 | 0.0 | -0.1 | -0.1 | 0.0 | -5.0 | -2.7 | -1.1 | -2.4 | -0.5 | -3.1 | 0.0 | -2.5 | -2.5 | -2.6 | -1.1 |
| | 12 | 0.0 | 0.1 | 0.0 | 0.1 | -0.9 | -0.1 | -0.1 | -0.1 | -0.1 | -0.1 | 0.0 | -0.1 | -0.1 | -0.1 | -0.1 |
| | 13 | 0.0 | 0.1 | 0.0 | 0.3 | -0.1 | 0.0 | 0.0 | 0.0 | 0.0 | 0.0 | 0.1 | 0.0 | 0.0 | 0.0 | 0.0 |
| | 14 | 0.0 | 0.4 | 0.1 | 0.8 | 0.0 | 0.0 | 0.0 | 0.0 | 0.0 | 0.0 | 0.4 | 0.0 | 0.0 | 0.0 | 0.0 |
| | 15 | 0.0 | 1.1 | 0.3 | 1.2 | 0.0 | 0.0 | 0.0 | 0.0 | 0.0 | 0.1 | 0.7 | 0.0 | 0.0 | 0.0 | 0.0 |
| | 16 | 0.7 | 3.3 | 1.0 | 3.2 | 0.1 | 0.2 | 0.2 | 0.2 | 0.2 | 0.3 | 1.0 | 0.2 | 0.2 | 0.2 | 0.2 |
| | 17 | 0.9 | 4.0 | 1.8 | 3.3 | 0.4 | 0.7 | 0.6 | 0.6 | 0.7 | 0.8 | 1.7 | 0.7 | 0.7 | 0.7 | 0.7 |
| | 18 | 3.9 | 4.2 | 3.6 | 3.5 | 1.2 | 2.7 | 2.3 | 2.4 | 2.8 | 3.7 | 4.0 | 2.6 | 2.5 | 3.1 | 2.8 |
| | 19 | 4.6 | 4.2 | 4.3 | 4.5 | 3.2 | 3.0 | 2.8 | 2.8 | 3.2 | 3.7 | 4.5 | 3.0 | 3.0 | 3.3 | 3.2 |
| | 20 | 6.1 | 4.8 | 5.1 | 5.1 | 4.8 | 3.9 | 3.6 | 3.7 | 4.1 | 4.7 | 4.7 | 3.9 | 3.8 | 4.3 | 4.2 |
| | 21 | 8.8 | 5.2 | 5.6 | 7.6 | 5.3 | 7.1 | 6.6 | 6.4 | 6.1 | 7.7 | 5.9 | 6.9 | 7.0 | 6.5 | 6.6 |
| | 22 | 9.3 | 6.2 | 5.7 | 13.5 | 5.7 | 8.7 | 6.9 | 6.9 | 12.4 | 13.3 | 6.5 | 8.5 | 7.8 | 12.6 | 7.0 |
| | 23 | 18.5 | 7.8 | 13.3 | 15.9 | 6.8 | 16.0 | 16.2 | 16.6 | 18.7 | 13.9 | 9.2 | 16.3 | 16.2 | 17.4 | 16.5 |
| | 24 | 24.6 | 12.6 | 17.7 | 19.1 | 13.4 | 18.9 | 17.0 | 16.6 | 24.0 | 19.7 | 17.6 | 18.8 | 18.1 | 22.8 | 17.3 |
| | 25 | 43.1 | 16.4 | 23.3 | 20.8 | 20.7 | 48.9 | 48.4 | 48.4 | 49.5 | 48.9 | 18.5 | 48.8 | 48.7 | 49.3 | 47.0 |
| | 26 | 96.4 | 62.5 | 58.9 | 54.0 | 62.4 | 52.2 | 49.7 | 51.3 | 55.9 | 56.7 | 60.2 | 52.2 | 51.5 | 56.1 | 50.5 |
| | 27 | — | 77.7 | 75.6 | 76.4 | 76.7 | 73.3 | 75.7 | 76.2 | 74.1 | 64.4 | 76.5 | 73.9 | 74.3 | 71.5 | 74.3 |
| Median S1–S9 | 28 | 0.0 | 0.4 | 0.1 | 0.8 | 0.0 | 0.0 | 0.0 | 0.0 | 0.0 | 0.0 | 0.4 | 0.0 | 0.0 | 0.0 | 0.0 |

Note 1: See Table 6.16 for descriptive labels for disease categories and subtotal groupings.

Source: Authors' calculations based on the percentage differences reported in the bottom panel of Table 7.2.

**Figure 7.1: Projections – Baseline and 27 Alternatives (Source: Authors' Calculations)**

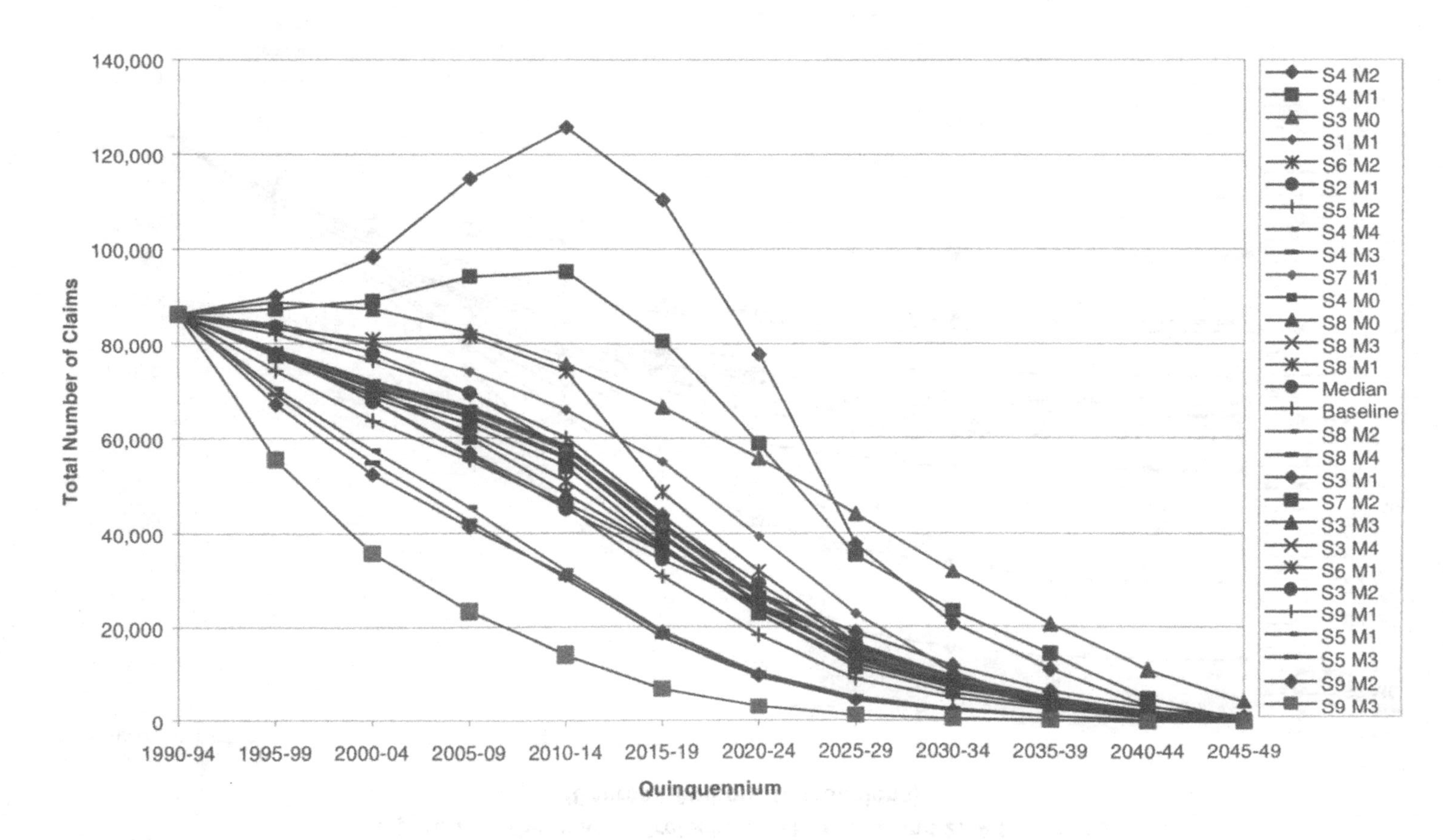

**Figure 7.2: Cumulative Projections – Baseline and 27 Alternatives (Source: Authors' Calculations)**

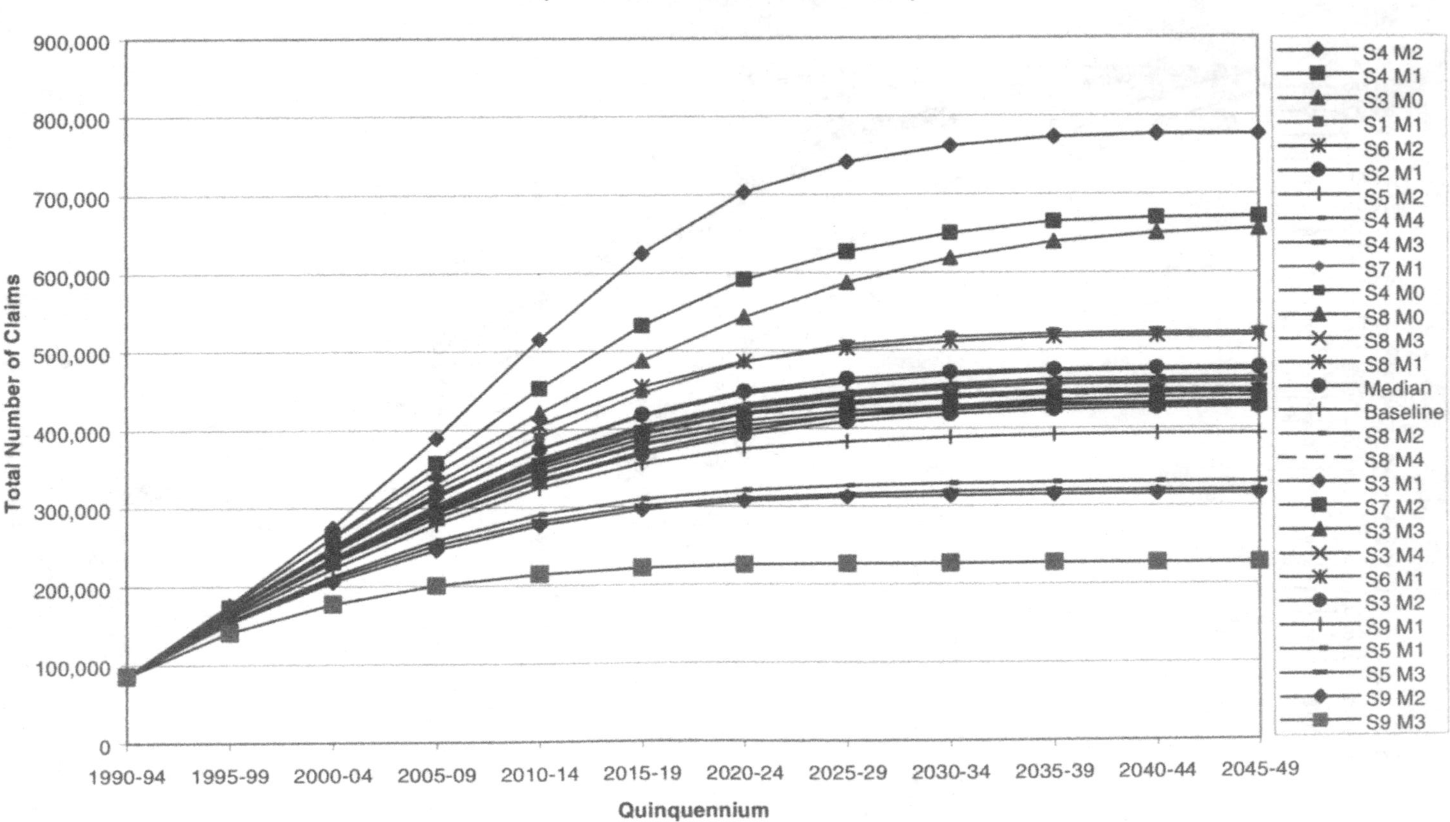

diseases is not good enough to select just one model for detailed simulation analysis.

More detailed evaluation of the level of uncertainty requires a joint analysis of the variation associated with each assumption or parameter of the model, to produce not just a single projected value at each future age and date but a distribution of projected values. This form of modeling permits one to compute the mean and variance of projected quantities. Our goal was to develop a basic projection model and to identify where it was sensitive to assumptions (e.g., the estimation of past exposure). Resolving this issue logically precedes the analysis of the joint variation associated with each assumption or parameter.

## 7.12 Analysis S10: Manville Trust Calibrations

This section presents a bridge model in which the SEER data in step 2 of the first-stage calibration were replaced with corresponding estimates from the Manville Trust claim data. The purpose of the bridge model was to allow us to replace the SEER data, which did not provide information on occupation or industry, with data that provided such information. This can only be done if we can demonstrate that this data interchange did not seriously distort the baseline projections developed in Chapter 6. We shall show that this was indeed the case.

The baseline projections were calibrated both to the SEER data 1975-1989 (first stage) and the Trust claims 1990-1992 (second stage). In this section, we evaluate three alternatives to the SEER data 1975-1989. First, in Model 1, we replaced the estimates in Table 6.3 with the diagnosis data for 1975-1989 in the Trust's files, displayed in Table 6.5. Recall that Table 6.5 showed claims of mesothelioma in the Trust's files, as alleged on the Proof of Claim (POC) Form, not necessarily diagnoses verified by the Manville Trust as mesothelioma. Second, in Model 2, we used only the data for 1985-1989 from Table 6.5, ignoring 1975-1984 in the first-stage calibration. In both models, the mesothelioma counts for 1975-1989 (or 1985-1989) were based on year of diagnosis, not year of claim. The progression of counts of diagnoses from 1975-1990 was 44 in 1975, 263 in 1980, 613 in 1985, 701 in 1986, 713 in 1987, 767 in 1988, 658 in 1989, and 675 in 1990. The peak was in 1988. The dropoff in 1989 and later likely reflected the lag between diagnosis and claim filing (see Section 6.5.3). Thus, the diagnosis count for 1985-1989 may be an underestimate.

Third, in Model 3, we changed the calibration period to 1990-1994 and used the Trust's claim counts for mesothelioma in Table 6.14 for calibration, along with the table (Table 7.10) of age and date of start of asbestos exposure for males filing claims in 1990-1992. Because the mesothelioma counts for 1990-1994 in Table 7.10 were based on year of claim, not year of diagnosis, the dropoff of diagnoses in 1989 and later was not a factor in this alternative calibration.

**Table 7.10: Observed Distribution of Start of Asbestos Exposure by Age at Start of Exposure and Date of Start of Exposure – Qualified Male Mesothelioma Claims Against the Manville Trust, 1990-1992**

| | Date of First Exposure to Asbestos | | | | | | | | | | | | | | | | |
|---|---|---|---|---|---|---|---|---|---|---|---|---|---|---|---|---|---|
| Age | 1915-1919 | 1920-1924 | 1925-1929 | 1930-1934 | 1935-1939 | 1940-1944 | 1945-1949 | 1950-1954 | 1955-1959 | 1960-1964 | 1965-1969 | 1970-1974 | 1975-1979 | 1980-1984 | 1985-1989 | Unknown | Total |
| 0-4 | 1 | 0 | 1 | 2 | 2 | 2 | 4 | 2 | 2 | 0 | 0 | 0 | 0 | 0 | 0 | 0 | 16 |
| 5-9 | 0 | 3 | 1 | 2 | 0 | 2 | 0 | 1 | 1 | 0 | 0 | 0 | 0 | 0 | 0 | 0 | 10 |
| 10-14 | 0 | 3 | 0 | 1 | 6 | 4 | 3 | 3 | 2 | 2 | 1 | 0 | 0 | 0 | 0 | 0 | 25 |
| 15-19 | 3 | 9 | 24 | 19 | 61 | 193 | 70 | 92 | 48 | 27 | 12 | 2 | 2 | 0 | 0 | 0 | 562 |
| 20-24 | 0 | 3 | 10 | 8 | 80 | 191 | 135 | 102 | 67 | 31 | 15 | 4 | 2 | 0 | 0 | 0 | 648 |
| 25-29 | 0 | 0 | 2 | 5 | 30 | 133 | 71 | 63 | 37 | 19 | 18 | 2 | 3 | 1 | 0 | 0 | 384 |
| 30-34 | 0 | 0 | 0 | 0 | 8 | 82 | 49 | 42 | 26 | 11 | 10 | 2 | 0 | 0 | 0 | 0 | 230 |
| 35-39 | 0 | 0 | 0 | 0 | 1 | 19 | 10 | 23 | 16 | 10 | 9 | 3 | 0 | 0 | 0 | 0 | 91 |
| 40-44 | 0 | 0 | 0 | 0 | 0 | 1 | 7 | 7 | 6 | 7 | 7 | 1 | 0 | 0 | 0 | 0 | 36 |
| 45-49 | 0 | 0 | 0 | 0 | 0 | 0 | 2 | 2 | 1 | 3 | 2 | 1 | 0 | 0 | 0 | 0 | 11 |
| 50-54 | 0 | 0 | 0 | 0 | 0 | 0 | 0 | 0 | 3 | 2 | 3 | 2 | 1 | 0 | 0 | 0 | 11 |
| 55-59 | 0 | 0 | 0 | 0 | 0 | 0 | 0 | 0 | 0 | 0 | 1 | 0 | 0 | 1 | 0 | 0 | 2 |
| 60-64 | 0 | 0 | 0 | 0 | 0 | 0 | 0 | 0 | 0 | 0 | 0 | 0 | 0 | 0 | 0 | 0 | 0 |
| 65-69 | 0 | 0 | 0 | 0 | 0 | 0 | 0 | 0 | 0 | 0 | 0 | 0 | 0 | 1 | 0 | 0 | 1 |
| Unknown | 0 | 0 | 0 | 0 | 1 | 1 | 2 | 0 | 1 | 0 | 0 | 0 | 0 | 0 | 0 | 82 | 87 |
| Total | 4 | 18 | 38 | 37 | 189 | 628 | 353 | 337 | 210 | 112 | 78 | 17 | 8 | 3 | 0 | 82 | 2,114 |

Source: Stallard and Manton (1993, Table 41).

The results are in Table 7.2. Compared with the baseline model, the three non-SEER mesothelioma models yielded substantially lower projected numbers of mesothelioma diagnoses. Because each diagnosis in the Trust data files corresponded to a filed claim, the ratios of the number of claims under Models 1-3 to the number in the baseline model can be interpreted as projections of the average propensity to sue (PTS). For the period 1990-2049, these PTS estimates were 39.9%, 50.0%, and 40.9% for Models 1-3, respectively. These can be compared with the PTS rates derived from Tables 6.3 and 6.5 (1975-1989) and Tables 6.11 and 6.14 (1990-1994): 9.5% for 1975-1979, 24.9% for 1980-1984, 40.0% for 1985-1989, and 40.3% for 1990-1994.

The historical PTS rates exhibited substantial increases over the period 1975-1989: This was reflected in a higher projection of mesothelioma and total claims in Model 1 (30.3% and 15.2% above baseline, respectively). The initial and projected PTS rates for Models 2 and 3 were close to 40%, except for Model 2, which projected a 50.0% PTS rate for 1990-2049. Compared with the baseline model 1990-2049, Model 2 projected 9.6% more mesothelioma and 0.3% more total claims, and Model 3 projected 5.5% more mesothelioma claims but 5.1% fewer total claims. Models 2 and 3 yielded projections well within the uncertainty intervals established for the baseline model in analyses S1-S9.

In general, the mesothelioma diagnoses (first stage) and claims (second stage) referred to data from different time periods so that they were not directly comparable. However, for Model 3, they referred to the same time period, and with the substitution of claims for diagnoses in step 3 of the first-stage calibration, they referred to the same set of counts. Thus, the difference between mesothelioma diagnoses and claims under Model 3 (27,734 vs. 25,693) reflected the differences between the parametric form of the Weibull hazard rate (first stage) and the nonparametric form of the CHR model (second stage). The Weibull model forced an age-independent continuing increase with TSFE, whereas the CHR model did not. The result was an excess of about 7.9% diagnoses versus claims under the Weibull assumption. Again, these differences were well within the uncertainty intervals established for the baseline model.

The analyses in this section indicated that the projections based on Model 3 were reasonably close to those of the baseline model. The substitution of the Trust claim data from 1990-1994 for the SEER diagnosis data 1975-1989 appears to be a reasonable method of introducing information on occupation and industry into the projection model.

## 7.13 Conclusions

We used the baseline model to project the number of claims against the Manville Trust, by age and date, for the period 1990-2049. Overall, we projected 445,751 claims among males, with 24,343 due to mesothelioma. We evaluated the sensitivity of these projections to various assumptions of the model and found the major effect to be in the estimation of the population of exposed workers. We did not include females in our detailed estimates.

Our projections face two problems. First, our definition of disease involved a specific hierarchy in which mesothelioma outranked all other diseases, lung cancer ranked second, and so forth (see Table 6.1). A disease was selected as the claimed injury only if it outranked all other reported diseases. The occurrence of multiple disease/injuries for any given claimant complicated the specification of disease incidence functions, because they were not independent. By selecting the highest ranked disease, we obtained a unique disease classification for each claimant, and one that was most likely to be considered by the Manville Trust in its claim settlement process. Nonetheless, the highest ranked disease/injury alleged on the POC Form was not necessarily the same as the disease/injury validated by the Manville Trust as part of the claim settlement process. We need to consider how to translate the projected counts of alleged diseases into corresponding counts of validated diseases.

Second, we defined the exposed population in terms of IWEs – a concept introduced by Walker (1982) to reflect the fact that these estimates were intermediate quantities that canceled out in the forecasting equations. We tested this assumption in Section 7.5 and found that it was not satisfied at the range of $b$ values estimated for insulation workers (Model 0 and baseline), but it was satisfied at $b$ values about one-quarter of that range (Models 3 and 4). In this latter range, there was an approximately linear dose-response relationship between the number of workers exposed and the projected number of mesothelioma diagnoses. An evaluation of the uncertainty associated with the linearity assumption was conducted in analysis S4, where it was shown to be small.

The IWE assumption might be satisfactory were it not for the complication that the $k$ parameter in the dose-response function for insulation workers was up to 1 unit higher than for the general population of exposed workers. Insulation workers had continual exposures over long durations of asbestos-related employment. Our approach involved a compromise wherein we selected the value $k = 3.2$ as a reasonable estimate of the value characterizing the general population of exposed workers. Sensitivity analysis S4 indicated the impact of alternative values: If $k = 5.3$, then the total claims 1990-2049 would increase from 445,751 claims to 777,038 – a 74.3% increase. This increase was sensitive to the estimates for the youngest cohorts, however, and dropped to 53.4% when the cohort aged 15-19 in 1970-1974 was deleted. We believe that the "correct" $k$ value for use in the general population of exposed workers is substantially below 4.2, and that is why we chose $k = 3.2$.

# 8

# Forecasts Based on a Hybrid Model

## 8.1 Introduction

This chapter introduces occupation as a stratifying dimension in our two-stage projection model. Because occupational information was not provided in the Surveillance, Epidemiology, and End Results Program (SEER) data files but was provided in the Manville Trust extract files, we used the results of Section 7.12 to justify replacing the SEER data in the first-stage calibration with the Trust data under the bridge model (Model 3) (i.e., using mesothelioma claim estimates for 1990-1994).

In making indirect estimates of past exposure to asbestos using data on claims against the Trust, we stratified the projection model by occupation, in addition to the standard covariates (age, sex, and time since first exposure). This allowed use of Selikoff's (1981) estimates of relative risk and average duration within occupation in the context of Walker's (1982) approach. In addition, with this hybrid approach, the estimates of past exposure are for real people, not insulation-worker equivalents (IWEs). Therefore, the projections to future years can accommodate depletions or decrements due to the rules against multiple claim filings. This hybrid model is the forecasting structure that the Rule 706 Panel found most credible.

The introduction of occupation as a stratifying dimension in our projection model offered the opportunity to restructure and simplify parts of the model. First, the use of an artificial measure of asbestos exposure (number of IWEs) was dropped in favor of estimates of the real number of exposed workers in each occupation. Second, estimates of relative risks of mesothelioma and associated durations of exposure were introduced for calibrating exposed workers in each occupation. Third, instead of approximating the effects of different durations of exposure to asbestos on the power parameter of the Peto et al. (1982) model, we used the OSHA (1983) form, which directly represented duration as a parameter of the model, along with a 10-year latency period during which no mesothelioma incidence could occur. Fourth, the calibration period was shortened from January 1, 1990 - November 30, 1992 to January 1, 1990

- June 30, 1992 to estimate better the distribution of time since first exposure for the entire 5-year period 1990-1994 and to eliminate any bias caused by the change in the Proof of Claim (POC) Forms in August 1992. Fifth, the use of Trust claims in the first-stage calibration provided direct estimates of the size of the potential claimant population and resolved sensitivity issues associated with the youngest cohorts in the SEER-based calibration. Sixth, with direct estimates of the size of the potential claimant population, it was possible to simplify the first-stage calibration by dropping steps 8-10. Seventh, with stratification of the model by occupation, it was possible to further simplify the first-stage model by dropping steps 5 and 6, so that the only step retained in the model estimation phase was step 7. Eighth, step 3 of the second-stage was modified to represent the effects of second-injury claim filing rules on the pool of potential claimants through new claim depletion rules.

These combined changes yielded a total of 365,615 claims [18.0% lower than the baseline claim hazard rate (CHR) model in Chapter 6] projected for 1990-2049. The 18.0% decrease represented the net effect of a number of significant changes. Without imposing the claim depletion rules, the projection was 440,375 claims (−1.2%). Thus, the claim depletion rules accounted for almost all of the overall decrease. The change in calibration period yielded a 0.4% increase in total claims for 1990-1994, but this was composed of an 8.6% increase in cancer claims and a 0.7% decrease in noncancer claims. Without claim depletion, the cancer projection for 1990-2049 increased 18.4%; with claim depletion, the increase was 12.1%. Without claim depletion, the noncancer projection for 1990-2049 decreased 4.7%; with claim depletion, the decrease was 23.4%.

All of these changes were within the uncertainty limits developed in Chapter 7. However, no single model in Chapter 7 came close to capturing the new pattern here of increasing cancer and decreasing noncancer claims. The best cancer models were only implicit in sensitivity analysis S4 with increases in the $k$ parameter and decreases in the $b$ parameter (implying better estimation of the actual number of workers). The best noncancer models were Models 1 and 2 from sensitivity analysis S9, with asbestos-related noncancer claim rates declining 3% per year and nonasbestos-related and other diseases declining 1% per year (implying better estimation of the rates of decline in claim filings). The preferred projection model in this chapter resolved the major issues identified in the sensitivity analyses of Chapter 7. The remaining sensitivities will be analyzed in Chapter 9.

## 8.2 Model Overview

### 8.2.1 First Stage

In the first stage, the distribution of all males exposed to Johns-Manville-produced asbestos and asbestos products, whose initial exposure was prior to

1975, was estimated by age (5-year groups), time since first exposure (TSFE, 5-year groups) to asbestos, and occupation (eight groups selected to match closely the categories used by Selikoff and co-workers – details on the construction of the occupation groups are in Section 8.3; age and TSFE were discussed in Chapter 6). This distribution was estimated from (1) the number of male mesothelioma claims against the Trust for 1990-1994 (using data for January 1, 1990 - June 30, 1992), stratified by age, TSFE, and occupation; (2) the mesothelioma incidence rate stratified by TSFE, for a population with known relative risk and exposure duration – estimated by refitting the OSHA (1983) model to insulation worker data from Selikoff and Seidman (1991) (see Table 7.4); and (3) the average duration of exposure to asbestos, and relative risk of mesothelioma by occupation, for equivalent exposure durations [estimated from Selikoff (1981)]. Details of this procedure are in Section 8.4.

The first stage consisted of three phases, as in Chapter 6: (1) data preparation; (2) model estimation; and (3) model projection.

### Data Preparation

1. *Nonmesothelioma mortality rates.* The hazard rates were shown in Table 6.2. We used the analysis in Section 6.5.1.
2. *Occupation groups with significant asbestos exposure.* To employ Selikoff's (1981) relative risk estimates in Table 3.1, we developed an industrial and occupational classification of the Trust data that closely matched Selikoff's (1981) classification. Because the industry/occupation groups differed by average exposure levels, this step obviated the need for steps 5 and 6 in Sections 6.5.5 and 6.6.1.
3. *Distribution of mesothelioma claim counts 1990-1994 by attained age at the time of claim and TSFE.* We estimated mesothelioma claim counts from Trust data for attained age groups 15-19, ..., 95-99 years and TSFE categories 2.5-7.5, ..., 92.5-97.5, stratified by occupation. This step replaced step 2 in Section 6.5.2, and is similar to the use of Table 6.14 in Model 3 in Section 7.12.
4. *Distribution of mesothelioma claim counts by age at start of exposure and date of first exposure (DOFE).* We estimated mesothelioma claim counts from Trust data for combinations of age and DOFE categories, stratified by occupation. This step replaced step 3 in Section 6.5.3, and was similar to the use of Table 7.10 in Model 3 in Section 7.12. As in Chapters 6 and 7, this step included smoothing of the exposure distribution, but here we modified the procedure to reflect the impact of WWII on the 1940-1944 estimates.
5. *Normalization of exposure.* We normalized the matrix from step 4 by cohort so that the cohort sum represented unit probability mass. This was similar to step 4 of Section 6.5.4.

### Model Estimation

6. *Estimation of the OSHA model for mesothelioma mortality.* The OSHA model generalized the Weibull incidence model used by Peto et al. (1982). We developed parameter estimates for this model that took into account the different relative risks and average exposure durations in the occupation groups. This did not correspond to any of the steps in Chapter 6. It was similar to the analyses in Section 7.5.
7. *Estimation of the population exposed to asbestos prior to 1975.* We calculated the size of the population exposed to asbestos prior to 1975 necessary to yield the Trust claim counts from step 3 for 1990-1994, stratified by occupation, using the OSHA model from step 6, under the assumption that the timing and number of mesothelioma deaths and claims were identical in the at-risk population during 1990-1994. This step was similar to step 7 in Section 6.6.2. Furthermore, because step 3 provided data on TSFE, this step obviated the need for steps 8-10 in Section 6.6. It could be argued that this step obviated the need for steps 4 and 5 in this chapter. These steps were retained primarily to accommodate the smoothing procedure and the special handling of WWII estimates.

### Model Projection

8. *First-stage calibration.* We used the modified forward projection equation to project the at-risk population from the age and date of initial exposure to the projection periods 1990-1994, 1995-1999, ..., 2045-2049, stratified by occupation. This step was similar to step 11 of Section 6.7.1.
9. *Forward projection of mesothelioma mortality.* We used the forward projection equation to project mesothelioma deaths throughout the period 1990-2049, stratified by occupation. This step was similar to step 12 of Section 6.7.2.

### 8.2.2 Second Stage

In the second stage, the claim filing rates for 1990-1994 were computed by age, TSFE, occupation, and type of disease, or injury, alleged to have resulted from exposure to Johns-Manville asbestos. The estimates were made for nine classes of alleged disease/injury identified on the POC Form. The male claim experience from January 1, 1990 to June 30, 1992 was extended to 1990-1994 by assuming the monthly claim filing counts were constant within disease and DOFE groups. For each of nine diseases, the claim counts were tabulated by age, TSFE, and occupation, and matched to the corresponding exposed population counts (available from the first-stage calibration). From these data, disease-specific claim filing rates were calculated by age, TSFE, and occupation for the period 1990-1994. The projection model exactly reproduced the age, occupation, and disease-specific male claim counts estimated for the period 1990-1994. This guaranteed that the initial conditions for the projections were based on plausible Trust claim estimates.

For each 5-year period starting with 1990-1994 and ending with 2045-2049, the disease-specific claim filing rates were multiplied by the surviving exposed population counts, specific to age, TSFE, and occupation to project the future number of claims against the Trust. The total number of male claims was obtained by summing over age, period, TSFE, occupation, and disease.

The second stage consisted of four steps that expanded the procedures used in Chapter 6 to reflect the process of claims depletion:

1. *Distribution of disease-specific claim counts 1990-1994 by attained age at the time of claim and TSFE.* We estimated disease-specific claim counts for combinations of attained age and TSFE categories, stratified by occupation. Three sets of criteria were used in assigning each claim to a disease category. This step expanded the procedures in step 1 of Section 6.8.1.
2. *Second-stage calibration.* We estimated CHRs for each disease defined in step 1, by attained age and TSFE, stratified by occupation. This step was similar to step 2 of Section 6.8.2.
3. *At-risk population projections.* Using the rules for filing multiple claims, we used the CHRs from step 2 to decrement the at-risk population for 1990-1994 estimated from step 9 of the first stage, to project the at-risk population for each type of claim for 1995-1999, ..., 2045-2049, stratified by occupation. Three sets of rules were used in these calculations. This step did not correspond to any of the steps in Chapter 6.
4. *Claim projections.* We applied the CHRs from step 2 to the appropriate at-risk population estimate from step 3 to project the disease-specific claims for 1990-1994, 1995-1999, ..., 2045-2049, by occupation. This step was similar to step 3 of Section 6.8.3.

## 8.3 Data Preparation

### 8.3.1 Step 1: Nonmesothelioma Mortality Rates

This step was described in Section 6.5.1 and the results for 1985-2049 were presented in Table 6.2. Sensitivity analyses in Sections 7.8 and 7.9 indicated little sensitivity to the assumptions about nonmesothelioma mortality rates. As a consequence, we assumed that the same rates applied to all occupations, and dispensed with Walker's (1982) assumption of 37% excess mortality among insulation workers. This was motivated, in part, by the standardized mortality ratios (SMRs) in Table 2.1 which indicated that mortality for noncancer causes was at most 1-9% higher than expected for insulation workers, but was at least 5-9% lower than expected when asbestosis and other noninfectious respiratory diseases were excluded. Because our projections account for the impact of asbestos-related diseases through the claim depletion calculations, it would be inappropriate to reflect the effects of these same diseases through the mortality calculations. The nonmesothelioma mortality rates reflect the mortality rates faced by the general population.

### 8.3.2 Step 2: Occupation Groups with Significant Asbestos Exposure

Our objective in this step was to develop a set of occupation codes that closely matched the categories of Selikoff (1981). To accomplish this, we employed the occupation and business-type codes for the first reported occupational exposure in the Manville Trust extract files (see Section 6.5.3). Those occupation codes were self-assigned by the claimant on the POC Form:

| Occupation | Description |
|---|---|
| A | Factory worker - asbestos as raw material |
| F | Friction materials worker |
| I | Insulator |
| S | Shipyard worker |
| R | Roofer |
| W | Factory worker - other |
| O | Other |
| U | Unknown |

Business type was indicated by the claimant in free format on the POC Form. Each entry was interpreted and classified by the Trust into one of 200 categories of business type. The joint frequencies of business type and occupation code are contained in Table 8.1.

Our analysis of Table 8.1 led to a sequence of three classification systems, or recodes, for occupation, designated R1, R2, and R3. All of the projections were based on the R3 recode which was derived from the R1 and R2 recodes.

To develop the R1 recode, we observed that the business types in Table 8.1 were initially classified by the Trust into 25 business categories. We collapsed those categories to a set of 11 industry groups plus a residual based on the descriptions provided by Selikoff (1981) (see Section 3.3.1). This yielded the R1 recode.

Table 8.1 shows that there was substantial occupational asbestos exposure in some lower-ranked R1 industries. Selikoff (1981) classified occupation and industry in a hierarchy that cut across both dimensions (see Section 3.3.1). We followed this strategy in our R2 recode of occupation using the following rule:

If occupation code is not any one of A, F, I, S, or R,then R2 = R1;
    else if occupation code = A then R2 = 1;
    else if occupation code = F then R2 = 2;

else if occupation code = I then R2 = 3;
else if occupation code = S then R2 = 4;
else if occupation code = R then R2 = 5.

The category names for the R2 recode were the same as for the R1 recode, but the classification system was better because it allowed variations in occupational exposures within each industry. The need for such supplemental information was described by Selikoff (1981). Our information was limited to that in Table 8.1, so the correspondence of the R2 recode to Selikoff's classification system was a rough approximation.

Figure 8.1 displays the distribution of claims by R2 occupation group and DOFE. Figure 8.2 displays the distribution of claims by disease (most severe alleged disease on the POC Form) and DOFE. Figure 8.3 displays the distribution of mesothelioma claims by R2 occupation group and DOFE. Figure 8.4 displays the distribution of mesothelioma claims filed in 1990-1992 by R2 occupation group and DOFE.

Our approach required one additional adjustment which created the R3 occupation recode: Groups 6, 7, 8, 9, and 11 of the R2 occupation recode were aggregated into one group (group 6) of the eight-group R3 occupation recode. Group 6 of the R3 recode was labeled Util/Trans/Chem/Longshore. This adjustment was required to increase the sample size for the first-stage calibration of the exposed population in this occupation group to a level comparable to the other groups: 12.0% of male mesothelioma claims in 1990-1994 fell into group 6 of the R3 recode.

Henceforth, all references to industries or occupation will be references to the R3 occupation groups.

The aggregate claims in Figure 8.4 were used to calibrate the bridge model projection (Model 3) in Section 7.12. The bridge model yielded 423,200 claims for 1990-2049, which compared well (−5.1%) with the 445,751 claims under the baseline projection in Chapter 6. With the mesothelioma claims for 1990-1994 in the bridge model stratified by occupation, we could produce projections of future claims by occupation.

### 8.3.3 Step 3: Distribution of Mesothelioma Claim Counts 1990-1994 by Attained Age at the Time of Claim and TSFE

This step generated estimates of mesothelioma claim counts 1990-1994 by attained age at the time of claim and TSFE. This differed from step 2 in Section 6.5.2 because the SEER data did not record TSFE. The information on TSFE in this step could be coordinated with the information on DOFE in the next step, if appropriate assumptions were made to extend the claim counts observed in the calibration period in 1990-1992 to the full 5 years 1990-1994.

To develop consistent estimates, we shortened the calibration period by 5 months and assumed that the filing counts for each cohort would be the same

**Table 8.1: Distribution of Qualified Male Claims Against the Manville Trust, 1988-1992**

| R1 | Manville Trust Business Classification System | | | R2 Recode 1 | 2 | 3 | 4 | 5 | =R1 | =R1 | =R1 | |
|---|---|---|---|---|---|---|---|---|---|---|---|---|
| | | | | Occupation Code[1] | | | | | | | | |
| Recode | Category | Type | Description | A | F | I | S | R | W | O | U | Total |
| 1-12 | | TOTAL | ALL CATEGORIES | 9,258 | 596 | 9,544 | 33,790 | 193 | 19,183 | 80,328 | 33,978 | 186,870 |
| 1 | | TOTAL | PRIMARY MANUFACTURING | **3,921** | 17 | 695 | 29 | 6 | 196 | 871 | 360 | 6,095 |
| | ASBPROD | Subtotal | Asbestos Abatement or Asbestos Products Manuf./Mining | **3,921** | 17 | 695 | 29 | 6 | 196 | 871 | 360 | 6,095 |
| | | 2 | Asbestos Abatement | **0** | 0 | 0 | 0 | 0 | 0 | 0 | 1 | 1 |
| | | 102 | Asbestos Abatement | **0** | 0 | 0 | 0 | 0 | 0 | 0 | 1 | 1 |
| | | 124 | Manville Asbestos Products Manufacturing/Mining | **0** | 0 | 0 | 0 | 0 | 0 | 0 | 8 | 8 |
| | | 125 | Non-Manville Asbestos Products Manufacturing/Mining | **0** | 0 | 0 | 0 | 0 | 0 | 1 | 40 | 41 |
| | | ASP | Asbestos Products Mfg. | **3,904** | 13 | 694 | 29 | 6 | 187 | 787 | 296 | 5,916 |
| | | GKM | Gasket Manufacturer | **8** | 0 | 0 | 0 | 0 | 1 | 9 | 1 | 19 |
| | | MIN | Mining | **9** | 4 | 1 | 0 | 0 | 8 | 74 | 13 | 109 |
| 2 | | TOTAL | SECONDARY MANUFACTURING | 2,958 | **63** | 548 | 560 | 2 | 12,820 | 11,784 | 1,445 | 30,180 |
| | AERO | Subtotal | Aerospace/Aviation | 24 | **2** | 4 | 3 | 0 | 76 | 184 | 21 | 314 |
| | | 1 | Aerospace/Aviation | 0 | **0** | 0 | 0 | 0 | 0 | 0 | 1 | 1 |
| | | 101 | Aerospace/Aviation | 0 | **0** | 0 | 0 | 0 | 0 | 0 | 3 | 3 |
| | | AIR | Aircraft Mfg./Repair/Airport Work | 24 | **2** | 4 | 3 | 0 | 76 | 184 | 17 | 310 |
| | BOILER | Subtotal | Boilermaking | 49 | **6** | 84 | 13 | 0 | 215 | 503 | 27 | 897 |
| | | 104 | Boilermaking | 0 | **0** | 0 | 0 | 0 | 0 | 0 | 8 | 8 |
| | | BMF | Boiler Mfg./Repair/Parts/etc. | 49 | **6** | 84 | 13 | 0 | 215 | 503 | 19 | 889 |
| | IRON | Subtotal | Iron/Steel | 1,098 | **18** | 252 | 514 | 0 | 4,004 | 8,100 | 1,020 | 15,006 |
| | | 108 | Iron/Steel | 0 | **0** | 0 | 0 | 0 | 0 | 21 | 631 | 652 |
| | | ALU | Aluminum Business | 55 | **6** | 28 | 19 | 0 | 533 | 243 | 30 | 914 |
| | | COP | Copper Company | 20 | **0** | 0 | 0 | 0 | 11 | 32 | 1 | 64 |
| | | FDY | Foundry | 28 | **0** | 9 | 4 | 0 | 56 | 123 | 11 | 231 |
| | | IRN | Iron Company | 4 | **0** | 7 | 14 | 0 | 33 | 189 | 9 | 256 |
| | | LED | Lead Company | 1 | **0** | 2 | 0 | 0 | 1 | 18 | 3 | 25 |

See footnotes at end of table.

(Continued)

**Table 8.1 (Continued)**

| R1 Recode | Manville Trust Business Classification System: Category | Type | Description | Occupation Code[1]: A | F | I | S | R | W | O | U | Total |
|---|---|---|---|---|---|---|---|---|---|---|---|---|
| | | SML | Smeltery | 5 | **0** | 2 | 0 | 0 | 10 | 51 | 1 | 69 |
| | | STM | Steel Manufacturer | 985 | **12** | 204 | 477 | 0 | 3,360 | 7,423 | 334 | 12,795 |
| | MANUF | Subtotal | Non-Asbestos Products Manufacturing | 1,238 | **27** | 160 | 28 | 2 | 1,538 | 2,724 | 268 | 5,985 |
| | | 113 | Non-Asbestos Products Manufacturing | 0 | **0** | 0 | 0 | 0 | 0 | 0 | 68 | 68 |
| | | ELO | Electronics | 3 | **0** | 1 | 4 | 0 | 4 | 31 | 3 | 46 |
| | | FBR | Fiberglass Producer/Plant/etc. | 4 | **0** | 3 | 1 | 0 | 1 | 7 | 1 | 17 |
| | | FLR | Flooring/Floor Covering | 16 | **1** | 1 | 1 | 0 | 2 | 79 | 3 | 103 |
| | | MAN | Manufacturer | 1,098 | **21** | 101 | 14 | 1 | 1,112 | 1,724 | 113 | 4,184 |
| | | OVN | Oven Manufacturing | 2 | **0** | 0 | 0 | 0 | 1 | 5 | 0 | 8 |
| | | PAP | Paper Products Business | 36 | **1** | 31 | 5 | 0 | 125 | 289 | 29 | 516 |
| | | PLC | Plastic/Plastic Mfg./etc. | 35 | **0** | 3 | 1 | 0 | 7 | 21 | 5 | 72 |
| | | PMP | Paper Manufacturing Plant | 20 | **0** | 6 | 1 | 0 | 39 | 115 | 12 | 193 |
| | | PPM | Papermill | 19 | **3** | 12 | 1 | 1 | 237 | 328 | 33 | 634 |
| | | RFT | Refractory | 5 | **1** | 2 | 0 | 0 | 10 | 125 | 1 | 144 |
| | TEXTILE | Subtotal | Textile | 26 | **1** | 31 | 1 | 0 | 97 | 136 | 42 | 334 |
| | | 121 | Textile | 0 | **0** | 0 | 0 | 0 | 0 | 0 | 12 | 12 |
| | | TXT | Textile/Dyeing/etc. | 26 | **1** | 31 | 1 | 0 | 97 | 136 | 30 | 322 |
| | TIRE | | Tire/Rubber | 523 | **9** | 17 | 1 | 0 | 6,890 | 137 | 67 | 7,644 |
| | | TRW | Tire Mfg.; Rubber Industry | 523 | **9** | 17 | 1 | 0 | 6,890 | 137 | 67 | 7,644 |
| 3 | | TOTAL | INSULATION WORK | 245 | 20 | **3,993** | 356 | 35 | 836 | 14,002 | 708 | 20,195 |
| | INSUL | Subtotal | Insulation - Installation/Repair/Ripout | 245 | 20 | **3,993** | 356 | 35 | 836 | 14,002 | 708 | 20,195 |
| | | 112 | Insulation - Installation/Repair/Ripout | 0 | 0 | **0** | 0 | 0 | 0 | 0 | 11 | 11 |
| | | AIC | Air Conditioning/Refrigeration | 1 | 0 | **5** | 0 | 0 | 2 | 75 | 1 | 84 |
| | | FIR | Fireproofing | 1 | 0 | **2** | 0 | 0 | 1 | 14 | 0 | 18 |
| | | HAV | Heating/Air Cond./Ventilation | 3 | 0 | **11** | 2 | 0 | 6 | 117 | 9 | 148 |
| | | HEV | Heating & Ventilating | 0 | 0 | **3** | 0 | 0 | 8 | 61 | 1 | 73 |
| | | HRC | Heating & Refrig. Contractor | 0 | 0 | **1** | 0 | 0 | 0 | 9 | 0 | 10 |
| | | HTG | Heating Co. (Home; Gas or Oil) | 0 | 0 | **7** | 0 | 0 | 2 | 55 | 4 | 68 |
| | | INS | Insulation Business | 143 | 1 | **2,676** | 51 | 2 | 18 | 548 | 181 | 3,620 |
| | | PIP | Pipe Coverer | 4 | 0 | **24** | 20 | 0 | 3 | 101 | 22 | 174 |
| | | UNI | Union | 93 | 19 | **1,264** | 283 | 33 | 796 | 13,022 | 479 | 15,989 |
| 4 | | TOTAL | SHIPBUILDING & REPAIR | 90 | 5 | 503 | **29,458** | 1 | 87 | 3,635 | 2,192 | 35,971 |

See footnotes at end of table.

(Continued)

**Table 8.1 (Continued)**

| R1 | Manville Trust Business Classification System | | | Occupation Code[1] | | | | | | | | |
|---|---|---|---|---|---|---|---|---|---|---|---|---|
| Recode | Category | Type | Description | A | F | I | S | R | W | O | U | Total |
| | SHIP | Subtotal | Shipyard - Construction/Repair | 90 | 5 | 503 | **29,458** | 1 | 87 | 3,635 | 2,192 | 35,971 |
| | | 120 | Shipyard - Construction/Repair | 0 | 0 | 0 | 0 | 0 | 0 | 14 | 265 | 279 |
| | | DRD | Dry Dock | 1 | 0 | 5 | **133** | 0 | 1 | 63 | 3 | 206 |
| | | SAS | Salvage Ships | 0 | 0 | 0 | **11** | 0 | 0 | 4 | 0 | 15 |
| | | SBR | Shipbuilding & Repair | 4 | 0 | 27 | **2,418** | 0 | 6 | 126 | 102 | 2,683 |
| | | SHB | Shipbuilding Business | 2 | 0 | 37 | **2,586** | 0 | 3 | 127 | 85 | 2,840 |
| | | SHC | Shipyard Construction | 0 | 0 | 2 | **34** | 0 | 1 | 6 | 7 | 50 |
| | | SHF | Shipfitting Business | 1 | 0 | 0 | **35** | 0 | 0 | 7 | 5 | 48 |
| | | SHH | Ship | 1 | 0 | 1 | **151** | 0 | 1 | 47 | 14 | 215 |
| | | SHI | Shipbuilding | 37 | 1 | 162 | **10,388** | 0 | 36 | 898 | 420 | 11,942 |
| | | SHR | Ship Repair | 3 | 2 | 13 | **782** | 0 | 2 | 79 | 103 | 984 |
| | | SPY | Shipyard(s) | 41 | 2 | 256 | **12,920** | 1 | 37 | 2,264 | 1,188 | 16,709 |
| 5 | | TOTAL | CONSTRUCTION TRADES | 136 | 27 | 1,477 | 387 | **120** | 314 | 13,904 | 885 | 17,250 |
| | CONSTR | Subtotal | Construction Trades | 89 | 24 | 1,234 | 268 | **26** | 235 | 10,147 | 568 | 12,591 |
| | | 107 | Construction Trades | 0 | 0 | 0 | 0 | **0** | 0 | 1 | 96 | 97 |
| | | CAR | Carpentry | 2 | 1 | 9 | 7 | **1** | 7 | 310 | 29 | 366 |
| | | COC | Concrete Business | 3 | 1 | 1 | 0 | **0** | 3 | 20 | 2 | 30 |
| | | CON | Construction | 26 | 9 | 612 | 88 | **19** | 128 | 3,652 | 208 | 4,742 |
| | | COT | Contractor | 15 | 5 | 578 | 47 | **6** | 33 | 4,746 | 117 | 5,547 |
| | | DRW | Dry Wall Business | 1 | 0 | 1 | 0 | **0** | 0 | 58 | 5 | 65 |
| | | ELE | Electric(al); Any Type (not Utility) | 35 | 5 | 28 | 35 | **0** | 39 | 911 | 68 | 1,121 |
| | | ERE | Erection | 0 | 0 | 1 | 0 | **0** | 0 | 3 | 0 | 4 |
| | | MAS | Masonry | 0 | 0 | 0 | 0 | **0** | 1 | 54 | 1 | 56 |
| | | PNT | Painting Business | 4 | 2 | 3 | 19 | **0** | 17 | 230 | 18 | 293 |
| | | SAN | Sandblasting | 1 | 1 | 0 | 14 | **0** | 0 | 8 | 2 | 26 |
| | | WEL | Welding Company | 2 | 0 | 1 | 41 | **0** | 4 | 79 | 12 | 139 |
| | | WLR | Welder | 0 | 0 | 0 | 17 | **0** | 3 | 71 | 10 | 101 |
| | | WOO | Woodcrafting | 0 | 0 | 0 | 0 | **0** | 0 | 4 | 0 | 4 |
| | PIPE | Subtotal | Pipefitting/Steamfitting/Plumbing | 14 | 0 | 210 | 77 | **0** | 40 | 2,138 | 221 | 2,700 |
| | | 115 | Pipefitting/Steamfitting/Plumbing | 0 | 0 | 0 | 0 | **0** | 0 | 4 | 104 | 108 |
| | | PBR | Plumber | 2 | 0 | 10 | 0 | **0** | 1 | 122 | 11 | 146 |
| | | PIF | Pipefitter | 3 | 0 | 81 | 19 | **0** | 17 | 471 | 22 | 613 |
| | | PLB | Plumbing; Heating | 4 | 0 | 52 | 10 | **0** | 14 | 975 | 56 | 1,111 |

See footnotes at end of table.

(Continued)

**Table 8.1 (Continued)**

| R1 | Manville Trust Business Classification System | | | Occupation Code[1] | | | | | | | | |
|---|---|---|---|---|---|---|---|---|---|---|---|---|
| Recode | Category | Type | Description | A | F | I | S | R | W | O | U | Total |
| | | PLM | Plumbing Supplies | 0 | 0 | 0 | 0 | **0** | 1 | 43 | 2 | 46 |
| | | PPF | Pipefitting Business | 3 | 0 | 59 | 45 | **0** | 6 | 352 | 11 | 476 |
| | | PPG | Plumbing/Pipefitting | 0 | 0 | 4 | 0 | **0** | 0 | 48 | 5 | 57 |
| | | STF | Steamfitting Business | 2 | 0 | 4 | 3 | **0** | 1 | 123 | 10 | 143 |
| | PLASTER | Subtotal | Plaster/Sheetrock | 0 | 2 | 5 | 1 | **0** | 3 | 214 | 7 | 232 |
| | | 116 | Plaster/Sheetrock | 0 | 0 | 0 | 0 | **0** | 0 | 0 | 1 | 1 |
| | | PLS | Plaster Business | 0 | 2 | 5 | 1 | **0** | 3 | 214 | 6 | 231 |
| | ROOFING | Subtotal | Roofing | 19 | 0 | 4 | 3 | **92** | 8 | 157 | 8 | 291 |
| | | 118 | Roofing | 0 | 0 | 0 | 0 | **0** | 0 | 0 | 1 | 1 |
| | | RFG | Roofing Business | 19 | 0 | 4 | 3 | **92** | 8 | 156 | 5 | 287 |
| | | SHG | Shingle Business | 0 | 0 | 0 | 0 | **0** | 0 | 1 | 2 | 3 |
| | SHEET | Subtotal | Sheetmetal | 14 | 1 | 24 | 38 | **2** | 28 | 1,248 | 81 | 1,436 |
| | | 119 | Sheetmetal | 0 | 0 | 0 | 0 | **0** | 0 | 0 | 8 | 8 |
| | | SMT | Sheet Metal/ Metal Business | 14 | 1 | 24 | 38 | **2** | 28 | 1,248 | 73 | 1,428 |
| 6 | | TOTAL | RAILROAD | 30 | 24 | 85 | 5 | 12 | 807 | 6,458 | 493 | 7,914 |
| | RAIL | Subtotal | Railroad | 30 | 24 | 85 | 5 | 12 | 807 | 6,458 | 493 | 7,914 |
| | | 117 | Railroad | 0 | 0 | 0 | 0 | 0 | 0 | 2 | 162 | 164 |
| | | MTA | Mass Transit (Trans. Author.) | 0 | 4 | 1 | 0 | 0 | 0 | 18 | 1 | 24 |
| | | RRD | Railroad Industry | 30 | 20 | 84 | 5 | 12 | 807 | 6,438 | 330 | 7,726 |
| 7 | | TOTAL | UTILITY SERVICES | 107 | 3 | 116 | 11 | 0 | 282 | 2,829 | 216 | 3,564 |
| | UTILITY | Subtotal | Utilities | 107 | 3 | 116 | 11 | 0 | 282 | 2,829 | 216 | 3,564 |
| | | 123 | Utilities | 0 | 0 | 0 | 0 | 0 | 0 | 0 | 41 | 41 |
| | | ATE | Atomic Energy Facility | 0 | 0 | 3 | 0 | 0 | 0 | 69 | 0 | 72 |
| | | CMI | Communcations Installation (Phones etc.) | 0 | 0 | 0 | 2 | 0 | 1 | 16 | 2 | 21 |
| | | GAS | Natural Gas Plant | 1 | 0 | 6 | 0 | 0 | 6 | 24 | 1 | 38 |
| | | POP | Power Plant | 42 | 0 | 42 | 2 | 0 | 73 | 1,038 | 89 | 1,286 |
| | | POW | Powerhouse | 3 | 0 | 19 | 1 | 0 | 26 | 179 | 8 | 236 |
| | | STH | Steam Heat Producer | 0 | 0 | 2 | 0 | 0 | 0 | 11 | 1 | 14 |
| | | UTL | Utility; Any Type | 61 | 3 | 44 | 6 | 0 | 176 | 1,492 | 74 | 1,856 |
| 8 | | TOTAL | CHEMICAL/PETROCHEMICAL | 824 | 13 | 931 | 95 | 3 | 1,550 | 6,860 | 518 | 10,794 |
| | CHEMICAL | Subtotal | Chemical | 586 | 3 | 418 | 24 | 0 | 1,019 | 2,373 | 211 | 4,634 |
| | | 106 | Chemical | 0 | 0 | 0 | 0 | 0 | 0 | 1 | 37 | 38 |

See footnotes at end of table.

(Continued)

**Table 8.1 (Continued)**

| R1 | Manville Trust Business Classification System | | | Occupation Code[1] | | | | | | | | |
|---|---|---|---|---|---|---|---|---|---|---|---|---|
| Recode | Category | Type | Description | A | F | I | S | R | W | O | U | Total |
| | | CHM | Chemical Mfg. and/or Refining | 586 | 3 | 418 | 24 | 0 | 1,019 | 2,372 | 174 | 4,596 |
| | PETRO | Subtotal | Petrochemical | 238 | 10 | 513 | 71 | 3 | 531 | 4,487 | 307 | 6,160 |
| | | 114 | Petrochemical | 0 | 0 | 0 | 0 | 0 | 0 | 0 | 14 | 14 |
| | | OIL | Oil Company | 28 | 4 | 38 | 5 | 0 | 12 | 421 | 55 | 563 |
| | | ORY | Oil Refinery | 50 | 5 | 234 | 2 | 1 | 232 | 1,645 | 118 | 2,287 |
| | | PCR | PetroChemical Refinery | 114 | 0 | 5 | 0 | 0 | 1 | 87 | 4 | 211 |
| | | PTR | Petroleum Industry | 1 | 0 | 6 | 0 | 0 | 1 | 50 | 7 | 65 |
| | | REF | Refinery | 45 | 1 | 230 | 64 | 2 | 285 | 2,284 | 109 | 3,020 |
| 9 | | TOTAL | MARITIME/LONGSHORE | 3 | 1 | 3 | 203 | 0 | 2 | 3,288 | 5,352 | 8,852 |
| | LONGSHOR | Subtotal | Longshore | 3 | 1 | 3 | 203 | 0 | 2 | 3,287 | 5,352 | 8,851 |
| | | 109 | Longshore | 0 | 0 | 0 | 0 | 0 | 0 | 0 | 1 | 1 |
| | | HAR | Harbor | 0 | 0 | 0 | 1 | 0 | 0 | 2 | 0 | 3 |
| | | LGS | Longshoreman | 3 | 1 | 3 | 202 | 0 | 2 | 3,285 | 5,351 | 8,847 |
| | MARITM | Subtotal | Maritime | 0 | 0 | 0 | 0 | 0 | 0 | 1 | 0 | 1 |
| | | MRN | Maritime Worker | 0 | 0 | 0 | 0 | 0 | 0 | 1 | 0 | 1 |
| 10 | | TOTAL | MILITARY | 52 | 63 | 127 | 1,633 | 0 | 50 | 4,285 | 738 | 6,948 |
| | MILIT | Subtotal | Military | 52 | 63 | 127 | 1,633 | 0 | 50 | 4,285 | 738 | 6,948 |
| | | 111 | Military | 0 | 0 | 0 | 0 | 0 | 0 | 0 | 6 | 6 |
| | | ARS | Arsenal | 5 | 2 | 7 | 4 | 0 | 20 | 220 | 9 | 267 |
| | | FRM | Foreign Military | 0 | 0 | 0 | 0 | 0 | 0 | 1 | 1 | 2 |
| | | MIS | Missile Building | 1 | 0 | 1 | 0 | 0 | 0 | 14 | 0 | 16 |
| | | UAF | U.S. Air Force | 15 | 11 | 5 | 5 | 0 | 5 | 210 | 21 | 272 |
| | | UCG | U.S. Coast Guard/C.G. Station | 0 | 0 | 0 | 34 | 0 | 1 | 72 | 17 | 124 |
| | | USA | U.S. Army | 5 | 21 | 5 | 21 | 0 | 3 | 369 | 40 | 464 |
| | | USG | U.S./State Government | 2 | 4 | 12 | 156 | 0 | 3 | 188 | 16 | 381 |
| | | USM | U.S. Military | 4 | 17 | 20 | 170 | 0 | 5 | 723 | 158 | 1,097 |
| | | USN | U.S. Navy | 20 | 8 | 77 | 1,243 | 0 | 13 | 2,488 | 470 | 4,319 |
| 11 | | TOTAL | MANUFACTURING/AUTOMOBILE | 112 | 199 | 19 | 2 | 0 | 130 | 842 | 79 | 1,383 |
| | AUTO | Subtotal | Automobile/Mechanical/Friction | 112 | 199 | 19 | 2 | 0 | 130 | 842 | 79 | 1,383 |
| | | 103 | Automobile/Mechanical/Friction | 0 | 0 | 0 | 0 | 0 | 0 | 0 | 7 | 7 |
| | | ATM | Auto Manufacturer | 73 | 26 | 13 | 0 | 0 | 108 | 453 | 7 | 680 |

See footnotes at end of table.

(Continued)

**Table 8.1 (Continued)**

| R1 | Manville Trust Business Classification System | | | Occupation Code[1] | | | | | | | | |
|---|---|---|---|---|---|---|---|---|---|---|---|---|
| Recode | Category | Type | Description | A | F | I | S | R | W | O | U | Total |
| | | ATR | Auto Repair and/or Service | 4 | 132 | 3 | 1 | 0 | 13 | 275 | 30 | 458 |
| | | AUD | Auto Dealership | 0 | 21 | 2 | 1 | 0 | 2 | 52 | 4 | 82 |
| | | BRK | Brake Lining Business | 13 | 19 | 1 | 0 | 0 | 6 | 52 | 3 | 94 |
| | | FRC | Friction Products Mfg. | 22 | 1 | 0 | 0 | 0 | 1 | 10 | 28 | 62 |
| 12 | | TOTAL | OTHER/UNKNOWN | 780 | 161 | 1,047 | 1,051 | 14 | 2,109 | 11,570 | 20,992 | 37,724 |
| | BYSTAND | Subtotal | Bystander/Building Occupant | 21 | 1 | 25 | 21 | 0 | 2 | 178 | 1,230 | 1,478 |
| | | BYS | Bystander (Housewife/Student/etc.) | 10 | 0 | 20 | 21 | 0 | 0 | 102 | 1,227 | 1,380 |
| | | EDU | Educational System | 0 | 1 | 0 | 0 | 0 | 0 | 1 | 0 | 2 |
| | | HSP | Hospital | 11 | 0 | 5 | 0 | 0 | 2 | 75 | 3 | 96 |
| | OTHER | Subtotal | Other (Specify) | 406 | 61 | 236 | 144 | 3 | 1,469 | 3,247 | 292 | 5,858 |
| | | 126 | Other (Specify) | 0 | 0 | 0 | 0 | 0 | 0 | 0 | 37 | 37 |
| | | ADV | Advertising | 0 | 0 | 0 | 0 | 0 | 1 | 4 | 0 | 5 |
| | | ALC | Alcohol: Distill, Brew, etc. | 3 | 0 | 4 | 0 | 0 | 2 | 24 | 0 | 33 |
| | | APP | Appliance Service & Repair | 0 | 0 | 0 | 0 | 0 | 0 | 9 | 0 | 9 |
| | | ASS | Assembling Business | 4 | 0 | 1 | 0 | 0 | 2 | 8 | 0 | 15 |
| | | BKP | Bakery Products | 0 | 0 | 0 | 0 | 0 | 3 | 15 | 4 | 22 |
| | | BOT | Bottling Company | 3 | 0 | 1 | 1 | 0 | 4 | 7 | 1 | 17 |
| | | BPM | Building Products Mfg. | 1 | 0 | 0 | 0 | 0 | 0 | 0 | 0 | 1 |
| | | BRM | Brick Making | 4 | 0 | 0 | 1 | 0 | 12 | 46 | 3 | 66 |
| | | BUR | Burner Service - Residential | 0 | 0 | 1 | 0 | 0 | 0 | 11 | 0 | 12 |
| | | CBM | Carbon Base Manufacturer | 5 | 0 | 5 | 0 | 0 | 67 | 112 | 3 | 192 |
| | | CEM | Cement Manufacturer | 5 | 0 | 0 | 1 | 0 | 13 | 33 | 2 | 54 |
| | | CIV | Civil Service | 0 | 1 | 0 | 3 | 0 | 0 | 67 | 11 | 82 |
| | | CLA | Clay Forming Business (Ceramics) | 1 | 0 | 0 | 0 | 0 | 67 | 53 | 1 | 122 |
| | | COF | Coffee Processing Plant | 0 | 0 | 0 | 0 | 0 | 4 | 5 | 1 | 10 |
| | | COM | Cosmetic Manufacturing | 0 | 0 | 0 | 0 | 0 | 0 | 1 | 0 | 1 |
| | | CRH | Craft Business | 0 | 0 | 0 | 0 | 0 | 0 | 1 | 0 | 1 |
| | | CTN | Cotton Mill | 0 | 0 | 0 | 1 | 0 | 3 | 5 | 2 | 11 |
| | | DAI | Dairy Manufacture | 0 | 1 | 0 | 0 | 0 | 3 | 11 | 2 | 17 |
| | | DEM | Demolition | 0 | 0 | 0 | 0 | 0 | 3 | 24 | 1 | 28 |
| | | DFC | Defense/Defense Contractor/etc. | 3 | 0 | 7 | 5 | 0 | 47 | 44 | 13 | 119 |
| | | DIS | Distributor | 3 | 0 | 4 | 1 | 0 | 1 | 16 | 4 | 29 |
| | | DME | Drilling Machinery/Equip. Mfg. | 0 | 0 | 1 | 0 | 0 | 1 | 8 | 2 | 12 |

See footnotes at end of table.

(Continued)

**Table 8.1 (Continued)**

| R1 | Manville Trust Business Classification System | | | Occupation Code[1] | | | | | | | | |
|---|---|---|---|---|---|---|---|---|---|---|---|---|
| Recode | Category | Type | Description | A | F | I | S | R | W | O | U | Total |
| | | DRB | Drug/Biological Company | 0 | 0 | 0 | 0 | 0 | 1 | 1 | 0 | 2 |
| | | DRE | Dredging Business | 1 | 1 | 1 | 3 | 0 | 0 | 13 | 0 | 19 |
| | | DRY | Drycleaning Business | 0 | 0 | 1 | 0 | 0 | 3 | 15 | 2 | 21 |
| | | DYE | Dye Cast Machining | 1 | 0 | 1 | 0 | 0 | 6 | 4 | 1 | 13 |
| | | ENG | Engineering | 3 | 0 | 37 | 13 | 0 | 5 | 209 | 20 | 287 |
| | | FAB | Fabricator | 6 | 1 | 1 | 1 | 0 | 34 | 39 | 2 | 84 |
| | | FAC | Factory | 186 | 3 | 6 | 1 | 0 | 78 | 70 | 7 | 351 |
| | | FAS | Fashion Industry | 0 | 0 | 0 | 0 | 0 | 2 | 2 | 0 | 4 |
| | | FIS | Fishing/Clamming (Commercial) | 0 | 0 | 0 | 3 | 0 | 0 | 4 | 1 | 8 |
| | | FPL | Food Processing; Any Type | 4 | 0 | 6 | 1 | 0 | 35 | 73 | 8 | 127 |
| | | FUR | Furnace Installation & Repair | 5 | 0 | 5 | 1 | 0 | 4 | 58 | 9 | 82 |
| | | GHA | Gov't Housekeeping Agency | 0 | 0 | 0 | 1 | 0 | 0 | 3 | 0 | 4 |
| | | GLM | Glass Manufacturing | 29 | 0 | 2 | 0 | 0 | 426 | 140 | 12 | 609 |
| | | GLS | Glassblowing | 0 | 0 | 0 | 0 | 0 | 17 | 3 | 0 | 20 |
| | | HON | Honey Processing Plant | 0 | 0 | 0 | 0 | 0 | 0 | 1 | 0 | 1 |
| | | HOS | Health Facility - Maintenance | 0 | 0 | 11 | 1 | 0 | 0 | 74 | 7 | 93 |
| | | HOT | Hotel | 0 | 0 | 2 | 0 | 0 | 2 | 14 | 0 | 18 |
| | | ICE | Ice Company | 0 | 0 | 0 | 0 | 0 | 1 | 4 | 1 | 6 |
| | | IND | Industry | 1 | 0 | 4 | 2 | 0 | 10 | 105 | 2 | 124 |
| | | INT | Installation Business | 0 | 0 | 14 | 1 | 0 | 1 | 46 | 2 | 64 |
| | | JAN | Janitor/Sanitary Engineer | 0 | 0 | 0 | 0 | 0 | 1 | 10 | 6 | 17 |
| | | LAU | Laundry Service | 0 | 0 | 2 | 0 | 0 | 5 | 22 | 3 | 32 |
| | | LUM | Lumber | 1 | 0 | 3 | 1 | 2 | 12 | 86 | 10 | 115 |
| | | MAC | Machining (e.g., Machine Shop) | 8 | 4 | 0 | 14 | 0 | 17 | 110 | 11 | 164 |
| | | MAT | Maintenance | 7 | 7 | 17 | 7 | 1 | 9 | 313 | 13 | 374 |
| | | MIL | Mill | 19 | 3 | 24 | 3 | 0 | 248 | 217 | 8 | 522 |
| | | MTC | Material Center | 1 | 0 | 0 | 0 | 0 | 0 | 6 | 0 | 7 |
| | | MTP | Meat Packing | 1 | 0 | 2 | 1 | 0 | 5 | 12 | 2 | 23 |
| | | MUN | Municipal Services | 1 | 2 | 1 | 0 | 0 | 1 | 77 | 2 | 84 |
| | | NUR | Nursery | 1 | 2 | 0 | 1 | 0 | 1 | 3 | 0 | 8 |
| | | PAK | Packing Company | 2 | 0 | 0 | 0 | 0 | 1 | 11 | 0 | 14 |
| | | PEN | Penal Institution | 0 | 0 | 0 | 0 | 0 | 0 | 3 | 0 | 3 |
| | | PHR | Pharmaceutical | 10 | 0 | 2 | 1 | 0 | 3 | 13 | 1 | 30 |

See footnotes at end of table.

(Continued)

**Table 8.1 (Continued)**

| R1 | Manville Trust Business Classification System | | | Occupation Code[1] | | | | | | | | |
|---|---|---|---|---|---|---|---|---|---|---|---|---|
| Recode | Category | Type | Description | A | F | I | S | R | W | O | U | Total |
| | | PLA | Plant | 76 | 1 | 60 | 0 | 0 | 93 | 356 | 23 | 609 |
| | | PRI | Printing Business | 1 | 0 | 0 | 0 | 0 | 1 | 17 | 0 | 19 |
| | | PUB | Publishing Co. | 0 | 1 | 0 | 0 | 0 | 4 | 16 | 0 | 21 |
| | | RSD | Research & Development | 0 | 0 | 2 | 3 | 0 | 1 | 18 | 2 | 26 |
| | | SCR | Scrapyard | 0 | 0 | 0 | 12 | 0 | 0 | 11 | 0 | 23 |
| | | SEC | Security (Guard, Nightwatch) | 0 | 0 | 0 | 0 | 0 | 0 | 3 | 0 | 3 |
| | | SHP | Shipping | 3 | 0 | 3 | 47 | 0 | 2 | 107 | 14 | 176 |
| | | SPF | Sports Facility | 0 | 0 | 1 | 0 | 0 | 0 | 3 | 1 | 5 |
| | | STO | Store (Retail or Other Type) | 1 | 1 | 1 | 0 | 0 | 1 | 31 | 5 | 40 |
| | | SWD | Sewage/Waste Disposal (Any Type) | 0 | 1 | 2 | 0 | 0 | 0 | 8 | 1 | 12 |
| | | TNK | Tank Co. (Water, Gas, etc.) | 0 | 0 | 0 | 0 | 0 | 0 | 5 | 1 | 6 |
| | | TRA | Transport Business | 2 | 22 | 1 | 7 | 0 | 7 | 265 | 13 | 317 |
| | | TRG | Trading (Any Type/Product) | 0 | 1 | 0 | 0 | 0 | 0 | 1 | 0 | 2 |
| | | TRK | Trucking | 2 | 8 | 0 | 2 | 0 | 197 | 99 | 9 | 317 |
| | | WAR | Warehousing | 2 | 1 | 0 | 4 | 0 | 2 | 42 | 6 | 57 |
| | UNK | Subtotal | Unknown | 353 | 99 | 786 | 886 | 11 | 638 | 8,145 | 19,470 | 30,388 |
| | | | Blank Business Type/Category | 80 | 22 | 254 | 498 | 2 | 130 | 1,358 | 12,189 | 14,533 |
| | | — | No Record in File | 0 | 0 | 0 | 0 | 0 | 0 | 0 | 6,294 | 6,294 |
| | | UNK | Unknown (Business Type not Clear) | 129 | 43 | 375 | 171 | 8 | 290 | 4,880 | 599 | 6,495 |
| | | UNS | Unspecified (not on Table) | 144 | 34 | 157 | 217 | 1 | 218 | 1,907 | 388 | 3,066 |

Note 1: Occupation Code – From information provided on the Proof of Claim form

A Factory worker – Asbestos as raw material
F Friction materials worker
I Insulator
S Shipyard worker
R Roofer
W Factory worker – Other
O Other
U Unknown

Source: Stallard and Manton (1994, Appendix A, Table 1).

**Figure 8.1: Qualified Male Claims, 1988-1992, by R2 Industry/Occupation Group and Date of First Exposure (Source: Authors' Calculations)**

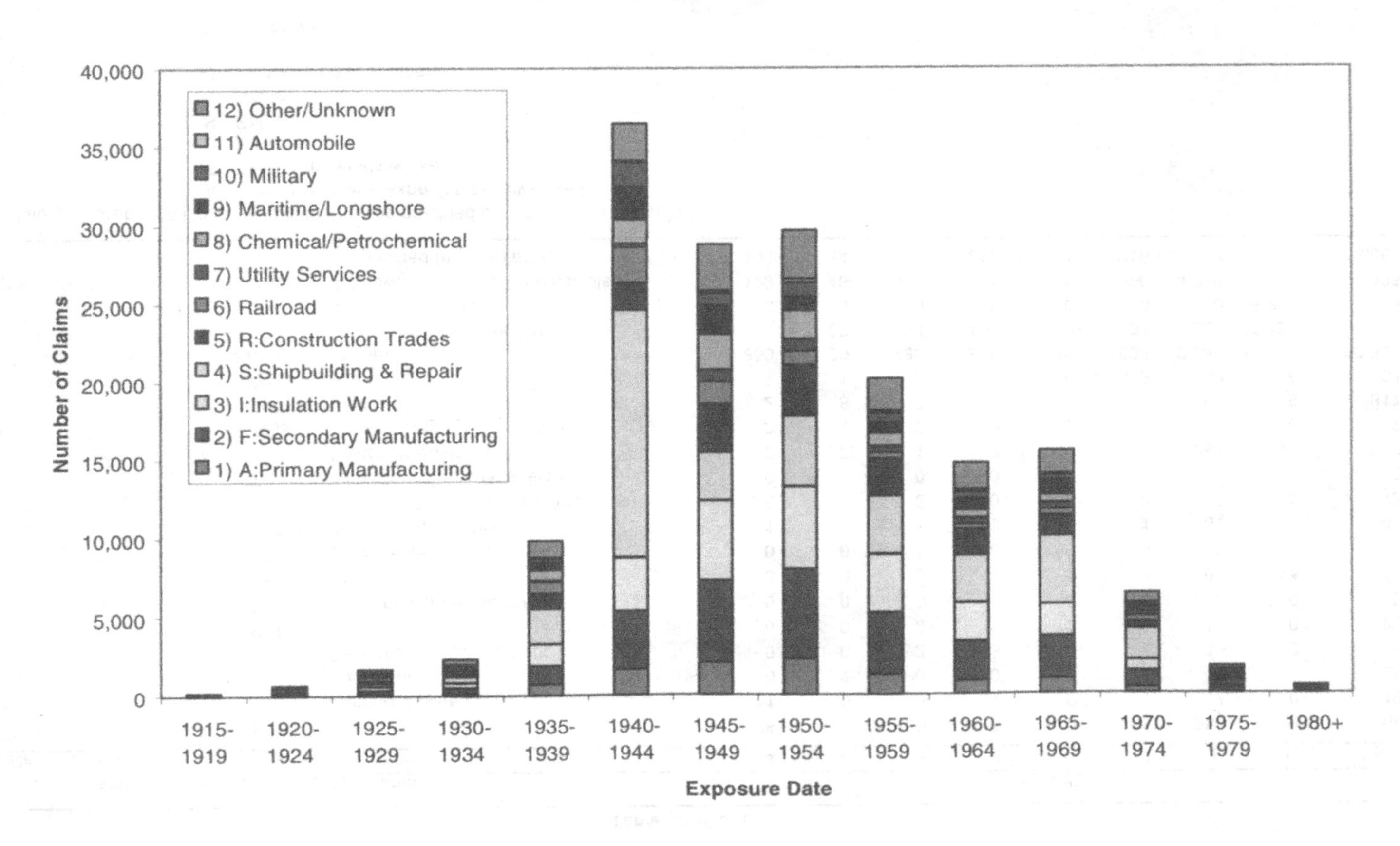

Figure 8.2: Qualified Male Claims, 1988-1992, by Disease Group and Date of First Exposure (Source: Authors' Calculations)

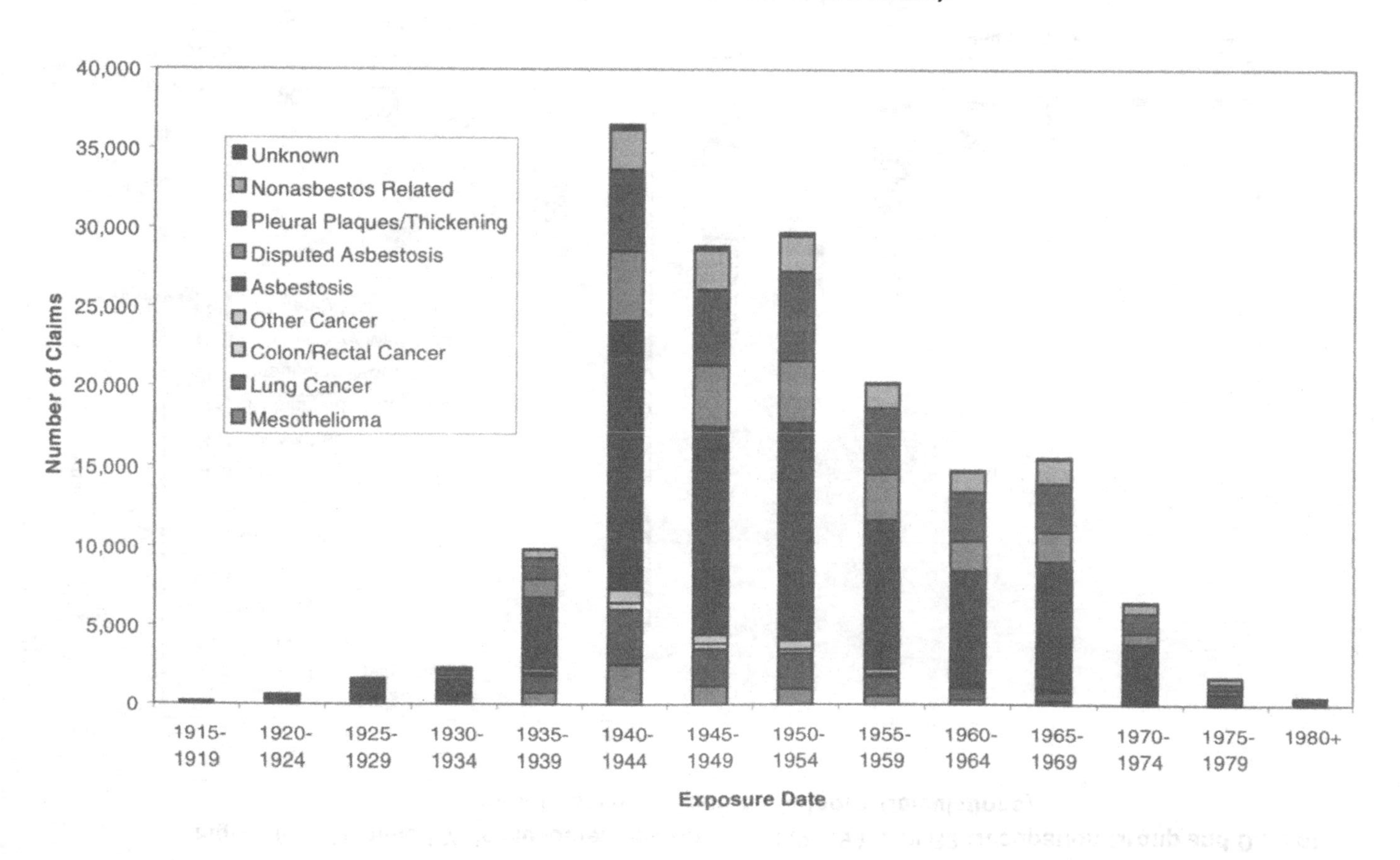

**Figure 8.3: Qualified Male Mesothelioma Claims, 1988-1992, by R2 Occupation Group and Date of First Exposure (Source: Authors' Calculations)**

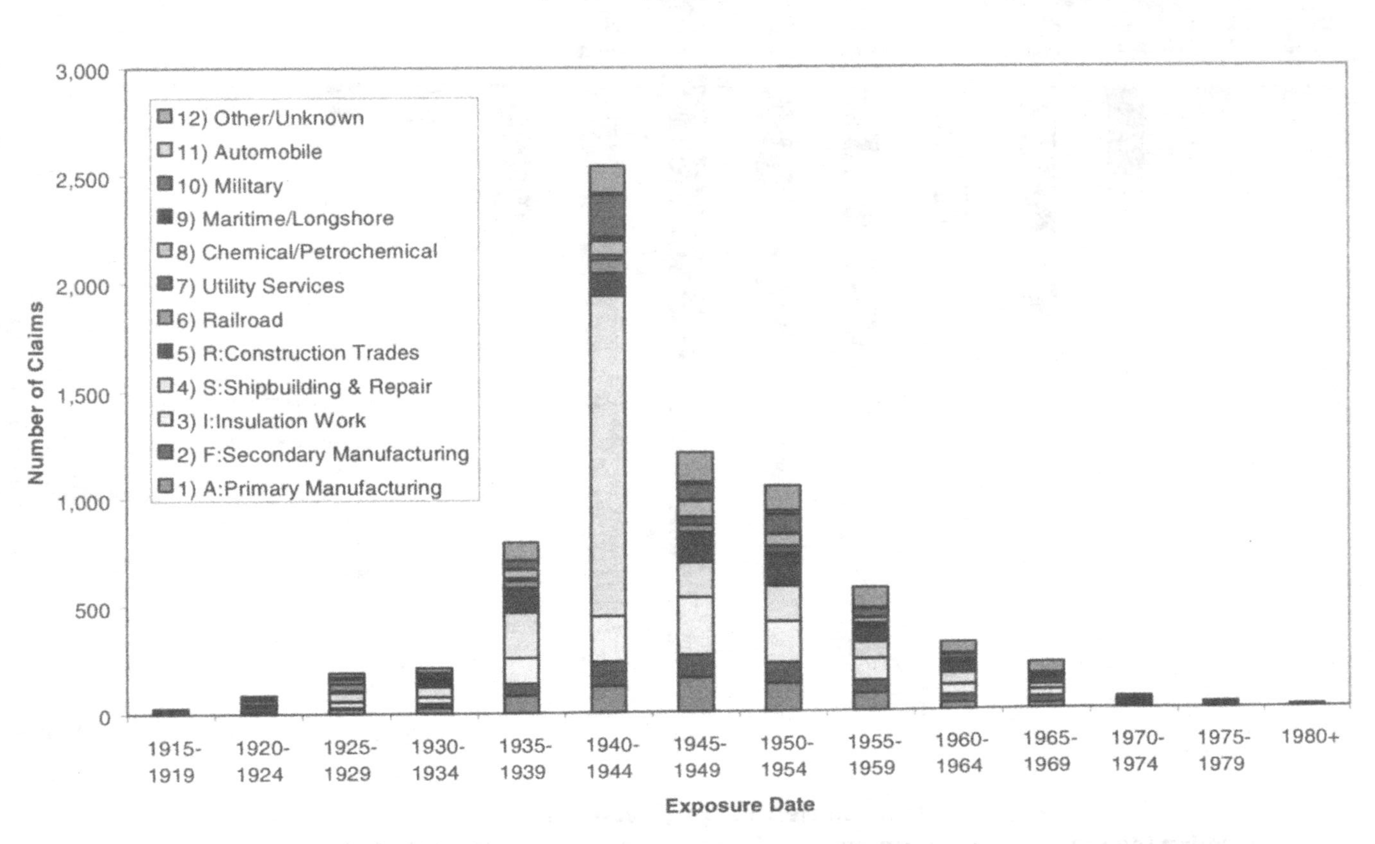

Figure 8.4: Qualified Male Mesothelioma Claims, 1990-1992, by R2 Occupation Group and Date of First Exposure (Source: Authors' Calculations)

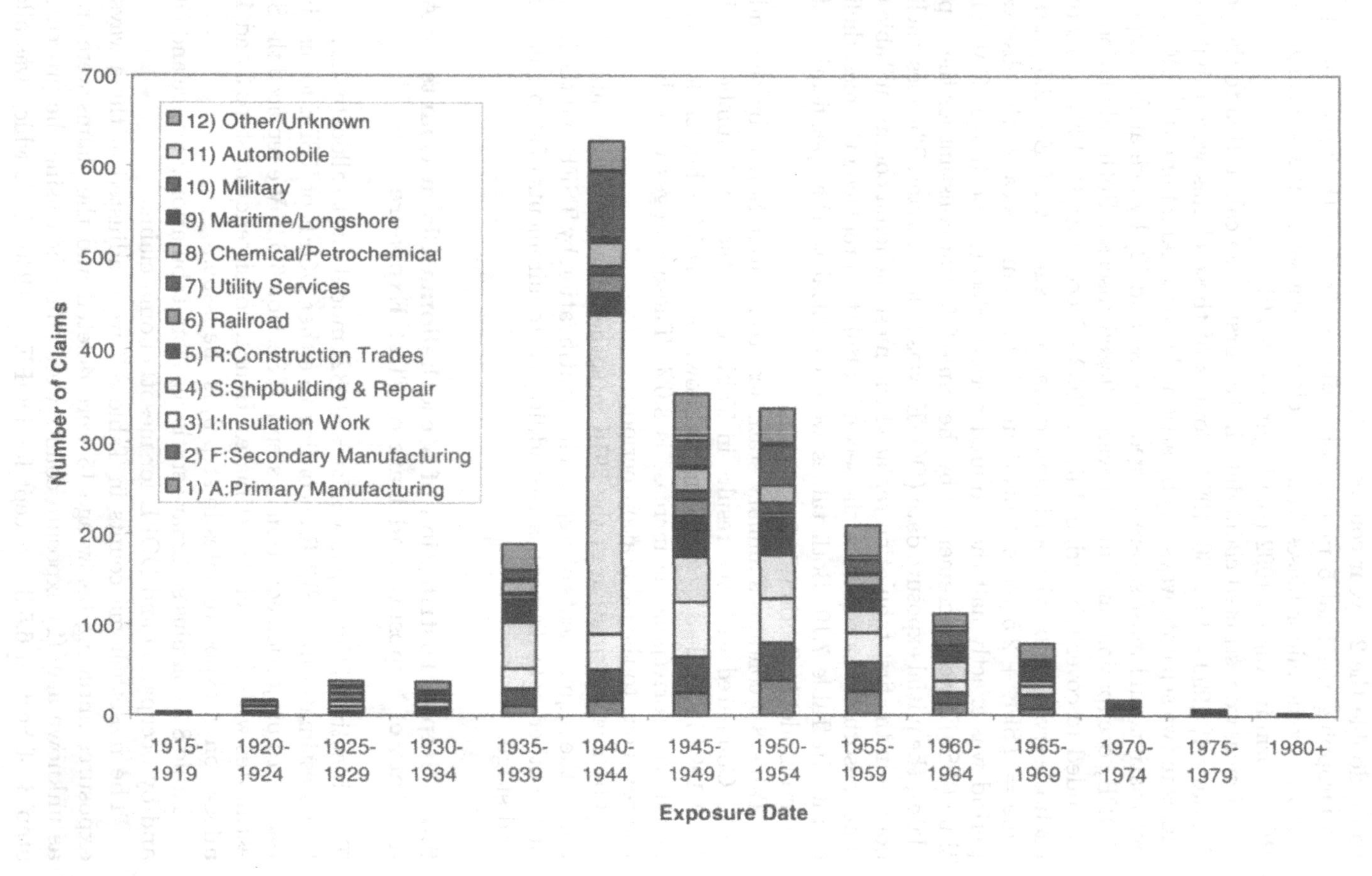

in 1990 to mid-1992 and mid-1992 to 1994. The 5-year counts were assumed to be double the 2.5-year counts.

Dropping the final 5 months of claims simplified the procedure. It also removed some minor biases from the claim data series due to changes in the POC Form in August 1992 (see Figures 6.2-6.6).

Hence, we restricted claim data in the first-stage calibration to the period January 1, 1990 - June 30, 1992. Counts of these claims were tabulated for males in two separate ways: (1) by attained age at claim filing (CAGE), TSFE, occupation, and most severe alleged disease and (2) by age at first exposure, DOFE, occupation, and most severe alleged disease. Both tabulations were extended to cover the period July 1, 1992 - December 31, 1994 by matching each observed claim with an imputed claim at age = CAGE + 2.5 years and time = TSFE + 2.5 years. This doubled the counts because the observation period was exactly half the required period. Because both age (CAGE) and time (TSFE) were incremented by the same 2.5-year constant for the imputed claim, the initial exposure date (DOFE) was held constant. The mesothelioma counts in the first tabulation replaced the mesothelioma counts in Table 6.14. The mesothelioma counts in the second tabulation replaced the mesothelioma counts in Table 7.10. Both tables were essential to the specification of the bridge model in Section 7.12.

Table 8.2 contains summary statistics from tabulation 1, by age and disease. Compared with the results in Table 6.14, the total number of claims 1990-1994 in Table 8.2 (excluding unknown age) is 1.7% higher; for mesothelioma, the corresponding increase is 8.0%. These changes were due solely to the revisions to the calibration period.

The age distribution of the 3916 mesothelioma claims in Table 8.2 completes the requirements of step 3. The stratification by TSFE and the inclusion of the other eight diseases satisfy additional requirements of step 1 in the second stage.

### 8.3.4 Step 4: Distribution of Mesothelioma Claim Counts by Age at Start of Exposure and Date of First Exposure

The objective of this step in Walker's (1982) model was to allocate the mesothelioma estimates from SEER according to either DOFE or TSFE for use in the backward projection equation in step 7 of Section 6.6.2. We replaced the SEER estimates with Trust claim estimates. Hence, consistency of steps 3 and 4, not allocation of exposure dates in step 3, is the issue.

Table 8.3 contains summary statistics from tabulation 2 by age and DOFE and by occupation and DOFE for mesothelioma claims.

The mesothelioma counts in Table 8.3 were adjusted in three ways: (1) exposures initiating below age 15 were deleted and the claims were treated as unknown age; (2) exposure counts were smoothed using the procedure in step 3 of Section 6.5.3 – except for DOFE = 1940-1944, which was allowed

**Table 8.2: Estimated Number of Qualified Male Claims Against the Manville Trust by Age or Occupation and Alleged Disease/Injury, 1990-1994, Based on Observed Claims, January 1990 to June 1992**

| | Most Severe Alleged Disease/Injury | | | | | | | | | |
|---|---|---|---|---|---|---|---|---|---|---|
| Age/ Occupation | Meso- thelioma | Lung Cancer | Colon/ Rectal Cancer | Other Cancer | Asbestosis | Disputed Asbestosis | Pleural Plaques/ Thickening | Nonasbestos Related Disease | Unknown Disease | Total |
| 15-19 | 0 | 0 | 0 | 0 | 0 | 0 | 0 | 0 | 0 | 0 |
| 20-24 | 0 | 0 | 0 | 0 | 0 | 0 | 1 | 0 | 2 | 3 |
| 25-29 | 0 | 0 | 0 | 1 | 7 | 4 | 4 | 17 | 28 | 61 |
| 30-34 | 6 | 4 | 0 | 1 | 109 | 72 | 29 | 175 | 206 | 602 |
| 35-39 | 14 | 14 | 0 | 2 | 672 | 248 | 181 | 363 | 291 | 1,785 |
| 40-44 | 44 | 35 | 3 | 10 | 1,825 | 493 | 518 | 493 | 320 | 3,741 |
| 45-49 | 90 | 133 | 15 | 15 | 2,588 | 869 | 945 | 517 | 250 | 5,422 |
| 50-54 | 214 | 247 | 17 | 25 | 3,254 | 1,298 | 1,366 | 730 | 237 | 7,388 |
| 55-59 | 332 | 516 | 57 | 95 | 4,384 | 1,751 | 1,879 | 931 | 306 | 10,251 |
| 60-64 | 556 | 1,012 | 109 | 175 | 5,973 | 2,326 | 2,496 | 1,527 | 448 | 14,622 |
| 65-69 | 765 | 1,344 | 190 | 229 | 7,082 | 2,723 | 2,657 | 2,070 | 461 | 17,521 |
| 70-74 | 801 | 1,203 | 151 | 193 | 5,314 | 1,978 | 1,880 | 1,472 | 479 | 13,471 |
| 75-79 | 628 | 892 | 108 | 125 | 3,327 | 1,143 | 1,035 | 834 | 336 | 8,428 |
| 80-84 | 311 | 408 | 66 | 65 | 1,461 | 435 | 376 | 424 | 220 | 3,766 |
| 85-89 | 125 | 123 | 13 | 18 | 471 | 133 | 111 | 115 | 83 | 1,192 |
| 90-94 | 22 | 34 | 2 | 3 | 96 | 20 | 24 | 28 | 16 | 245 |
| 95-99 | 0 | 5 | 1 | 3 | 17 | 3 | 2 | 4 | 7 | 42 |
| Unknown Age | 8 | 8 | 0 | 8 | 90 | 46 | 42 | 56 | 22 | 280 |
| 1. A: Primary Manufacturing | 272 | 420 | 42 | 52 | 2,264 | 484 | 1,178 | 110 | 38 | 4,860 |
| 2. F: Secondary Manufacturing | 376 | 960 | 166 | 202 | 4,980 | 2,488 | 2,824 | 330 | 126 | 12,452 |
| 3. I: Insulation Work | 452 | 790 | 120 | 106 | 4,318 | 1,512 | 1,696 | 392 | 56 | 9,442 |
| 4. S: Shipbuilding and Repair | 1,042 | 1,094 | 100 | 156 | 7,386 | 1,214 | 1,696 | 460 | 178 | 13,326 |
| 5. R: Construction Trades | 352 | 494 | 46 | 80 | 2,154 | 802 | 926 | 350 | 28 | 5,232 |
| 6. Util/Trans/Chem/Longshore | 466 | 1,096 | 150 | 188 | 9,674 | 3,054 | 2,922 | 4,602 | 182 | 22,334 |
| 7. Military | 364 | 246 | 24 | 28 | 946 | 266 | 430 | 80 | 16 | 2,400 |
| 8. Other/Unknown | 592 | 878 | 84 | 156 | 4,948 | 3,722 | 1,874 | 3,432 | 3,088 | 18,774 |
| Total | 3,916 | 5,978 | 732 | 968 | 36,670 | 13,542 | 13,546 | 9,756 | 3,712 | 88,820 |

Source: Top panel – Stallard and Manton (1994, Appendix B, Report 1); bottom panel – authors' calculations.

**Table 8.3: Distribution of Date of Start of Asbestos Exposure by Age or Occupation at Start of Exposure – Based on Qualified Male Mesothelioma Claims Against the Manville Trust, January 1990 to June 1992**

| | Date of First Exposure to Asbestos | | | | | | | | | | | | | | | | |
|---|---|---|---|---|---|---|---|---|---|---|---|---|---|---|---|---|---|
| Age/ Occupation | 1915-1919 | 1920-1924 | 1925-1929 | 1930-1934 | 1935-1939 | 1940-1944 | 1945-1949 | 1950-1954 | 1955-1959 | 1960-1964 | 1965-1969 | 1970-1974 | 1975-1979 | 1980-1984 | 1985-1989 | Unknown | Total |
| 0-4 | 2 | 0 | 2 | 4 | 4 | 4 | 8 | 4 | 2 | 0 | 0 | 0 | 0 | 0 | 0 | 0 | 30 |
| 5-9 | 0 | 6 | 2 | 4 | 0 | 4 | 0 | 2 | 2 | 0 | 0 | 0 | 0 | 0 | 0 | 0 | 20 |
| 10-14 | 0 | 6 | 0 | 2 | 12 | 4 | 6 | 6 | 4 | 4 | 2 | 0 | 0 | 0 | 0 | 0 | 46 |
| 15-19 | 6 | 18 | 44 | 36 | 116 | 352 | 128 | 170 | 86 | 52 | 24 | 4 | 2 | 0 | 0 | 0 | 1,038 |
| 20-24 | 0 | 4 | 20 | 16 | 144 | 358 | 246 | 192 | 124 | 52 | 30 | 6 | 4 | 0 | 0 | 0 | 1,196 |
| 25-29 | 0 | 0 | 4 | 8 | 52 | 244 | 134 | 126 | 72 | 38 | 36 | 4 | 4 | 0 | 0 | 0 | 722 |
| 30-34 | 0 | 0 | 0 | 0 | 16 | 150 | 92 | 72 | 50 | 20 | 18 | 4 | 0 | 0 | 0 | 0 | 422 |
| 35-39 | 0 | 0 | 0 | 0 | 2 | 34 | 20 | 42 | 32 | 16 | 16 | 4 | 0 | 0 | 0 | 0 | 166 |
| 40-44 | 0 | 0 | 0 | 0 | 0 | 2 | 12 | 12 | 10 | 14 | 14 | 2 | 0 | 0 | 0 | 0 | 66 |
| 45-49 | 0 | 0 | 0 | 0 | 0 | 0 | 4 | 2 | 2 | 6 | 4 | 2 | 0 | 0 | 0 | 0 | 20 |
| 50-54 | 0 | 0 | 0 | 0 | 0 | 0 | 0 | 0 | 6 | 4 | 6 | 4 | 0 | 0 | 0 | 0 | 20 |
| 55-59 | 0 | 0 | 0 | 0 | 0 | 0 | 0 | 0 | 0 | 0 | 2 | 0 | 0 | 2 | 0 | 0 | 4 |
| 60-64 | 0 | 0 | 0 | 0 | 0 | 0 | 0 | 0 | 0 | 0 | 0 | 0 | 0 | 0 | 0 | 0 | 0 |
| 65-69 | 0 | 0 | 0 | 0 | 0 | 0 | 0 | 0 | 0 | 0 | 0 | 0 | 0 | 2 | 0 | 0 | 2 |
| Unknown Age | 0 | 0 | 0 | 0 | 2 | 2 | 4 | 0 | 2 | 0 | 0 | 0 | 0 | 0 | 0 | 154 | 164 |
| 1. A: Primary Manufacturing | 0 | 2 | 6 | 4 | 22 | 32 | 50 | 70 | 52 | 18 | 14 | 2 | 0 | 0 | 0 | 0 | 272 |
| 2. F: Secondary Manufacturing | 0 | 6 | 2 | 4 | 34 | 58 | 72 | 78 | 60 | 24 | 28 | 8 | 0 | 2 | 0 | 0 | 376 |
| 3. I: Insulation Work | 0 | 2 | 8 | 6 | 42 | 74 | 116 | 92 | 62 | 26 | 18 | 4 | 2 | 0 | 0 | 0 | 452 |
| 4. S: Shipbuilding and Repair | 4 | 10 | 12 | 14 | 96 | 628 | 96 | 82 | 44 | 40 | 12 | 0 | 2 | 0 | 0 | 2 | 1,042 |
| 5. R: Construction Trades | 2 | 2 | 4 | 4 | 44 | 46 | 76 | 82 | 34 | 22 | 30 | 6 | 0 | 0 | 0 | 0 | 352 |
| 6. Util/Trans/Chem/Longshore | 0 | 8 | 24 | 10 | 42 | 116 | 110 | 74 | 50 | 20 | 10 | 0 | 2 | 0 | 0 | 0 | 466 |
| 7. Military | 0 | 2 | 6 | 8 | 14 | 136 | 52 | 84 | 24 | 30 | 4 | 4 | 0 | 0 | 0 | 0 | 364 |
| 8. Other/Unknown | 2 | 2 | 10 | 20 | 54 | 64 | 82 | 66 | 66 | 26 | 36 | 6 | 4 | 2 | 0 | 152 | 592 |
| Total | 8 | 34 | 72 | 70 | 348 | 1,154 | 654 | 628 | 392 | 206 | 152 | 30 | 10 | 4 | 0 | 154 | 3,916 |

Source: Stallard and Manton (1994, Appendix B, Report 2 – all counts were doubled in the above table).

to retain its excess counts, compared to the average of 1945-1949 and 1950-1954, to avoid distorting the WWII exposures in the shipbuilding and repair industry (see Figure 8.4); and (3) exposures initiating after 1974 were deleted and the claims were treated as unknown DOFE.

### 8.3.5 Step 5: Normalization of Exposure

We converted the exposure counts from step 4 for each occupation to the relative frequencies within each cohort along the diagonals of the age by DOFE table, with all probability mass restricted to ages 15 and older in the intervals 1915-1919 to 1970-1974. This followed the procedure in step 4 of Section 6.5.4. The normalized value for each DOFE represented the probability that a mesothelioma claim filed in 1990-1994 had first exposure to Johns-Manville asbestos at that date and age (note: age at first exposure plus TSFE equals attained age when filing claim).

Table 8.4 contains the marginal counts of male mesothelioma claims in 1990-1994, by attained age and occupation, obtained from applying the normalization factors to the age-specific mesothelioma claim counts for 1990-1994 obtained in step 3. Comparison with Table 8.2 shows that 25 mesothelioma claims (0.64%) were deleted from the analysis, primarily as a result of the restrictions on age at first exposure ($\geq$15) and DOFE ($<$1975).

Because the mesothelioma distribution was used to infer the size of the exposed population, it was necessary to impose similar restrictions on the ages and dates of first exposures for the other diseases. The revised counts for the other diseases are displayed in Table 8.4. Compared with Table 8.2, the declines were larger than for mesothelioma: lung cancer, $-0.6\%$; colon/rectal cancer, $-0.5\%$; other cancer, $-2.3\%$; asbestosis, $-1.4\%$; disputed asbestosis, $-1.5\%$; pleural plaques/thickening, $-1.3\%$; nonasbestos-related diseases, $-6.2\%$; unknown diseases (missing diagnosis/no record), $-16.8\%$.

Overall, 2214 claims (2.5%) initially estimated for 1990-1994 were deleted, with 1746 too young (age $<$ 35) to satisfy the exposure restriction; 1224 were nonasbestos-related or unknown disease; and 1087 were of other or unknown occupation. We chose not to restore these claims by pro rata increase in the retained claims because the deleted claims were not a random sample; they were a subgroup with less reliable information. The 86,606 claims retained were 0.4% higher than the number (86,248) retained for 1990-1994 in Table 6.16. These retained claims formed the basic counts used in step 1 of the second stage.

## 8.4 Model Estimation

### 8.4.1 Step 6: Estimation of the OSHA Model for Mesothelioma

The general expression for the rate of mortality due to mesothelioma, as a function of $t$ = TSFE, $d$ = duration of exposure (assumed continuous at a

**Table 8.4: Estimated Number of Qualified Male Claims Against the Manville Trust by Alleged Disease/Injury and Age at Claim 1990-1994 or Occupation at Start of Exposure – Based on Observed Claims, January 1990 to June 1992**

| Age/ Occupation | Most Severe Alleged Disease/Injury | | | | | | | | | |
|---|---|---|---|---|---|---|---|---|---|---|
| | Meso-thelioma | Lung Cancer | Colon/ Rectal Cancer | Other Cancer | Asbestosis | Disputed Asbestosis | Pleural Plaques/ Thickening | Nonasbestos Related Disease | Unknown Disease | Total |
| 15-19 | 0 | 0 | 0 | 0 | 0 | 0 | 0 | 0 | 0 | 0 |
| 20-24 | 0 | 0 | 0 | 0 | 0 | 0 | 0 | 0 | 0 | 0 |
| 25-29 | 0 | 0 | 0 | 0 | 0 | 0 | 0 | 0 | 0 | 0 |
| 30-34 | 0 | 0 | 0 | 0 | 0 | 0 | 0 | 0 | 0 | 0 |
| 35-39 | 4 | 10 | 0 | 0 | 395 | 178 | 98 | 14 | 6 | 705 |
| 40-44 | 43 | 35 | 3 | 6 | 1,825 | 493 | 518 | 493 | 320 | 3,736 |
| 45-49 | 90 | 133 | 15 | 15 | 2,588 | 869 | 945 | 517 | 250 | 5,422 |
| 50-54 | 214 | 247 | 15 | 25 | 3,254 | 1,298 | 1,366 | 730 | 237 | 7,386 |
| 55-59 | 332 | 516 | 57 | 95 | 4,384 | 1,751 | 1,879 | 931 | 306 | 10,251 |
| 60-64 | 556 | 1,012 | 109 | 175 | 5,973 | 2,326 | 2,496 | 1,527 | 448 | 14,622 |
| 65-69 | 765 | 1,344 | 190 | 229 | 7,082 | 2,723 | 2,657 | 2,070 | 461 | 17,521 |
| 70-74 | 801 | 1,203 | 151 | 193 | 5,314 | 1,978 | 1,880 | 1,472 | 479 | 13,471 |
| 75-79 | 628 | 892 | 108 | 125 | 3,327 | 1,143 | 1,035 | 834 | 336 | 8,428 |
| 80-84 | 311 | 408 | 66 | 64 | 1,461 | 433 | 376 | 423 | 218 | 3,760 |
| 85-89 | 125 | 123 | 13 | 18 | 471 | 133 | 111 | 115 | 23 | 1,132 |
| 90-94 | 22 | 18 | 1 | 1 | 71 | 12 | 13 | 28 | 6 | 172 |
| 95-99 | 0 | 0 | 0 | 0 | 0 | 0 | 0 | 0 | 0 | 0 |
| 1. A: Primary Manufacturing | 270 | 414 | 42 | 44 | 2,214 | 478 | 1,160 | 105 | 32 | 4,759 |
| 2. F: Secondary Manufacturing | 376 | 952 | 165 | 195 | 4,950 | 2,479 | 2,802 | 325 | 126 | 12,370 |
| 3. I: Insulation Work | 448 | 784 | 120 | 104 | 4,244 | 1,479 | 1,660 | 387 | 55 | 9,281 |
| 4. S: Shipbuilding and Repair | 1,038 | 1,090 | 99 | 156 | 7,357 | 1,210 | 1,688 | 450 | 172 | 13,260 |
| 5. R: Construction Trades | 351 | 494 | 46 | 79 | 2,144 | 802 | 926 | 347 | 28 | 5,217 |
| 6. Util/Trans/Chem/Longshore | 466 | 1,092 | 150 | 187 | 9,388 | 2,964 | 2,860 | 4,373 | 180 | 21,660 |
| 7. Military | 363 | 245 | 24 | 27 | 939 | 259 | 422 | 79 | 14 | 2,372 |
| 8. Other/Unknown | 579 | 870 | 82 | 154 | 4,909 | 3,666 | 1,856 | 3,088 | 2,483 | 17,687 |
| Total | 3,891 | 5,941 | 728 | 946 | 36,145 | 13,337 | 13,374 | 9,154 | 3,090 | 86,606 |

Source: Stallard and Manton (1994, Appendix B, Report 3).

constant intensity from start to end), and $w$ = latency (the time during which no incidence can occur – this is the time for the disease to undergo preclinical development, and, in the case of a mortality model, it includes the period in which the disease is clinically manifest), was given by OSHA (1983, p. 51,124) and EPA (1986, p. 85), and was presented in Section 2.4. Generalizing the Weibull parameterization in Section 6.6.2 to the OSHA hazard rate form, we obtain an equivalent expression:

$$I_t = \begin{cases} 0 & (t \leq w) \\ b(t-w)^k & (w < t \leq w+d) \\ b(t-w)^k - b(t-w-d)^k & (w+d < t), \end{cases}$$

where $b$ is a scale parameter and $k$ is a shape parameter. The OSHA model fixed the latency time $w$ at 10 years and the exponent $k$ equal to 3. Peto et al. (1982) considered the three-parameter form of the Weibull model,

$$I_t = b(t-w)^k,$$

with $w = 10$ years and found that it fit their data better than their basic two-parameter Weibull model (with $w = 0$ years) up to 15 years after first exposure, and that it fit equally well thereafter. Peto et al. (1982) also discussed the impact of duration on the incidence function, although they did not present a fully parameterized model such as provided by OSHA (1983).

Setting $w = 0$ and $d > t$ in the OSHA model retrieves the basic Weibull hazard form used by Peto et al. (1982) (i.e., the hazard rate at time $t$ for exposure initiated at time 0). The expression following the minus sign in the third line of the OSHA equation represents a hazard rate at time $t$ for exposure initiated at time $d$, rather than at time 0. Termination of exposure at time $d$ reduces the hazard at $t$ associated with continuous constant exposure over the interval (0, $t$) by the hazard at $t$ associated with continuous constant exposure over the subinterval $(d, t)$. The third line of the OSHA equation represents the hazard at $t$ associated with continuous constant exposure over the complementary subinterval $(0, d)$.

Three approximations are of interest. First, for short-duration exposures, we used a Taylor series expansion about $d = 0$, retaining terms up to the first-order derivative, to obtain

$$I_t^{(1)}(d) \approx dbk(t-w)^{k-1} \qquad (t > w),$$

which shows that the impact of short-duration exposures is approximately proportional to duration of exposure with an increase, for $w = 0$, as the $k-1$ power of TSFE. This is similar to the assumption used by Selikoff (1981) (see Section 3.3.4b), except that $k-1$ replaces $k$ in the exponent.

Second, for intermediate-duration exposures, using a mean-value approximation, we obtain

$$I_t^{(2)}(d) \approx dbk(t-w-d/2)^{k-1} \qquad (t > w + d/2),$$

which better approximates the OSHA model than the Taylor series approximation. However, the hazard is no longer strictly proportional to duration of exposure. The approximation works best when $t - w$ is large, say, at least twice $d$.

Third, for very long-duration exposures, the OSHA hazard rate can be approximated with a Weibull hazard form:

$$I_t^{(3)}(d) \approx b(t-w)^k \qquad (t > w).$$

For this approximation, the relative error is

$$R_t = \begin{cases} 0 & (t \le w + d) \\ \left(\dfrac{t-w-d}{t-w}\right)^k & (t > w + d)\,. \end{cases}$$

This approximation can be useful for analyzing data on insulation workers.

Selikoff and Seidman (1991) presented data on the timing of 458 mesothelioma deaths among 17,800 North American insulation workers in 1967-1986. These data, reproduced in Table 7.4, included a subset of 409 deaths with TSFE ranging from 15 to 49 years. Section 7.5 discussed parameter estimation for the Weibull model; parameter estimates and goodness-of-fit statistics were presented in Table 7.5 for the Peto et al. (1982) form of the model with $w = 0$. The EPA (1986) selected this same study for detailed analysis and assumed average duration of employment ($d$) to be 25 years. Selikoff and Seidman (1991) did not provide this value nor did the EPA explain its calculation. A similar 25-year figure was cited by Selikoff (1981) (see Section 3.3.4b) in conjunction with his mesothelioma projection model using the ratio of employment time to 25 years as a multiplicative factor applied to the insulator risk function. However, this approach was not equivalent to the OSHA model because it did not reduce the exponent in the first or second approximation from $k$ to $k - 1$.

To fit the OSHA model to the data in Table 7.4, with $w = 10$ years by assumption, we required the average duration of exposure for all the TSFE categories with exposure terminating, on average, more than 10 years prior to the midpoint of the TSFE category. If exposure terminates within the interval $(t - w,\, t)$, then the OSHA model retains a simple Weibull form:

$$I_t = b(t-w)^k, \qquad (0 < t - w \le d)$$

and the value of $d$ is not needed if $t - d \le w$.

Because Selikoff and Seidman (1991) did not provide duration data, we developed our own estimates. Selikoff (1979, Table 10) provided the joint distribution of age and TSFE for the 17,800 North American insulation workers enrolled in the study. These were used in Table 8.5 to estimate the average hiring age and average duration of exposure for this cohort. The average hiring age ranged from 22.0 to 26.1 years, with an overall average of 23.9 years and with the highest age corresponding to WWII.

**Table 8.5: Age and TSFE of 17,800 North American Insulation Workers at Initial Observation in 1967, Prior to 20-Year Follow-up**

| Age | Midpoint/ | TSFE Group/Midpoint | | | | | | | | | Total | Average | Average | Retirement |
|---|---|---|---|---|---|---|---|---|---|---|---|---|---|---|
| | | 0-9 | 10-14 | 15-19 | 20-24 | 25-29 | 30-34 | 35-39 | 40-49 | 50-54 | | | | |
| Group | Average | 5.00 | 12.50 | 17.50 | 22.50 | 27.50 | 32.50 | 37.50 | 45.00 | 52.50 | 16.19 | TSFE | Duration | Adjuster |
| | | Number of persons | | | | | | | | | | | | |
| 15 - 19 | 17.50 | 244 | | | | | | | | | 244 | 5.00 | 5.00 | 0 |
| 20 - 24 | 22.50 | 1,695 | | | | | | | | | 1,695 | 5.00 | 5.00 | 0 |
| 25 - 29 | 27.50 | 2,066 | 345 | 1 | | | | | | | 2,412 | 6.08 | 6.08 | 0 |
| 30 - 34 | 32.50 | 1,065 | 1,356 | 341 | | | | | | | 2,762 | 10.23 | 10.23 | 0 |
| 35 - 39 | 37.50 | 313 | 1,141 | 1,342 | 192 | | | | | | 2,988 | 14.60 | 14.60 | 0 |
| 40 - 44 | 42.50 | 79 | 424 | 1,026 | 591 | 139 | 1 | | | | 2,260 | 18.05 | 18.05 | 0 |
| 45 - 49 | 47.50 | 49 | 131 | 433 | 442 | 487 | 47 | | | | 1,589 | 21.60 | 21.60 | 0 |
| 50 - 54 | 52.50 | 27 | 88 | 214 | 332 | 377 | 182 | 77 | | | 1,297 | 24.38 | 24.38 | 0 |
| 55 - 59 | 57.50 | 13 | 49 | 129 | 206 | 176 | 146 | 193 | 72 | | 984 | 28.08 | 28.08 | 0 |
| 60 - 64 | 62.50 | 1 | 21 | 59 | 131 | 126 | 87 | 99 | 179 | | 703 | 31.73 | 31.73 | 0 |
| 65 - 69 | 67.50 | | 6 | 18 | 41 | 58 | 45 | 29 | 201 | 21 | 419 | 37.24 | 34.74 | 2.5 |
| 70 - 74 | 72.50 | | | 6 | 14 | 22 | 21 | 16 | 105 | 71 | 255 | 42.20 | 34.70 | 7.5 |
| 75 - 79 | 77.50 | | 1 | | 4 | 8 | 4 | 7 | 37 | 50 | 111 | 45.09 | 32.59 | 12.5 |
| 80 - 84 | 82.50 | | | | | 2 | 1 | 2 | 16 | 31 | 52 | 48.27 | 30.77 | 17.5 |
| 85 - 89 | 87.50 | | | | | | | 2 | 7 | 20 | 29 | 49.66 | 27.16 | 22.5 |
| Total | 40.05 | 5,552 | 3,562 | 3,569 | 1,953 | 1,395 | 534 | 425 | 617 | 193 | 17,800 | 16.19 | 15.86 | 0.33 |
| 15 - 64 | 38.43 | 5,552 | 3,555 | 3,545 | 1,894 | 1,305 | 463 | 369 | 251 | 0 | 16,934 | 14.94 | 14.94 | 0.00 |
| 65 - 89 | 71.82 | 0 | 7 | 24 | 59 | 90 | 71 | 56 | 366 | 193 | 866 | 40.79 | 33.96 | 6.82 |
| Average Hiring Age | | 22.64 | 23.95 | 24.42 | 26.08 | 24.92 | 24.83 | 22.09 | 21.95 | 23.91 | 23.86 | | | |
| TSFE Midpoint | | 5.00 | 12.50 | 17.50 | 22.50 | 27.50 | 32.50 | 37.50 | 45.00 | 52.50 | 16.19 | | | |
| Average Duration | | 5.00 | 12.49 | 17.47 | 22.37 | 27.18 | 31.87 | 36.65 | 41.45 | 41.09 | 15.86 | | | |
| Avg. Time Since Separation | | 0.00 | 0.01 | 0.03 | 0.13 | 0.32 | 0.63 | 0.85 | 3.55 | 11.41 | 0.33 | | | |

Source: Data from Selikoff et al. (1979, Table 10).

To compute duration of employment, we made two assumptions:

(1) All union members below age 65 were actively employed in insulation work.
(2) All union members at age 65 and older were retired.

These assumptions were reasonable because workers who had permanently left insulation work were unlikely to maintain their union membership. The use of age 65 as the average retirement age in Assumptions 1 and 2 is consistent with Selikoff (1981, Table 2-3). Male labor force participation rates in the late 1960s were still close to 70% at age 64 (Quinn and Burkhauser, 1994, p. 53). It is unlikely that a significant number would have retired before age 65.

For each age group of 65 or more years, we defined the retirement adjuster in Table 8.5 as the difference between the midpoint of each indicated age group and the assumed retirement age (65). This value was subtracted from the average TSFE to obtain the average duration of exposure for the age group. We defined the average time since separation (TSS) as the weighted average of the retirement adjusters using the age-specific number of insulation workers within each indicated TSFE group as the weights. The average TSS was subtracted from the TSFE midpoint to obtain the average duration for the TSFE group.

Table 8.5 shows that the overall average TSS was 0.33 years and, within the range 15-49 years TSFE, the average TSS was in the range 0.03-3.55 years. Thus, the simple Weibull form of the OSHA model could be used for modeling the incidence of mesothelioma in this cohort for $w \geq 3.55$ years.

However, some union members may have voluntarily separated from insulation work prior to retirement during the 20-year epidemiologic follow-up. Unfortunately, we lacked information on separation rates by age and TSFE for this cohort. To deal with the effect of separations, we considered the distribution of person-years of observation by age and TSFE reported in Table 8.6 for this cohort (Selikoff and Seidman, 1991, Table 1).

We made two additional assumptions:

(3) Active workers voluntarily separated from insulation work at an annualized rate of 6% per year.
(4) The person-years of observation within age and TSFE categories in Table 8.5 were concentrated at the midpoint age, but were uniformly distributed along TSFE and time of follow-up (TOFU).

The choice of a 6% voluntary separation rate was based on aggregate data in Selikoff (1981, Table 2-3), in which 11,700 of the 17,800 members separated from the union during 1967-1980, implying an average total separation hazard rate of 8.2% per year. With an estimated 3075 deaths during 1967-1980 (interpolating from 2271 deaths in 1967-1976 to 4951 deaths in 1967-1986), the voluntary separation rate must have been about 6%.

The assumption of a uniform distribution of TOFU was based on the observations that, compared with Table 8.5, the average hiring age in Table 8.6

**Table 8.6: Age and TSFE of 17,800 North American Insulation Workers During 20-Year Follow-up**

| Age Group | Midpoint/ Average | TSFE Group/Midpoint 0-14 7.50 | 15-19 17.50 | 20-24 22.50 | 25-29 27.50 | 30-34 32.50 | 35-39 37.50 | 40-44 42.50 | 45-49 47.50 | 50-54 52.50 | Total 23.37 | Average TSFE | Within Range 20-54 This Table | Table 8.5 |
|---|---|---|---|---|---|---|---|---|---|---|---|---|---|---|
| | | Number of person-years | | | | | | | | | | | | |
| 15 - 19 | 17.50 | 183 | | | | | | | | | 183 | 7.50 | 0.00 | 0.00 |
| 20 - 24 | 22.50 | 4,886 | | | | | | | | | 4,886 | 7.50 | 0.00 | 0.00 |
| 25 - 29 | 27.50 | 15,452 | 12 | | | | | | | | 15,463 | 7.51 | 0.00 | 0.00 |
| 30 - 34 | 32.50 | 23,020 | 5,248 | 14 | | | | | | | 28,282 | 9.36 | 22.50 | 0.00 |
| 35 - 39 | 37.50 | 12,274 | 23,037 | 6,095 | 14 | | | | | | 41,421 | 15.28 | 22.51 | 22.50 |
| 40 - 44 | 42.50 | 3,819 | 15,631 | 24,018 | 5,205 | 17 | | | | | 48,690 | 20.26 | 23.40 | 23.46 |
| 45 - 49 | 47.50 | 967 | 5,570 | 16,507 * | 19,676 | 3,817 | 9 | | | | 46,548 | 24.53 | 25.92 | 25.48 |
| 50 - 54 | 52.50 | 557 | 1,599 | 6,418 | 14,700 | 14,217 | 2,170 | 6 | | | 39,666 | 28.35 | 29.12 | 27.52 |
| 55 - 59 | 57.50 | 335 | 872 | 2,364 | 6,066 | 10,674 | 8,004 | 1,259 | 6 | | 29,580 | 31.73 | 32.46 | 31.14 |
| 60 - 64 | 62.50 | 133 | 483 | 1,327 | 2,714 | 4,673 | 5,502 | 4,274 | 761 | 3 | 19,871 | 34.73 | 35.35 | 33.77 |
| 65 - 69 | 67.50 | 29 | 204 | 581 | 1,397 | 2,227 | 2,599 | 2,643 | 2,504 | 483 | 12,666 | 38.03 | 38.44 | 38.52 |
| 70 - 74 | 72.50 | | 49 | 221 | 529 | 1,055 | 1,251 | 1,158 | 1,282 | 1,903 | 7,447 | 41.84 | 42.00 | 42.79 |
| 75 - 79 | 77.50 | 1 | 3 | 47 | 188 | 358 | 572 | 544 | 426 | 1,712 | 3,850 | 44.83 | 44.85 | 45.39 |
| 80 - 84 | 82.50 | | 2 | 3 | 29 | 111 | 165 | 221 | 185 | 1,079 | 1,794 | 47.65 | 47.68 | 48.27 |
| 85 - 89 | 87.50 | | | 0 | | 18 | 68 | 95 | 94 | 971 | 1,246 | 50.25 | 50.25 | 49.66 |
| Total | 47.30 | 61,655 | 52,710 | 57,595 | 50,519 | 37,166 | 20,340 | 10,200 | 5,256 | 6,151 | 301,593 | 23.37 | 30.24 | 30.00 |
| 15 - 64 | 44.85 | 61,626 | 52,452 | 56,744 | 48,376 | 33,397 | 15,686 | 5,539 | 767 | 3 | 274,590 | 21.61 | 28.36 | 27.72 |
| 65 - 89 | 72.22 | 30 | 257 | 852 | 2,142 | 3,769 | 4,654 | 4,662 | 4,490 | 6,147 | 27,003 | 41.25 | 41.52 | 41.69 |
| Average Hiring Age | | 25.16 | 23.21 | 23.49 | 23.89 | 23.99 | 23.88 | 23.27 | 22.18 | 25.12 | 23.94 | | | |
| TSFE Midpoint | | 7.50 | 17.50 | 22.50 | 27.50 | 32.50 | 37.50 | 42.50 | 47.50 | 52.50 | 23.37 | | | |
| Avg. Duration: Best Case | | 7.50 | 17.48 | 22.43 | 27.30 | 31.95 | 36.15 | 39.74 | 42.45 | 39.88 | 22.72 | | | |
| Avg. Duration: 6% Case | | 6.55 | 15.05 | 19.42 | 24.32 | 29.07 | 33.48 | 37.45 | 40.79 | 39.41 | 20.37 | | | |
| Avg. Duration: Worst Case | | 2.50 | 8.69 | 12.49 | 17.46 | 22.38 | 27.18 | 31.81 | 36.27 | 37.65 | 14.41 | | | |
| Avg. TSS: Best Case | | 0.00 | 0.02 | 0.07 | 0.20 | 0.55 | 1.35 | 2.76 | 5.05 | 12.62 | 0.65 | | | |
| Avg. TSS: 6% Case | | 0.95 | 2.45 | 3.08 | 3.18 | 3.43 | 4.02 | 5.05 | 6.71 | 13.09 | 2.99 | | | |
| Avg. TSS: Worst Case | | 5.00 | 8.81 | 10.01 | 10.04 | 10.12 | 10.32 | 10.69 | 11.23 | 14.85 | 8.96 | | | |

*Note: This entry was incorrectly listed as 16,567 in Selikoff and Seidman (1991).

Source: Data from Selikoff and Seidman (1991, Table 1).

remained quite near to 23.9 years, the average TSFE among retirees increased only slightly from 40.8 to 41.3 years, and the average TSFE among insulation workers with TSFE in the range 20-54 years increased only slightly from 30.0 to 30.2 years. Thus, for each combination of age and TSFE (beyond the first two TSFE groups, 0-19 years TSFE), there was an approximate equilibrium between persons entering and exiting the cell over the 20-year follow-up.

With the above assumptions, the average TSS within a typical cell of Table 8.6, with age $< 65$ and TSFE $\geq 20$ years, was

$$\mathrm{E(TSS)} = \frac{1}{20}\int_0^{20}\left[y - \frac{1}{0.06}(1 - e^{-0.06y})\right] dy = 3.04,$$

where the second term in the integrand is the mean work time under a truncated exponential distribution with maximum value equal to $y$ (i.e., as measured from the time of initial observation in Table 8.5 to the time for which person-years of exposure were recorded in Table 8.6). For ages above age 65, the maximum value was reduced by the retirement adjuster in Table 8.5. For TSFE values below 20 years, the TSS was further restricted not to exceed TSFE. The constant 20 was derived from the uniform distribution of TOFU in Assumption 4 over the 20-year follow-up. For each value of TOFU, the difference between the maximum possible work time and the mean work time was equal to the mean time since separation (TSS), as illustrated in the above integrand.

Table 8.6 shows that the overall average TSS was 2.99 years, and within the range 15-49 years of TSFE, the average TSS was in the range 2.45-6.71 years. Thus, the simple Weibull form of the OSHA model applied for $w \geq 6.71$ years.

Table 8.6 also presents average TSS estimates under "best case" and "worst case" scenarios. The best case assumed no voluntary separation prior to retirement; the worst case assumed all active workers separate immediately following the initiation of the study.

The overall average TSS in the worst case was 8.96 years, and within the range 15-49 years of TSFE, the average TSS was in the range 8.81-11.23 years. For the worst case, the use of the simple Weibull model could require the use of the third approximation, $I_t^{(3)}(d)$, developed earlier. With the latency constant $w$ equal to 10 years in the OSHA model, the relative error in the worst case was at TSFE group 45-49 years, for example,

$$R_{47.5} = \left(\frac{47.5 - 10 - 36.27}{47.5 - 10}\right)^k = \left(\frac{1.23}{37.5}\right)^k < 0.0011 \qquad (k \geq 2).$$

Even with the latency constant set to zero, the relative error in the worst case was trivial for large enough $k$, for example,

$$R_{47.5} = \left(\frac{47.5 - 36.27}{47.5}\right)^k = \left(\frac{11.23}{47.5}\right)^k < 0.0031 \qquad (k \geq 4).$$

**Table 8.7: Parameter Estimates and Standard Errors for Mesothelioma Deaths Among 17,800 North American Insulation Workers**

| Model | k | Std. Error k | b | Std. Error b | w | Chi-Squared |
|---|---|---|---|---|---|---|
| | | | Estimates based on TSFEs 15-49 | | | |
| 0 | 3.20 | — | 3.58E-08 | 1.8E-09 | 0 | 40.43 |
| 1 | 4.18 | 0.208 | 1.14E-09 | 8.5E-10 | 0 | 17.08 |
| 2 | 4.00 | — | 2.15E-09 | 1.1E-10 | 0 | 17.82 |
| 3 | 3.00 | — | 2.18E-07 | 1.1E-08 | 10 | 11.06 |
| 4 | 2.83 | 0.148 | 3.77E-07 | 1.8E-07 | 10 | 9.74 |
| 5 | 2.00 | — | 5.13E-06 | 2.5E-07 | 10 | 44.11 |

Source: Stallard and Manton (1994, Appendix B, Table 1).

These results indicated that the analysis of Selikoff and Seidman's (1991) insulation worker data for 15-49 years TSFE can be conducted using a fully parameterized form of the OSHA (1983) model, which can be approximated to a high degree of accuracy by the Weibull hazard rate model as specified in Section 7.5 for the no-latency case. The accuracy of the Weibull approximation improves for the 10-year latency case, which was recommended by Peto et al. (1982) and OSHA (1983). Thus, it is important to determine whether the 10-year latency model provides a better fit to the data than the no-latency model.

The relative error in the Weibull approximation increased substantially for TSFE $>$ 50 years. Given the inconsistencies noted for this range of TSFE in Section 7.5, we restricted our current analysis to the range 15-49 years TSFE.

Six models were estimated, based on the Weibull approximation to the OSHA (1983) model. The parameter estimates and test statistics are presented in Table 8.7.

Models 0-2 are no-latency models; Models 3-5 are 10-year latency models. Models 1 and 4 permit $k$ to attain its unrestricted MLE value; Models 2, 3 and 5 restrict $k$ to integer values. For Models 0 and 5, $k$ was set to the values recommended by Peto et al. (1982, p. 132) for no-latency and 10-year latency models, respectively. For Model 2, $k$ was set to the upper limit of the range cited by Peto et al. (1982); for Model 3, $k$ was set to the value cited by OSHA (1983). Models 0 and 1 are the same as Models 0 and 1 in Table 7.5.

Of the six models, only Models 3 and 4 provided acceptable statistical fits to these data ($p > 0.05$). Model 3 was the form reported by the EPA (1986, p. 90), except that they reported $b = 2.25 \times 10^{-7}$ (+3.2%; our notation, where $b = K_M \times f$ in their notation – see Section 2.4). Model 3 was also the basis of the Resource Planning Corporation (RPC) (1993) model, except that they assumed that $b = 4.37 \times 10^{-8}$ ($-80.1\%$) would be appropriate for the average worker exposed to Johns-Manville asbestos. Model 4 was the best-fitting model; its estimates of $b$ and $k$ were used in all further calculations in this chapter. The larger standard errors in Model 4 (and 1) were due to the extremely large sampling correlation of $-0.995$ between $b$ and $k$. When $k$ was fixed (as in Models 2, 3, and 5), the $b$ estimate was quite precise.

Model 5 was the latency form preferred by Peto et al. (1982). Although the fit to the Selikoff and Seidman (1991) data was poor, as was the fit of Peto et al.'s (1982) no-latency form (Model 0), the fit of each model to their own data was excellent (i.e., $\chi^2 \leq 4.06$; 5 d.f.).

The parameters in Table 8.7 were modified to reflect the relative intensity and absolute duration of exposures in the eight occupation groups defined in step 2. To do this, we relied extensively on estimates produced by Selikoff (1981) (see Sections 3.3.1-3.3.2). The results are in Table 8.8.

The estimates of relative risks of mesothelioma in Table 8.8 were taken directly from Selikoff (1981, Table 2-16) for occupation groups 1-5. The estimates of average duration of exposure for occupation groups 1-5 were taken from Selikoff (1981, Table 2-13), using the average of the two values reported for 1950-1959 and 1960-1969.

Occupation group 6 required special treatment because it was an aggregate of five groups in the R2 occupation classification (see step 2), and these did not completely match Selikoff's groups (see footnotes 3 and 4 in Table 8.8). The weighted average relative risk was 10% and the average duration was rounded to 7 years because the occupational match was only approximate. As noted in footnote 5 in Table 8.8, the implied intensity of exposure at 10% relative risk is approximately equal to the 1976 OSHA standard for permissible exposure: 2 f/ml (OSHA, 1983, p. 51,087).

Occupation groups 7 and 8 had no correspondence in Selikoff's (1981) classification. It was necessary to make some assumptions to proceed with the model specification.

We were guided by two considerations. First, the average cumulative exposure in occupation groups 1-5 should have been higher than in groups 6-8, because the former were the main groups with significant occupational exposure (Selikoff, 1981) (see Section 3.3.1). Our classification reinforced this difference by including the self-reported occupation codes (A, F, I, S, and R; see step 2) with the first five R1-industry groups to form the R2 (and R3) classifications. Second, to the extent that the cumulative exposure in occupation groups 6-8 was substantially lower, the sensitivity to additional proportional changes in relative risk was minimal (see Section 7.5). The remaining sensitivity reflected changes in the depletion of the pool of surviving exposed workers as the initial pool was altered.

We approximated the relative risk for occupation groups 7 and 8 as 10% of the risk of insulation workers. This level roughly approximated the 1976 OSHA standard for ambient asbestos concentrations (Table 8.8, footnote 5). The average duration of exposure for occupation group 7 (Military) was assumed

**Table 8.8: Estimates of Relative Risk and Average Duration of Employment in Asbestos-Exposed Workers**

| | Relative Risk of Mesothelioma | | Average Duration of Occupational Exposure | |
|---|---|---|---|---|
| R3 Occupation | From Selikoff (1981) | From Stallard and Manton (1994) | From Selikoff (1981) | From Stallard and Manton (1994) |
| 1. A: Primary Manufacturing | 1.00 | 1.00 | 3.65 | 3.65 |
| 2. F: Secondary Manufacturing | 0.50 | 0.50 | 3.75 | 3.75 |
| 3. I: Insulation Work | 1.00 | 1.00 | 14.15 | 14.15 |
| 4. S: Shipbuilding and Repair | 0.50 | 0.50 | 4.75 | 4.75 |
| 5. R: Construction Trades | 0.15-0.25[1] | 0.20 | 7.90 | 7.90 |
| 6. Util/Trans/Chem/Longshore | 0.10[2] | 0.10 | 6.76[2] | 7.00[6] |
| R2-6: Railroad Engine Repair | 0.20 | — | 7.70 | — |
| R2-7: Utility Services | 0.18[3] | — | 6.00 | — |
| R2-8: Chemical/Petrochemical | 0.15 | — | 8.05 | — |
| R2-9: Maritime/Longshore | 0.10[4] | — | 7.60 | — |
| R2-11: Automobile Maintenance/Repair | 0.04 | — | 6.85 | — |
| 7. Military | — | 0.10[5] | — | 4.00[7] |
| 8. Other/Unknown | — | 0.10[5] | — | 7.00[8] |

Notes –

1. Higher risk refers to years 1958-1972 when the use of sprayed asbestos-fireproofing was common.
2. Weighted average of R2 Industry/Occupation groups 6-9 and 11.
3. Estimate is weighted average of values for (1) stationary engineers and firemen and (2) utility services – see Table 3.1 of Chapter 3.
4. Selikoff's estimate refers to Marine Engine Room Personnel (except U.S. Navy); our estimate refers almost totally to Longshoremen – see Table 8.1.
5. Estimate is based on ratio of the 1976 OSHA standard of 2 f/ml to our estimate of 20 f/ml for unit relative risk – the rough average of 20-40 f/ml in primary manufacturing and 15 f/ml in insulation work (Selikoff, 1981, Table 2-14).
6. Rounded up from 6.76 years.
7. Approximation based on 4-year enlistment.
8. Approximation based on R3 Industry/Occupation group 6.

Source: Adapted from Selikoff (1981, Tables 2-13 and 2-16).

to be 4 years; and for group 8, it was 7 years. These assumptions implied average cumulative exposures of 8-14 f-yr/ml.

The relative risk estimates in Table 8.8 were used to rescale the parameter $b$ in the OSHA model to the appropriate levels for each occupation. The duration estimates were used in place of the parameter $d$ in the OSHA model.

### 8.4.2 Step 7: Estimation of the Population Exposed to Asbestos Prior to 1975

This step was similar to step 7 in Section 6.6.2. Both steps focused on the estimation of the distribution of the exposed population by age and date of first exposure (DOFE). Two differences were important. First, in this chapter, we developed estimates of actual numbers of exposed workers, not IWEs. Second, to do this, we conducted the calculations separately for each occupation group. The notation was modified to reflect this change.

Let $E_{A,Y}^{Y_2,i}$ be the number of mesothelioma claims during calendar period $Y_2$ in occupation $i$ and at TSFE = $Y_2 - Y$ with initial exposure at age $A$ in calendar period $Y$. Following the convention introduced in Section 6.5, (1) lowercase age, year, and time indexes and subscripts referred to annual time units and (2) uppercase age, year, and time indexes and subscripts referred to quinquennial time units. We also followed the convention in Figure 6.1 that time subscripts 0, 1, and 2 referred to initial exposure, diagnosis, and claim filing dates, respectively. Thus, $Y_2$ referred to a 5-year claim filing period.

Similarly, let $M_{A_2,Y_2}^i$ be the total number of mesothelioma claims filed in occupation $i$, at age $A_2$, in period $Y_2$. It follows that

$$M_{A_2,Y_2}^i = \sum_{T=1} E_{A_2-T,Y_2-T}^{Y_2,i} ,$$

where the summation starts with $T = 1$ to reflect a minimum latency time of at least 5 years. Alternatively, we could have started the summation at $T = 2$ to reflect the 10-year latency of the OSHA model or at $T = 4$ to reflect the assumption that the latest period in which exposure initiated was 1970-1974.

The $E_{A,Y}^{Y_2,i}$ values were obtained from Table 8.3 using the procedures in steps 4 and 5. The $M_{A_2,Y_2}^i$ values resulting from those procedures are presented in Table 8.4.

Let $I_T^i$ be the mesothelioma quinquennial hazard rate in occupation $i$ at TSFE $t_T = 5T$, where the conversion from annual to quinquennial time units affected only the scale parameter of the OSHA model; that is,

$$B_i = b_i \times 5^{k+1}.$$

We defined the $T$-quinquennial survival probability for occupation $i$ as

$$S_{A,Y}^{T,i} = \exp\left\{ -\sum_{N=0}^{T-1} h_{A+N,Y+N} - \int_0^T I_N^i \, dN \right\},$$

where the nonmesothelioma hazard rates were obtained from step 1.

We approximated the probability of a mesothelioma claim in the same exposure group using

$$Q_{A,Y}^{T,i} = S_{A,Y}^{T,i} \times I_T^i,$$

and we estimated the initial size of this exposure group using

$$N^i_{A,Y} = E^{Y_2,i}_{A,Y} / Q^{(Y_2-Y),i}_{A,Y},$$

which is similar to the first-stage projection calibration equation in Section 6.6.2.

The use of mortality parameters for the mesothelioma hazard rates in the above calculations assumed that, on average, the claims were filed at or about the time of death. This would imply that about 50% of the mesothelioma claimants were alive at the date of claim filing. The actual number in the Trust data was 23.6%, lower than expected, suggesting that the average claim was filed after death. However, Section 6.3 presented results indicating that the median lag from diagnosis to claim filing date was only 1.06 years, the median lag from diagnosis to date of death was 0.73 years, and the resulting 0.33-year discrepancy was small enough not to be of concern. Given (1) the use of 5-year age and date intervals in our projections, (2) the small relative shift of the claim versus death date, and (3) the self-correcting nature of the two-stage calibration procedure, we assumed that the mortality parameters were acceptable for the first-stage calibration based on mesothelioma claims.

One limitation of this approach was that it implicitly assumed that the propensity to sue was 100% in 1990-1994 in the case of death due to mesothelioma. To the extent that the propensity to sue was lower, the size of the exposed population would be underestimated, whereas the claim hazard rates would be overestimated by an equivalent, compensating amount. This approach yielded the smallest possible estimates of the at-risk population consistent with the Trust claims for mesothelioma during 1990-1994.

Table 8.9 presents the estimates of $N^i_{A,Y}$ for the eight occupational groups. Overall, we estimated that 2.61 million persons were initially exposed to Johns-Manville asbestos during 1940-1974, about 10.1% of Selikoff's (1981; his Table 2-12) estimate for all types of exposures beginning in the same period. To improve the validity of the comparison, Selikoff's estimate should be adjusted downward to correct his overprediction of mesothelioma deaths ($\times$ 0.651; Table 6.13), his inclusion of persons with very low exposures ($\times$ 0.68; Selikoff, 1981, Table 2-18), and his inclusion of non-Manville exposures [$\times$ 0.30; based on market shares (Hersch, 1992)]. With these adjustments, Selikoff's estimate was only about 1.32 times larger than ours. Alternatively, the propensity to sue could have been as low as 75.8% under Selikoff's definition of the population at risk. More significantly, our 1940-1974 estimate in Table 8.9 was 3.37 times larger than the corresponding national IWE estimate in Table 6.8, and 12.7 times larger than the corresponding Manville IWE estimate under Model 3 in Section 7.12. These comparisons suggested that the current procedure produced estimates of the population at risk that were reasonable for forecasting claims.

Figure 8.5 displays the distribution of initial exposures by DOFE and occupation group.

**Table 8.9: Estimates of Number of Male Workers Exposed to Manville Asbestos by Age or Occupation at Start of Exposure and Date of Start of Exposure**

| | Date of First Exposure to Asbestos | | | | | | | | | | | | |
|---|---|---|---|---|---|---|---|---|---|---|---|---|---|
| Age/ Occupation | 1915-1919 | 1920-1924 | 1925-1929 | 1930-1934 | 1935-1939 | 1940-1944 | 1945-1949 | 1950-1954 | 1955-1959 | 1960-1964 | 1965-1969 | 1970-1974 | Total |
| 15-19 | 5,872 | 18,718 | 20,920 | 25,538 | 38,392 | 119,360 | 86,732 | 79,245 | 88,770 | 58,017 | 51,138 | 44,474 | 637,178 |
| 20-24 | 0 | 8,419 | 19,043 | 24,569 | 54,218 | 146,794 | 84,540 | 110,017 | 72,393 | 68,597 | 59,829 | 67,179 | 715,599 |
| 25-29 | 0 | 0 | 6,407 | 14,288 | 33,968 | 117,016 | 72,357 | 65,903 | 67,617 | 46,365 | 63,405 | 52,610 | 539,937 |
| 30-34 | 0 | 0 | 0 | 5,735 | 26,275 | 73,397 | 61,047 | 60,156 | 45,905 | 39,058 | 36,457 | 24,398 | 372,427 |
| 35-39 | 0 | 0 | 0 | 0 | 9,946 | 52,215 | 40,735 | 44,377 | 39,726 | 26,609 | 40,445 | 25,761 | 279,814 |
| 40-44 | 0 | 0 | 0 | 0 | 0 | 13,320 | 25,257 | 22,745 | 21,653 | 26,260 | 32,448 | 39,537 | 181,220 |
| 45-49 | 0 | 0 | 0 | 0 | 0 | 0 | 15,452 | 10,957 | 9,087 | 9,943 | 23,405 | 16,403 | 85,248 |
| 50-54 | 0 | 0 | 0 | 0 | 0 | 0 | 0 | 5,842 | 8,208 | 9,160 | 16,145 | 41,950 | 81,306 |
| 55-59 | 0 | 0 | 0 | 0 | 0 | 0 | 0 | 0 | 2,809 | 2,213 | 7,754 | 18,041 | 30,817 |
| 60-64 | 0 | 0 | 0 | 0 | 0 | 0 | 0 | 0 | 0 | 0 | 2,891 | 0 | 2,891 |
| 65-69 | 0 | 0 | 0 | 0 | 0 | 0 | 0 | 0 | 0 | 0 | 0 | 0 | 0 |
| 1. A: Primary Manufacturing | 0 | 529 | 1,106 | 1,091 | 2,790 | 4,522 | 7,366 | 11,688 | 11,470 | 7,498 | 7,361 | 2,818 | 58,238 |
| 2. F: Secondary Manufacturing | 0 | 1,197 | 872 | 3,011 | 9,369 | 17,332 | 21,721 | 26,675 | 25,843 | 19,299 | 26,301 | 26,358 | 177,981 |
| 3. I: Insulation Work | 0 | 91 | 382 | 794 | 2,157 | 4,253 | 5,880 | 6,005 | 5,345 | 4,754 | 6,051 | 5,707 | 41,418 |
| 4. S: Shipbuilding and Repair | 4,107 | 4,618 | 4,717 | 10,154 | 21,050 | 170,849 | 27,835 | 22,019 | 19,122 | 17,585 | 10,686 | 3,207 | 315,949 |
| 5. R: Construction Trades | 1,765 | 2,085 | 2,167 | 5,873 | 17,864 | 21,999 | 33,495 | 37,920 | 28,176 | 27,235 | 48,606 | 39,189 | 266,373 |
| 6. Util/Trans/Chem/Longshore | 0 | 10,652 | 18,363 | 17,753 | 41,566 | 102,221 | 100,086 | 90,747 | 73,488 | 44,800 | 39,994 | 24,565 | 564,234 |
| 7. Military | 0 | 2,987 | 8,907 | 10,270 | 20,708 | 118,854 | 76,511 | 91,448 | 72,216 | 51,577 | 29,431 | 31,374 | 514,282 |
| 8. Other/Unknown | 0 | 4,976 | 9,858 | 21,184 | 47,296 | 82,071 | 113,227 | 112,742 | 120,509 | 113,475 | 165,485 | 197,137 | 987,959 |
| Total | 5,872 | 27,137 | 46,371 | 70,130 | 162,799 | 522,102 | 386,120 | 399,242 | 356,169 | 286,222 | 333,915 | 330,354 | 2,926,435 |
| | Cumulative Total (in 000's) | | | | | | | | | | | | |
| | 2,926 | 2,921 | 2,893 | 2,847 | 2,777 | 2,614 | 2,092 | 1,706 | 1,307 | 950 | 664 | 330 | |

Source: Authors' calculations.

**Figure 8.5: Estimated Number of Exposed Male Workers by Occupation and Date of Start of Exposure (Source: Authors' Calculations)**

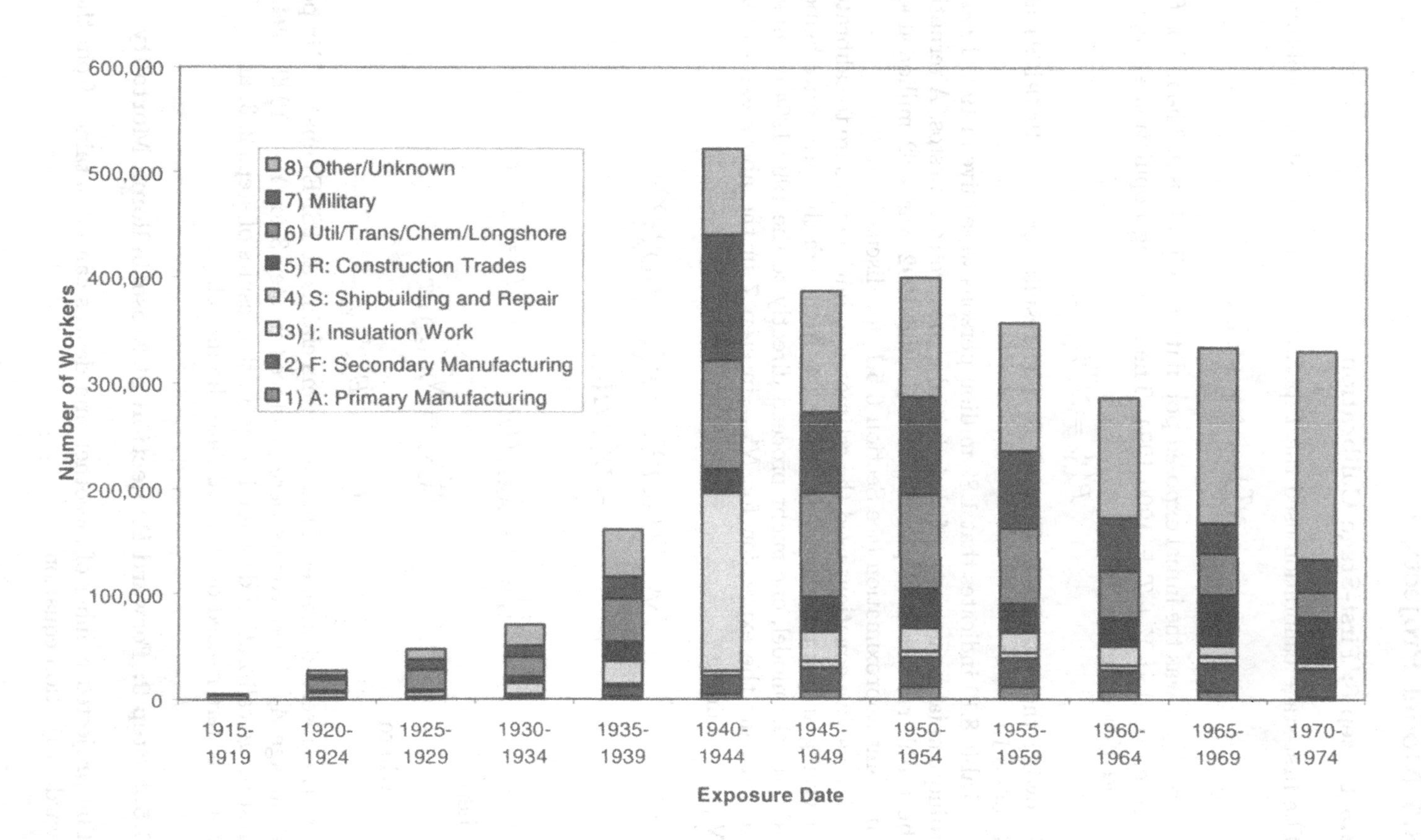

## 8.5 Model Projection

### 8.5.1 Step 8: First-Stage Calibration

The first-stage calibration used the $T$-period survival equation to obtain

$$N^{T,i}_{A,Y} = N^{i}_{A,Y} \times S^{T,i}_{A,Y},$$

where $N^{i}_{A,Y}$ was the initial exposed population in Table 8.9. The index $T$ was selected so that $Y + T =$ 1990-1994. The surviving population at age $A$ in period $Y$ was defined by

$$P^{T,i}_{A,Y} = N^{T,i}_{A\text{-}T,Y\text{-}T},$$

following the notation in Chapter 6. The results for $Y =$ 1990-1994 are in Table 8.10.

Table 8.10 indicates that 1.82 million persons were alive in 1990-1994 on a quinquennial anniversary of their initial exposure to asbestos. Alternatively, the number of exposed persons alive in mid-1992 was 1.82 million under a midpoint approximation (see Section 6.5.1, for discussion).

The above equations used the estimates of initial exposure to calibrate the surviving at-risk population in 1990-1994. Although this procedure helped to validate the model, one might proceed directly to the 1990-1994 estimate by substituting the expression for $N^{i}_{A,Y}$ in step 7 in the above expression for $N^{i}_{A,Y}$, obtaining

$$\begin{aligned} N^{(Y_2\text{-}Y),i}_{A,Y} &= E^{Y_2,i}_{A,Y} \times S^{(Y_2\text{-}Y),i}_{A,Y} / Q^{(Y_2\text{-}Y),i}_{A,Y} \\ &= E^{Y_2,i}_{A,Y} / I^{i}_{Y_2\text{-}Y} \end{aligned}$$

or

$$P^{T,i}_{A_2,Y_2} = M^{T,i}_{A_2,Y_2} / I^{i}_{T} \qquad (T = Y_2 - Y)$$

where

$$P^{T,i}_{A_2,Y_2} = N^{T,i}_{A_2\text{-}T,Y_2\text{-}T}$$

and where

$$M^{T,i}_{A_2,Y_2} = E^{Y_2,i}_{A_2\text{-}T,Y_2\text{-}T}$$

was the observed number of mesothelioma claims at TSFE equal to $T$ in period $Y_2$ at age $A_2$. These counts were produced in step 3 for $Y_2 =$ 1990-1994. The first-stage calibration depended only on the results of steps 2, 3, and 6. Steps 4 and 5 were needed only to validate the model.

### 8.5.2 Step 9: Forward Projection of Mesothelioma Mortality

The projected number of mesothelioma deaths can be obtained from the forward projection equation

**Table 8.10: Estimated Number of Male Workers Exposed to Manville Asbestos Surviving to 1990-1994 by Attained Age or Occupation and Time Since First Exposure**

| Age/ | Time Since First Exposure | | | | | | | | | | | | |
|---|---|---|---|---|---|---|---|---|---|---|---|---|---|
| Occupation | 20 | 25 | 30 | 35 | 40 | 45 | 50 | 55 | 60 | 65 | 70 | 75 | Total |
| 35-39 | 42,823 | 0 | 0 | 0 | 0 | 0 | 0 | 0 | 0 | 0 | 0 | 0 | 42,823 |
| 40-44 | 64,405 | 48,524 | 0 | 0 | 0 | 0 | 0 | 0 | 0 | 0 | 0 | 0 | 112,929 |
| 45-49 | 49,955 | 56,235 | 54,045 | 0 | 0 | 0 | 0 | 0 | 0 | 0 | 0 | 0 | 160,235 |
| 50-54 | 22,615 | 58,185 | 62,381 | 80,051 | 0 | 0 | 0 | 0 | 0 | 0 | 0 | 0 | 223,232 |
| 55-59 | 22,846 | 31,924 | 40,222 | 62,200 | 67,509 | 0 | 0 | 0 | 0 | 0 | 0 | 0 | 224,701 |
| 60-64 | 32,628 | 32,755 | 31,233 | 53,517 | 86,181 | 67,335 | 0 | 0 | 0 | 0 | 0 | 0 | 303,649 |
| 65-69 | 12,119 | 23,274 | 18,721 | 31,876 | 45,213 | 57,300 | 79,875 | 0 | 0 | 0 | 0 | 0 | 268,378 |
| 70-74 | 26,243 | 13,983 | 15,226 | 22,599 | 33,727 | 39,987 | 79,873 | 20,578 | 0 | 0 | 0 | 0 | 252,216 |
| 75-79 | 8,740 | 7,265 | 4,259 | 9,000 | 18,072 | 24,427 | 45,847 | 20,913 | 9,663 | 0 | 0 | 0 | 148,186 |
| 80-84 | 0 | 2,327 | 2,547 | 2,407 | 5,834 | 10,196 | 17,837 | 8,118 | 5,757 | 4,800 | 0 | 0 | 59,823 |
| 85-89 | 0 | 478 | 327 | 1,126 | 1,432 | 3,184 | 6,361 | 3,124 | 1,651 | 2,157 | 2,054 | 0 | 21,894 |
| 90-94 | 0 | 0 | 0 | 153 | 294 | 736 | 608 | 438 | 244 | 265 | 335 | 217 | 3,290 |
| 95-99 | 0 | 0 | 0 | 0 | 0 | 0 | 0 | 0 | 0 | 0 | 0 | 0 | 0 |
| 1. A: Primary Manufacturing | 2,652 | 6,418 | 5,227 | 8,502 | 7,950 | 4,295 | 2,100 | 959 | 296 | 109 | 26 | 0 | 38,534 |
| 2. F: Secondary Manufacturing | 23,508 | 22,711 | 15,987 | 18,432 | 16,835 | 11,417 | 7,737 | 3,194 | 781 | 174 | 127 | 0 | 120,903 |
| 3. I: Insulation Work | 5,065 | 4,367 | 3,201 | 3,856 | 4,107 | 3,098 | 1,685 | 647 | 184 | 55 | 7 | 0 | 26,272 |
| 4. S: Shipbuilding and Repair | 3,061 | 9,662 | 14,794 | 15,222 | 14,231 | 12,985 | 65,110 | 6,354 | 2,055 | 619 | 357 | 150 | 144,600 |
| 5. R: Construction Trades | 34,110 | 37,293 | 19,846 | 17,828 | 22,324 | 15,376 | 9,471 | 4,397 | 1,096 | 295 | 180 | 67 | 162,284 |
| 6. Util/Trans/Chem/Longshore | 22,519 | 33,773 | 35,932 | 49,295 | 53,206 | 51,172 | 42,437 | 12,189 | 4,218 | 2,664 | 1,052 | 0 | 308,458 |
| 7. Military | 30,125 | 27,602 | 46,991 | 63,710 | 73,019 | 56,701 | 68,427 | 10,248 | 3,562 | 1,803 | 336 | 0 | 382,525 |
| 8. Other/Unknown | 161,333 | 133,124 | 86,984 | 86,084 | 66,590 | 48,119 | 33,433 | 15,182 | 5,122 | 1,502 | 305 | 0 | 637,779 |
| Total | 282,374 | 274,950 | 228,961 | 262,929 | 258,262 | 203,165 | 230,401 | 53,171 | 17,315 | 7,222 | 2,389 | 217 | 1,821,354 |
| | Cumulative Total | | | | | | | | | | | | |
| | 1,821,354 | 1,538,980 | 1,264,031 | 1,035,070 | 772,141 | 513,879 | 310,715 | 80,314 | 27,143 | 9,828 | 2,606 | 217 | |

Source: Authors' calculations.

$$E_{A,Y}^{Y+T,i} = N_{A,Y}^{i} \times Q_{A,Y}^{T,i},$$

where $A$ and $Y$ refer to age and date of first exposure, respectively. Equivalently,

$$M_{A,Y}^{T,i} = P_{A,Y}^{T,i} \times I_{T}^{i},$$

where $A$ and $Y$ refer to attained age and date of the projection.

The projected mesothelioma deaths by calendar period for 1970-2049 are presented in Table 8.11. The total for 1970-1989 of 9500 mesothelioma deaths was 1.67 times larger than the 5673 claims filed against the Johns-Manville Corporation or the Manville Trust prior to 1990. This is consistent with the finding in Section 7.12 that the Trust diagnosis counts prior to 1990 were low compared to SEER diagnosis counts.

## 8.6 Second Stage: CHR Forecasting Model

### 8.6.1 Step 1: Distribution of Disease-Specific Claim Counts for 1990-1994 by Attained Age and TSFE

The first step was to estimate claim counts for 1990-1994 for each of the nine disease/injury categories used by the Trust, for combinations of age and TSFE. These were developed using the tabulation procedures described in step 3 of the first stage, with initial marginal counts presented in Table 8.2. Additional restrictions were imposed in step 5 of the first stage, with final marginal counts presented in Table 8.4.

### 8.6.2 Step 2: Second-Stage Calibration

The second step was to develop estimates of claim hazard rates (CHRs) combining the claim counts from step 1 with the estimates of the at-risk population from step 8 of the first stage. Let $C_{A,Y_2}^{T,i,d}$ be the claim count for age group $A$, $Y_2$ =1990-1994, $T$ = TSFE, $i$ = occupation group, and $d$ = disease/injury category. Let $P_{A,Y_2}^{T,i}$ be the corresponding at-risk population estimate. Then, following the procedure in Section 6.8.2, the CHR was defined as the ratio of the claim counts to the population at risk:

$$H_{A,Y}^{T,i,d} = C_{A,Y_2}^{T,i,d} / P_{A,Y_2}^{T,i} \qquad (Y \geq Y_2 = 1990\text{-}1994),$$

where the variance can be approximated by

$$\text{var}[H_{A,Y}^{T,i,d}] \approx 2H_{A,Y}^{T,i,d} / P_{A,Y_2}^{T,i},$$

conditional on $P_{A,Y_2}^{T,i}$. The constant 2 on the right-hand side of the variance equation results from the doubling of claims in Section 8.3.3. These equations imply a total of 72 CHR tables similar in form to those in Table 8.12, with one table for each combination of occupation group and disease/injury category.

**Table 8.11: Estimated and Projected Mesothelioma Mortality Among Male Workers Exposed to Manville Asbestos by Attained Age or Occupation and Date of Death, 1970-2049**

| Age/ Occupation | Date of Death 1970-1974 | 1975-1979 | 1980-1984 | 1985-1989 | 1990-1994 | 1995-1999 | 2000-2004 | 2005-2009 | 2010-2014 | 2015-2019 | 2020-2024 | 2025-2029 | 2030-2034 | 2035-2039 | 2040-2044 | 2045-2049 | Total |
|---|---|---|---|---|---|---|---|---|---|---|---|---|---|---|---|---|---|
| 30-34 | 3 | 2 | 2 | 1 | 0 | 0 | 0 | 0 | 0 | 0 | 0 | 0 | 0 | 0 | 0 | 0 | 9 |
| 35-39 | 18 | 19 | 13 | 14 | 7 | 0 | 0 | 0 | 0 | 0 | 0 | 0 | 0 | 0 | 0 | 0 | 72 |
| 40-44 | 64 | 56 | 56 | 41 | 44 | 18 | 0 | 0 | 0 | 0 | 0 | 0 | 0 | 0 | 0 | 0 | 278 |
| 45-49 | 178 | 143 | 121 | 122 | 90 | 92 | 32 | 0 | 0 | 0 | 0 | 0 | 0 | 0 | 0 | 0 | 779 |
| 50-54 | 264 | 306 | 258 | 215 | 214 | 159 | 156 | 50 | 0 | 0 | 0 | 0 | 0 | 0 | 0 | 0 | 1,622 |
| 55-59 | 314 | 414 | 461 | 401 | 332 | 325 | 241 | 230 | 69 | 0 | 0 | 0 | 0 | 0 | 0 | 0 | 2,787 |
| 60-64 | 253 | 448 | 574 | 624 | 556 | 456 | 440 | 326 | 305 | 88 | 0 | 0 | 0 | 0 | 0 | 0 | 4,068 |
| 65-69 | 200 | 327 | 566 | 717 | 765 | 693 | 564 | 539 | 400 | 369 | 104 | 0 | 0 | 0 | 0 | 0 | 5,244 |
| 70-74 | 88 | 226 | 373 | 638 | 801 | 843 | 774 | 626 | 597 | 444 | 404 | 113 | 0 | 0 | 0 | 0 | 5,927 |
| 75-79 | 0 | 84 | 224 | 371 | 628 | 787 | 821 | 765 | 618 | 591 | 442 | 399 | 112 | 0 | 0 | 0 | 5,842 |
| 80-84 | 0 | 0 | 69 | 186 | 311 | 528 | 666 | 695 | 660 | 536 | 518 | 392 | 353 | 100 | 0 | 0 | 5,013 |
| 85-89 | 0 | 0 | 0 | 45 | 125 | 213 | 366 | 471 | 496 | 483 | 398 | 392 | 303 | 273 | 80 | 0 | 3,645 |
| 90-94 | 0 | 0 | 0 | 0 | 22 | 64 | 112 | 199 | 264 | 285 | 288 | 243 | 247 | 196 | 179 | 54 | 2,151 |
| 95-99 | 0 | 0 | 0 | 0 | 0 | 8 | 24 | 45 | 83 | 115 | 129 | 137 | 119 | 127 | 105 | 97 | 987 |
| 1. A: Primary Manufacturing | 73 | 115 | 166 | 221 | 270 | 304 | 317 | 307 | 276 | 230 | 178 | 128 | 84 | 50 | 28 | 13 | 2,760 |
| 2. F: Secondary Manufacturing | 105 | 164 | 233 | 307 | 376 | 427 | 450 | 442 | 407 | 349 | 278 | 205 | 137 | 81 | 42 | 18 | 4,019 |
| 3. I: Insulation Work | 109 | 180 | 268 | 362 | 448 | 510 | 535 | 521 | 473 | 401 | 318 | 233 | 157 | 92 | 48 | 19 | 4,674 |
| 4. S: Shipbuilding and Repair | 533 | 725 | 895 | 1,009 | 1,040 | 984 | 859 | 697 | 531 | 382 | 261 | 169 | 101 | 55 | 19 | 5 | 8,262 |
| 5. R: Construction Trades | 113 | 167 | 228 | 292 | 352 | 398 | 421 | 418 | 391 | 346 | 290 | 227 | 166 | 111 | 62 | 26 | 4,008 |
| 6. Util/Trans/Chem/Longshore | 182 | 263 | 347 | 420 | 466 | 478 | 453 | 398 | 325 | 246 | 173 | 113 | 67 | 37 | 18 | 8 | 3,994 |
| 7. Military | 101 | 158 | 226 | 297 | 363 | 412 | 436 | 429 | 394 | 336 | 266 | 194 | 130 | 80 | 42 | 21 | 3,884 |
| 8. Other/Unknown | 165 | 252 | 355 | 468 | 581 | 673 | 726 | 733 | 695 | 620 | 519 | 407 | 292 | 191 | 104 | 42 | 6,823 |
| Total | 1,381 | 2,024 | 2,718 | 3,376 | 3,895 | 4,184 | 4,197 | 3,944 | 3,491 | 2,910 | 2,283 | 1,676 | 1,133 | 696 | 363 | 151 | 38,423 |
| Cumulative Total | 38,423 | 37,042 | 35,017 | 32,300 | 28,923 | 25,028 | 20,844 | 16,647 | 12,703 | 9,212 | 6,302 | 4,019 | 2,342 | 1,209 | 514 | 151 | |

Source: Authors' calculations.

**Table 8.12: Estimated Claim Hazard Rates (per 1000 Workers ) for Total Claims, 1990-1994, by Attained Age or Occupation and Time Since First Exposure**

| Age/ | Time Since First Exposure | | | | | | | | | | | | |
|---|---|---|---|---|---|---|---|---|---|---|---|---|---|
| Occupation | 20 | 25 | 30 | 35 | 40 | 45 | 50 | 55 | 60 | 65 | 70 | 75 | Total |
| 35-39 | 16.46 | 0.00 | 0.00 | 0.00 | 0.00 | 0.00 | 0.00 | 0.00 | 0.00 | 0.00 | 0.00 | 0.00 | 16.46 |
| 40-44 | 30.02 | 37.14 | 0.00 | 0.00 | 0.00 | 0.00 | 0.00 | 0.00 | 0.00 | 0.00 | 0.00 | 0.00 | 33.08 |
| 45-49 | 24.70 | 41.94 | 33.85 | 0.00 | 0.00 | 0.00 | 0.00 | 0.00 | 0.00 | 0.00 | 0.00 | 0.00 | 33.84 |
| 50-54 | 20.36 | 35.45 | 41.56 | 28.36 | 0.00 | 0.00 | 0.00 | 0.00 | 0.00 | 0.00 | 0.00 | 0.00 | 33.09 |
| 55-59 | 24.38 | 43.88 | 52.54 | 53.15 | 42.57 | 0.00 | 0.00 | 0.00 | 0.00 | 0.00 | 0.00 | 0.00 | 45.62 |
| 60-64 | 12.80 | 44.00 | 47.10 | 54.07 | 60.52 | 47.26 | 0.00 | 0.00 | 0.00 | 0.00 | 0.00 | 0.00 | 48.15 |
| 65-69 | 11.54 | 38.15 | 53.06 | 53.82 | 79.13 | 93.35 | 60.83 | 0.00 | 0.00 | 0.00 | 0.00 | 0.00 | 65.29 |
| 70-74 | 7.66 | 23.07 | 34.01 | 39.99 | 61.21 | 87.49 | 56.65 | 69.88 | 0.00 | 0.00 | 0.00 | 0.00 | 53.41 |
| 75-79 | 6.16 | 23.14 | 43.41 | 37.68 | 52.25 | 78.98 | 58.57 | 79.56 | 47.55 | 0.00 | 0.00 | 0.00 | 56.88 |
| 80-84 | 0.00 | 20.32 | 47.35 | 37.95 | 48.18 | 71.13 | 66.65 | 89.24 | 48.55 | 62.83 | 0.00 | 0.00 | 62.85 |
| 85-89 | 0.00 | 5.83 | 59.07 | 49.67 | 36.10 | 42.71 | 51.40 | 72.96 | 54.77 | 54.50 | 50.34 | 0.00 | 51.70 |
| 90-94 | 0.00 | 0.00 | 0.00 | 39.95 | 33.93 | 40.34 | 49.72 | 75.06 | 54.77 | 86.77 | 67.22 | 20.23 | 52.32 |
| 95-99 | 0.00 | 0.00 | 0.00 | 0.00 | 0.00 | 0.00 | 0.00 | 0.00 | 0.00 | 0.00 | 0.00 | 0.00 | 0.00 |
| 1. A: Primary Manufacturing | 59.28 | 69.59 | 129.03 | 86.44 | 139.22 | 187.80 | 252.46 | 242.04 | 139.61 | 219.36 | 193.87 | 0.00 | 123.51 |
| 2. F: Secondary Manufacturing | 26.86 | 58.00 | 82.23 | 111.10 | 146.85 | 201.69 | 184.69 | 190.93 | 210.21 | 367.16 | 133.67 | 0.00 | 102.31 |
| 3. I: Insulation Work | 73.24 | 236.57 | 394.18 | 419.40 | 452.85 | 538.55 | 567.90 | 601.99 | 355.93 | 876.72 | 1350.76 | 0.00 | 353.26 |
| 4. S: Shipbuilding and Repair | 442.10 | 170.94 | 80.61 | 76.90 | 99.28 | 114.48 | 61.77 | 125.24 | 57.52 | 65.49 | 36.35 | 20.97 | 91.70 |
| 5. R: Construction Trades | 6.18 | 12.94 | 36.60 | 60.16 | 49.40 | 56.33 | 49.22 | 49.90 | 35.58 | 77.53 | 40.29 | 18.58 | 32.15 |
| 6. Util/Trans/Chem/Longshore | 51.71 | 91.21 | 63.21 | 43.26 | 66.25 | 88.10 | 85.21 | 77.24 | 51.91 | 55.60 | 51.38 | 0.00 | 70.22 |
| 7. Military | 2.05 | 5.58 | 4.39 | 4.14 | 6.11 | 7.31 | 9.53 | 10.04 | 10.62 | 15.50 | 11.79 | 0.00 | 6.20 |
| 8. Other/Unknown | 10.86 | 17.50 | 25.23 | 29.53 | 46.45 | 58.12 | 58.10 | 52.41 | 30.76 | 44.38 | 53.43 | 0.00 | 27.73 |
| Total | 20.20 | 38.17 | 42.99 | 44.05 | 58.15 | 73.09 | 59.09 | 76.87 | 48.67 | 61.21 | 52.71 | 20.23 | 47.55 |

Source: Authors' calculations.

Because the claim counts were fixed, the cell sizes were reduced approximately by a factor of 8 compared to the cell sizes in Section 6.8.2, and this introduced additional concerns about the credibility of these estimates. To address these concerns, we calculated the CHRs with a smoothing procedure that exactly reproduced the age, occupation, and disease-specific male claim counts estimated for the period 1990-1994. Without smoothing, the total claim count projected for 1990-2049 was 387,110; with smoothing, it was 365,615 (−5.6%). The decrease was identifiable with certain classes. Unknown occupation decreased its claims by 10,087; asbestos-related noncancer decreased its claims by 15,569. On the other hand, mesothelioma (which was treated the same as the other diseases in the second-stage calibration) decreased its claims by 22 (from 26,071 to 26,049).

To conduct the smoothing, each rate table of age ($A$) by TSFE ($T$), within occupation ($i$) and disease ($d$), was represented as the product of a TSFE rate schedule and an age multiplier. Using plus (+) to indicate summation over the indicated subscript, and asterisk (*) to indicate ratios of summed counts, we defined the marginal TSFE rate schedule for disease $d$ in occupation $i$:

$$H^{T,i,d}_{*,Y} = C^{T,i,d}_{+,Y_2} / P^{T,i}_{+,Y_2} \qquad (Y \geq Y_2 = 1990\text{-}1994).$$

The expected number of claims at age $A$ was then

$$\widehat{C}^{T,i,d}_{A,Y} = P^{T,i}_{A,Y} \times H^{T,i,d}_{*,Y}.$$

The age multiplier was determined as

$$R^{i,d}_{A,Y} = C^{+,i,d}_{A,Y_2} / \widehat{C}^{+,i,d}_{A,Y_2}.$$

The final estimator was

$$\begin{aligned}\widetilde{C}^{T,i,d}_{A,Y} &= R^{i,d}_{A,Y} \times \widehat{C}^{T,i,d}_{A,Y} \\ &= P^{T,i}_{A,Y} \times R^{i,d}_{A,Y} \times H^{T,i,d}_{*,Y} \\ &= P^{T,i}_{A,Y} \times \widehat{H}^{T,i,d}_{A,Y},\end{aligned}$$

where

$$\widehat{H}^{T,i,d}_{A,Y} = R^{i,d}_{A,Y} \times H^{T,i,d}_{*,Y}$$

and $\widehat{H}^{T,i,d}_{A,Y}$ was the smoothed CHR estimator. Motivation for this particular form of smoothing followed from its impact on $\widetilde{C}^{T,i,d}_{A,Y_2}$. Specifically, it yielded

$$\begin{aligned}\widetilde{C}^{T,i,d}_{A,Y_2} &= \widehat{C}^{T,i,d}_{A,Y_2} \times R^{i,d}_{A,Y_2} \\ &= \widehat{C}^{T,i,d}_{A,Y_2} \times C^{+,i,d}_{A,Y_2} / \widehat{C}^{+,i,d}_{A,Y_2}.\end{aligned}$$

Summing over $T$, we obtained

$$\widetilde{C}^{+,i,d}_{A,Y_2} = C^{+,i,d}_{A,Y_2},$$

which implied that the smoothing procedure exactly reproduced the age-specific claim counts in the calibration period. The procedure was more general than the Weibull model used for smoothing in Section 7.4. It was also easier to implement.

We required estimates for 13 ages (35-99) and 12 TSFEs (20, 25, ...,75). With the triangular form of the table, there were 90 cells per table, or 6480 cells over all 72 tables – about 6.7 claims per cell. With smoothing, 25 parameters were estimated per table, or 1800 parameters over all 72 tables – about 24.1 claims per parameter. The effect was to increase the stability of the CHRs by a factor of 3.6, which could be a significant improvement for cells with the least reliable information.

### 8.6.3 Step 3: At-Risk Population Projections

The third step was to develop projections of the population at risk of various types of claim. These projections modified the projections developed in step 8 of the first stage. We begin by considering these projections in detail.

Table 8.10 displays the distribution of the at-risk population, $P^{T,i}_{A,Y}$, by TSFE for $Y = 1990\text{-}1994$, stratified by attained age and occupation (i.e., $P^{T,+}_{A,Y}$ and $P^{T,i}_{+,Y}$, respectively). Table 8.13 displays the distribution of these estimates by quinquennia for $Y$ in the range 1990-2049, stratified by attained age and occupation (i.e., $P^{+,+}_{A,Y}$ and $P^{+,i}_{+,Y}$, respectively). Following the convention in Section 6.5.1, we can multiply each person-count in Table 8.13 by 5 years to obtain the corresponding person-years at risk. Thus, the table summarizes a total of 39.5 million (5 × 7,909,756) person-years of exposure over the period 1990-2049. The modal age group was 70-74 years and the modal occupation group was other/unknown.

The projection can be represented by a modified form of the $T$-period survival equation:

$$\begin{aligned} P^{T,i}_{A,Y} &= N^{i}_{A-T,Y-T} \times S^{T,i}_{A-T,Y-T} \\ &= P^{T_2,i}_{A_2,Y_2} \times S^{T,i}_{A-T,Y-T} / S^{T_2,i}_{A_2-T_2,Y_2-T_2} \\ &= P^{T_2,i}_{A_2,Y_2} \times S^{T_2,T,i}_{A_2,Y_2}, \end{aligned}$$

where

$$S^{T_2,T,i}_{A_2,Y_2} = \exp\left\{ -\sum_{N=0}^{T-T_2-1} h_{A_2+N,Y_2+N} - \int_{T_2}^{T} I^{i}_{N}\, dN \right\}.$$

The first-stage projection reflected the impact of nonmesothelioma and mesothelioma mortality in reducing the size of the at-risk population from one quinquennium to the next. Because the only part of the depletion that depended on asbestos exposure was the mesothelioma hazard, we characterized this projection as the mesothelioma decrement (MDEC) model.

**Table 8.13: Estimated and Projected Number of Male Workers Exposed to Manville Asbestos by Attained Age or Occupation and Date, 1990-2049, Mesothelioma Decrement Model**

| Age/ Occupation | Date of Projection | | | | | | | | | | | | Total | |
|---|---|---|---|---|---|---|---|---|---|---|---|---|---|---|
| | 1990-1994 | 1995-1999 | 2000-2004 | 2005-2009 | 2010-2014 | 2015-2019 | 2020-2024 | 2025-2029 | 2030-2034 | 2035-2039 | 2040-2044 | 2045-2049 | Persons | Person-Years |
| | Number of persons surviving to midpoint of quinquennium | | | | | | | | | | | | | |
| 35-39 | 42,823 | 0 | 0 | 0 | 0 | 0 | 0 | 0 | 0 | 0 | 0 | 0 | 42,823 | 214,117 |
| 40-44 | 112,929 | 42,299 | 0 | 0 | 0 | 0 | 0 | 0 | 0 | 0 | 0 | 0 | 155,228 | 776,142 |
| 45-49 | 160,234 | 110,916 | 41,591 | 0 | 0 | 0 | 0 | 0 | 0 | 0 | 0 | 0 | 312,742 | 1,563,708 |
| 50-54 | 223,232 | 155,648 | 107,892 | 40,529 | 0 | 0 | 0 | 0 | 0 | 0 | 0 | 0 | 527,301 | 2,636,503 |
| 55-59 | 224,703 | 212,929 | 148,841 | 103,375 | 38,936 | 0 | 0 | 0 | 0 | 0 | 0 | 0 | 728,785 | 3,643,924 |
| 60-64 | 303,648 | 208,355 | 198,198 | 139,040 | 96,837 | 36,616 | 0 | 0 | 0 | 0 | 0 | 0 | 982,694 | 4,913,472 |
| 65-69 | 268,375 | 270,059 | 186,216 | 178,032 | 125,478 | 87,714 | 33,342 | 0 | 0 | 0 | 0 | 0 | 1,149,216 | 5,746,080 |
| 70-74 | 252,217 | 223,331 | 226,394 | 157,087 | 151,167 | 107,193 | 75,300 | 28,825 | 0 | 0 | 0 | 0 | 1,221,515 | 6,107,573 |
| 75-79 | 148,184 | 190,807 | 170,642 | 174,820 | 122,419 | 118,926 | 85,079 | 60,207 | 23,279 | 0 | 0 | 0 | 1,094,362 | 5,471,810 |
| 80-84 | 59,824 | 97,920 | 128,389 | 116,724 | 121,619 | 86,449 | 85,261 | 61,862 | 44,314 | 17,395 | 0 | 0 | 819,757 | 4,098,783 |
| 85-89 | 21,896 | 32,665 | 54,987 | 74,080 | 69,043 | 73,760 | 53,620 | 54,077 | 40,062 | 29,232 | 11,725 | 0 | 515,146 | 2,575,732 |
| 90-94 | 3,288 | 9,074 | 14,118 | 24,752 | 34,679 | 33,507 | 37,099 | 27,858 | 29,007 | 22,141 | 16,595 | 6,859 | 258,975 | 1,294,876 |
| 95-99 | 0 | 973 | 2,853 | 4,698 | 8,701 | 12,848 | 13,031 | 15,133 | 11,870 | 12,897 | 10,245 | 7,962 | 101,211 | 506,057 |
| 1. A: Primary Manufacturing | 38,532 | 33,217 | 27,526 | 21,814 | 16,463 | 11,775 | 7,948 | 5,020 | 2,930 | 1,550 | 753 | 313 | 167,840 | 839,199 |
| 2. F: Secondary Manufacturing | 120,903 | 104,961 | 87,778 | 70,353 | 53,749 | 38,902 | 26,454 | 16,783 | 9,768 | 5,112 | 2,350 | 899 | 538,013 | 2,690,063 |
| 3. I: Insulation Work | 26,272 | 22,215 | 18,057 | 14,083 | 10,513 | 7,485 | 5,041 | 3,168 | 1,839 | 940 | 418 | 144 | 110,175 | 550,877 |
| 4. S: Shipbuilding and Repair | 144,600 | 113,035 | 84,293 | 60,076 | 41,002 | 26,717 | 16,575 | 9,662 | 5,150 | 2,467 | 809 | 202 | 504,586 | 2,522,932 |
| 5. R: Construction Trades | 162,284 | 138,895 | 115,102 | 92,098 | 71,060 | 52,600 | 37,148 | 24,794 | 15,401 | 8,822 | 4,304 | 1,563 | 724,070 | 3,620,352 |
| 6. Util/Trans/Chem/Longshore | 308,457 | 252,253 | 196,922 | 146,020 | 102,619 | 67,971 | 42,264 | 24,549 | 13,120 | 6,428 | 2,775 | 1,070 | 1,164,447 | 5,822,236 |
| 7. Military | 382,525 | 341,740 | 294,521 | 243,305 | 191,527 | 142,796 | 100,302 | 65,851 | 39,923 | 22,247 | 10,776 | 4,761 | 1,840,276 | 9,201,382 |
| 8. Other/Unkn. | 637,779 | 548,659 | 455,922 | 365,390 | 281,946 | 208,766 | 147,000 | 98,136 | 60,401 | 34,100 | 16,380 | 5,869 | 2,860,348 | 14,301,739 |
| Total | 1,821,354 | 1,554,975 | 1,280,121 | 1,013,138 | 768,879 | 557,012 | 382,731 | 247,962 | 148,533 | 81,665 | 38,565 | 14,821 | 7,909,756 | 39,548,779 |

Source: Authors' calculations.

Two other types of decrement were of interest. First, we were interested in adjusting the at-risk population so that it contained only surviving persons who were first exposed to asbestos prior to 1975 but who had not filed any claim by period $Y >$ 1990-1994. We characterized this projection as the total decrement (TDEC) model because it reflected the total impact of all types of claim. The survival function reflected both the nonmesothelioma mortality hazard rates as well as the smoothed CHRs:

$$S_{A_2,Y_2}^{T_2,T,i} = \exp\left\{-\sum_{N=0}^{T-T_2-1} h_{A_2+N,Y_2+N} - \sum_{N=T_2}^{T-1} \widetilde{H}_{A_2+N-T_2,Y_2+N-T_2}^{T,i,+}\right\},$$

where, under a midpoint approximation,

$$\widetilde{H}_{A,Y}^{T,i,d} = \frac{1}{2}\widehat{H}_{A,Y}^{T,i,d} + \frac{1}{2}\widehat{H}_{A+1,Y+1}^{T+1,i,d}$$

and

$$\widetilde{H}_{A,Y}^{T,i,+} = \sum_{d=1}^{9} \widetilde{H}_{A,Y}^{T,i,d}.$$

The nonparametric CHRs for mesothelioma were included in the TDEC model, whereas the parametric hazard function, $I_T^i$, was deleted.

Table 8.14 displays the TDEC projection for 1990-2049, stratified by attained age and occupation. The estimates for 1990-1994 were unchanged. The total person-years at risk dropped from 39.5 to 35.5 million (−10.2%). This may seem like a modest drop until one realizes that the reason for the drop was that claimants now were being removed from the at-risk population. The exact number and timing of affected claims can be determined using the methods in step 4.

Second, we were interested in adjusting the at-risk population so that it contained only surviving persons who were first exposed to asbestos prior to 1975 but had not filed a cancer claim by period $Y >$ 1990-1994. We focused on cancer claims because cancer claims were very expensive to settle (about $113,500 per claim; Table 6.1), cancer injuries were usually lethal (so that cancer decrements approximated the excess mortality associated with asbestos exposure), and cancer claims were treated differently from noncancer claims under the Trust Distribution Process. We characterized this projection as the cancer decrement (CDEC) model because it reflected the impact of all types of cancer claims. The survival function was

$$S_{A_2,Y_2}^{T_2,T,i} = \exp\left\{-\sum_{N=0}^{T-T_2-1} h_{A_2+N,Y_2+N} - \sum_{N=T_2}^{T-1} \widetilde{H}_{A_2+N-T_2,Y_2+N-T_2}^{T,i,\mathrm{CAN}}\right\},$$

where

$$\widetilde{H}_{A,Y}^{T,i,\mathrm{CAN}} = \sum_{d=1}^{4} \widetilde{H}_{A,Y}^{T,i,d},$$

**Table 8.14: Estimated and Projected Number of Male Workers Exposed to Manville Asbestos by Attained Age or Occupation and Date, 1990-2049, Total Decrement Model**

| Age/ | Date of Projection | | | | | | | | | | | | Total | |
|---|---|---|---|---|---|---|---|---|---|---|---|---|---|---|
| Occupation | 1990-1994 | 1995-1999 | 2000-2004 | 2005-2009 | 2010-2014 | 2015-2019 | 2020-2024 | 2025-2029 | 2030-2034 | 2035-2039 | 2040-2044 | 2045-2049 | Persons | Person-Years |
| | Number of persons surviving to midpoint of quinquennium | | | | | | | | | | | | | |
| 35-39 | 42,823 | 0 | 0 | 0 | 0 | 0 | 0 | 0 | 0 | 0 | 0 | 0 | 42,823 | 214,117 |
| 40-44 | 112,929 | 41,265 | 0 | 0 | 0 | 0 | 0 | 0 | 0 | 0 | 0 | 0 | 154,195 | 770,974 |
| 45-49 | 160,234 | 106,354 | 39,546 | 0 | 0 | 0 | 0 | 0 | 0 | 0 | 0 | 0 | 306,134 | 1,530,669 |
| 50-54 | 223,232 | 150,653 | 98,715 | 37,636 | 0 | 0 | 0 | 0 | 0 | 0 | 0 | 0 | 510,236 | 2,551,181 |
| 55-59 | 224,703 | 205,098 | 138,457 | 90,354 | 34,950 | 0 | 0 | 0 | 0 | 0 | 0 | 0 | 693,562 | 3,467,810 |
| 60-64 | 303,648 | 198,775 | 182,723 | 123,288 | 80,326 | 31,506 | 0 | 0 | 0 | 0 | 0 | 0 | 920,267 | 4,601,333 |
| 65-69 | 268,375 | 256,424 | 168,161 | 156,138 | 105,388 | 68,416 | 27,502 | 0 | 0 | 0 | 0 | 0 | 1,050,404 | 5,252,019 |
| 70-74 | 252,217 | 210,659 | 204,062 | 134,294 | 126,048 | 85,325 | 55,423 | 22,928 | 0 | 0 | 0 | 0 | 1,090,956 | 5,454,779 |
| 75-79 | 148,184 | 181,308 | 153,094 | 150,572 | 99,824 | 94,973 | 64,782 | 42,406 | 18,032 | 0 | 0 | 0 | 953,175 | 4,765,877 |
| 80-84 | 59,824 | 92,590 | 116,049 | 99,897 | 100,193 | 67,368 | 65,394 | 45,225 | 30,065 | 13,091 | 0 | 0 | 689,695 | 3,448,477 |
| 85-89 | 21,896 | 30,827 | 49,303 | 63,990 | 56,522 | 58,248 | 40,020 | 39,992 | 28,207 | 19,140 | 8,570 | 0 | 416,714 | 2,083,572 |
| 90-94 | 3,288 | 8,605 | 12,657 | 21,233 | 28,822 | 26,521 | 28,339 | 20,173 | 20,963 | 15,203 | 10,578 | 4,971 | 201,351 | 1,006,756 |
| 95-99 | 0 | 955 | 2,638 | 4,136 | 7,341 | 10,538 | 10,222 | 11,452 | 8,549 | 9,300 | 7,014 | 5,059 | 77,203 | 386,015 |
| 1. A: Primary Manufacturing | 38,532 | 29,039 | 20,554 | 13,897 | 8,957 | 5,379 | 2,992 | 1,600 | 807 | 343 | 143 | 60 | 122,304 | 611,519 |
| 2. F: Secondary Manufacturing | 120,903 | 94,305 | 69,874 | 48,976 | 32,286 | 19,848 | 11,354 | 6,052 | 2,935 | 1,257 | 480 | 165 | 408,437 | 2,042,184 |
| 3. I: Insulation Work | 26,272 | 15,367 | 8,224 | 4,193 | 2,029 | 911 | 374 | 147 | 58 | 17 | 5 | 2 | 57,599 | 287,996 |
| 4. S: Shipbuild-ing and Repair | 144,600 | 102,878 | 70,046 | 45,891 | 28,683 | 17,088 | 9,684 | 5,179 | 2,560 | 1,145 | 317 | 52 | 428,125 | 2,140,626 |
| 5. R: Construct-ion Trades | 162,284 | 134,344 | 106,747 | 81,252 | 59,552 | 41,920 | 28,193 | 17,958 | 10,637 | 5,819 | 2,731 | 966 | 652,403 | 3,262,014 |
| 6. Util/Trans/Ch-em/Longshore | 308,457 | 234,895 | 170,765 | 118,216 | 77,372 | 47,539 | 27,321 | 14,683 | 7,279 | 3,326 | 1,329 | 489 | 1,011,671 | 5,058,357 |
| 7. Military | 382,525 | 339,828 | 291,093 | 238,946 | 186,858 | 138,393 | 96,564 | 62,987 | 37,973 | 21,038 | 10,124 | 4,461 | 1,810,791 | 9,053,956 |
| 8. Other/Unkn. | 637,779 | 532,855 | 428,102 | 330,166 | 243,677 | 171,816 | 115,201 | 73,570 | 43,565 | 23,788 | 11,031 | 3,834 | 2,615,386 | 13,076,929 |
| Total | 1,821,354 | 1,483,513 | 1,165,404 | 881,538 | 639,414 | 442,895 | 291,682 | 182,177 | 105,816 | 56,734 | 26,161 | 10,029 | 7,106,716 | 35,533,580 |
| % Change from Table 8.13 | 0.0 | -4.6 | -9.0 | -13.0 | -16.8 | -20.5 | -23.8 | -26.5 | -28.8 | -30.5 | -32.2 | -32.3 | -10.2 | -10.2 |

Source: Authors' calculations.

where the index $d$ referred to the four cancer categories used by the Trust.

Table 8.15 displays the CDEC projection for the period 1990-2049, stratified by attained age and occupation. The estimates for 1990-1994 were unchanged while the total person-years at risk increased to 39.0 million – just 0.5 million below the MDEC projection. The differences between these two projections reflected the impact of lung cancer, colon/rectal cancer, and other cancer.

### 8.6.4 Step 4: Claim Projections

For each 5-year period starting with 1990-1994 and ending with 2045-2049, the disease-specific claim filing rates were multiplied by the surviving exposed population counts, specific to age, TSFE, and occupation, to project the future number of claims against the Trust.

The projected claims were based on a multiplicative CHR model:

$$\widetilde{C}_{A,Y}^{T,i,d} = P_{A,Y}^{T,i} \times \widehat{H}_{A,Y}^{T,i,d},$$

where the smoothed CHRs were obtained from step 2 and the at-risk population counts from step 3. The total number of male claims was obtained by summing over age, period, TSFE, occupation, and disease.

Each of the three alternative projections of the at-risk population in step 3 gave a corresponding projection of claims against the Trust in step 4.

The first projection was based on the MDEC model in Table 8.13. The claim projections are summarized in Table 8.16. The total projected number of claims was 440,375 for the period 1990-2049, 1.2% less than the 445,751 claims under the baseline CHR model in Chapter 6 (Table 6.17). This apparent agreement hid divergences between the cancer and noncancer projections. Table 8.16 implies a total of 80,463 cancer claims (18.4% higher than in Table 6.17) and 359,912 noncancer claims (4.7% lower than in Table 6.17). The change in calibration period in step 3 yielded an 8.6% increase in cancer claims and a 0.7% decrease in noncancer claims. The residual 9.8% cancer increase and 4.0% noncancer decrease were attributable to the combined effects of all revisions to the model structure in this chapter, except the introduction of claim depletion.

One of these revisions was to drop the SEER data from the first-stage calibration, as in the bridge model developed in Section 7.12 (Table 7.2, analysis S10, Model 3). Compared to the bridge model, the MDEC model for 1990-2049 projected 4.1% more total claims, 14.7% more cancer claims, and 2.0% more noncancer claims. The largest relative increase was for cancer claims, primarily due to the change in the calibration period in step 3. The remaining differences were small.

The second projection was based on the TDEC model in Table 8.14. The claim projections are summarized in Table 8.17. The total projected number of claims was 349,237, 20.7% lower than in Table 8.16. Relative to this model,

**Table 8.15: Estimated and Projected Number of Male Workers Exposed to Manville Asbestos by Attained Age or Occupation and Date, 1990-2049, Cancer Decrement Model**

| Age/ | Date of Projection | | | | | | | | | | | | Total | |
|---|---|---|---|---|---|---|---|---|---|---|---|---|---|---|
| Occupation | 1990-1994 | 1995-1999 | 2000-2004 | 2005-2009 | 2010-2014 | 2015-2019 | 2020-2024 | 2025-2029 | 2030-2034 | 2035-2039 | 2040-2044 | 2045-2049 | Persons | Person-Years |
| | Number of persons surviving to midpoint of quinquennium | | | | | | | | | | | | | |
| 35-39 | 42,823 | 0 | 0 | 0 | 0 | 0 | 0 | 0 | 0 | 0 | 0 | 0 | 42,823 | 214,117 |
| 40-44 | 112,929 | 42,280 | 0 | 0 | 0 | 0 | 0 | 0 | 0 | 0 | 0 | 0 | 155,209 | 776,046 |
| 45-49 | 160,234 | 110,778 | 41,535 | 0 | 0 | 0 | 0 | 0 | 0 | 0 | 0 | 0 | 312,547 | 1,562,736 |
| 50-54 | 223,232 | 155,449 | 107,515 | 40,424 | 0 | 0 | 0 | 0 | 0 | 0 | 0 | 0 | 526,620 | 2,633,100 |
| 55-59 | 224,703 | 212,431 | 148,259 | 102,668 | 38,727 | 0 | 0 | 0 | 0 | 0 | 0 | 0 | 726,789 | 3,633,944 |
| 60-64 | 303,648 | 207,499 | 196,894 | 137,836 | 95,602 | 36,225 | 0 | 0 | 0 | 0 | 0 | 0 | 977,704 | 4,888,519 |
| 65-69 | 268,375 | 268,619 | 184,275 | 175,777 | 123,555 | 85,836 | 32,770 | 0 | 0 | 0 | 0 | 0 | 1,139,207 | 5,696,036 |
| 70-74 | 252,217 | 221,704 | 223,695 | 154,225 | 148,148 | 104,723 | 72,937 | 28,139 | 0 | 0 | 0 | 0 | 1,205,788 | 6,028,941 |
| 75-79 | 148,184 | 189,419 | 168,080 | 171,471 | 119,151 | 115,639 | 82,452 | 57,747 | 22,574 | 0 | 0 | 0 | 1,074,718 | 5,373,589 |
| 80-84 | 59,824 | 97,022 | 126,384 | 113,953 | 118,256 | 83,291 | 82,141 | 59,395 | 42,085 | 16,696 | 0 | 0 | 799,048 | 3,995,242 |
| 85-89 | 21,896 | 32,345 | 53,947 | 72,269 | 66,694 | 70,960 | 50,992 | 51,522 | 38,045 | 27,387 | 11,128 | 0 | 497,185 | 2,485,926 |
| 90-94 | 3,288 | 9,005 | 13,877 | 24,118 | 33,606 | 32,171 | 35,455 | 26,309 | 27,487 | 20,914 | 15,452 | 6,506 | 248,189 | 1,240,946 |
| 95-99 | 0 | 974 | 2,832 | 4,630 | 8,497 | 12,481 | 12,565 | 14,515 | 11,269 | 12,286 | 9,729 | 7,472 | 97,251 | 486,255 |
| 1. A: Primary Manufacturing | 38,532 | 32,685 | 26,498 | 20,463 | 15,019 | 10,398 | 6,733 | 4,101 | 2,326 | 1,139 | 523 | 224 | 158,641 | 793,206 |
| 2. F: Secondary Manufacturing | 120,903 | 103,678 | 85,428 | 67,283 | 50,355 | 35,539 | 23,481 | 14,447 | 8,131 | 4,101 | 1,827 | 694 | 515,866 | 2,579,329 |
| 3. I: Insulation Work | 26,272 | 21,276 | 16,453 | 12,158 | 8,543 | 5,668 | 3,519 | 2,036 | 1,095 | 501 | 214 | 79 | 97,814 | 489,071 |
| 4. S: Shipbuild-ing and Repair | 144,600 | 111,570 | 82,182 | 57,967 | 39,085 | 25,169 | 15,441 | 8,913 | 4,723 | 2,263 | 741 | 184 | 492,838 | 2,464,188 |
| 5. R: Construct-ion Trades | 162,284 | 138,307 | 114,029 | 90,693 | 69,506 | 51,073 | 35,782 | 23,692 | 14,613 | 8,312 | 4,036 | 1,472 | 713,799 | 3,568,996 |
| 6. Util/Trans/Ch-em/Longshore | 308,457 | 250,978 | 194,822 | 143,592 | 100,259 | 65,967 | 40,733 | 23,484 | 12,452 | 6,056 | 2,595 | 998 | 1,150,393 | 5,751,965 |
| 7. Military | 382,525 | 341,460 | 293,971 | 242,524 | 190,620 | 141,875 | 99,469 | 65,177 | 39,450 | 21,939 | 10,593 | 4,675 | 1,834,277 | 9,171,384 |
| 8. Other/Unkn. | 637,779 | 547,574 | 453,909 | 362,691 | 278,850 | 205,638 | 144,154 | 95,777 | 58,672 | 32,974 | 15,780 | 5,653 | 2,839,451 | 14,197,257 |
| Total | 1,821,354 | 1,547,527 | 1,267,293 | 997,371 | 752,236 | 541,327 | 369,312 | 237,627 | 141,461 | 77,284 | 36,309 | 13,979 | 7,803,079 | 39,015,396 |
| % Change from Table 8.13 | 0.0 | -0.5 | -1.0 | -1.6 | -2.2 | -2.8 | -3.5 | -4.2 | -4.8 | -5.4 | -5.8 | -5.7 | -1.3 | -1.3 |

Source: Authors' calculations.

**Table 8.16: Projected Number of Qualified Male Claims Against the Manville Trust by Most Severe Alleged Disease/Injury or Occupation and Quinquennial Dates of Claim, 1990-2049, Mesothelioma Decrement Model**

| Item | Date of Claim | | | | | | | | | | | | | | % Change |
|---|---|---|---|---|---|---|---|---|---|---|---|---|---|---|---|
| | 1990-1994 | 1995-1999 | 2000-2004 | 2005-2009 | 2010-2014 | 2015-2019 | 2020-2024 | 2025-2029 | 2030-2034 | 2035-2039 | 2040-2044 | 2045-2049 | Total | % | From Table 6.17 |
| Most Severe Alleged Disease/Injury | | | | | | | | | | | | | | | |
| 1. Mesothelioma | 3,891 | 4,524 | 4,026 | 3,923 | 3,417 | 2,723 | 2,056 | 1,381 | 820 | 519 | 138 | 2 | 27,419 | 6.2 | 12.6 |
| 2. Lung cancer | 5,941 | 6,800 | 6,396 | 5,954 | 5,180 | 4,190 | 3,126 | 1,955 | 1,178 | 815 | 134 | 0 | 41,668 | 9.5 | 24.2 |
| 3. Colon/rectal cancer | 728 | 856 | 841 | 754 | 673 | 553 | 382 | 230 | 124 | 40 | 0 | 0 | 5,182 | 1.2 | 21.0 |
| 4. Other cancer | 946 | 1,038 | 972 | 923 | 770 | 575 | 443 | 280 | 134 | 79 | 35 | 0 | 6,194 | 1.4 | 6.9 |
| 5. Asbestosis | 36,145 | 35,202 | 28,577 | 23,386 | 18,162 | 13,153 | 8,574 | 4,715 | 2,628 | 1,531 | 343 | 1 | 172,416 | 39.2 | -11.9 |
| 6. Disputed asbestosis | 13,337 | 13,195 | 11,003 | 9,098 | 6,994 | 4,918 | 3,071 | 1,538 | 872 | 443 | 35 | 0 | 64,504 | 14.6 | 5.8 |
| 7. Pleural plaques/thickening | 13,374 | 13,395 | 10,904 | 8,574 | 6,512 | 4,395 | 2,695 | 1,459 | 833 | 456 | 62 | 0 | 62,660 | 14.2 | -5.6 |
| 8. Nonasbestos-related disease | 9,154 | 8,766 | 7,018 | 5,732 | 4,626 | 3,413 | 1,999 | 1,059 | 562 | 346 | 244 | 11 | 42,929 | 9.7 | 7.8 |
| 9. Unknown | 3,090 | 3,079 | 2,836 | 2,377 | 2,176 | 1,777 | 1,138 | 765 | 89 | 42 | 34 | 0 | 17,403 | 4.0 | 17.6 |
| Total | 86,606 | 86,854 | 72,573 | 60,721 | 48,509 | 35,696 | 23,484 | 13,382 | 7,240 | 4,271 | 1,025 | 14 | 440,375 | 100.0 | -1.2 |
| Cancer (1-4) | 11,506 | 13,217 | 12,235 | 11,553 | 10,039 | 8,041 | 6,007 | 3,846 | 2,257 | 1,453 | 307 | 2 | 80,463 | 18.3 | 18.4 |
| Noncancer (5-9) | 75,100 | 73,637 | 60,338 | 49,168 | 38,470 | 27,655 | 17,477 | 9,536 | 4,983 | 2,817 | 718 | 12 | 359,912 | 81.7 | -4.7 |
| Asbestos-related noncancer (5-7) | 62,856 | 61,792 | 50,484 | 41,058 | 31,668 | 22,466 | 14,340 | 7,712 | 4,333 | 2,429 | 441 | 1 | 299,580 | 68.0 | -7.3 |
| Nonasbestos-related & unknown (8-9) | 12,244 | 11,845 | 9,854 | 8,110 | 6,802 | 5,189 | 3,137 | 1,824 | 650 | 388 | 278 | 11 | 60,332 | 13.7 | 10.4 |
| Occupation | | | | | | | | | | | | | | | |
| 1. A: Primary Manufacturing | 4,759 | 5,543 | 4,961 | 3,980 | 3,174 | 2,598 | 1,773 | 756 | 507 | 420 | 11 | 0 | 28,482 | 6.5 | |
| 2. F: Secondary Manufacturing | 12,370 | 13,225 | 12,725 | 11,555 | 9,579 | 7,553 | 5,117 | 3,189 | 1,838 | 914 | 147 | 0 | 78,213 | 17.8 | |
| 3. I: Insulation Work | 9,281 | 9,950 | 8,636 | 6,728 | 5,411 | 4,054 | 2,896 | 1,537 | 845 | 768 | 0 | 0 | 50,106 | 11.4 | |
| 4. S: Shipbuilding and Repair | 13,260 | 12,478 | 6,903 | 5,136 | 3,263 | 1,960 | 1,111 | 553 | 217 | 72 | 12 | 3 | 44,969 | 10.2 | |
| 5. R: Construction Trades | 5,217 | 5,617 | 5,913 | 5,049 | 3,674 | 2,622 | 1,749 | 1,125 | 775 | 427 | 97 | 11 | 32,276 | 7.3 | |
| 6. Util/Trans/Chem/Longshore | 21,660 | 19,064 | 13,540 | 10,069 | 7,435 | 4,996 | 2,969 | 1,541 | 759 | 292 | 199 | 0 | 82,524 | 18.7 | |
| 7. Military | 2,372 | 2,501 | 2,378 | 2,174 | 1,855 | 1,423 | 1,043 | 632 | 329 | 197 | 66 | 0 | 14,968 | 3.4 | |
| 8. Other/Unknown | 17,687 | 18,478 | 17,516 | 16,030 | 14,118 | 10,489 | 6,826 | 4,049 | 1,969 | 1,181 | 494 | 0 | 108,838 | 24.7 | |
| Total | 86,606 | 86,854 | 72,573 | 60,721 | 48,509 | 35,696 | 23,484 | 13,382 | 7,240 | 4,271 | 1,025 | 14 | 440,375 | 100.0 | |

Source: Authors' calculations.

the estimate produced from the IWE model (Table 6.17) was 27.7% too high due to the inability of that model to reduce the at-risk population as claims were filed and claimants were removed.

The effects were not evenly distributed over cancer and noncancer claims. Table 8.17 implies a total of 59,813 cancer claims (25.7% lower than in Table 8.16 and 12.0% lower than in Table 6.17) and 289,424 noncancer claims (19.6% lower than in Table 8.16 and 23.4% lower than in Table 6.17).

Strictly speaking, the TDEC model projected claimants, not claims. For valuation purposes, claims could be handled by embedding a submodel of the claim process that assigned a value to each claimant at the time of his initial claim filing. In this chapter, we employed an alternative strategy exemplified by the CDEC model in Table 8.15. Under this strategy, we projected cancer claims with a claim depletion of the at-risk population due only to cancer claims. Noncancer claims were assumed not to reduce the at-risk population. The CDEC model correctly projected cancer claims and claimants, but produced an excess of noncancer claims.

The claim projections for the CDEC model are summarized in Table 8.18. The total projected number of claims was 424,323, 3.7% lower than in Table 8.16. Table 8.18 indicates a total of 76,191 cancer claims (5.3% lower than in Table 8.16 but 27.4% higher than in Table 8.17) and 348,132 noncancer claims (3.3% lower than in Table 8.16 but 20.3% higher than in Table 8.17). The 27.4% increase in cancer claimants from Table 8.17 to 8.18 reflected second-order claims among persons who were noncancer claimants under the TDEC model, under the assumption that noncancer claimants were not at increased risk of death. Such claims were permitted under the Trust Distribution Process. Alternatively, the 20.3% increase in noncancer claims from Table 8.17 to 8.18 reflected second- or higher-order claims among persons who were noncancer claimants under the TDEC model. Such claims, however, were not allowed under the Trust Distribution Process.

There was an essential asymmetry between cancer and noncancer claims. Cancer claims were of interest regardless of whether or not they are preceded by a noncancer claim. Under the Trust Distribution Process, second injury claims for cancer were treated as independent claims and the presence of a prior noncancer asbestos injury could be used to support the veracity of the cancer claim. The CDEC model is the appropriate model for projecting cancer claimants.

On the other hand, cancer and noncancer claims were competing risks and noncancer claims were relevant only if they preceded a cancer claim (or death). The TDEC model is the appropriate model for projecting noncancer claimants.

Tables 8.19-8.21 display the combined CDEC/TDEC claim projections, under which cancer claims were obtained from the CDEC model and noncancer claims from the TDEC model. A total of 365,615 claims were projected for 1990-2049 under the combined CDEC/TDEC model. This was 4.7% higher

**Table 8.17: Projected Number of Qualified Male Claims Against the Manville Trust by Most Severe Alleged Disease/Injury or Occupation and Quinquennial Dates of Claim, 1990-2049, Total Decrement (TDEC) Model**

| Item | Date of Claim | | | | | | | | | | | | Total | % | % Change From |
|---|---|---|---|---|---|---|---|---|---|---|---|---|---|---|---|
| | 1990-1994 | 1995-1999 | 2000-2004 | 2005-2009 | 2010-2014 | 2015-2019 | 2020-2024 | 2025-2029 | 2030-2034 | 2035-2039 | 2040-2044 | 2045-2049 | | | Table 6.17 |
| Most Severe Alleged Disease/Injury | | | | | | | | | | | | | | | |
| 1. Mesothelioma | 3,891 | 4,054 | 3,255 | 2,913 | 2,340 | 1,718 | 1,201 | 773 | 430 | 249 | 77 | 0 | 20,901 | 6.0 | -14.1 |
| 2. Lung cancer | 5,941 | 5,977 | 4,995 | 4,213 | 3,358 | 2,460 | 1,683 | 1,023 | 584 | 351 | 101 | 0 | 30,686 | 8.8 | -8.5 |
| 3. Colon/rectal cancer | 728 | 740 | 642 | 511 | 415 | 302 | 187 | 100 | 44 | 10 | 0 | 0 | 3,681 | 1.1 | -14.0 |
| 4. Other cancer | 946 | 916 | 763 | 648 | 486 | 326 | 235 | 133 | 50 | 34 | 8 | 0 | 4,546 | 1.3 | -21.5 |
| 5. Asbestosis | 36,145 | 31,113 | 22,724 | 16,960 | 12,003 | 7,819 | 4,581 | 2,357 | 1,230 | 587 | 178 | 0 | 135,697 | 38.9 | -30.7 |
| 6. Disputed asbestosis | 13,337 | 11,852 | 8,976 | 6,837 | 4,855 | 3,083 | 1,724 | 759 | 396 | 164 | 11 | 0 | 51,995 | 14.9 | -14.7 |
| 7. Pleural plaques/thickening | 13,374 | 11,880 | 8,696 | 6,249 | 4,273 | 2,543 | 1,344 | 659 | 322 | 137 | 25 | 0 | 49,502 | 14.2 | -25.4 |
| 8. Nonasbestos-related disease | 9,154 | 8,118 | 6,089 | 4,717 | 3,621 | 2,523 | 1,395 | 704 | 350 | 230 | 160 | 7 | 37,068 | 10.6 | -6.9 |
| 9. Unknown | 3,090 | 2,914 | 2,553 | 2,020 | 1,765 | 1,387 | 827 | 534 | 43 | 13 | 16 | 0 | 15,162 | 4.3 | 2.5 |
| Total | 86,606 | 77,564 | 58,692 | 45,069 | 33,115 | 22,162 | 13,178 | 7,042 | 3,450 | 1,776 | 576 | 7 | 349,237 | 100.0 | -21.7 |
| Cancer (1-4) | 11,506 | 11,687 | 9,654 | 8,286 | 6,599 | 4,806 | 3,307 | 2,029 | 1,109 | 644 | 186 | 0 | 59,813 | 17.1 | -12.0 |
| Noncancer (5-9) | 75,100 | 65,877 | 49,038 | 36,783 | 26,516 | 17,356 | 9,871 | 5,013 | 2,342 | 1,132 | 390 | 7 | 289,424 | 82.9 | -23.4 |
| Asbestos-related noncancer (5-7) | 62,856 | 54,844 | 40,396 | 30,046 | 21,130 | 13,445 | 7,649 | 3,775 | 1,949 | 888 | 214 | 0 | 237,193 | 67.9 | -26.6 |
| Nonasbestos-related & unknown (8-9) | 12,244 | 11,033 | 8,642 | 6,737 | 5,386 | 3,910 | 2,222 | 1,238 | 393 | 244 | 175 | 7 | 52,231 | 15.0 | -4.4 |
| Occupation | | | | | | | | | | | | | | | |
| 1. A: Primary Manufacturing | 4,759 | 4,790 | 3,639 | 2,480 | 1,706 | 1,188 | 667 | 240 | 135 | 88 | 2 | 0 | 19,694 | 5.6 | |
| 2. F: Secondary Manufacturing | 12,370 | 11,593 | 9,767 | 7,753 | 5,603 | 3,759 | 2,123 | 1,113 | 527 | 215 | 28 | 0 | 54,853 | 15.7 | |
| 3. I: Insulation Work | 9,281 | 6,657 | 3,776 | 1,899 | 981 | 457 | 199 | 67 | 22 | 11 | 0 | 0 | 23,350 | 6.7 | |
| 4. S: Shipbuilding and Repair | 13,260 | 11,175 | 5,714 | 3,886 | 2,215 | 1,206 | 625 | 270 | 101 | 27 | 3 | 1 | 38,482 | 11.0 | |
| 5. R: Construction Trades | 5,217 | 5,387 | 5,451 | 4,437 | 3,059 | 2,078 | 1,315 | 806 | 524 | 281 | 61 | 7 | 28,621 | 8.2 | |
| 6. Util/Trans/Chem/Longshore | 21,660 | 17,663 | 11,705 | 8,134 | 5,590 | 3,466 | 1,905 | 919 | 420 | 149 | 92 | 0 | 71,702 | 20.5 | |
| 7. Military | 2,372 | 2,485 | 2,348 | 2,132 | 1,807 | 1,378 | 1,003 | 604 | 313 | 186 | 62 | 0 | 14,690 | 4.2 | |
| 8. Other/Unknown | 17,687 | 17,814 | 16,292 | 14,347 | 12,154 | 8,630 | 5,340 | 3,024 | 1,409 | 820 | 329 | 0 | 97,845 | 28.0 | |
| Total | 86,606 | 77,564 | 58,692 | 45,069 | 33,115 | 22,162 | 13,178 | 7,042 | 3,450 | 1,776 | 576 | 7 | 349,237 | 100.0 | |

Source: Authors' calculations.

**Table 8.18: Projected Number of Qualified Male Claims Against the Manville Trust by Most Severe Alleged Disease/Injury or Occupation and Quinquennial Dates of Claim, 1990-2049, Cancer Decrement (CDEC) Model**

| Item | Date of Claim | | | | | | | | | | | | | | % Change |
|---|---|---|---|---|---|---|---|---|---|---|---|---|---|---|---|
| | 1990-1994 | 1995-1999 | 2000-2004 | 2005-2009 | 2010-2014 | 2015-2019 | 2020-2024 | 2025-2029 | 2030-2034 | 2035-2039 | 2040-2044 | 2045-2049 | Total | % | From Table 6.17 |
| Most Severe Alleged Disease/Injury | | | | | | | | | | | | | | | |
| 1. Mesothelioma | 3,891 | 4,445 | 3,890 | 3,730 | 3,201 | 2,506 | 1,858 | 1,238 | 719 | 440 | 128 | 2 | 26,049 | 6.1 | 7.0 |
| 2. Lung cancer | 5,941 | 6,667 | 6,152 | 5,630 | 4,824 | 3,825 | 2,791 | 1,729 | 1,025 | 667 | 129 | 0 | 39,381 | 9.3 | 17.4 |
| 3. Colon/rectal cancer | 728 | 837 | 807 | 711 | 627 | 505 | 342 | 202 | 105 | 33 | 0 | 0 | 4,897 | 1.2 | 14.4 |
| 4. Other cancer | 946 | 1,019 | 937 | 874 | 716 | 525 | 398 | 245 | 111 | 66 | 28 | 0 | 5,864 | 1.4 | 1.2 |
| 5. Asbestosis | 36,145 | 34,705 | 27,748 | 22,377 | 17,122 | 12,151 | 7,725 | 4,181 | 2,277 | 1,232 | 315 | 1 | 165,979 | 39.1 | -15.2 |
| 6. Disputed asbestosis | 13,337 | 13,033 | 10,735 | 8,773 | 6,665 | 4,600 | 2,806 | 1,368 | 762 | 367 | 29 | 0 | 62,476 | 14.7 | 2.4 |
| 7. Pleural plaques/thickening | 13,374 | 13,217 | 10,612 | 8,222 | 6,138 | 4,040 | 2,383 | 1,260 | 693 | 341 | 54 | 0 | 60,333 | 14.2 | -9.1 |
| 8. Nonasbestos-related disease | 9,154 | 8,705 | 6,916 | 5,614 | 4,507 | 3,306 | 1,918 | 1,005 | 525 | 331 | 233 | 10 | 42,224 | 10.0 | 6.0 |
| 9. Unknown | 3,090 | 3,061 | 2,805 | 2,336 | 2,128 | 1,731 | 1,095 | 733 | 79 | 32 | 32 | 0 | 17,121 | 4.0 | 15.7 |
| Total | 86,606 | 85,690 | 70,602 | 58,266 | 45,928 | 33,189 | 21,316 | 11,962 | 6,295 | 3,509 | 948 | 13 | 424,323 | 100.0 | -4.8 |
| Cancer (1-4) | 11,506 | 12,968 | 11,786 | 10,945 | 9,367 | 7,362 | 5,389 | 3,414 | 1,960 | 1,207 | 284 | 2 | 76,191 | 18.0 | 12.1 |
| Noncancer (5-9) | 75,100 | 72,721 | 58,815 | 47,322 | 36,560 | 25,828 | 15,926 | 8,548 | 4,335 | 2,303 | 663 | 11 | 348,132 | 82.0 | -7.8 |
| Asbestos-related noncancer (5-7) | 62,856 | 60,955 | 49,094 | 39,371 | 29,926 | 20,791 | 12,913 | 6,809 | 3,731 | 1,940 | 399 | 1 | 288,787 | 68.1 | -10.6 |
| Nonasbestos-related & unknown (8-9) | 12,244 | 11,766 | 9,721 | 7,950 | 6,635 | 5,037 | 3,013 | 1,738 | 604 | 363 | 265 | 10 | 59,345 | 14.0 | 8.6 |
| Occupation | | | | | | | | | | | | | | | |
| 1. A: Primary Manufacturing | 4,759 | 5,436 | 4,746 | 3,703 | 2,886 | 2,304 | 1,510 | 621 | 403 | 303 | 8 | 0 | 26,679 | 6.3 | |
| 2. F: Secondary Manufacturing | 12,370 | 13,000 | 12,289 | 10,958 | 8,924 | 6,862 | 4,512 | 2,734 | 1,522 | 735 | 113 | 0 | 74,019 | 17.4 | |
| 3. I: Insulation Work | 9,281 | 9,463 | 7,789 | 5,714 | 4,338 | 3,024 | 1,981 | 970 | 481 | 392 | 0 | 0 | 43,434 | 10.2 | |
| 4. S: Shipbuilding and Repair | 13,260 | 12,313 | 6,737 | 4,955 | 3,110 | 1,844 | 1,033 | 510 | 199 | 66 | 11 | 3 | 44,040 | 10.4 | |
| 5. R: Construction Trades | 5,217 | 5,587 | 5,857 | 4,974 | 3,594 | 2,546 | 1,683 | 1,074 | 734 | 403 | 90 | 10 | 31,770 | 7.5 | |
| 6. Util/Trans/Chem/Longshore | 21,660 | 18,964 | 13,389 | 9,897 | 7,267 | 4,857 | 2,864 | 1,475 | 720 | 275 | 185 | 0 | 81,553 | 19.2 | |
| 7. Military | 2,372 | 2,498 | 2,372 | 2,166 | 1,845 | 1,414 | 1,034 | 626 | 326 | 194 | 65 | 0 | 14,912 | 3.5 | |
| 8. Other/Unknown | 17,687 | 18,428 | 17,423 | 15,900 | 13,964 | 10,339 | 6,697 | 3,952 | 1,911 | 1,141 | 474 | 0 | 107,916 | 25.4 | |
| Total | 86,606 | 85,690 | 70,602 | 58,266 | 45,928 | 33,189 | 21,316 | 11,962 | 6,295 | 3,509 | 948 | 13 | 424,323 | 100.0 | |

Source: Authors' calculations.

than under the TDEC model. The entire excess was attributable to cancer claims.

Table 8.19 presents the CDEC/TDEC projection stratified by most severe alleged disease. All four cancers were projected to peak in 1995-1999; noncancers and total claims peaked in 1990-1994. Mesothelioma accounted for 4.5% of claims in 1990-1994, but 7.1% in 1990-2049. Lung cancer accounted for 6.9% of claims in 1990-1994, but 10.8% in 1990-2049. Both major cancer injuries increased their share of claims over time.

Table 8.19 also presents the CDEC/TDEC projection stratified by occupation. There was substantial variation in the timing of claims (e.g., about 33% of claims for insulation work and shipbuilding and repair, but less than 18% of construction trades and other/unknown, were projected to occur in 1990-1994). Over half (51.3%) of the claims came from three low-risk occupation groups (6-8).

Table 8.20 presents the CDEC/TDEC projection stratified by date of first exposure. The largest number of claims for 1990-2049 (18.2%) were projected to arise from exposure 1965-1969, followed by 1970-1974 (16.7%). These results contrast with claims for the baseline period 1990-1994, where 12.1% of claims arose from exposure 1965-1969 and 6.6% from 1970-1974. The exposure period 1950-1964 accounted for 43.5% of claims in 1990-2049 and 42.1% in 1990-1994.

Table 8.21 presents the CDEC/TDEC projection stratified by attained age and cohort (based on attained age in 1990-1994). Compared with Table 6.18, the peak age was the same, but the peak cohort was 15 years older. The relative frequency of claims by attained age decreased up to 74 years and increased thereafter. Similarly, the relative frequency of claims by cohort decreased for the three youngest cohorts and increased for the others. Three factors were primarily responsible:

1. The tabulations in Section 8.3.3 increased the ages of the 50% of claims imputed for mid-1992 to 1994 by 2.5 years relative to the corresponding tabulations in Section 6.8.1.
2. The data for the first-stage calibration in the CDEC/TDEC model were approximately 20 years removed from the last exposure period (1970-1974) compared with 5, 10, or 15 years in the IWE model. This greater lag increased the credibility of the first-stage calibration and allowed us to drop adjustments in steps 8-10 of Section 6.6 that differentially affected the exposure estimates for the youngest cohorts.
3. The claim depletion rules in the CDEC/TDEC model had a greater impact on the younger cohorts due to the greater length of time over which their claims developed.

In summary, from the 1,821,354 men estimated to be occupationally exposed to Manville asbestos, we projected that 349,237 would file a claim against the Manville Trust in the period 1990-2049. Assuming a noncancer claimant may later file a cancer claim, we projected a total of 365,615 claims

**Table 8.19: Projected Number of Qualified Male Claims Against the Manville Trust by Most Severe Alleged Disease/Injury or Occupation and Quinquennial Dates of Claim, 1990-2049, Combined Cancer and Total Decrement (CDEC/TDEC) Model**

| Item | Date of Claim | | | | | | | | | | | | Total | % | % Change From Table 6.17 |
|---|---|---|---|---|---|---|---|---|---|---|---|---|---|---|---|
| | 1990-1994 | 1995-1999 | 2000-2004 | 2005-2009 | 2010-2014 | 2015-2019 | 2020-2024 | 2025-2029 | 2030-2034 | 2035-2039 | 2040-2044 | 2045-2049 | | | |
| Most Severe Alleged Disease/Injury | | | | | | | | | | | | | | | |
| 1. Mesothelioma | 3,891 | 4,445 | 3,890 | 3,730 | 3,201 | 2,506 | 1,858 | 1,238 | 719 | 440 | 128 | 2 | 26,049 | 7.1 | 7.0 |
| 2. Lung cancer | 5,941 | 6,667 | 6,152 | 5,630 | 4,824 | 3,825 | 2,791 | 1,729 | 1,025 | 667 | 129 | 0 | 39,381 | 10.8 | 17.4 |
| 3. Colon/rectal cancer | 728 | 837 | 807 | 711 | 627 | 505 | 342 | 202 | 105 | 33 | 0 | 0 | 4,897 | 1.3 | 14.4 |
| 4. Other cancer | 946 | 1,019 | 937 | 874 | 716 | 525 | 398 | 245 | 111 | 66 | 28 | 0 | 5,864 | 1.6 | 1.2 |
| 5. Asbestosis | 36,145 | 31,113 | 22,724 | 16,960 | 12,003 | 7,819 | 4,581 | 2,357 | 1,230 | 587 | 178 | 0 | 135,697 | 37.1 | -30.7 |
| 6. Disputed asbestosis | 13,337 | 11,852 | 8,976 | 6,836 | 4,855 | 3,083 | 1,724 | 759 | 396 | 164 | 11 | 0 | 51,995 | 14.2 | -14.7 |
| 7. Pleural plaques/thickening | 13,374 | 11,880 | 8,696 | 6,249 | 4,273 | 2,543 | 1,344 | 659 | 322 | 137 | 25 | 0 | 49,502 | 13.5 | -25.4 |
| 8. Nonasbestos-related disease | 9,154 | 8,118 | 6,089 | 4,717 | 3,621 | 2,523 | 1,395 | 704 | 350 | 230 | 160 | 7 | 37,068 | 10.1 | -6.9 |
| 9. Unknown | 3,090 | 2,914 | 2,553 | 2,020 | 1,764 | 1,387 | 827 | 534 | 43 | 13 | 16 | 0 | 15,162 | 4.1 | 2.5 |
| Total | 86,606 | 78,845 | 60,824 | 47,728 | 35,883 | 24,717 | 15,260 | 8,427 | 4,302 | 2,338 | 674 | 9 | 365,615 | 100.0 | -18.0 |
| Cancer (1-4) | 11,506 | 12,968 | 11,786 | 10,945 | 9,367 | 7,362 | 5,389 | 3,414 | 1,960 | 1,207 | 284 | 2 | 76,191 | 20.8 | 12.1 |
| Noncancer (5-9) | 75,100 | 65,877 | 49,038 | 36,783 | 26,516 | 17,356 | 9,871 | 5,013 | 2,342 | 1,132 | 390 | 7 | 289,424 | 79.2 | -23.4 |
| Asbestos-related noncancer (5-7) | 62,856 | 54,844 | 40,396 | 30,046 | 21,130 | 13,445 | 7,649 | 3,775 | 1,949 | 888 | 214 | 0 | 237,193 | 64.9 | -26.6 |
| Nonasbestos-related & unknown (8-9) | 12,244 | 11,033 | 8,642 | 6,737 | 5,386 | 3,910 | 2,222 | 1,238 | 393 | 244 | 175 | 7 | 52,231 | 14.3 | -4.4 |
| Occupation | | | | | | | | | | | | | | | |
| 1. A: Primary Manufacturing | 4,759 | 4,913 | 3,890 | 2,807 | 2,042 | 1,531 | 964 | 386 | 235 | 207 | 2 | 0 | 21,736 | 5.9 | |
| 2. F: Secondary Manufacturing | 12,370 | 11,827 | 10,219 | 8,367 | 6,288 | 4,450 | 2,693 | 1,525 | 791 | 352 | 55 | 0 | 58,935 | 16.1 | |
| 3. I: Insulation Work | 9,281 | 7,148 | 4,542 | 2,746 | 1,796 | 1,142 | 705 | 366 | 181 | 116 | 0 | 0 | 28,023 | 7.7 | |
| 4. S: Shipbuilding and Repair | 13,260 | 11,372 | 5,940 | 4,163 | 2,475 | 1,403 | 766 | 360 | 145 | 46 | 9 | 2 | 39,941 | 10.9 | |
| 5. R: Construction Trades | 5,217 | 5,425 | 5,530 | 4,551 | 3,195 | 2,213 | 1,439 | 900 | 591 | 331 | 75 | 7 | 29,475 | 8.1 | |
| 6. Util/Trans/Chem/Longshore | 21,660 | 17,787 | 11,915 | 8,390 | 5,854 | 3,689 | 2,081 | 1,040 | 494 | 182 | 110 | 0 | 73,200 | 20.0 | |
| 7. Military | 2,372 | 2,489 | 2,357 | 2,146 | 1,824 | 1,396 | 1,020 | 617 | 321 | 192 | 65 | 0 | 14,799 | 4.0 | |
| 8. Other/Unknown | 17,687 | 17,884 | 16,433 | 14,558 | 12,411 | 8,894 | 5,592 | 3,233 | 1,545 | 913 | 357 | 0 | 99,505 | 27.2 | |
| Total | 86,606 | 78,845 | 60,824 | 47,728 | 35,883 | 24,717 | 15,260 | 8,427 | 4,302 | 2,338 | 674 | 9 | 365,615 | 100.0 | |

Source: Stallard and Manton (1994, Tables D and F).

**Table 8.20: Projected Number of Qualified Male Claims Against the Manville Trust by Date of First Exposure and Quinquennial Dates of Claim, 1990-2049, Combined Cancer and Total Decrement (CDEC/TDEC) Model**

| | Date of Claim | | | | | | | | | | | | | |
|---|---|---|---|---|---|---|---|---|---|---|---|---|---|---|
| Date of First Exposure | 1990-1994 | 1995-1999 | 2000-2004 | 2005-2009 | 2010-2014 | 2015-2019 | 2020-2024 | 2025-2029 | 2030-2034 | 2035-2039 | 2040-2044 | 2045-2049 | Total | % |
| 1915-1919 | 4 | 0 | 0 | 0 | 0 | 0 | 0 | 0 | 0 | 0 | 0 | 0 | 4 | 0.0 |
| 1920-1924 | 126 | 4 | 0 | 0 | 0 | 0 | 0 | 0 | 0 | 0 | 0 | 0 | 130 | 0.0 |
| 1925-1929 | 442 | 187 | 2 | 0 | 0 | 0 | 0 | 0 | 0 | 0 | 0 | 0 | 632 | 0.2 |
| 1930-1934 | 843 | 645 | 246 | 5 | 0 | 0 | 0 | 0 | 0 | 0 | 0 | 0 | 1,739 | 0.5 |
| 1935-1939 | 4,087 | 1,644 | 1,216 | 445 | 8 | 0 | 0 | 0 | 0 | 0 | 0 | 0 | 7,400 | 2.0 |
| 1940-1944 | 13,615 | 11,296 | 4,171 | 2,981 | 1,078 | 40 | 0 | 0 | 0 | 0 | 0 | 0 | 33,181 | 9.1 |
| 1945-1949 | 14,850 | 9,388 | 6,408 | 2,743 | 2,027 | 761 | 12 | 0 | 0 | 0 | 0 | 0 | 36,188 | 9.9 |
| 1950-1954 | 15,019 | 14,403 | 9,257 | 6,466 | 2,817 | 2,205 | 830 | 10 | 0 | 0 | 0 | 0 | 51,006 | 14.0 |
| 1955-1959 | 11,582 | 12,908 | 12,119 | 8,151 | 5,770 | 2,591 | 2,112 | 795 | 18 | 0 | 0 | 0 | 56,046 | 15.3 |
| 1960-1964 | 9,842 | 8,641 | 9,482 | 9,115 | 6,048 | 4,360 | 2,011 | 1,657 | 678 | 14 | 0 | 0 | 51,849 | 14.2 |
| 1965-1969 | 10,494 | 10,929 | 9,524 | 10,053 | 9,635 | 6,534 | 4,720 | 2,002 | 1,803 | 754 | 17 | 0 | 66,465 | 18.2 |
| 1970-1974 | 5,703 | 8,801 | 8,400 | 7,768 | 8,500 | 8,226 | 5,575 | 3,963 | 1,802 | 1,570 | 657 | 9 | 60,975 | 16.7 |
| Total | 86,606 | 78,845 | 60,824 | 47,728 | 35,883 | 24,717 | 15,260 | 8,427 | 4,302 | 2,338 | 674 | 9 | 365,615 | 100.0 |

Source: Stallard and Manton (1994, Table E).

**Table 8.21: Projected Number of Qualified Male Claims Against the Manville Trust by Attained Age or Age in 1990-1994 and Quinquennial Dates of Claim, 1990-2049, Combined Cancer and Total Decrement (CDEC/TDEC) Model**

| Item | Date of Claim | | | | | | | | | | | | | | % Change From |
|---|---|---|---|---|---|---|---|---|---|---|---|---|---|---|---|
| | 1990-94 | 1995-99 | 2000-04 | 2005-09 | 2010-14 | 2015-19 | 2020-24 | 2025-29 | 2030-34 | 2035-39 | 2040-44 | 2045-49 | Total | % | Table 6.18 |
| Attained Age | | | | | | | | | | | | | | | |
| 35-39 | 705 | | | | | | | | | | | | 705 | 0.2 | -66.2 |
| 40-44 | 3,736 | 1,344 | | | | | | | | | | | 5,080 | 1.4 | -13.4 |
| 45-49 | 5,422 | 5,578 | 854 | | | | | | | | | | 11,854 | 3.2 | -5.0 |
| 50-54 | 7,386 | 5,049 | 4,480 | 1,056 | | | | | | | | | 17,972 | 4.9 | -46.4 |
| 55-59 | 10,251 | 9,182 | 6,857 | 4,778 | 1,549 | | | | | | | | 32,617 | 8.9 | -42.8 |
| 60-64 | 14,622 | 10,454 | 8,776 | 6,498 | 5,057 | 1,522 | | | | | | | 46,928 | 12.8 | -34.2 |
| 65-69 | 17,521 | 15,571 | 11,013 | 9,495 | 7,021 | 5,395 | 1,330 | | | | | | 67,347 | 18.4 | -32.1 |
| 70-74 | 13,471 | 12,570 | 10,786 | 7,778 | 6,818 | 4,851 | 3,548 | 923 | | | | | 60,745 | 16.6 | -6.5 |
| 75-79 | 8,428 | 10,390 | 8,161 | 7,802 | 5,561 | 4,810 | 3,407 | 2,389 | 578 | | | | 51,526 | 14.1 | 4.2 |
| 80-84 | 3,760 | 6,105 | 6,432 | 5,825 | 5,540 | 4,004 | 3,497 | 2,478 | 1,668 | 672 | | | 39,982 | 10.9 | 34.2 |
| 85-89 | 1,132 | 2,016 | 2,759 | 3,362 | 3,020 | 3,093 | 2,368 | 1,961 | 1,500 | 1,227 | 316 | | 22,755 | 6.2 | 65.8 |
| 90-94 | 172 | 587 | 706 | 1,132 | 1,318 | 1,042 | 1,109 | 676 | 556 | 440 | 358 | 9 | 8,104 | 2.2 | 47.5 |
| 95-99 | 0 | 0 | 0 | 0 | 0 | 0 | 0 | 0 | 0 | 0 | 0 | 0 | 0 | 0.0 | -100.0 |
| Total | 86,606 | 78,845 | 60,824 | 47,728 | 35,883 | 24,717 | 15,260 | 8,427 | 4,302 | 2,338 | 674 | 9 | 365,615 | 100.0 | -18.0 |
| Age in 1990-1994 | | | | | | | | | | | | | | | |
| 35-39 | 705 | 1,344 | 854 | 1,056 | 1,549 | 1,522 | 1,330 | 923 | 578 | 672 | 316 | 9 | 10,858 | 3.0 | -73.6 |
| 40-44 | 3,736 | 5,578 | 4,480 | 4,778 | 5,057 | 5,395 | 3,548 | 2,389 | 1,668 | 1,227 | 358 | 0 | 38,214 | 10.5 | -52.0 |
| 45-49 | 5,422 | 5,049 | 6,857 | 6,498 | 7,021 | 4,851 | 3,407 | 2,478 | 1,500 | 440 | 0 | | 43,523 | 11.9 | -49.5 |
| 50-54 | 7,386 | 9,182 | 8,776 | 9,495 | 6,818 | 4,810 | 3,497 | 1,961 | 556 | 0 | | | 52,480 | 14.4 | 4.1 |
| 55-59 | 10,251 | 10,454 | 11,013 | 7,778 | 5,561 | 4,004 | 2,368 | 676 | 0 | | | | 52,107 | 14.3 | 3.6 |
| 60-64 | 14,622 | 15,571 | 10,786 | 7,802 | 5,540 | 3,093 | 1,109 | 0 | | | | | 58,523 | 16.0 | 3.5 |
| 65-69 | 17,521 | 12,570 | 8,161 | 5,825 | 3,020 | 1,042 | 0 | | | | | | 48,139 | 13.2 | 24.9 |
| 70-74 | 13,471 | 10,390 | 6,432 | 3,362 | 1,318 | 0 | | | | | | | 34,974 | 9.6 | 36.4 |
| 75-79 | 8,428 | 6,105 | 2,759 | 1,132 | 0 | | | | | | | | 18,423 | 5.0 | 52.3 |
| 80-84 | 3,760 | 2,016 | 706 | 0 | | | | | | | | | 6,482 | 1.8 | 60.1 |
| 85-89 | 1,132 | 587 | 0 | | | | | | | | | | 1,719 | 0.5 | 59.3 |
| 90-94 | 172 | 0 | | | | | | | | | | | 172 | 0.0 | 2.4 |
| 95-99 | 0 | | | | | | | | | | | | 0 | 0.0 | -100.0 |
| Total | 86,606 | 78,845 | 60,824 | 47,728 | 35,883 | 24,717 | 15,260 | 8,427 | 4,302 | 2,338 | 674 | 9 | 365,615 | 100.0 | -18.0 |

Source: Authors' calculations.

in 1990-2049, of which 76,191 would be for cancer, 237,193 for asbestos-related noncancer, and 52,231 for nonasbestos-related (or unknown) disease.

## 8.7 Conclusions

The development of projections is generally an iterative process, with each projection addressing a specific question. The projections and the sensitivity they exhibit as assumptions are varied may suggest additional questions or issues to be addressed by further projections.

The projections in this chapter were based on a hybrid form of the methods used by Walker (1982) and Selikoff (1981).

As discussed in Chapters 3-5, those two projections used different assumptions and produced different results. Table 6.13 showed that the national projection of mesothelioma for 1990-2009 produced by Selikoff was 4.9 times that of Walker (58,880 vs. 12,000). Section 6.7 showed that this discrepancy was associated with scale factors that could make both projections consistent with the SEER data and with the baseline IWE projection in Chapter 6.

Sensitivity analyses in Section 7.6 showed that the most unstable part of the baseline IWE projection was the first-stage calibration of the initial IWE population. This instability was greatly reduced when the elapsed time between the last exposure period and the observed manifestation (i.e., diagnosis, death, or claim) of mesothelioma increased to 15+ years. With initial calibration for first exposures prior to 1975 based on Trust data for 1990-1994, the total projected number of claims 1990-2049 in Section 7.12 was within 5.1% of the baseline IWE projection of 445,751 claims. This justified our decision to drop the SEER data and perform the first-stage calibration using only mesothelioma claim data for 1990-1994. This step allowed us to use occupation in the projections. Occupation was recorded in the Trust files, but not in the SEER files.

With occupation known, we used relative risk and duration of exposure estimates from Selikoff (1981) to estimate the number of surviving exposed men alive in 1990-1994, not just the number of IWEs. This allowed us to determine the effect of claim filings on the pool of men who had never filed a claim or, in the case of cancer, who had never filed a cancer claim.

The preferred projection for 1990-2049 (CDEC/TDEC) in this chapter was 18.0% lower than the baseline IWE projection (Table 6.17) and 13.6% lower than the bridge model projection (Table 7.2, analysis S10, Model 3). The latter projection was also calibrated using Trust mesothelioma claims, whereas the former used SEER mesothelioma incidence data. Compared to the bridge model projection, our preferred cancer projection was 8.6% higher (12.1% higher than the baseline IWE projection) and our preferred noncancer projection was 18.0% lower (23.4% lower than the baseline IWE projection). These changes were attributable to changes in the calibration period and in the second-injury claim filing rules. In our opinion, these changes were improvements. The fact that one can alter the model structure and assumptions, however, should alert the reader to the possibility of further alterations and, consequently, further revision of the projections – especially as more data become available.

The projections in this chapter resolved important issues raised in evaluations of Walker (1982) and Selikoff (1981). Although the projections here were improvements over those in Chapter 6, the sensitivity analyses in Chapters 5 and 7 indicated that there was substantial uncertainty in these types of projections, so, despite the desire for point estimates, we caution the reader that any point estimate of the future number of claims is highly uncertain. Deviations of $\pm 10\%$ are very likely; deviations up or down by a factor of 2 are possible, as determined through sensitivity analyses involving 27 alternative projections (Table 7.9). Nothing in this chapter diminishes the uncertainty established in Chapter 7. Thus, the reduction in the preferred CDEC/TDEC projection serves only to shift the baseline downward, not to remove the intrinsic uncertainty. Consequently, any decision based on projections must plan for the contingency of substantial variation in outcome. We investigate this further in Chapter 9.

# 9

# Uncertainty in Forecasts Based on a Hybrid Model

## 9.1 Introduction

We evaluated the sensitivity of the hybrid model by varying sets of assumptions across plausible ranges, as in Chapter 7. We selected for sensitivity analyses characteristics of the model that we felt could have the greatest influence on the hybrid model. Factors identified as being of minor influence in Chapter 7 are not reanalyzed in this chapter.

We begin by summarizing the impact of claim filing rules on the baseline projection. Next, we summarize the results of the baseline cancer decrement/total decrement (CDEC/TDEC) projection model in Chapter 8. In the remainder of the chapter, we examine how different that projection could be under plausible modifications of the model.

Table 9.1 summarizes the characteristics of the baseline and alternative projections. The sensitivity analyses were grouped into seven areas:

S1. *Validated disease.* The baseline projection was based on analysis of alleged disease/injury. In these projections, we considered the impact of switching to validated disease.

S2. *Multiple disease.* The baseline projection was based on claim hazard rates (CHRs) for the most severe alleged disease/injury. When a claimant alleged multiple disease/injuries, the lower-ranked disease/injuries were ignored in calibrating the CHRs for the baseline projection. In these projections, we considered the impact of these additional lower-ranked diseases on future claim filings.

S3. *CHR smoothing.* Step 2 of the second-stage baseline projection used a multiplicative form of the CHR with factors representing age effects and time since first exposure (TSFE) effects. In these projections, we reverted to the crude rates for combinations of age and TSFE.

S4. *Exposure smoothing.* In step 4 of the first-stage baseline projection, we smoothed estimates of the distribution of mesothelioma claims by age and date of first exposure. In these projections, we reverted to the exposure

**Table 9.1: Sensitivity Analyses Conducted to Assess the Uncertainty in Projections of Qualified Male Claims Against the Manville Trust, 1990-2049**

| Analysis | Model | # | Description |
|---|---|---|---|
| S0 | | 0 | Baseline hybrid model in Section 8.6 |
| S1 | | | Convert from alleged to validated disease |
| | 1 | 1 | Transitions based on Table 6.1 |
| | 2 | 2 | Transitions based on Table 9.3 |
| S2 | | | Impact of claims with multiple diseases |
| | 1 | 3 | Project all disease mentions |
| | 2 | 4 | Eliminate 50% of cancer in TDEC projection |
| | 3 | 5 | Reallocate 50% of cancer in TDEC projection |
| S3 | 1 | 6 | Impact of smoothing in Section 8.6, Step 2 |
| S4 | 1 | 7 | Impact of smoothing in Section 8.3, Step 4 |
| S5 | | | Parametric models of claim hazard rates |
| | 1 | 8 | $k = 2$ (and $b = 5.13 \quad 10^{-6}$) |
| | 2 | 9 | $k = 3$ (and $b = 2.18 \quad 10^{-7}$) |
| S6 | | | Relative risks in Table 8.8 |
| | 1 | 10 | Reduced by 50% |
| | 2 | 11 | Increased by 50% |
| S7 | | | Average durations in Table 8.8 |
| | 1 | 12 | Reduced by 50% |
| | 2 | 13 | Increased by 50% |

Source: S0–S2, Stallard and Manton (1994); S3–S7, authors' calculations.

data obtained in step 3 stratified by age and TSFE, and used those in place of the smoothed estimates obtained in step 4.

S5. *Weibull k parameter*. In Section 7.5, we saw that the insulation-worker equivalent (IWE) model was sensitive to changes in the $k$ parameter of the Weibull hazard function. In these projections, we evaluated the impact of setting $k = 2$ (Peto et al., 1982) and $k = 3$ (OSHA, 1983).

S6. *Relative risks of mesothelioma*. Selikoff's (1981) estimates of relative risks in Tables 3.1 and 8.8 were presented without standard errors. In addition, the correspondence between the Trust occupation groups and Selikoff's groups was only approximate. In these projections, we evaluated the impact of increasing or decreasing the relative risks in Table 8.8 by 50%.

S7. *Duration of exposure*. Selikoff's (1981) estimates of duration of exposure (Table 8.8) were only approximations to the durations experienced by persons exposed to Manville asbestos. In these projections, we evaluated the impact of increasing or decreasing the average durations in Table 8.8 by 50%.

## 9.2 Impact of Claim Filing Rules

Application of the claim filing rates to the surviving exposed population in 1990-2049 required us to define (1) rules for handling claims with multiple diseases or injuries and (2) rules to decrement or deplete this population.

Because multiple diseases were often alleged by a given claimant, three criteria were evaluated. First, in our baseline model, we selected the most severe disease (SDIS criterion) as indicative of his injury. The hierarchy used was the same as that used by the Trust settlement negotiation process: (1) mesothelioma; (2) lung cancer; (3) colon/rectal cancer; (4) other cancer; (5) asbestosis; (6) disputed asbestosis; (7) pleural plaques/thickening; (8) nonasbestos-related disease; and (9) unknown disease. Disease groups were frequently used: (a) "cancer" referred to 1, 2, 3, and 4; (b) "noncancer" referred to 5, 6, 7, 8, and 9; (c) "asbestos-related disease" referred to 1, 2, 3, 4, 5, 6, and 7; (d) "nonasbestos-related and unknown disease" referred to 8 and 9; and (e) "asbestos-related noncancer disease" referred to 5, 6, and 7. Section 6.2 presented results indicating that among claims settled postpetition, 93.4% of claims with cancer as the most serious alleged condition were evaluated as cancer, whereas 95.6% of claims with noncancer as the most serious alleged condition were evaluated as noncancer. Use of the most severe alleged injury as the basis for the CDEC/TDEC projections was reasonable.

Second, under the validated disease (VDIS) criterion, the transition rates from most severe alleged disease to final validated disease classification determined by the Trust settlement process (Table 6.1) were applied to the claim filing rates obtained under the SDIS criterion. This change in procedure was the basis of sensitivity analysis S1, which evaluated the trade-off between the uncertainty in the transition matrix used to convert from alleged to validated disease and the uncertainty in decrementing the population of claims using alleged disease when the actual process was based on validated diseases. The projections using the VDIS criterion were also rerun using a transition matrix from alleged to validated disease prepared by RPC (1993).

Third, rules for decrementing or depleting the surviving exposed population depended on the Trust Distribution Process (TDP) rules regarding second-injury claims. The TDP rules allowed the Trust to (1) give limited releases that allowed noncancer living claimants to pursue second-injury claims for cancer (87% of claims in 1990-1994 were noncancer) and (2) settle some claims (initially, about 3.0% of settled claims) for zero dollars accompanied by the right to file a claim at a later date. In the baseline model, the occurrence of a cancer claim as a second-injury claim is represented through the CDEC projection where the noncancer claims are removed as a competing risk. A further sensitivity analysis was developed using the MDIS criterion in which all claims that mentioned a specific disease/injury were recognized in the CHR calculations regardless of its ranking in the SDIS hierarchy. The MDIS criterion was stratified to reflect cancer versus noncancer claims, consistent with the rules for second-injury claim filings. Projections of disease/injuries can be an important supplement to projections of claimants if there are significant improvements in the treatment of specific diseases that could increase or decrease the costs associated with those diseases. Such projections also permit consideration of the potential impact of a change to multiple, single-claim filings in future years for claimants who currently were filing a single claim with multiple listed diseases.

## 9.3 Baseline Model: SDIS Criterion

The baseline projections developed in Section 8.6 addressed two questions: (1) how many persons would file claims of any type against the Trust? and (2) how many persons would file cancer claims against the Trust? To facilitate discussion of the sensitivity of these projections, we briefly review the assumptions and findings of the baseline model.

First, to estimate how many men would ever file a claim against the Trust, we decremented (TDEC projection) the surviving exposed population for (1) all nonasbestos-related deaths in the 5 years between the midpoints of each projection period 1990-1994, 1995-1999, ..., 2045-2049 (estimated using projected death rates for the total U.S. male population) and (2) all claims filed in the corresponding 5-year intervals. The decrement calculations treated the 9 diseases (defined by the SDIS criterion) and nonasbestos-related death as a set of 10 independent competing risks such that for any given man, only one outcome occurred. This gave rise to an exponential survival function in which all active risks were represented in the summation in the exponent of the survival function as indicated in Section 8.6.3. The total number of outcomes of each type was estimated and subtracted from the initial exposed population in each projection period to obtain the surviving exposed population for the next period. Under TDEC, we projected 349,237 men would file at least one claim in 1990-2049 (Table 9.2); 59,813 men filing a first claim would allege a cancer disease/injury in 1990-2049.

Second, to determine how many men would ever file a cancer claim against the Trust, we decremented (CDEC projection) the surviving exposed population for (1) all nonasbestos-related deaths in the 5 years between the midpoints of each projection period (same as TDEC) and (2) all cancer claims (based on the SDIS criterion) filed in the corresponding 5-year intervals. Noncancer claims were not decremented in this projection; that is, the noncancer hazard rates were removed from the summation in the exponent of the exponential survival functions. Under CDEC, we projected 76,191 men would file a cancer claim in 1990-2049 (Table 9.2).

Comparison of the CDEC and TDEC projections suggests that 16,378 (25.3% of 64,685) cancer claims in 1995-2049 would be second-injury claims. Implicit in this comparison was the assumption that the 289,424 men filing a noncancer claim in 1990-2049 were at the same risk of filing a subsequent cancer claim as men who never filed a noncancer claim. If the risk of first

**Table 9.2: Analysis S0, Model 1 – Baseline Projections of Qualified Male Claims Against the Manville Trust, 1990-2049**

| | Projection Interval | | |
|---|---|---|---|
| Projection Type | 1990-1994 | 1995-2049 | 1990-2049 |
| TDEC | | | |
| Total | 86,606 | 262,631 | 349,237 |
| Cancer | 11,506 | 48,307 | 59,813 |
| Noncancer | 75,100 | 214,324 | 289,424 |
| Asbestos-related noncancer | 62,856 | 174,337 | 237,193 |
| Nonasbestos-related & unknown | 12,244 | 39,987 | 52,231 |
| CDEC | | | |
| Total | 86,606 | 337,717 | 424,323 |
| Cancer | 11,506 | 64,685 | 76,191 |
| Noncancer | 75,100 | 273,032 | 348,132 |
| Asbestos-related noncancer | 62,856 | 225,931 | 288,787 |
| Nonasbestos-related & unknown | 12,244 | 47,101 | 59,345 |
| CDEC/TDEC | | | |
| Total | 86,606 | 279,009 | 365,615 |
| Cancer | 11,506 | 64,685 | 76,191 |
| Noncancer | 75,100 | 214,324 | 289,424 |
| Asbestos-related noncancer | 62,856 | 174,337 | 237,193 |
| Nonasbestos-related & unknown | 12,244 | 39,987 | 52,231 |

Source: Stallard and Manton (1994, Table A).

cancer filing for everybody is lower than the risk of second cancer filing for men who previously filed a noncancer claim, then the TDEC projection will (a) overestimate the number filing a first claim with cancer and (b) underestimate the percent of cancer claims that are second-injury claims.

Combining 76,191 cancer (CDEC) and 289,424 noncancer (TDEC) claims, we projected a total of 365,615 claims in 1990-2049 (CDEC/TDEC in Table 9.2). The difference between the CDEC and TDEC projections was that TDEC depleted the surviving exposed population as noncancer claims were filed, and this additional depletion caused the composite projection (CDEC/TDEC) of total claims to be lower than in the CDEC projection.

## 9.4 Analysis S1: Validated Disease

The conversion from alleged to validated disease served two purposes. First, because claims were settled on the basis of validated diseases, not alleged diseases, projections of validated disease could be used to analyze monetary liabilities of the Trust. Second, because the rules for second-injury claim filing were based on the occurrence of a prior validated cancer claim, accurate implementation of these rules required the use of validated disease/injury data.

The drawback to the use of validated disease was that this information was developed as part of the settlement process and, in the case of the Trust extract made available to us, the information simply was not available for most claims filed during the baseline calibration period from January 1, 1990 to June 30, 1992. Consequently, we used alleged disease/injuries in the baseline model.

To evaluate the potential sensitivity of our results to this choice, we used two estimates of the transition matrices from alleged to validated disease: one from Table 6.1 and the other from RPC's (1993) analysis of Trust claim data (displayed in Table 9.3). RPC (1993) did not include disputed asbestosis or colon/rectal cancer as distinct categories, either alleged or validated. In constructing Table 9.3, we replicated RPC's (1993) transition rates for other cancers in the row for colon/rectal cancer and RPC's (1993) transition rates for asbestosis in the row for disputed asbestosis. However, the column entries for these two diseases were set to zero because RPC (1993) did not include them among validated diseases.

Comparison of Tables 9.3 and 6.1 revealed numerous differences. Except for lung cancer, the diagonals of comparable categories were generally quite different. The off-diagonals also exhibited significant differences. For example, 5.8% of nonasbestos-related diseases converted to lung cancer in Table 6.1 but only 0.9% in Table 9.3. Nonetheless, the estimated average settlement amounts derived from the two transition matrices were within 1%, suggesting that the conversion from alleged to validated disease may not be a major source of sensitivity with respect to the aggregate costs of the claims.

In Table 9.4, VTDEC denotes the TDEC projection with all disease counts converted from the SDIS to VDIS criterion using the Table 6.1 transition matrix. This conversion did not change the total number of claims filed under the TDEC rules. The 59,813 alleged cancer claims (under the SDIS criterion) in the TDEC projection for 1990-2049 converted to 70,227 validated cancer claims (under the VDIS criterion; +17.4%). The other 279,010 validated noncancer claimants (VDIS criterion) were free to file a cancer claim later.

To estimate how many men would ever file a validated cancer claim against the Trust, in a second projection (VCDEC), we decremented the surviving exposed population for (1) all nonasbestos-related deaths in the 5 years between the midpoints of each projection period (same as TDEC and VTDEC) and (2) all validated cancer claims (VDIS criterion) filed in the corresponding intervals. Validated noncancer claims were not decremented in this projection.

Under VCDEC, we projected 87,451 validated cancer claims in 1990-2049: 70,227 first claims and 17,224 second or subsequent claims. We assumed that a man was free to file cancer claims until his injury was validated as cancer. The separation into first and second or subsequent injury claims assumed that men filing a noncancer claim were at the same risk of filing a subsequent cancer claim as men who never filed a noncancer claim. Combined with the 279,010 validated noncancer claims in the VTDEC projection, we projected 366,461 validated claims in 1990-2049 (VCDEC/VTDEC in Table 9.4): 100.2%

**Table 9.3: Resource Planning Corporation's Estimates of the Distribution of Validated Disease/Injury by Most Severe Alleged Disease/Injury, Based on 15,000 Post–Bankruptcy-Petition Closed Claims and 3000 Randomly Sampled Open Claims Against the Manville Trust**

| Most Severe Alleged Disease/Injury | Validated Disease/Injury: Percent Distribution 1 | 2 | 3 | 4 | 5 | 6 | 7 | 8 | Number of Claims in Table 6.1 | Average Settlement Amount ($000s): Estimated[1] | Average Settlement Amount ($000s): Estimated in Table 6.1 | Ratio of Estimated Amounts This Table vs. Table 6.1 |
|---|---|---|---|---|---|---|---|---|---|---|---|---|
| 1. Mesothelioma | 87.5 | 5.7 | 0.0 | 0.0 | 3.4 | 0.0 | 0.7 | 2.7 | 1,404 | 181.2 | 190.3 | 0.95 |
| 2. Lung cancer | 1.4 | 90.9 | 0.0 | 1.4 | 2.0 | 0.0 | 1.3 | 3.0 | 1,717 | 75.4 | 76.0 | 0.99 |
| 3. Colon/rectal cancer | 2.0 | 10.9 | 0.0 | 64.5 | 13.0 | 0.0 | 3.9 | 5.7 | 211 | 45.3 | 37.5 | 1.21 |
| 4. Other cancer | 2.0 | 10.9 | 0.0 | 64.5 | 13.0 | 0.0 | 3.9 | 5.7 | 448 | 45.3 | 52.0 | 0.87 |
| 5. Asbestosis | 0.1 | 0.4 | 0.0 | 0.2 | 65.7 | 0.0 | 30.5 | 3.1 | 6,179 | 37.1 | 38.4 | 0.97 |
| 6. Disputed asbestosis | 0.1 | 0.4 | 0.0 | 0.2 | 65.7 | 0.0 | 30.5 | 3.1 | 1,678 | 37.1 | 24.6 | 1.51 |
| 7. Pleural plaques/thickening | 0.0 | 0.2 | 0.0 | 0.1 | 17.1 | 0.0 | 77.8 | 4.8 | 1,008 | 18.1 | 24.3 | 0.75 |
| 8. Nonasbestos-related disease | 0.2 | 0.9 | 0.0 | 0.4 | 55.7 | 0.0 | 20.6 | 22.2 | 586 | 31.6 | 28.2 | 1.12 |
| 9. Unknown | 1.8 | 10.3 | 0.0 | 2.5 | 5.8 | 0.0 | 10.4 | 69.2 | 208 | 16.8 | 11.3 | 1.48 |
| Total[2] | 9.5 | 13.2 | 0.0 | 3.5 | 43.4 | 0.0 | 25.2 | 5.2 | 13,439 | 55.5 | 55.9 | 0.99 |
| Cancer (1-4)[2] | 33.5 | 45.3 | 0.0 | 11.9 | 4.4 | 0.0 | 1.5 | 3.4 | 3,780 | 109.5 | 113.5 | 0.96 |
| Noncancer (5-9)[2] | 0.1 | 0.6 | 0.0 | 0.3 | 58.7 | 0.0 | 34.4 | 5.9 | 9,659 | 34.3 | 33.3 | 1.03 |
| Asbestos-related noncancer (5-7)[2] | 0.1 | 0.4 | 0.0 | 0.2 | 60.1 | 0.0 | 35.9 | 3.3 | 8,865 | 34.9 | 34.2 | 1.02 |
| Nonasbestos-related & unknown (8-9)[2] | 0.6 | 3.4 | 0.0 | 1.0 | 42.6 | 0.0 | 17.9 | 34.5 | 794 | 27.7 | 23.8 | 1.17 |

| Item | Validated Disease/Injury -- Trust Distribution Process 1 | 2 | 3 | 4 | 5 | 6 | 7 | 8 | Total |
|---|---|---|---|---|---|---|---|---|---|
| Scheduled Settlement Amount ($000s)[3] | 200.0 | 78.0 | 40.0 | 40.0 | 50.0 | 25.0 | 12.0 | 0.0 | 55.5 |
| Number of Claims[2] | 1,278 | 1,773 | 0 | 474 | 5,838 | 0 | 3,383 | 693 | 13,439 |

Note 1: Based on scheduled settlement amount for validated disease/injury.
Note 2: Based on disease-specific number of claims in Table 6.1
Note 3: The $78,000 amount for lung cancer is based on relative frequencies of the two types of lung cancer claims filed during 1995-1999.

Source: Resource Planning Corporation (1993, Table 4) and authors' calculations.

**Table 9.4: Analysis S1, Model 1 – Projections of Qualified Male Claims Against the Manville Trust, 1990-2049, with Validated Diseases Based on Transitions in Table 6.1**

| Projection Type | Projection Interval 1990-1994 | 1995-2049 | 1990-2049 | % Change From Table 9.2 |
|---|---|---|---|---|
| VTDEC | | | | |
| Total | 86,606 | 262,631 | 349,237 | 0.0 |
| Cancer | 14,391 | 55,836 | 70,227 | 17.4 |
| Noncancer | 72,215 | 206,795 | 279,010 | -3.6 |
| Asbestos-related noncancer | 66,850 | 188,180 | 255,030 | 7.5 |
| Nonasbestos-related & unknown | 5,365 | 18,615 | 23,980 | -54.1 |
| VCDEC | | | | |
| Total | 86,606 | 334,453 | 421,059 | -0.8 |
| Cancer | 14,391 | 73,060 | 87,451 | 14.8 |
| Noncancer | 72,215 | 261,393 | 333,608 | -4.2 |
| Asbestos-related noncancer | 66,850 | 239,468 | 306,318 | 6.1 |
| Nonasbestos-related & unknown | 5,365 | 21,925 | 27,290 | -54.0 |
| VCDEC/VTDEC | | | | |
| Total | 86,606 | 279,855 | 366,461 | 0.2 |
| Cancer | 14,391 | 73,060 | 87,451 | 14.8 |
| Noncancer | 72,215 | 206,795 | 279,010 | -3.6 |
| Asbestos-related noncancer | 66,850 | 188,180 | 255,030 | 7.5 |
| Nonasbestos-related & unknown | 5,365 | 18,615 | 23,980 | -54.1 |
| VADEC | | | | |
| Total | 86,606 | 265,870 | 352,476 | — |
| Cancer | 14,391 | 56,542 | 70,933 | — |
| Noncancer | 72,215 | 209,328 | 281,543 | — |
| Asbestos-related noncancer | 66,850 | 190,409 | 257,259 | — |
| Nonasbestos-related & unknown | 5,365 | 18,919 | 24,284 | — |
| Asbestos-related | 81,241 | 246,951 | 328,192 | — |
| VCDEC/VADEC | | | | |
| Total | 86,606 | 282,388 | 368,994 | — |
| Cancer | 14,391 | 73,060 | 87,451 | — |
| Noncancer | 72,215 | 209,328 | 281,543 | — |
| Asbestos-related noncancer | 66,850 | 190,409 | 257,259 | — |
| Nonasbestos-related & unknown | 5,365 | 18,919 | 24,284 | — |
| Asbestos-related | 81,241 | 263,469 | 344,710 | — |

Source: Authors' calculations.

of the total of 365,615 claims in the CDEC/TDEC projection. The VCDEC projection of validated cancer claims was 114.8% of the CDEC value, and the VCDEC projection of validated noncancer claims was 95.8% of the CDEC value. These differences suggested that conversion from alleged to evaluated disease could significantly change the disease mix of the baseline projection.

Because the 23,980 nonasbestos-related disease claims were expected to receive no payment, they retained the right to file a claim at a later date. To estimate how many men would ever file a validated asbestos-related disease claim, we produced a third projection (VADEC), where we decremented the surviving exposed population for (1) all nonasbestos-related deaths in the 5 years between the midpoints of each projection period (same as TDEC and VTDEC) and (2) all validated asbestos-related disease claims (VDIS criterion) filed in the corresponding intervals. Occurrences of validated nonasbestos-related disease were not decremented in this projection.

Under VADEC, we projected that 328,192 claims would be validated as asbestos-related disease in 1990-2049, with 257,259 validated as asbestos-related noncancer. Combined with 87,451 cancer claims in 1990-2049, under VCDEC/VADEC, we projected a total of 344,710 validated asbestos-related disease claims in 1990-2049.

Table 9.5 presents the VTDEC projection with all disease counts converted from the SDIS to VDIS criterion using RPC's transition matrix (Table 9.3). The 59,813 alleged cancer claims (SDIS criterion) in the TDEC projection for 1990-2049 converted to 58,835 validated cancer claims (VDIS criterion; −1.6%).

Under VCDEC, we projected 74,665 validated cancer claims in 1990-2049: 58,835 first claims and 15,830 second (or subsequent) claims. Combined with the 290,402 validated noncancer claims in the VTDEC projection, we projected 365,067 claims in 1990-2049 (VCDEC/VTDEC in Table 9.5): 99.9% of the total of 365,615 claims in the CDEC/TDEC projection. The VCDEC projection of validated cancer claims was 98.0% of the CDEC value, and the VCDEC projection of validated noncancer claims was 100.7% of the CDEC value. Conversion from alleged to evaluated disease did not significantly change the hybrid projection.

Under VADEC, we projected that 324,431 claims would be validated as asbestos-related disease in 1990-2049, with 264,730 validated as asbestos-related noncancer. Combined with 74,665 cancer claims in 1990-2049, under VCDEC/VADEC, we projected a total of 339,395 validated asbestos-related disease claims in 1990-2049.

Comparison of the results under the VCDEC/VADEC model for the two alternative transition matrices revealed a 0.1% difference in total claims, a 1.6% difference in asbestos-related disease claims, and a 17.1% difference in cancer claims. The cancer differences were important because the settlement costs for cancer claims were more than three times those of noncancer claims (Table 6.1). The difference in cancer claims between the two alternative transition matrices suggested that the conversion from alleged to validated claims could be a major additional source of uncertainty in projecting the total liability associated with these claims.

Given the range of uncertainty of these projections, the precision gained in switching from the VTDEC to the VADEC projection was small. More important was the difference in treatment of cancer versus noncancer claims in the decrement criteria of the VCDEC/VADEC projection, namely that

**Table 9.5: Analysis S1, Model 2 – Projections of Qualified Male Claims Against the Manville Trust, 1990-2049, with Validated Diseases Based on Transitions in Table 9.3**

| Projection Type | Projection Interval 1990-1994 | 1995-2049 | 1990-2049 | % Change From Table 9.2 |
|---|---|---|---|---|
| VTDEC | | | | |
| Total | 86,606 | 262,631 | 349,237 | 0.0 |
| Cancer | 11,465 | 47,370 | 58,835 | -1.6 |
| Noncancer | 75,141 | 215,261 | 290,402 | 0.3 |
| Asbestos-related noncancer | 68,413 | 193,108 | 261,521 | 10.3 |
| Nonasbestos-related & unknown | 6,728 | 22,153 | 28,881 | -44.7 |
| VCDEC | | | | |
| Total | 86,606 | 338,528 | 425,134 | 0.2 |
| Cancer | 11,465 | 63,200 | 74,665 | -2.0 |
| Noncancer | 75,141 | 275,328 | 350,469 | 0.7 |
| Asbestos-related noncancer | 68,413 | 248,326 | 316,739 | 9.7 |
| Nonasbestos-related & unknown | 6,728 | 27,002 | 33,730 | -43.2 |
| VCDEC/VTDEC | | | | |
| Total | 86,606 | 278,461 | 365,067 | -0.1 |
| Cancer | 11,465 | 63,200 | 74,665 | -2.0 |
| Noncancer | 75,141 | 215,261 | 290,402 | 0.3 |
| Asbestos-related noncancer | 68,413 | 193,108 | 261,521 | 10.3 |
| Nonasbestos-related & unknown | 6,728 | 22,153 | 28,881 | -44.7 |
| VADEC | | | | |
| Total | 86,606 | 267,099 | 353,705 | — |
| Cancer | 11,465 | 48,236 | 59,701 | — |
| Noncancer | 75,141 | 218,863 | 294,004 | — |
| Asbestos-related noncancer | 68,413 | 196,317 | 264,730 | — |
| Nonasbestos-related & unknown | 6,728 | 22,546 | 29,274 | — |
| Asbestos-related | 79,878 | 244,553 | 324,431 | — |
| VCDEC/VADEC | | | | |
| Total | 86,606 | 282,063 | 368,669 | — |
| Cancer | 11,465 | 63,200 | 74,665 | — |
| Noncancer | 75,141 | 218,863 | 294,004 | — |
| Asbestos-related noncancer | 68,413 | 196,317 | 264,730 | — |
| Nonasbestos-related & unknown | 6,728 | 22,546 | 29,274 | — |
| Asbestos-related | 79,878 | 259,517 | 339,395 | — |

Source: Stallard and Manton (1994, Table B).

noncancer claimants could claim a cancer injury at a later date, but not the reverse.

## 9.5 Analysis S2: Multiple Diseases

Projections based on the SDIS criterion did not provide complete estimates of the number of claims by disease, except for the most serious disease,

**Table 9.6: Analysis S2, Model 1 – Projections of Multiple Diseases Among Qualified Male Claims Against the Manville Trust, 1990-2049**

| Projection Type | Projection Interval | | | % Change From |
|---|---|---|---|---|
| | 1990-1994 | 1995-2049 | 1990-2049 | Table 9.2 |
| MDIS-TDEC | | | | |
| Total | 122,876 | 381,274 | 504,150 | 44.4 |
| Cancer | 12,172 | 51,179 | 63,351 | 5.9 |
| Noncancer | 110,704 | 330,095 | 440,799 | 52.3 |
| Asbestos-related noncancer | 86,523 | 251,801 | 338,324 | 42.6 |
| Nonasbestos-related & unknown | 24,181 | 78,294 | 102,475 | 96.2 |
| NCMDIS-TDEC | | | | |
| Total | 104,050 | 302,096 | 406,146 | — |
| Cancer | 0 | 0 | 0 | — |
| Noncancer | 104,050 | 302,096 | 406,146 | — |
| Asbestos-related noncancer | 81,528 | 231,178 | 312,706 | — |
| Nonasbestos-related & unknown | 22,522 | 70,918 | 93,440 | — |
| CMDIS-CDEC | | | | |
| Total | 18,812 | 106,439 | 125,251 | — |
| Cancer | 12,172 | 68,460 | 80,632 | — |
| Noncancer | 6,640 | 37,979 | 44,619 | — |
| Asbestos-related noncancer | 4,987 | 28,182 | 33,169 | — |
| Nonasbestos-related & unknown | 1,653 | 9,797 | 11,450 | — |
| MDIS-CDEC/TDEC | | | | |
| Total | 122,862 | 408,535 | 531,397 | 45.3 |
| Cancer | 12,172 | 68,460 | 80,632 | 5.8 |
| Noncancer | 110,690 | 340,075 | 450,765 | 55.7 |
| Asbestos-related noncancer | 86,515 | 259,360 | 345,875 | 45.8 |
| Nonasbestos-related & unknown | 24,175 | 80,715 | 104,890 | 100.8 |

Source: Stallard and Manton (1994, Table C).

mesothelioma. Consequently, we produced projections based on multiple diseases (MDIS criterion) where the disease-specific claim filing rates were based on the number of claims in 1990-1994 that mentioned the disease, regardless of severity ranking. This provided complete estimates for each disease. They could be used to determine settlement values if each disease had a fixed dollar value.

We considered three projections. First, to estimate how many diseases would be claimed by men at the time of first claim filing, we used the TDEC projection of the surviving exposed population. Applying the claim filing rates based on the MDIS criterion, we projected 504,150 disease occurrences in 1990-2049 and 381,274 in 1995-2049 (Table 9.6). Dividing by the 262,631 men projected to file in 1995-2049 produced an average of 1.452 diseases per first claim.

Second, to estimate how many disease/injuries would be claimed by men at the time of noncancer claim filing, we used the TDEC projection of the surviving exposed population. The claim filing rates based on the MDIS criterion were revised to include in their numerators only claims in which cancer was not mentioned (which defines the NCMDIS criterion). With these modified claim filing rates, we projected 406,146 disease/injuries in 1990-2049 and 302,096 in 1995-2049. Dividing by the 214,324 men projected to file noncancer claims in 1995-2049 produced an average of 1.410 disease/injuries per first noncancer claim.

Third, to estimate how many disease/injuries would be claimed by men at the time of cancer claim filing, we used the CDEC projection of the surviving exposed population. The claim filing rates based on the MDIS criterion were modified to reflect only claims in which cancer was mentioned (which defines the CMDIS criterion). Using these modified claim filing rates, we projected 125,251 disease/injuries in 1990-2049 and 106,439 in 1995-2049. Dividing by the 64,685 men projected to file cancer claims in 1995-2049 gave an average of 1.645 disease/injuries per first cancer claim.

Dividing the 37,979 noncancers in 1995-2049 in the CMDIS-CDEC projection by the 64,685 men projected (CDEC projection) to file cancer claims, we obtain an average of 0.587 noncancers per cancer claim. In 1990-1994, 41.8% of all male cancer claims mentioned a noncancer injury and 35.1% an asbestos-related noncancer injury. Among the 41.8%, the average number of noncancers was 1.38.

The second and third projections were combined to form the MDIS-CDEC/TDEC projection in Table 9.6 – producing 531,397 disease/injuries in 1990-2049 and 408,535 in 1995-2049. Dividing by the 279,009 claims projected in 1995-2049, we found an average of 1.464 disease/injuries per claim. This compared with the average of 1.452 disease/injuries per first claim in the first projection for 1995-2049.

Table 9.7 displays the MDIS-CDEC/TDEC projection by alleged disease/injury and occupation, by claim date in the period 1990-2049. This is comparable to the CDEC/TDEC projection in Table 8.19. Only mesothelioma remained unchanged; the unknown category was almost the same; and lung cancer increased modestly (+2.7%). Large increases were seen for other cancer, disputed asbestosis, pleural plaques/thickening, and nonasbestos-related disease. Colon/rectal cancer and asbestosis increased moderately (+11.2% and +12.8%, respectively).

The CDEC/TDEC projection assumed that the claim filing rates observed in 1990-1994 continued unchanged in 1995-2049 by age, occupation, TSFE, and disease (using either the SDIS or MDIS criteria). The MDIS results, however, allowed us to speculate how rates might change. For example, with data on multiple disease/injuries per claim, we generated claim filing rates for (1) cancer claims that mentioned a noncancer and (2) cancer claims that did not mention a noncancer. For joint mentions of cancer and noncancer, we might assume that (1) the noncancer was sufficient to motivate a potential

**Table 9.7: Analysis S2, Model 1 – Projected Number of Qualified Male Claims Against the Manville Trust by Multiple Alleged Disease/Injury or Occupation and Quinquennial Dates of Claim, 1990-2049, Combined Cancer and Total Decrement (MDIS-CDEC/TDEC) Model**

| | Date of Claim | | | | | | | | | | | | | | % Change |
|---|---|---|---|---|---|---|---|---|---|---|---|---|---|---|---|
| Item | 1990-1994 | 1995-1999 | 2000-2004 | 2005-2009 | 2010-2014 | 2015-2019 | 2020-2024 | 2025-2029 | 2030-2034 | 2035-2039 | 2040-2044 | 2045-2049 | Total | % | From Table 8.19 |
| Multiple Alleged Disease/Injury | | | | | | | | | | | | | | | |
| 1. Mesothelioma | 3,891 | 4,445 | 3,890 | 3,730 | 3,201 | 2,506 | 1,858 | 1,238 | 719 | 440 | 128 | 2 | 26,049 | 4.9 | 0.0 |
| 2. Lung cancer | 6,085 | 6,842 | 6,304 | 5,791 | 4,961 | 3,936 | 2,879 | 1,775 | 1,053 | 695 | 129 | 0 | 40,450 | 7.6 | 2.7 |
| 3. Colon/rectal cancer | 807 | 929 | 896 | 795 | 692 | 561 | 382 | 224 | 125 | 33 | 0 | 0 | 5,445 | 1.0 | 11.2 |
| 4. Other cancer | 1,389 | 1,524 | 1,409 | 1,272 | 1,072 | 799 | 579 | 353 | 171 | 92 | 27 | 0 | 8,688 | 1.6 | 48.1 |
| 5. Asbestosis | 38,816 | 34,189 | 25,616 | 19,467 | 14,162 | 9,526 | 5,782 | 3,029 | 1,575 | 707 | 208 | 1 | 153,077 | 28.8 | 12.8 |
| 6. Disputed asbestosis | 19,478 | 17,558 | 13,388 | 10,500 | 7,649 | 5,069 | 3,019 | 1,563 | 866 | 391 | 104 | 1 | 79,587 | 15.0 | 53.1 |
| 7. Pleural plaques/thickening | 28,221 | 25,399 | 19,173 | 14,655 | 10,805 | 7,076 | 4,037 | 2,070 | 1,093 | 530 | 151 | 1 | 113,211 | 21.3 | 128.7 |
| 8. Nonasbestos-related disease | 21,053 | 18,982 | 14,733 | 11,702 | 9,120 | 6,358 | 3,831 | 1,994 | 1,111 | 581 | 187 | 5 | 89,656 | 16.9 | 141.9 |
| 9. Unknown | 3,122 | 2,887 | 2,581 | 2,040 | 1,779 | 1,388 | 829 | 536 | 42 | 13 | 16 | 0 | 15,234 | 2.9 | 0.5 |
| Total | 122,862 | 112,754 | 87,989 | 69,954 | 53,440 | 37,220 | 23,197 | 12,783 | 6,755 | 3,483 | 950 | 9 | 531,397 | 100.0 | 45.3 |
| Cancer (1-4) | 12,172 | 13,740 | 12,499 | 11,590 | 9,926 | 7,803 | 5,698 | 3,591 | 2,068 | 1,260 | 284 | 2 | 80,631 | 15.2 | 5.8 |
| Noncancer (5-9) | 110,690 | 99,014 | 75,490 | 58,364 | 43,515 | 29,417 | 17,499 | 9,192 | 4,687 | 2,223 | 666 | 8 | 450,765 | 84.8 | 55.7 |
| Asbestos-related noncancer (5-7) | 86,515 | 77,145 | 58,176 | 44,622 | 32,616 | 21,671 | 12,839 | 6,662 | 3,534 | 1,628 | 463 | 3 | 345,875 | 65.1 | 45.8 |
| Nonasbestos-related & unkn. (8-9) | 24,175 | 21,869 | 17,314 | 13,742 | 10,899 | 7,746 | 4,660 | 2,530 | 1,153 | 594 | 203 | 5 | 104,890 | 19.7 | 100.8 |
| Occupation | | | | | | | | | | | | | | | |
| 1. A: Primary Manufacturing | 6,491 | 6,842 | 5,557 | 4,111 | 3,014 | 2,269 | 1,455 | 564 | 347 | 254 | 2 | 0 | 30,905 | 5.8 | 42.2 |
| 2. F: Secondary Manufacturing | 16,836 | 16,446 | 14,752 | 12,491 | 9,633 | 7,118 | 4,549 | 2,677 | 1,492 | 712 | 74 | 0 | 86,781 | 16.3 | 47.2 |
| 3. I: Insulation Work | 12,214 | 9,591 | 6,342 | 4,004 | 2,755 | 1,835 | 1,121 | 586 | 295 | 185 | 0 | 0 | 38,929 | 7.3 | 38.9 |
| 4. S: Shipbuilding and Repair | 18,834 | 16,565 | 8,655 | 6,046 | 3,737 | 2,143 | 1,147 | 541 | 234 | 77 | 17 | 6 | 58,002 | 10.9 | 45.2 |
| 5. R: Construction Trades | 7,090 | 7,411 | 7,575 | 6,229 | 4,393 | 3,060 | 1,976 | 1,184 | 691 | 359 | 70 | 4 | 40,043 | 7.5 | 35.9 |
| 6. Util/Trans/Chem/Longshore | 32,335 | 26,861 | 18,171 | 12,969 | 9,145 | 5,844 | 3,374 | 1,707 | 827 | 311 | 180 | 0 | 111,724 | 21.0 | 52.6 |
| 7. Military | 3,232 | 3,483 | 3,443 | 3,133 | 2,691 | 2,070 | 1,479 | 882 | 492 | 257 | 81 | 0 | 21,243 | 4.0 | 43.5 |
| 8. Other/Unknown | 25,830 | 25,555 | 23,495 | 20,971 | 18,072 | 12,880 | 8,095 | 4,641 | 2,377 | 1,329 | 525 | 0 | 143,769 | 27.1 | 44.5 |
| Total | 122,862 | 112,754 | 87,989 | 69,954 | 53,440 | 37,220 | 23,197 | 12,783 | 6,755 | 3,483 | 950 | 9 | 531,397 | 100.0 | 45.3 |

Source: Stallard and Manton (1994, Table H) and authors' calculations.

**Table 9.8: Analysis S2, Model 2 – Projections of Qualified Male Claims Against the Manville Trust, 1990-2049, with 50% of Cancer Claims Eliminated from the TDEC Projection for 1995-2049**

| Projection Type | Projection Interval 1990-1994 | Projection Interval 1995-2049 | Projection Interval 1990-2049 | % Change From Table 9.2 |
|---|---|---|---|---|
| TDEC | | | | |
| Total | 86,606 | 243,006 | 329,612 | -5.6 |
| Cancer | 11,506 | 24,808 | 36,314 | -39.3 |
| Noncancer | 75,100 | 218,198 | 293,298 | 1.3 |
| Asbestos-related noncancer | 62,856 | 177,786 | 240,642 | 1.5 |
| Nonasbestos-related & unknown | 12,244 | 40,412 | 52,656 | 0.8 |
| CDEC | | | | |
| Total | 86,606 | 337,717 | 424,323 | 0.0 |
| Cancer | 11,506 | 64,685 | 76,191 | 0.0 |
| Noncancer | 75,100 | 273,032 | 348,132 | 0.0 |
| Asbestos-related noncancer | 62,856 | 225,931 | 288,787 | 0.0 |
| Nonasbestos-related & unknown | 12,244 | 47,101 | 59,345 | 0.0 |
| CDEC/TDEC | | | | |
| Total | 86,606 | 282,883 | 369,489 | 1.1 |
| Cancer | 11,506 | 64,685 | 76,191 | 0.0 |
| Noncancer | 75,100 | 218,198 | 293,298 | 1.3 |
| Asbestos-related noncancer | 62,856 | 177,786 | 240,642 | 1.5 |
| Nonasbestos-related & unknown | 12,244 | 40,412 | 52,656 | 0.8 |

Source: Authors' calculations.

claimant to file a claim and (2) the subsequent occurrence of a cancer diagnosis resulted in a second-injury claim. Then, assuming claims with joint mentions of cancer and noncancer injuries represented a backlog of claims, we identified two ways that these types of claim might be handled in the future.

First, we assumed that the claim filing rates for joint mentions can be set to zero because they reflect a backlog that will not recur; that is, in future years, the cancers now included in joint-mention claims will be filed as second-injury cancer claims among men who previously filed first-injury noncancer claims. To obtain an estimate of this effect, we arbitrarily assumed that 50% of cancer claims could be eliminated from the claim filing rates in the TDEC projection for 1995-2049 due to their joint occurrence with noncancer disease/injuries. The 50% rate was selected after observing above that 41.8% of all male claims in 1990-1994 mentioned a noncancer injury – a rate that may be regarded as a lower bound for the proportion of future cancer claimants who could file a prior noncancer claim. The results are summarized in Table 9.8.

Under these assumptions, we projected 243,006 claims (−7.5%) in 1995-2049, 24,808 cancer (−48.6%), 177,786 asbestos-related noncancer (+2.0%), and 40,412 nonasbestos-related disease (+1.1%). The only important change was a decline in first-injury cancer claims. This decline was close to the assumed percentage of joint-mention claims (48.6% vs. 50.0%). Also, because total cancer claim rates were not changed (only their timing as first vs. second injuries), the CDEC projection was unaffected. However, the percentage of second injury cancer claims in 1995-2049 increased from 25.3% to 61.7%.

Despite the rather dramatic changes in the number of claims in which cancer was filed as the first disease/injury, the net impact of this scenario was small. Under the modified CDEC/TDEC projection, the total number of claims for 1990-2049 in Table 9.8 was 369,489 - just 1.1% higher than in Table 9.2.

Second, we assumed that filing rates for joint-mention claims remained at their 1990-1994 levels but such claims became temporally separated: The first-injury claim was filed as a noncancer disease/injury and the second-injury claim filed as a cancer disease/injury. This overrode the SDIS criterion and resulted in some proportion of first-injury cancer claims being reassigned as noncancer claims during the period 1995-2049. Because the total claim filing rate was unchanged by reassigning the component rates, the TDEC projection of the surviving exposed population was unchanged. The results are summarized in Table 9.9. Assuming 50% of cancer claims were reassigned, with 42% reassigned to asbestos-related noncancer and 8% to nonasbestos-related disease (i.e., proportional to the 1990-1994 data), we projected 262,631 claims (no change) in 1995-2049: 24,153 cancer (−50.0%), 194,626 asbestos-related noncancer (+11.6%), and 43,852 nonasbestos-related disease (+9.7%). These were upper bounds to the TDEC effect. Also, because total cancer claim rates were unchanged, the CDEC projection was unaffected. However, the percentage of second-injury cancer claims in 1995-2049 increased from 25.3% to 62.7%. In addition, the combined CDEC/TDEC total increased to 389,769 claims (+6.6%) in 1990-2049 and 303,163 claims (+8.7%) in 1995-2049. Given the extreme nature of the assumptions about cancer claim filing rates, the effects were not large. The assumptions underlying the baseline CDEC/TDEC projections in Chapter 8 seemed reasonable because results were insensitive to alternative treatments of claims with multiple diseases.

## 9.6 Analysis S3: CHR Smoothing

The claim filing rates in the second-stage calibration were recomputed without the CHR smoothing procedure in Section 8.6.2. Smoothing was introduced because the introduction of occupation into the model reduced the previous cell sizes roughly by a factor of 8. The results are summarized in Table 9.10. As with the baseline projection, the modified projection model exactly reproduced the age, occupation, and disease-specific male claim counts estimated for the period 1990-1994. Total claims in the CDEC/TDEC projection for 1990-2049 were 365,615 with smoothing (Table 9.2); without smoothing, the total was 387,110 (+5.9%).

**Table 9.9: Analysis S2, Model 3 – Projections of Qualified Male Claims Against the Manville Trust, 1990-2049, with 50% of Cancer Claims Allocated to Noncancers in the TDEC Projection for 1995-2049**

| Projection Type | Projection Interval 1990-1994 | 1995-2049 | 1990-2049 | % Change From Table 9.2 |
|---|---|---|---|---|
| TDEC | | | | |
| Total | 86,606 | 262,631 | 349,237 | 0.0 |
| Cancer | 11,506 | 24,153 | 35,659 | -40.4 |
| Noncancer | 75,100 | 238,478 | 313,578 | 8.3 |
| Asbestos-related noncancer | 62,856 | 194,626 | 257,482 | 8.6 |
| Nonasbestos-related & unknown | 12,244 | 43,852 | 56,096 | 7.4 |
| CDEC | | | | |
| Total | 86,606 | 337,717 | 424,323 | 0.0 |
| Cancer | 11,506 | 64,685 | 76,191 | 0.0 |
| Noncancer | 75,100 | 273,032 | 348,132 | 0.0 |
| Asbestos-related noncancer | 62,856 | 225,931 | 288,787 | 0.0 |
| Nonasbestos-related & unknown | 12,244 | 47,101 | 59,345 | 0.0 |
| CDEC/TDEC | | | | |
| Total | 86,606 | 303,163 | 389,769 | 6.6 |
| Cancer | 11,506 | 64,685 | 76,191 | 0.0 |
| Noncancer | 75,100 | 238,478 | 313,578 | 8.3 |
| Asbestos-related noncancer | 62,856 | 194,626 | 257,482 | 8.6 |
| Nonasbestos-related & unknown | 12,244 | 43,852 | 56,096 | 7.4 |

Source: Authors' calculations.

The increase was identifiable with certain classes (not shown in Table 9.10). Claims with unknown occupation increased by 10,087 (+10.1%). Claims with asbestos-related noncancer increased by 15,569 (+6.6%). Claims with mesothelioma (which was treated the same as the other diseases in the second-stage calibration) increased by 22 (from 26,049 to 26,071). All occupation categories except primary manufacturing exhibited increased claims. However, the increases for insulation workers (+1.1%) and military (+0.7%) were small. The largest increase was for unknown occupation (+10.1%). All DOFE periods except 1970-1974 exhibited increased claims. The relative increases were small for periods after 1950 (+6.7%, +5.4%, +5.2%, and +3.1%, respectively for 1950-1954, 1955-1959, 1960-1964, and 1965-1969). Prior to 1950, the relative increases were all above 13.5% (1940-1944), with the largest relative increase (36.7%) for 1925-1929. The largest post-WWII increase was 20.7% for 1945-1949.

**Table 9.10: Analysis S3, Model 1 – Projections of Qualified Male Claims Against the Manville Trust, 1990-2049, Without CHR Smoothing**

| Projection Type | Projection Interval | | | % Change From |
|---|---|---|---|---|
| | 1990-1994 | 1995-2049 | 1990-2049 | Table 9.2 |
| TDEC | | | | |
| Total | 86,606 | 281,887 | 368,493 | 5.5 |
| Cancer | 11,506 | 48,533 | 60,039 | 0.4 |
| Noncancer | 75,100 | 233,354 | 308,454 | 6.6 |
| Asbestos-related noncancer | 62,856 | 189,906 | 252,762 | 6.6 |
| Nonasbestos-related & unknown | 12,244 | 43,448 | 55,692 | 6.6 |
| CDEC | | | | |
| Total | 86,606 | 377,766 | 464,372 | 9.4 |
| Cancer | 11,506 | 67,150 | 78,656 | 3.2 |
| Noncancer | 75,100 | 310,616 | 385,716 | 10.8 |
| Asbestos-related noncancer | 62,856 | 258,240 | 321,096 | 11.2 |
| Nonasbestos-related & unknown | 12,244 | 52,376 | 64,620 | 8.9 |
| CDEC/TDEC | | | | |
| Total | 86,606 | 300,504 | 387,110 | 5.9 |
| Cancer | 11,506 | 67,150 | 78,656 | 3.2 |
| Noncancer | 75,100 | 233,354 | 308,454 | 6.6 |
| Asbestos-related noncancer | 62,856 | 189,906 | 252,762 | 6.6 |
| Nonasbestos-related & unknown | 12,244 | 43,448 | 55,692 | 6.6 |

Source: Authors' calculations.

## 9.7 Analysis S4: Exposure Smoothing

Step 4 of the first-stage baseline projection (Section 8.3.4) was recomputed without smoothing the estimates of the distribution of mesothelioma claims by age and date of first exposure (DOFE). In so doing, we replaced the exposure data in Table 8.3 with a corresponding table based on the distribution of mesothelioma claims stratified by age and TSFE as tabulated in step 3. This replacement was possible because the approximations assumed that tabulations based on TSFE and DOFE were interchangeable using the midpoint of the relevant TSFE or DOFE categories to convert between the two time measures. This replacement also allowed the normalization in step 5 to be simplified using TSFE in place of DOFE. The results are summarized in Table 9.11. As with the baseline projection, the modified projection model exactly reproduced the age, occupation, and disease specific male claim counts estimated for the period 1990-1994. With smoothing, the total number of claims in the CDEC/TDEC projection for 1990-2049 was 365,615 (Table 9.2); without exposure smoothing, it was 368,946 (+0.9%). The direction of change differed by type of disease: decreasing for cancer (−1.2%) and increasing for noncancer claims (1.5%). The magnitude of these differences was small: The smoothing of exposure had minor impact.

**Table 9.11: Analysis S4, Model 1 – Projections of Qualified Male Claims Against the Manville Trust, 1990-2049, Without Exposure Smoothing**

| Projection Type | Projection Interval | | | % Change From Table 9.2 |
|---|---|---|---|---|
| | 1990-1994 | 1995-2049 | 1990-2049 | |
| TDEC | | | | |
| Total | 86,578 | 265,721 | 352,299 | 0.9 |
| Cancer | 11,491 | 47,134 | 58,625 | -2.0 |
| Noncancer | 75,087 | 218,587 | 293,674 | 1.5 |
| Asbestos-related noncancer | 62,845 | 177,612 | 240,457 | 1.4 |
| Nonasbestos-related & unknown | 12,242 | 40,975 | 53,217 | 1.9 |
| CDEC | | | | |
| Total | 86,578 | 343,268 | 429,846 | 1.3 |
| Cancer | 11,491 | 63,781 | 75,272 | -1.2 |
| Noncancer | 75,087 | 279,487 | 354,574 | 1.9 |
| Asbestos-related noncancer | 62,845 | 231,022 | 293,867 | 1.8 |
| Nonasbestos-related & unknown | 12,242 | 48,465 | 60,707 | 2.3 |
| CDEC/TDEC | | | | |
| Total | 86,578 | 282,368 | 368,946 | 0.9 |
| Cancer | 11,491 | 63,781 | 75,272 | -1.2 |
| Noncancer | 75,087 | 218,587 | 293,674 | 1.5 |
| Asbestos-related noncancer | 62,845 | 177,612 | 240,457 | 1.4 |
| Nonasbestos-related & unknown | 12,242 | 40,975 | 53,217 | 1.9 |

Source: Authors' calculations.

## 9.8 Analysis S5: Weibull $k$ Parameter

In Section 7.5, we found that the IWE model was sensitive to changes in the $k$ parameter of the Weibull hazard function for mesothelioma mortality. We considered $k$ values of 3.2 and 4.2 for the no-latency form of the Peto et al. (1982) model. The estimate $k = 3.2$ was obtained from Peto et al.'s (1982) analysis of the North American insulation worker data. The estimate $k = 4.2$ was obtained from our reanalysis (Table 7.5) of data from the same North American insulation workers, extended to 20 years of follow-up. Our results indicated that Peto et al.'s (1982) estimate of $k$ was too low by 1 unit. However, although our estimate $k = 4.2$ was appropriate for modeling mesothelioma mortality among insulation workers, an estimate $k = 3.2$ was appropriate for use with the general population of asbestos-exposed workers, based on the same arguments used in obtaining the first two approximations in Section 8.4.1. The projections with $k = 4.2$ were 50.5% higher for 1990-2049 than those with $k = 3.2$. We noted that without more detailed information on occupation, it would be difficult to develop more accurate estimates of $k$.

The baseline model in Chapter 8 provided the necessary information on occupation. In Section 8.4.1, we described the estimation and specification of the OSHA (1983) form of the mesothelioma mortality model. In Table 8.7,

**Table 9.12: Analysis S5, Model 1 – Projections of Qualified Male Claims Against the Manville Trust, 1990-2049, with $k = 2$**

| Projection Type | Projection Interval 1990-1994 | 1995-2049 | 1990-2049 | % Change From Table 9.2 |
|---|---|---|---|---|
| TDEC | | | | |
| Total | 86,606 | 180,689 | 267,295 | -23.5 |
| Cancer | 11,506 | 34,215 | 45,721 | -23.6 |
| Noncancer | 75,100 | 146,474 | 221,574 | -23.4 |
| Asbestos-related noncancer | 62,856 | 120,802 | 183,658 | -22.6 |
| Nonasbestos-related & unknown | 12,244 | 25,672 | 37,916 | -27.4 |
| CDEC | | | | |
| Total | 86,606 | 211,408 | 298,014 | -29.8 |
| Cancer | 11,506 | 40,582 | 52,088 | -31.6 |
| Noncancer | 75,100 | 170,826 | 245,926 | -29.4 |
| Asbestos-related noncancer | 62,856 | 142,447 | 205,303 | -28.9 |
| Nonasbestos-related & unknown | 12,244 | 28,379 | 40,623 | -31.5 |
| CDEC/TDEC | | | | |
| Total | 86,606 | 187,056 | 273,662 | -25.2 |
| Cancer | 11,506 | 40,582 | 52,088 | -31.6 |
| Noncancer | 75,100 | 146,474 | 221,574 | -23.4 |
| Asbestos-related noncancer | 62,856 | 120,802 | 183,658 | -22.6 |
| Nonasbestos-related & unknown | 12,244 | 25,672 | 37,916 | -27.4 |

Source: Authors' calculations.

we showed that the 10-year latency form produced a better fit to the North American insulation worker data than the no-latency model and that Peto et al.'s (1982) estimate of $k$ for the 10-year latency form again appeared to be too low, this time by 0.8. Compared with the OSHA (1983) form of the 10-year latency model, Peto et al.'s (1982) estimate of $k$ was too low by 1 unit (2.0 vs. 3.0). Because our estimate of $k = 2.83$ fell between these two alternatives, it was of interest to evaluate the impact of setting $k = 2$ and $k = 3$. The results are summarized in Tables 9.12 and 9.13.

With $k = 2$ (−29.3%), the total number of claims in the CDEC/TDEC baseline projection for 1990-2049 decreased 25.2% from 365,615 to 273,662. With $k = 3$ (+6.0%), the total number increased 7.5% to 393,088. The reduction in cancer claims for $k = 2$ was 31.6% and the increase for $k = 3$ was 9.8%, with each relative change being larger than the corresponding relative change for noncancer claims (−23.4% and +6.9%, respectively).

Table 8.7 indicated that the best $k$ estimate was somewhere near 3, and likely closer to our estimate of 2.83. The chi-squared tests in Table 8.7 rejected the Peto et al. (1982) estimate of $k = 2$ but not the OSHA (1983) estimate of $k = 3$. The 7.5% difference in projected total claims between the baseline and OSHA (1983) models was relatively small. OSHA (1986, p. 22,639) viewed the

**Table 9.13: Analysis S5, Model 2 – Projections of Qualified Male Claims Against the Manville Trust, 1990-2049, with $k = 3$**

| Projection Type | Projection Interval 1990-1994 | 1995-2049 | 1990-2049 | % Change From Table 9.2 |
|---|---|---|---|---|
| TDEC | | | | |
| Total | 86,606 | 286,754 | 373,360 | 6.9 |
| Cancer | 11,506 | 52,438 | 63,944 | 6.9 |
| Noncancer | 75,100 | 234,316 | 309,416 | 6.9 |
| Asbestos-related noncancer | 62,856 | 189,969 | 252,825 | 6.6 |
| Nonasbestos-related & unknown | 12,244 | 44,347 | 56,591 | 8.3 |
| CDEC | | | | |
| Total | 86,606 | 376,566 | 463,172 | 9.2 |
| Cancer | 11,506 | 72,166 | 83,672 | 9.8 |
| Noncancer | 75,100 | 304,400 | 379,500 | 9.0 |
| Asbestos-related noncancer | 62,856 | 251,367 | 314,223 | 8.8 |
| Nonasbestos-related & unknown | 12,244 | 53,033 | 65,277 | 10.0 |
| CDEC/TDEC | | | | |
| Total | 86,606 | 306,482 | 393,088 | 7.5 |
| Cancer | 11,506 | 72,166 | 83,672 | 9.8 |
| Noncancer | 75,100 | 234,316 | 309,416 | 6.9 |
| Asbestos-related noncancer | 62,856 | 189,969 | 252,825 | 6.6 |
| Nonasbestos-related & unknown | 12,244 | 44,347 | 56,591 | 8.3 |

Source: Authors' calculations.

estimate of $k = 3$ as reasonable, especially given OSHA's mandate "to make assumptions which err on the side of overprotection of workers." We agree.

Thus, although there remained considerable sensitivity to the estimate of $k$, the value $k = 2.83$ used in our baseline projection was optimal for the 10-year latency form of the OSHA (1983) model, and it yielded claim projections only moderately lower than those obtained with $k = 3$. The use of Peto et al.'s (1982) estimate of $k = 2$ yielded claim projections that were reduced an additional 25.2%. We believe that this widely used estimate of $k$ was too low and produced biased projections.

## 9.9 Analysis S6: Relative Risks of Mesothelioma

Selikoff's (1981) estimates of relative risks in Tables 3.1 and 8.8 were presented without standard errors. We dealt with this in Section 3.3.3 by estimating the standard errors of comparable relative risk estimates reported by McDonald and McDonald (1980). We found that three of the four relative risk estimates reported by Selikoff (1981) were within one standard error of McDonald and McDonald's (1980) estimates. These standard errors were converted to coefficients of variation in the range 25-44%, with a mean of 33%. We took 33%

**Table 9.14: Analysis S6, Model 1 – Projections of Qualified Male Claims Against the Manville Trust, 1990-2049, with Relative Risks Reduced by 50%**

| Projection Type | Projection Interval | | | % Change From |
|---|---|---|---|---|
| | 1990-1994 | 1995-2049 | 1990-2049 | Table 9.2 |
| TDEC | | | | |
| Total | 86,606 | 301,370 | 387,976 | 11.1 |
| Cancer | 11,506 | 56,852 | 68,358 | 14.3 |
| Noncancer | 75,100 | 244,518 | 319,618 | 10.4 |
| Asbestos-related noncancer | 62,856 | 200,788 | 263,644 | 11.2 |
| Nonasbestos-related & unknown | 12,244 | 43,730 | 55,974 | 7.2 |
| CDEC | | | | |
| Total | 86,606 | 348,742 | 435,348 | 2.6 |
| Cancer | 11,506 | 67,642 | 79,148 | 3.9 |
| Noncancer | 75,100 | 281,100 | 356,200 | 2.3 |
| Asbestos-related noncancer | 62,856 | 233,296 | 296,152 | 2.6 |
| Nonasbestos-related & unknown | 12,244 | 47,804 | 60,048 | 1.2 |
| CDEC/TDEC | | | | |
| Total | 86,606 | 312,160 | 398,766 | 9.1 |
| Cancer | 11,506 | 67,642 | 79,148 | 3.9 |
| Noncancer | 75,100 | 244,518 | 319,618 | 10.4 |
| Asbestos-related noncancer | 62,856 | 200,788 | 263,644 | 11.2 |
| Nonasbestos-related & unknown | 12,244 | 43,730 | 55,974 | 7.2 |

Source: Authors' calculations.

as a reasonable approximation to the average coefficient of variation of the relative risks in Tables 3.1 and 8.8.

Another source of uncertainty in the baseline model related to the approximate correspondence between the Trust's occupation groups and Selikoff's groups (discussed in Section 8.3.2). Given the different methods of occupational coding in the two analyses, it was impossible to quantify the additional uncertainty introduced into the model by these differences. However, the resulting average coefficient of variation should be at least 33%, based on the estimate derived earlier.

Given this uncertainty, we evaluated the impact of increasing or decreasing each of the relative risks in Table 8.8 by 50%. The results are summarized in Tables 9.14 and 9.15.

With a 50% reduction in relative risks, the total number of claims in the CDEC/TDEC baseline projection for 1990-2049 increased from 365,615 to 398,766 (+9.1%). With 50% increase in relative risks, the total number decreased to 340,473 (−6.9%). The increase in cancer claims for reduced relative risks was 3.9% and the reduction for increased relative risks was 3.4%, with each relative change being smaller than the corresponding relative change for noncancer claims (+10.4% and –7.8%, respectively).

**Table 9.15: Analysis S6, Model 2 – Projections of Qualified Male Claims Against the Manville Trust, 1990-2049, with Relative Risks Increased by 50%**

| Projection Type | Projection Interval | | | % Change From |
|---|---|---|---|---|
| | 1990-1994 | 1995-2049 | 1990-2049 | Table 9.2 |
| TDEC | | | | |
| Total | 86,606 | 234,040 | 320,646 | -8.2 |
| Cancer | 11,506 | 42,303 | 53,809 | -10.0 |
| Noncancer | 75,100 | 191,737 | 266,837 | -7.8 |
| Asbestos-related noncancer | 62,856 | 154,882 | 217,738 | -8.2 |
| Nonasbestos-related & unknown | 12,244 | 36,855 | 49,099 | -6.0 |
| CDEC | | | | |
| Total | 86,606 | 328,031 | 414,637 | -2.3 |
| Cancer | 11,506 | 62,130 | 73,636 | -3.4 |
| Noncancer | 75,100 | 265,901 | 341,001 | -2.0 |
| Asbestos-related noncancer | 62,856 | 219,445 | 282,301 | -2.2 |
| Nonasbestos-related & unknown | 12,244 | 46,456 | 58,700 | -1.1 |
| CDEC/TDEC | | | | |
| Total | 86,606 | 253,867 | 340,473 | -6.9 |
| Cancer | 11,506 | 62,130 | 73,636 | -3.4 |
| Noncancer | 75,100 | 191,737 | 266,837 | -7.8 |
| Asbestos-related noncancer | 62,856 | 154,882 | 217,738 | -8.2 |
| Nonasbestos-related & unknown | 12,244 | 36,855 | 49,099 | -6.0 |

Source: Authors' calculations.

Given the large sizes of the change in the assumed relative risks and the relatively smaller sizes of the change in the projected number of claims, we concluded that uncertainty in the relative risk estimates was not a major source of uncertainty in the projections. On the other hand, the changes in the projections were large enough to falsify the assumption of the IWE models in Chapters 4-7 that proportional changes in the relative risk of mesothelioma would cancel out in the first-stage calibration, leaving the second-stage projections unaltered.

## 9.10 Analysis S7: Duration of Exposure

Selikoff's (1981) estimates of average durations of exposures in Table 8.8 were presented without standard errors. These estimates were derived using the workforce turnover model described in Section 3.3.2, so the variability in the average durations depended on the variability of the estimates of the number of new entrants in asbestos-exposed occupations. Selikoff's (1981) estimates of new entrants in 1950-1969 were 3.0 times those of similar estimates prepared by Nicholson et al. (1981a), implying average durations two-thirds lower. Even allowing for the improvements in Selikoff's (1981) methods, these differences

**Table 9.16: Analysis S7, Model 1 – Projections of Qualified Male Claims Against the Manville Trust, 1990-2049, with Average Durations of Exposure Reduced by 50%**

| Projection Type | Projection Interval | | | % Change From Table 9.2 |
|---|---|---|---|---|
| | 1990-1994 | 1995-2049 | 1990-2049 | |
| TDEC | | | | |
| Total | 86606 | 271222 | 357828 | 2.5 |
| Cancer | 11506 | 51508 | 63014 | 5.4 |
| Noncancer | 75100 | 219714 | 294814 | 1.9 |
| Asbestos-related noncancer | 62856 | 180677 | 243533 | 2.7 |
| Nonasbestos-related & unknown | 12244 | 39037 | 51281 | -1.8 |
| CDEC | | | | |
| Total | 86606 | 316981 | 403587 | -4.9 |
| Cancer | 11506 | 61764 | 73270 | -3.8 |
| Noncancer | 75100 | 255217 | 330317 | -5.1 |
| Asbestos-related noncancer | 62856 | 212203 | 275059 | -4.8 |
| Nonasbestos-related & unknown | 12244 | 43014 | 55258 | -6.9 |
| CDEC/TDEC | | | | |
| Total | 86,606 | 281,478 | 368,084 | 0.7 |
| Cancer | 11,506 | 61,764 | 73,270 | -3.8 |
| Noncancer | 75,100 | 219,714 | 294,814 | 1.9 |
| Asbestos-related noncancer | 62,856 | 180,677 | 243,533 | 2.7 |
| Nonasbestos-related & unknown | 12,244 | 39,037 | 51,281 | -1.8 |

Source: Authors' calculations.

indicated that the variability of the estimates of duration of exposure must be large. In addition, given that Selikoff (1981) and we used different methods of coding occupation, the estimates in Table 8.8 were at best only approximations to the durations experienced by workers exposed to Manville asbestos.

The impact of increasing or decreasing each of the average duration estimates in Table 8.8 by 50% is summarized in Tables 9.16 and 9.17.

With the 50% reduction in average durations of exposure, the total number of claims in the CDEC/TDEC baseline projection for 1990-2049 increased from 365,615 to 368,084 (+0.7%). With the 50% increase in average durations of exposure, the total number also increased, to 372,009 (+1.7%). For cancer, however, the reduction in average durations of exposure led to a corresponding 3.8% reduction in claims, whereas the increase in average durations of exposure led to a 4.9% increase. For noncancer claims, both changes led to small increases in claims (+1.9% and +0.9%, respectively).

These results were consistent with the comparisons of Selikoff (1981) and Nicholson et al. (1981a) in Table 3.4 in which the shorter durations implicit in Selikoff's (1981) model resulted in a 0.8% decrease in the projected number of cancer deaths for 1979-1999.

**Table 9.17: Analysis S7, Model 2 – Projections of Qualified Male Claims Against the Manville Trust, 1990-2049, with Average Durations of Exposure Increased by 50%**

| Projection Type | Projection Interval | | | % Change From |
|---|---|---|---|---|
| | 1990-1994 | 1995-2049 | 1990-2049 | Table 9.2 |
| TDEC | | | | |
| Total | 86,606 | 264,412 | 351,018 | 0.5 |
| Cancer | 11,506 | 47,452 | 58,958 | -1.4 |
| Noncancer | 75,100 | 216,960 | 292,060 | 0.9 |
| Asbestos-related noncancer | 62,856 | 174,906 | 237,762 | 0.2 |
| Nonasbestos-related & unknown | 12,244 | 42,054 | 54,298 | 4.0 |
| CDEC | | | | |
| Total | 86,606 | 362,034 | 448,640 | 5.7 |
| Cancer | 11,506 | 68,443 | 79,949 | 4.9 |
| Noncancer | 75,100 | 293,591 | 368,691 | 5.9 |
| Asbestos-related noncancer | 62,856 | 241,632 | 304,488 | 5.4 |
| Nonasbestos-related & unknown | 12,244 | 51,959 | 64,203 | 8.2 |
| CDEC/TDEC | | | | |
| Total | 86,606 | 285,403 | 372,009 | 1.7 |
| Cancer | 11,506 | 68,443 | 79,949 | 4.9 |
| Noncancer | 75,100 | 216,960 | 292,060 | 0.9 |
| Asbestos-related noncancer | 62,856 | 174,906 | 237,762 | 0.2 |
| Nonasbestos-related & unknown | 12,244 | 42,054 | 54,298 | 4.0 |

Source: Authors' calculations.

As large changes in the average durations of exposure led to small changes in the projected number of claims, we concluded that uncertainty in the average durations was a minor source of uncertainty in the projections. The impact was substantially less than that of comparable changes in the relative risks in Section 9.9. To understand why, consider the impact of duration in the three approximations shown in Section 8.4.1. In the first approximation, duration behaved like relative risk; in the second, duration affected both relative risk and time, but in opposite ways; and in the third, duration had no effect. The pattern of changes in Tables 9.16 and 9.17 were consistent with the second and third approximations.

## 9.11 Overall Sensitivity: Analyses S1-S7

Comparisons of sensitivity analyses S1-S7 yielded additional insight into the uncertainty of the baseline model. Table 9.18 compares the projected total number of claims for 1990-2049, sorted in ascending size within each individual sensitivity analysis. The totals ranged from 273,662 to 531,397 claims with a median of 369,218 claims and a relative range of –25.2% to +45.3%.

**Table 9.18: Comparisons of Alternative Projections of the Total Number of Qualified Male Claims Against the Manville Trust, 1990-2049, with Models Sorted by Increasing Impact Within Senitivity Analysis Groupings**

| Analysis | Model | # | Total Claims 1990-2049 | Difference (%) of Total Claims in Model from Total Claims in Baseline Analysis |
|---|---|---|---|---|
| Baseline S0 | | 0 | 365,615 | 0.0 |
| S1 | 2 | 2 | 365,067 | -0.1 |
| | 1 | 1 | 366,461 | 0.2 |
| S2 | 2 | 4 | 369,489 | 1.1 |
| | 3 | 5 | 389,769 | 6.6 |
| | 1 | 3 | 531,397 | 45.3 |
| S3 | 1 | 6 | 387,110 | 5.9 |
| S4 | 1 | 7 | 368,946 | 0.9 |
| S5 | 1 | 8 | 273,662 | -25.2 |
| | 2 | 9 | 393,088 | 7.5 |
| S6 | 2 | 11 | 340,473 | -6.9 |
| | 1 | 10 | 398,766 | 9.1 |
| S7 | 1 | 12 | 368,084 | 0.7 |
| | 2 | 13 | 372,009 | 1.7 |
| Median S1–S7 | | 14 | 369,218 | 1.0 |

Source: Authors' calculations.

The lowest projection was analysis S5, Model 1, with $k = 2$ – the value reported by Peto et al. (1982) for the 10-year latency form of the Weibull model. This projection was substantially lower than any other projection and was based on a $k$ value that we judged to be implausibly low.

The highest projection was analysis S2, Model 1, which was based on summary totals of the number of mentions of each disease, with multiple mentions per claim counted as separate disease occurrences in the tabulation of the Manville Trust calibration data for 1990-1994. This projection was substantially higher than any other projection and was believed to be an upper bound to the number of claims that would be filed in future years under the assumption that each occurrence of an asbestos-related disease would lead to a distinct claim.

The remaining 11 projections fell within 9% of the baseline projection. These projections included reasonable variations in conversion from alleged to validated disease, smoothing procedures, and estimated parameter values, including the Weibull shape parameter $k$, the relative risks, and the average durations of exposure to asbestos. These factors were the most critical parts of the model. Assessing their influence on the projections was important to the evaluation of the performance of the model and to the acceptance of the model structure underlying the baseline projections in Chapter 8. The low degree of sensitivity associated with these factors suggested that the constraint on the first-stage calibration to exactly reproduce the total number of claims estimated for 1990-1994 successfully reduced the overall variability of the projections to a modest level.

Figure 9.1 displays the distribution of total claims by quinquennium for the full set of models. Except for the above-noted two extreme models, the remaining 11 projections were tightly clustered from the beginning to the end of the projection. Figure 9.2 shows the cumulative number of claims at each quinquennium according to the same projections. Again, the clustering of the 11 central projections is apparent.

The sensitivity analyses exhibited variability in the mix of cancer and noncancer claims, even where the total number of claims was unchanged from the baseline model. Figures 9.3 and 9.4 display the quinquennial and cumulative distributions, respectively, of cancer claims for the 13 models. Figures 9.5 and 9.6 display the corresponding results for noncancer claims.

Except for analyses S1, Model 1 (validated diseases based on Table 6.1), and S5, Model 1 ($k = 2$), all of the cancer projections, including the multiple mentions projection in analysis S2, Model 1, were clustered relatively tightly. From 1990-1994 to 2005-2009, analysis S1, Model 1, produced the highest numbers of cancer claims. Beginning in 2010-2014, analysis S5, Model 2 ($k = 3$), produced the highest quinquennial numbers of cancer claims and, beginning in 2015-2019, the second highest cumulative numbers of cancer claims. These results indicated that the cancer projections were more stable than the projections of total claims.

In contrast, the extremes of the noncancer projections exhibited greater variability, primarily due to the multiple-mentions projection in analysis S2, Model 1. The rate of decrease of the noncancer projections was more rapid than that of the cancer projections. For example, the median cancer projection in 2020-2024 was 46.8% of the size of the median in 1990-1994, whereas the corresponding noncancer ratio was 13.6%.

## 9.12 Conclusions

The Rule 706 Panel concluded that the structure and assumptions of the baseline hybrid model in Chapter 8 improved the structure and assumptions of the baseline model in Chapter 6.

The high level of uncertainty in Chapters 6 and 7 about how best to approximate the parameter $k$ in applications to asbestos-exposed workers was

Figure 9.1: Projections of Total Claims – Baseline and 13 Alternatives (Source: Authors' Calculations)

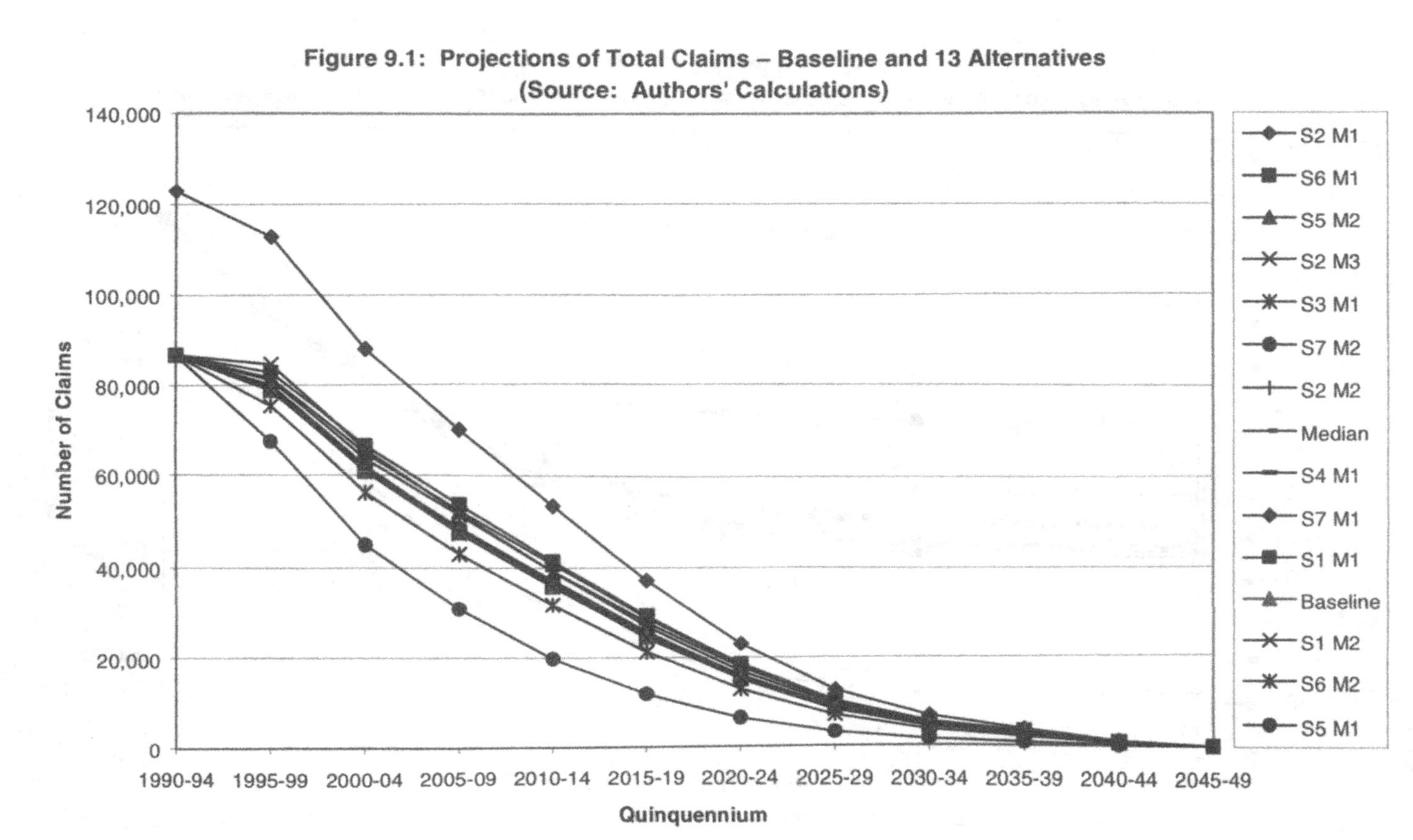

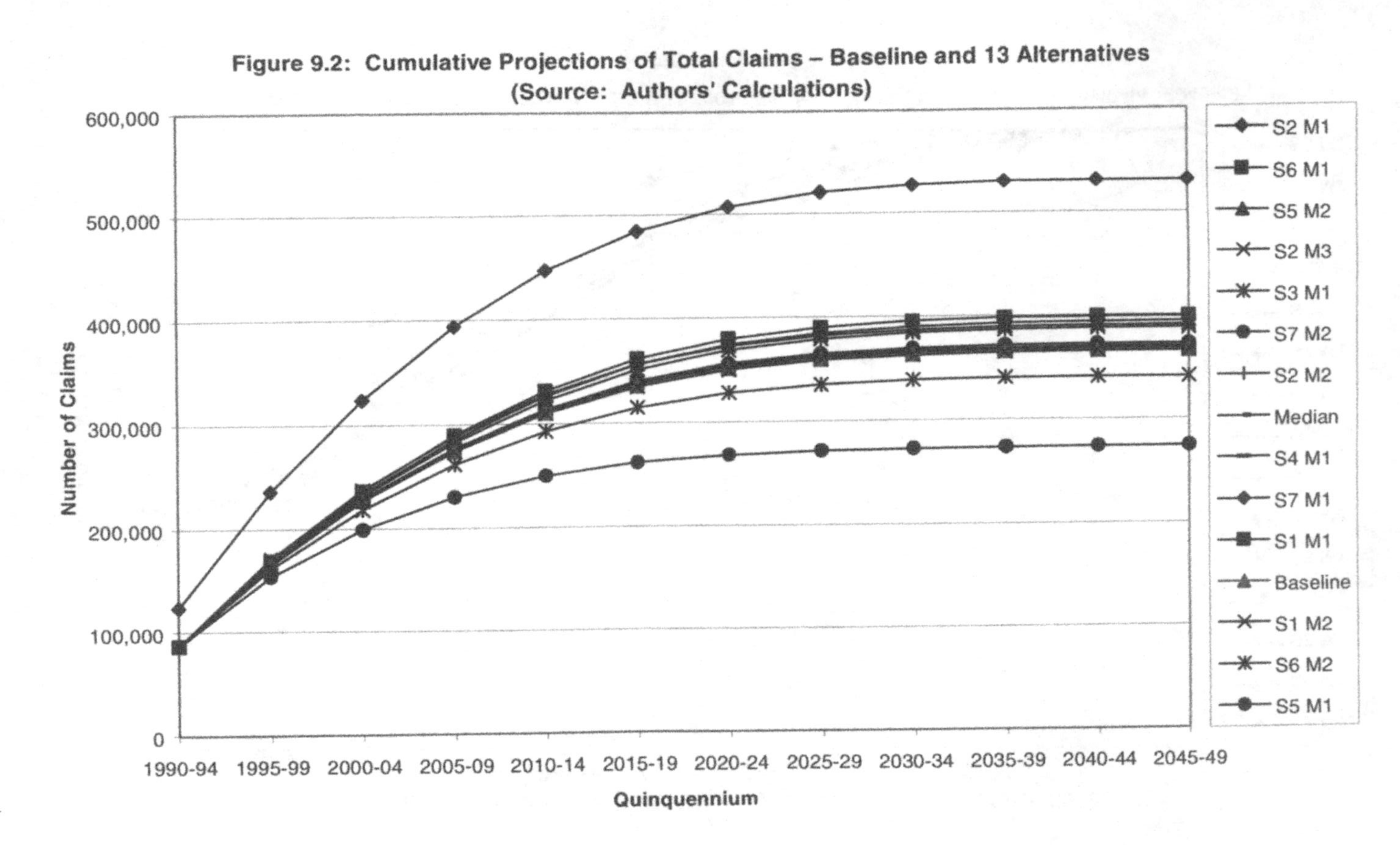


Figure 9.2: Cumulative Projections of Total Claims – Baseline and 13 Alternatives (Source: Authors' Calculations)

Figure 9.3: Projections of Cancer Claims – Baseline and 13 Alternatives (Source: Authors' Calculations)

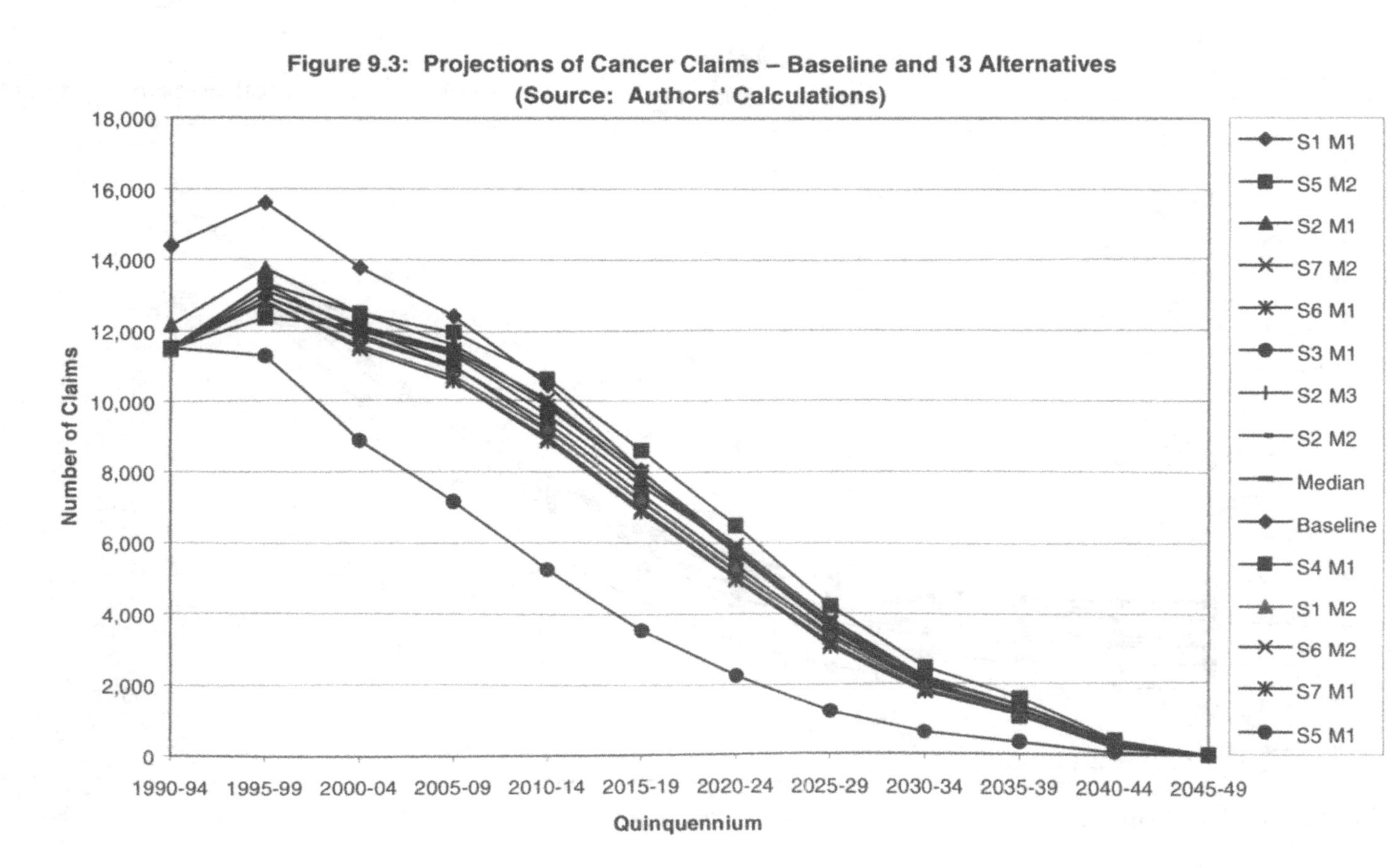

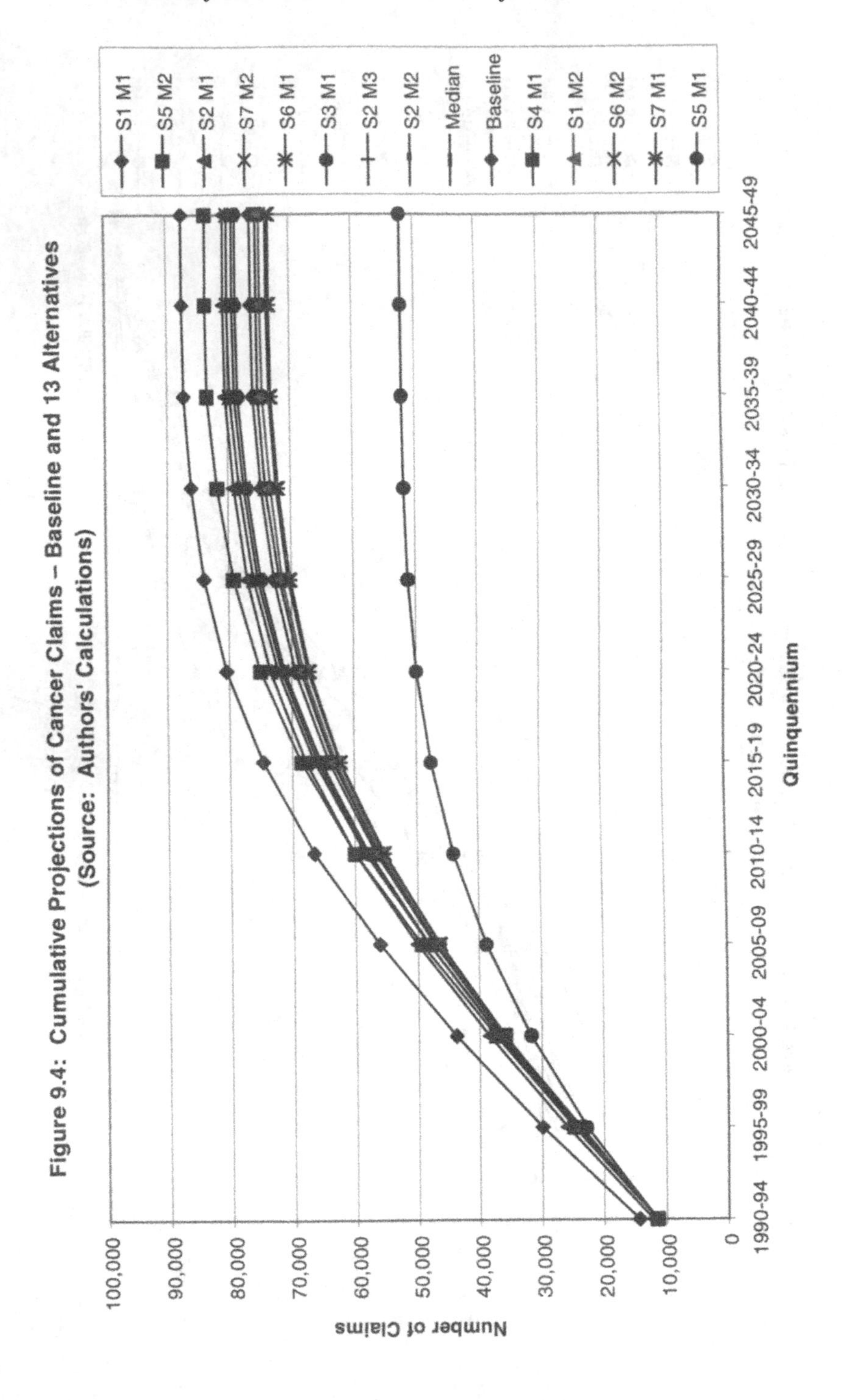


Figure 9.4: Cumulative Projections of Cancer Claims – Baseline and 13 Alternatives (Source: Authors' Calculations)

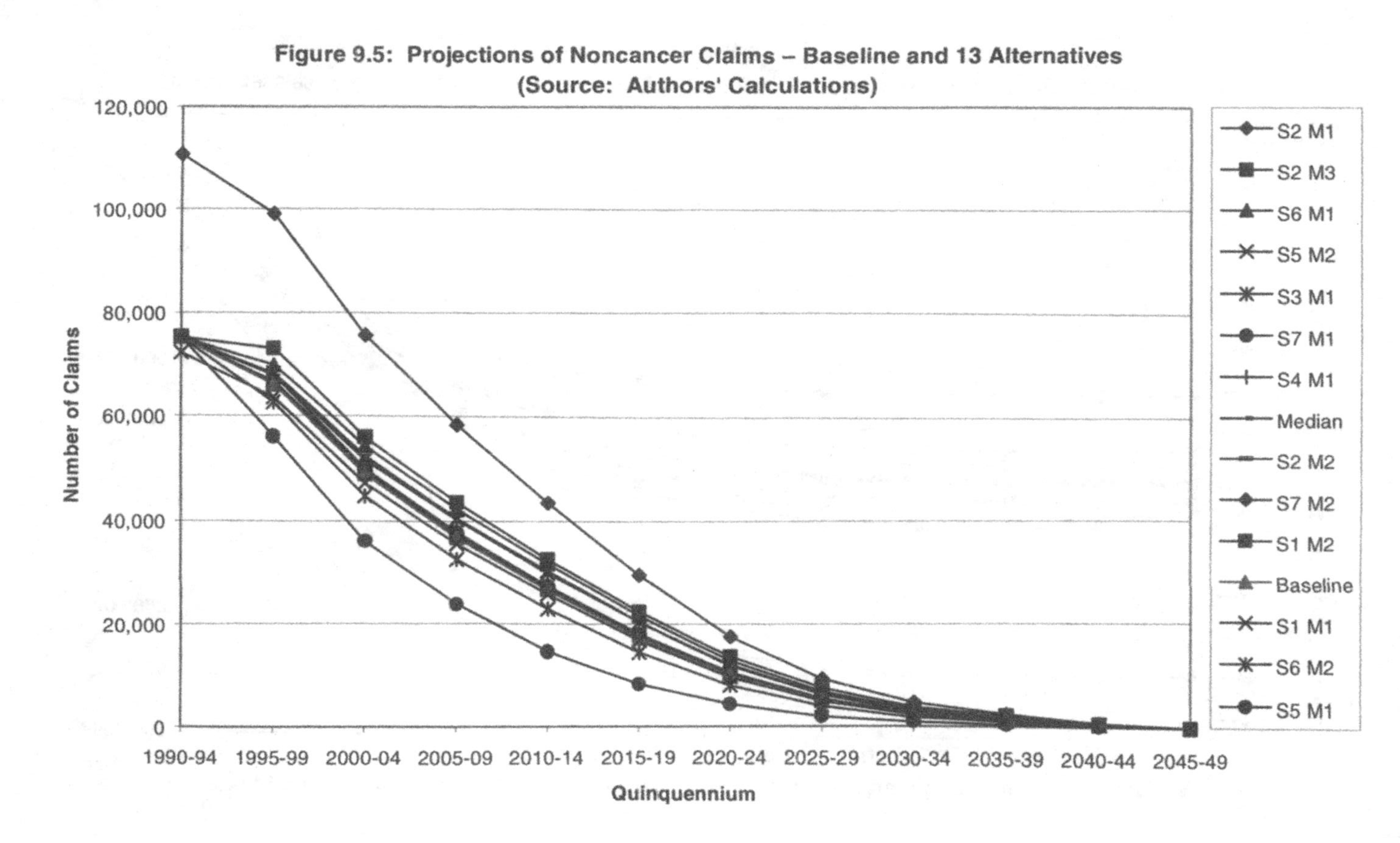


Figure 9.5: Projections of Noncancer Claims – Baseline and 13 Alternatives (Source: Authors' Calculations)

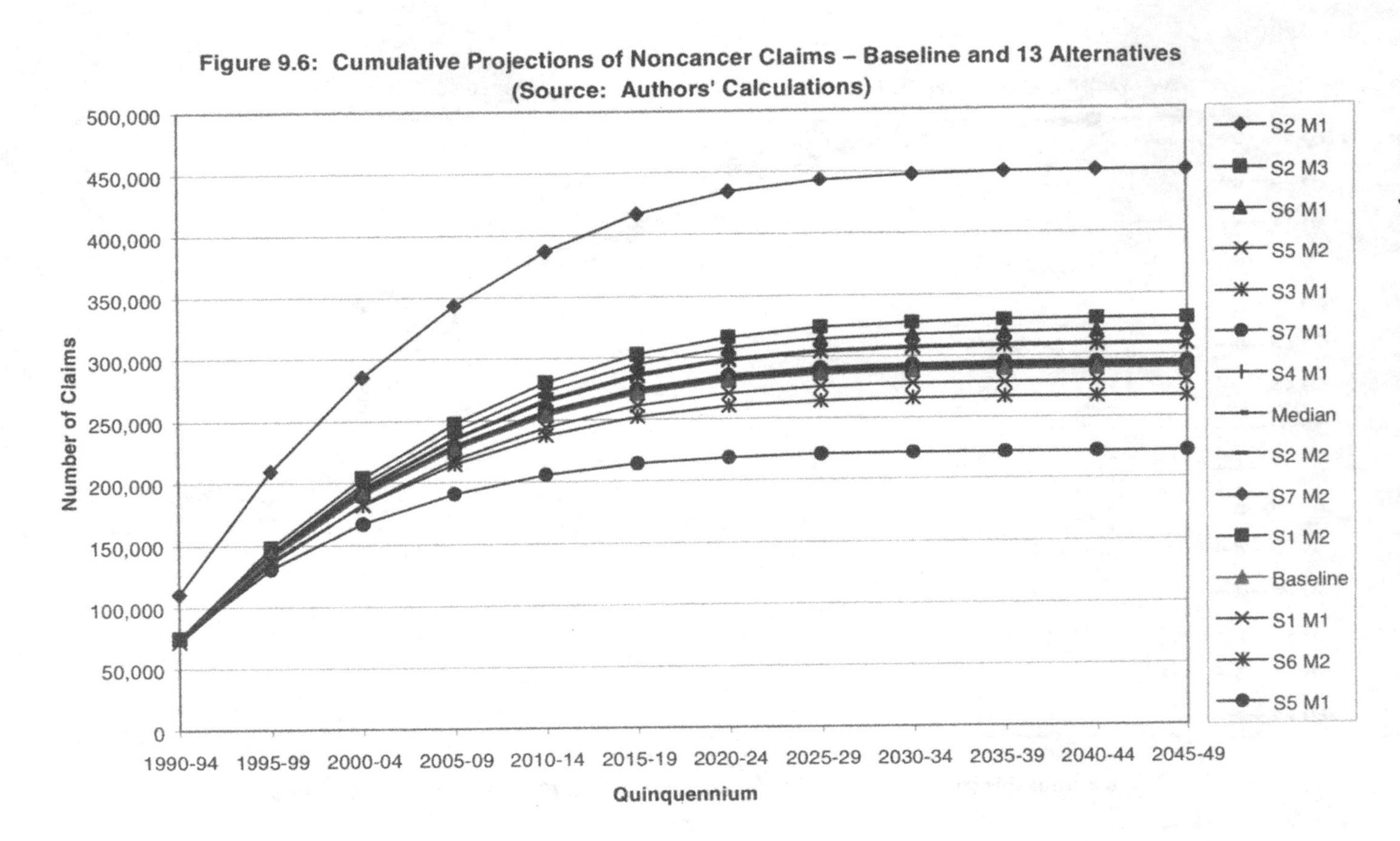


Figure 9.6: Cumulative Projections of Noncancer Claims – Baseline and 13 Alternatives (Source: Authors' Calculations)

largely resolved by considering occupation in the model in Chapter 8. This change yielded the largest impact of the variations tested in Chapter 7. The sensitivity to the value of $k$ continued in the analyses conducted in this chapter. The difference was that the reductions in the value of $k$ by up to 1 unit depending on the average durations of exposure over the various occupations was automatically handled by the OSHA (1983) form of the mesothelioma incidence function. The first-stage calibration based on the mesothelioma incidence rates was more accurate. In addition, it yielded estimates of the actual number of real workers exposed to asbestos at various times in the past, rather than the numbers of hypothetical IWEs in the prior models.

To implement this model, we required estimates of the relative risks of mesothelioma and average durations of exposure by occupation. Estimates of these parameters were provided by Selikoff (1981) in a form that roughly corresponded to the occupations coded by the Manville Trust. The analyses in this chapter indicated that the sensitivity to these parameter estimates was relatively low and was especially low for average duration of exposure.

We tested changes in relative risks of 50%. Although this range may seem large, in Section 2.4 we saw that the EPA (1986) determined that the relative risks varied by a factor of 5, with a 95% confidence interval ranging from 0.2 to 5.0. In addition, in Table 8.7 we found that the proportionality constant $b$ in the Weibull hazard model had a relative standard error of 5% for fixed values of $k$, which increased to 48% for variable values of $k$. Both of these results indicated that the potential variability of the relative risks may be substantially larger than the 50% considered here.

One factor considered in Chapter 7 (analysis S9), but not here, was the impact of decreases in the propensity to sue for asbestos-related diseases. The motivation for the analysis in Chapter 7 was the possibility that the claim filing rates would decline as the assets of defendants were depleted. That analysis indicated that the cumulative total number of claims could be 49% lower than the baseline projection. Counterbalancing this effect was the finding in this chapter (analysis S2, Model 1) that the cumulative total number of claims could be 45% higher than the baseline projection if each individual disease generated a claim.

Under this hypothetical modification of the claim process, the number of "claims" in the initial calibration period 1990-1994 was 42% higher than the baseline number. Thus, this analysis also illustrated the impact of an immediate 42% increase in the claim filing rates in the first-stage calibration. All of the other analyses were based on the assumption that the claim filing rates estimated for 1990-1994 would continue to apply post-settlement. Although there was concern about the validity of this assumption, no data were available to determine the size or direction of any change that might occur, and there were arguments made for changes in both directions. To deal with this issue, we examined the monthly claim data in Figures 6.2-6.6 to assess the impact of the judicial stay issued in July 1990 suspending all Trust payments except in a limited number of special cases. This stay was cited by those arguing for consideration of additional adjustments in the projection as an important example of the types of change that could occur. Our evaluation of the monthly claim data indicated that there was no obvious decrease in the claim filing rates at that time; if anything, there was a small increase. The argument that the legal proceedings suppressed the claim filing rates in the calibration period was not compelling. The Rule 706 Panel decided to make no specific adjustment to the projections on this account.

On the other hand, the Rule 706 Panel concluded that the previously established uncertainty limits of 50% should be maintained to account for the various sources of uncertainty described in this chapter and in Chapter 7. The Panel believed it was reasonable to expect that the actual deviations of future claims from the baseline projection could be substantially larger than shown in the central projections in Figures 9.1-9.6, possibly extending beyond the levels indicated by the two most extreme projections.

We shall evaluate the performance of the Rule 706 Panel's models in comparison with recent claims experience in Chapter 10.

# 10

# Conclusions and Implications

## 10.1 Introduction

The official work of the Rule 706 Panel ended with the presentation of the analyses reported in Chapters 6-9 as oral and written court testimony in March and May 1994. At that time, we expected a long trial during which Panel members would be subjected to intense direct questioning and cross-examination. Instead, a negotiated agreement was reached in July 1994. The litigation was settled in December 1994 (Weinstein, 1994), final approval of a new Trust Distribution Process (1995 TDP) was given by Judge Weinstein on January 19, 1995, and the 1995 TDP went into effect on February 21, 1995 and remained in effect until June 19, 2001, at which time the pro rata share (payment rate) was cut in half. A revised Trust Distribution Process (2002 TDP) was implemented on August 28, 2002, effective with claims filed on or after January 2, 2003. The Trust anticipated raising the pro rata share in mid-2003 after the 2002 TDP had been in effect long enough to assess its impact on the filing of claims.

The charge to the Rule 706 Panel was to project the number of claims for asbestos-related injuries against the Manville Trust by disease category, over time, by age of claimant, for as long as there were asbestos-related injuries due to Manville asbestos. Because the payment schedules and pro rata payment rates were contested, the Panel did not report monetary values of the projected claims using the settlement amounts in Table 6.1 or any other payment schedules. Nonetheless, in his oral testimony on March 15, 1994, Joel E. Cohen described how one *could* use the projections to compute present values of current and future liabilities for the purpose of setting pro rata payment rates. This method was effectively implemented by the Manville Trust using the Rule 706 Panel projections and the Trust's own projections in the witness statement of Mark E. Lederer, Manville Trust CFO, on May 5, 1994 (Lederer, 1994).

Lederer (1994) combined a range of claim projections with a range of economic assumptions about the cash flows that could be generated from the

Trust's assets. He obtained sets of pro rata shares ranging from 9.1% to 20.2%. The Rule 706 Panel's baseline model led to pro rata shares in the range 13.4-15.0%. Lederer (1994) recommended that the pro rata share for the 1995 TDP be set at 10.0% to account for uncertainty in the combined projections and the asymmetry of the Trust's future ability to compensate for error in the initial pro rata share value. If the value were set too high, the Trust would be unable to recapture the excess funds that had been paid out. If the value were set too low, the Trust would be able to supplement payments to Trust beneficiaries with liquidated claims who had been underpaid. Both the 1995 and 2002 TDPs required that the Trust reestimate at least every 3 years the values of its assets and liabilities to determine if a revision of the pro rata share value was warranted.

The Trust retained the initial 10% pro rata share for the first six years of its renewed operations, indicating that the initial projections were reasonably satisfactory. However, the pro rata share was cut in half in 2001. The 2002 TDP contained substantial revisions to the scheduled payment values. These changes suggested that the claim filing experience was developing in ways that were previously unanticipated.

In this chapter, we use the Rule 706 Panel's baseline hybrid projections to evaluate the Trust's claim filing experience for 1995-1999. Because the Panel's projections provided a set of expectations for the detailed characteristics of the claim filing process, the deviations from those expectations were analyzed for systematic patterns. These analyses provided further insight into the modeling and forecasting processes, given that 5 years were sufficient for the vagaries of the claim filing and Trust payment processes to have emerged.

We also compared the Trust's claim filing experience for 1995-1999 with the corresponding values from the extreme upper and lower projections in Chapters 7 and 9. These comparisons indicated that the deviations of actual from projected values were larger than expected under the uncertainty analyses described in these chapters, providing further evidence that the claim filing experience was developing in ways that were previously unanticipated.

We began by analyzing the differences between the original claim data provided to us in 1992 and the updated data on claim filings through the end of 2000. The most significant finding was that a substantial reclassification of unknown diseases resulted in a 27.8% increase in cancer claims and a 4.0% decrease in noncancer claims, compared to the 1990-1992 calibration data used in Chapters 6-9 (see Table 10.3).

The actual numbers of claims in the periods 1990-1994 and 1995-1999 exceeded the projected numbers by 1.2% and 103.6%, respectively. Cancer claims exceeded projections by 13.1% and 45.1%. Noncancer claims differed from projections by −0.6% and 115.1%. Given the substantial reclassification of disease claims in the calibration period, we adjusted the comparisons for 1995-1999 to remove the effects of the observed differences in 1990-1994. These adjustments indicated that there was an excess of 28.3% cancer and 116.3%

noncancer claims in 1995-1999 over what might be projected using the updated data for 1990-1994 (see Table 10.4).

It was known for several years that the total claim filings substantially exceeded the Rule 706 Panel's baseline projections. We now report our understanding of the reasons for these discrepancies. Two processes govern the filing of claims: (1) a stable, predictable biological process, interacting with (2) a complex and evolving litigation process that links the diagnosis of an asbestos-related disease to the decision to file a legal claim against one or more asbestos defendants.

Evidence that the biological processes underlying the mesothelioma and lung cancer rates were stable was provided by Surveillance, Epidemiology, and End Results Program (SEER) data 1973-1999, and national mortality data for 1999.

Evidence for the changing nature of the litigation was provided by the accelerated pace of bankruptcy filings (Figure 10.1). Following a 4-year period with only two bankruptcies, 1998 marked the start of a new round of bankruptcies. By year-end 1999, 6 more companies had filed for bankruptcy. By year-end 2002, another 28 companies had filed (Biggs, 2003). In less than 5 years, the number of bankruptcies had more than doubled (increasing from 31 to 65 companies).

The observation of relatively modest adjusted increases for cancer claims in the face of vastly increasing overall claims (28.3% vs. 116.3%) supported our model of distinct processes governing the biology and the litigation.

We conclude the chapter by discussing the use of these and similar projections in economic and actuarial cost calculations. We indicate how our forecasting model for the Manville Trust might be applied to other asbestos product liability cases. We briefly consider how these methods and results might be used in developing global settlements for asbestos-related injuries, in the context of continuing changes in litigation.

## 10.2 Data

The Manville Trust continued to maintain data on the claim settlement process and prepared an extract of all qualified claims filed against the Trust from the start of operations on November 28, 1988, through December 21, 2000. An initial extract of these data, prepared on December 24, 2000, was updated on February 11, 2001.

Summary counts of all claims, in total and by gender, for each year and each disease, show substantial variation from year to year (e.g., 1996 and 2000), with a generally upward trend from the early 1990s onward (Table 10.1).

When the numbers of claims were plotted by year and disease (Figure 10.2), most claims were for asbestosis and pleural plaques. To discern the cancer claims, Figure 10.3 displays the numbers of claims by year for the four

**Figure 10.1: Cumulative Distribution of Bankruptcies Among 63 Asbestos Defendants, 1976-2002 (Source: Data from Biggs, 2003, Ref. List 2 – Two Bankruptcies Without Dates Omitted)**

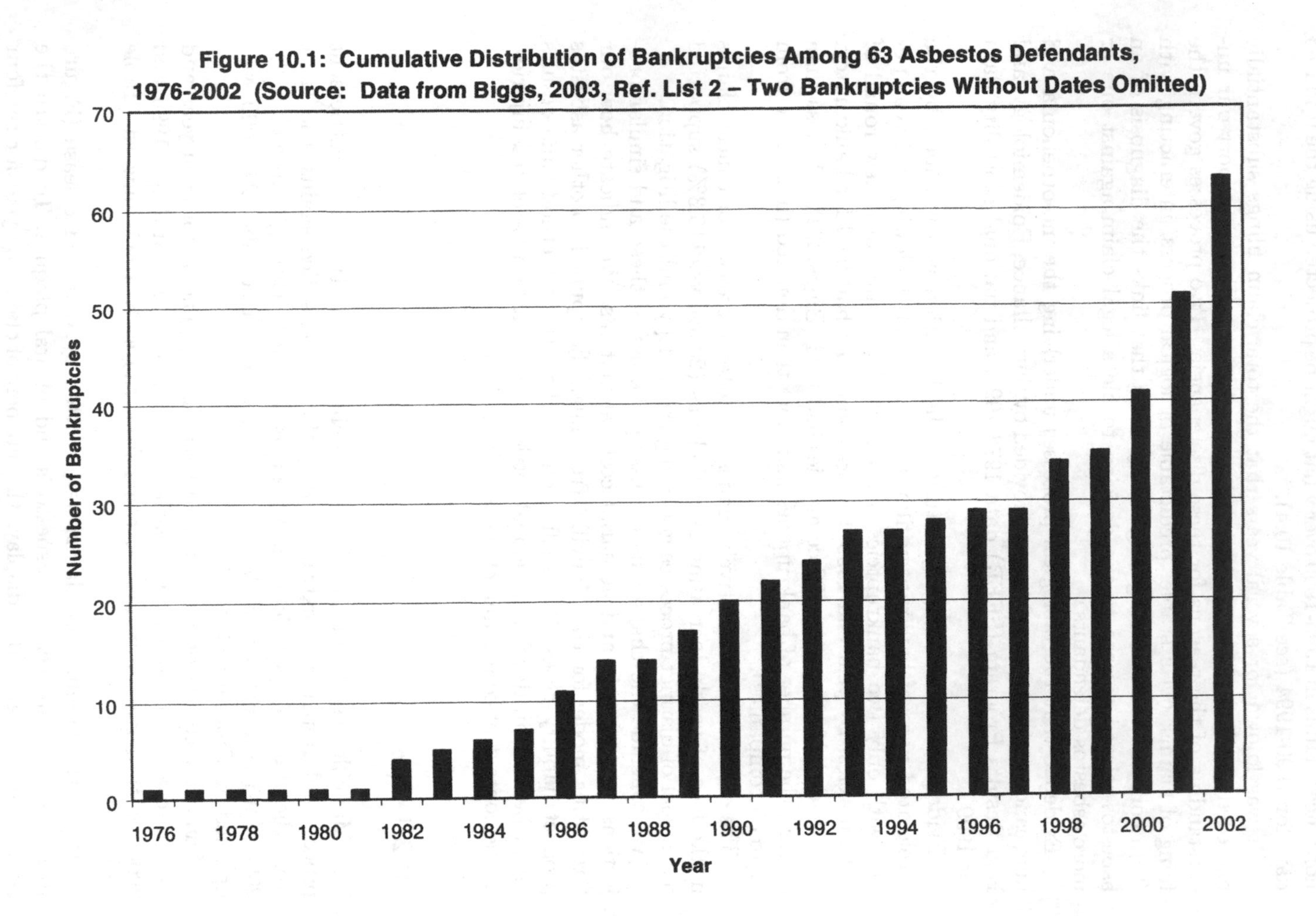

Table 10.1: Number and Type of Qualified Claims Filed Against the Manville Trust by Gender, 1988-2000

| Alleged Disease | Claim Filing Year | | | | | | | | | | | | | Total |
|---|---|---|---|---|---|---|---|---|---|---|---|---|---|---|
| | 1988 | 1989 | 1990 | 1991 | 1992 | 1993 | 1994 | 1995 | 1996 | 1997 | 1998 | 1999 | 2000* | |
| | | | | | | | Both Sexes | | | | | | | |
| 1. Mesothelioma | 1,507 | 5,005 | 1,120 | 873 | 752 | 523 | 1,293 | 1,399 | 1,377 | 1,225 | 1,174 | 1,532 | 1,912 | 19,692 |
| 2. Lung cancer | 2,311 | 9,279 | 1,541 | 1,454 | 1,398 | 794 | 1,926 | 2,441 | 2,336 | 1,937 | 1,647 | 2,101 | 2,871 | 32,036 |
| 3. Colon/rectal cancer | 322 | 1,201 | 169 | 220 | 217 | 124 | 238 | 354 | 418 | 300 | 286 | 302 | 451 | 4,602 |
| 4. Other cancer | 982 | 1,915 | 248 | 206 | 154 | 98 | 218 | 305 | 264 | 164 | 151 | 146 | 234 | 5,085 |
| 5. Asbestosis | 22,039 | 55,731 | 8,378 | 9,368 | 10,640 | 9,233 | 16,855 | 23,806 | 41,086 | 14,831 | 21,158 | 20,584 | 42,243 | 295,952 |
| 6. Pleural plaques/thickening | 3,028 | 18,955 | 2,624 | 2,335 | 2,168 | 3,335 | 3,920 | 3,221 | 5,587 | 5,215 | 4,997 | 7,062 | 8,746 | 71,193 |
| 7. Nonasbestos-related disease | 2,233 | 7,049 | 897 | 648 | 293 | 0 | 0 | 0 | 9 | 16 | 15 | 24 | 247 | 11,431 |
| 8. Unknown | 84 | 1,198 | 6,582 | 212 | 137 | 5 | 0 | 0 | 0 | 0 | 0 | 5 | 173 | 8,396 |
| Total | 32,506 | 100,333 | 21,559 | 15,316 | 15,759 | 14,112 | 24,450 | 31,526 | 51,077 | 23,688 | 29,428 | 31,756 | 56,876 | 448,386 |
| | | | | | | | Males | | | | | | | |
| 1. Mesothelioma | 1,438 | 4,738 | 1,068 | 818 | 710 | 486 | 1,182 | 1,277 | 1,270 | 1,129 | 1,090 | 1,413 | 1,713 | 18,332 |
| 2. Lung cancer | 2,263 | 9,061 | 1,505 | 1,413 | 1,368 | 765 | 1,855 | 2,347 | 2,249 | 1,850 | 1,584 | 2,011 | 2,752 | 31,023 |
| 3. Colon/rectal cancer | 317 | 1,164 | 165 | 216 | 212 | 116 | 235 | 345 | 407 | 283 | 274 | 291 | 434 | 4,459 |
| 4. Other cancer | 953 | 1,873 | 240 | 201 | 149 | 94 | 211 | 289 | 254 | 160 | 147 | 143 | 227 | 4,941 |
| 5. Asbestosis | 21,364 | 53,975 | 8,076 | 9,082 | 10,331 | 8,834 | 16,047 | 22,952 | 39,214 | 14,078 | 20,181 | 19,704 | 38,436 | 282,274 |
| 6. Pleural plaques/thickening | 2,982 | 18,519 | 2,567 | 2,295 | 2,133 | 3,184 | 3,861 | 3,143 | 5,433 | 5,050 | 4,900 | 6,955 | 8,591 | 69,613 |
| 7. Nonasbestos-related disease | 2,165 | 6,742 | 859 | 638 | 286 | 0 | 0 | 0 | 9 | 16 | 14 | 23 | 236 | 10,988 |
| 8. Unknown | 72 | 1,019 | 6,143 | 199 | 134 | 5 | 0 | 0 | 0 | 0 | 0 | 5 | 159 | 7,736 |
| Total | 31,554 | 97,091 | 20,623 | 14,862 | 15,323 | 13,484 | 23,391 | 30,353 | 48,836 | 22,566 | 28,190 | 30,545 | 52,549 | 429,367 |
| | | | | | | | Females | | | | | | | |
| 1. Mesothelioma | 65 | 265 | 51 | 55 | 42 | 37 | 111 | 122 | 107 | 96 | 84 | 119 | 198 | 1,352 |
| 2. Lung cancer | 44 | 209 | 36 | 41 | 30 | 29 | 71 | 94 | 87 | 87 | 63 | 90 | 117 | 998 |
| 3. Colon/rectal cancer | 5 | 37 | 4 | 4 | 5 | 8 | 3 | 9 | 11 | 17 | 12 | 11 | 16 | 142 |
| 4. Other cancer | 23 | 39 | 8 | 5 | 5 | 4 | 7 | 16 | 10 | 4 | 4 | 3 | 7 | 135 |
| 5. Asbestosis | 537 | 1,571 | 294 | 279 | 307 | 399 | 808 | 854 | 1,872 | 753 | 977 | 880 | 3,805 | 13,336 |
| 6. Pleural plaques/thickening | 39 | 353 | 51 | 38 | 35 | 151 | 59 | 78 | 154 | 165 | 97 | 107 | 153 | 1,480 |
| 7. Nonasbestos-related disease | 57 | 237 | 32 | 10 | 7 | 0 | 0 | 0 | 0 | 0 | 1 | 1 | 11 | 356 |
| 8. Unknown | 10 | 167 | 367 | 11 | 3 | 0 | 0 | 0 | 0 | 0 | 0 | 0 | 13 | 571 |
| Total | 780 | 2,878 | 843 | 443 | 434 | 628 | 1,059 | 1,173 | 2,241 | 1,122 | 1,238 | 1,211 | 4,322 | 18,372 |
| | | | | | | | Unknown Sex | | | | | | | |
| Total | 172 | 364 | 93 | 11 | 2 | 0 | 0 | 0 | 0 | 0 | 0 | 0 | 5 | 647 |

*Note: Includes pro rata allocations for 705 claims filed during December 22-31, 2000.

Source: Authors' tabulations of Manville Trust data.

cancer categories only. These figures reveal that the large increase in total claims in 1996 comprised a modest decrease in the number of cancer claims and a large increase in the number of noncancer claims.

The claim filing rates in these figures may have been affected by one or more of the following characteristics of the litigation environment:

1990–1992: The calibration period for the Rule 706 Panel's model – A judicial stay was issued in July 1990; a new Proof of Claim Form was issued in September 1992.

1993–1994: The time during which the Rule 706 Panel's model was developed – The Manville Trust case was effectively resolved by mid-1994.

1995–1996: First 2 years of operation of the 1995 TDP – The increase in total claims in 1996 was not accompanied by a corresponding increase in cancer claims.

1997–1999: Total claims were close to or lower than their 1995 levels; cancer claims were lower.

2000: Filing rates for both total claims and cancer claims increased significantly. At the same time, similar increases were experienced by a large number of solvent asbestos defendants, reflecting a major change in the external litigation environment; see Figure 10.1.

2001–2002: The Trust's 2001 and 2002 Annual Reports indicated that the number of qualified claims increased to 89,426 in 2001 and decreased to 53,487 in 2002. The 1995 TDP was replaced with the 2002 TDP. No breakdown by disease category was provided; this period was not included in our analysis.

## 10.3 Comparisons of Original and Updated Data

Before we could confidently compare the actual and projected number of claims in the period 1995-1999, we had to validate the new data by comparing the actual number of claims in the period 1988-1992 based on the original 1992 data described in Sections 6.2 and 6.5 with the 2000 data described in Section 10.2.

The top panel of Table 10.2 shows that 14,219 of 206,810 claims (6.9%) in the original 1992 file were disqualified at that time, yielding the total of 192,591 claims reported in Table 6.4. The 2000 update of the Trust files contained substantial revisions to this information, reflecting the acquisition of more accurate and complete information during the claim settlement process. Of the 14,219 disqualified claims in 1992, 7802 were requalified in 2000; however, another 15,879 claims originally qualified were later disqualified. A total of 22,296 claims (10.8%) were disqualified in the 2000 data. The highest relative rates of disqualification for previously qualified claims were for claims filed in 1991 and 1992.

The top panel of Table 10.2 shows the distribution of qualified and disqualified claims by claim filing year, as recorded in the 2000 data. Comparison

**Figure 10.2: Disease-Specific Numbers of Qualified Claims Filed Against the Manville Trust, 1990-2000 (Source: Authors' Calculations)**

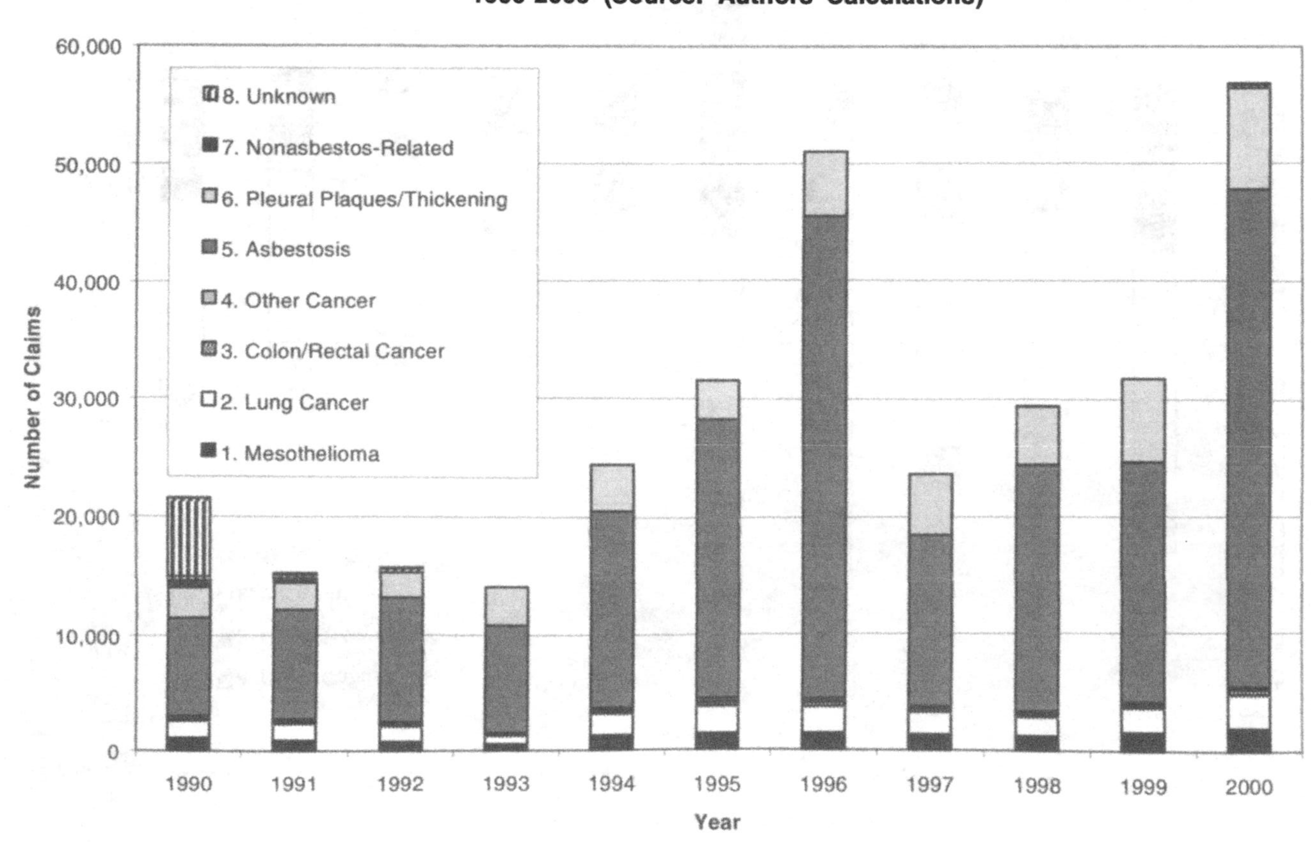

**Figure 10.3: Cancer-Specific Numbers of Qualified Claims Filed Against the Manville Trust, 1990-2000 (Source: Authors' Calculations)**

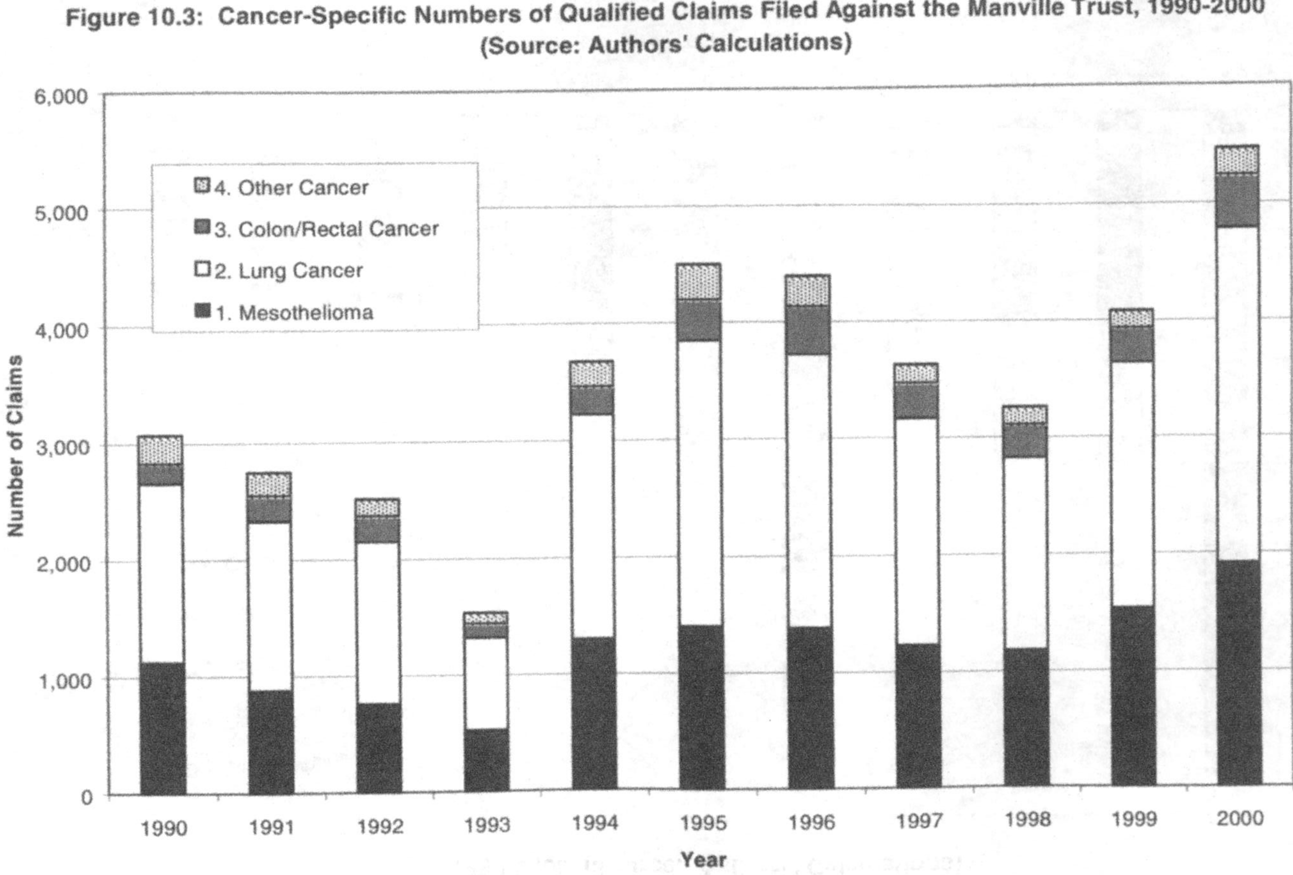

**Table 10.2: Number of Claims Filed Against the Manville Trust, 1988-1992, by Qualification Status in 1992 and Percentages by Final Qualification Status in 2000**

| Status in 1992 | Claim Filing Year | Status in 2000 Qualified | Status in 2000 Disqualified | Total | Status in 2000 Qualified | Status in 2000 Disqualified |
|---|---|---|---|---|---|---|
| Qualified | 1988 | 32,388 | 103 | 32,491 | 99.7% | 0.3% |
| | 1989 | 99,229 | 8,892 | 108,121 | 91.8% | 8.2% |
| | 1990 | 15,395 | 958 | 16,353 | 94.1% | 5.9% |
| | 1991 | 15,077 | 3,441 | 18,518 | 81.4% | 18.6% |
| | 1992 | 14,623 | 2,485 | 17,108 | 85.5% | 14.5% |
| | Subtotal | 176,712 | 15,879 | 192,591 | 91.8% | 8.2% |
| Disqualified | 1988 | 118 | 465 | 583 | 20.2% | 79.8% |
| | 1989 | 1,103 | 4,392 | 5,495 | 20.1% | 79.9% |
| | 1990 | 6,164 | 615 | 6,779 | 90.9% | 9.1% |
| | 1991 | 236 | 536 | 772 | 30.6% | 69.4% |
| | 1992 | 181 | 404 | 585 | 30.9% | 69.1% |
| | Unknown | 0 | 5 | 5 | 0.0% | 100.0% |
| | Subtotal | 7,802 | 6,417 | 14,219 | 54.9% | 45.1% |
| Total | | 184,514 | 22,296 | 206,810 | 89.2% | 10.8% |
| **Status in 1992** | **Sex** | **Status in 2000 Qualified** | **Status in 2000 Disqualified** | **Total** | **Status in 2000 Qualified** | **Status in 2000 Disqualified** |
| Qualified | Male | 171,162 | 15,723 | 186,885 | 91.6% | 8.4% |
| | Female | 4,986 | 147 | 5,133 | 97.1% | 2.9% |
| | Unknown | 564 | 9 | 573 | 98.4% | 1.6% |
| | Subtotal | 176,712 | 15,879 | 192,591 | 91.8% | 8.2% |
| Disqualified | Male | 7,354 | 5,976 | 13,330 | 55.2% | 44.8% |
| | Female | 370 | 207 | 577 | 64.1% | 35.9% |
| | Unknown | 78 | 234 | 312 | 25.0% | 75.0% |
| | Subtotal | 7,802 | 6,417 | 14,219 | 54.9% | 45.1% |
| Total | | 184,514 | 22,296 | 206,810 | 89.2% | 10.8% |
| **Male Subtotals for 1990-1992** | **Status in 1992** | **Status in 2000 Qualified** | **Status in 2000 Disqualified** | **Total** | **Status in 2000 Qualified** | **Status in 2000 Disqualified** |
| | Qualified | 43,706 | 6,818 | 50,524 | 86.5% | 13.5% |
| | Disqualified | 6,166 | 1,487 | 7,653 | 80.6% | 19.4% |
| | Total | 49,872 | 8,305 | 58,177 | 85.7% | 14.3% |

Source: Authors' tabulations of Manville Trust data.

with Table 6.4 shows small changes for each year, the most significant being a reduction of 327 claims in 1990 and a roughly compensating increase of 338 claims in 1989.

The middle panel of Table 10.2 shows the distribution of qualified and disqualified claims by sex, as recorded in the 2000 data. Comparison with Table 6.4 shows minor changes in classification by sex: 16 fewer cases with unknown sex, 15 additional males, and 1 additional female. Females had a higher percentage of claims that either remained qualified or became qualified if originally disqualified.

The bottom panel of Table 10.2 shows the distribution of qualified and disqualified male claims filed during 1990-1992, using the date recorded in the 2000 data. Comparison with Table 6.4 shows a reduction of 406 claims (0.8%), which was primarily due to revisions in the 1990 filing dates. The most notable findings in this panel were that 80.6% of originally disqualified claims were reclassified as qualified in the 2000 data and 13.5% of originally qualified claims were disqualified in the 2000 data. These changes were important because qualified male claims filed in 1990-1992 were used in calibrating our models in Chapters 6-9. Changes in the composition of this group could affect our projections.

Table 10.3 shows the distribution of the 49,872 qualified male claims in the 2000 data files by the original and revised classification of most severe alleged disease. The claims classified as cancer in the 1992 data files continued to be classified as cancer in the 2000 data files, but a substantial number of noncancer claims in the 1992 data files were reclassified as cancer in the 2000 data files. The net impact of these changes is shown in the row with the disease-specific ratios of the 2000 classifications to the 1992 classifications. Mesothelioma, lung cancer, and colon/rectal cancers increased by 20.3%, 32.5%, and 45.0%, respectively, with an overall increase of 27.8% for cancer claims. Asbestosis remained nearly the same (+1.9%). The remaining three categories (pleural plaques, nonasbestos-related diseases, and unknown diseases) declined 11-16%. Overall, noncancer claims decreased 4.0%.

The primary increments to mesothelioma and lung cancer in the 2000 data were 528 claims classified as "unknown disease" in the 1992 data, 91.9% of which were filed in 1990 without diagnostic information. These changes were important because qualified male mesothelioma claims filed in 1990-1992 were the primary data used in calibrating our models in Chapters 6-9. Changes to the initial number of claims classified as due to this disease could have substantial effects on our projections.

## 10.4 Comparisons of Actual and Projected Numbers of Claims

We grouped the observed claims in the 2000 data into two filing periods for detailed analysis: 1990-1994 and 1995-1999. For both filing periods, we compared the actual claim counts with the baseline hybrid model projection summarized in Table 8.19.

Table 10.4 shows that the actual number of claims filed in 1990-1994 was 1.2% higher than projected. This discrepancy was within the tolerances used for the generation of the baseline calibration data tables in Section 8.3 (e.g., compare Tables 8.2 and 8.4). Thus the use of the claim data for the period

**Table 10.3: Distribution of Qualified Male Claims Against the Manville Trust Filed in 1990-1992, Cross-Classified by Most Severe Alleged Disease, Using the Classifications in the 1992 Data File vs. the 2000 Data File**

| Most Severe Alleged Disease: 1992 Data File | Most Severe Alleged Disease: 2000 Data File | | | | | | | | | |
|---|---|---|---|---|---|---|---|---|---|---|
| | 1. Mesothelioma | 2. Lung Cancer | 3. Colon/ Rectal Cancer | 4. Other Cancer | 5. Asbestosis | 7. Pleural Plaques/ Thickening | 8. Nonasbestos-Related Disease | 9. Unknown | Total |
| 1. Mesothelioma | 2,133 | 0 | 0 | 0 | 1 | 2 | 0 | 0 | 2,136 |
| 2. Lung cancer | 6 | 3,190 | 0 | 1 | 4 | 0 | 0 | 0 | 3,201 |
| 3. Colon/rectal cancer | 2 | 8 | 381 | 3 | 8 | 0 | 0 | 0 | 402 |
| 4. Other cancer | 0 | 23 | 3 | 468 | 6 | 1 | 1 | 0 | 502 |
| 5. Asbestosis | 49 | 489 | 103 | 45 | 20,163 | 7 | 0 | 1 | 20,857 |
| 6. Disputed asbestosis | 17 | 117 | 28 | 14 | 5,294 | 4 | 0 | 0 | 5,474 |
| 7. Pleural plaques/thickening | 18 | 147 | 39 | 29 | 766 | 6,620 | 1 | 0 | 7,620 |
| 8. Nonasbestos-related disease | 12 | 73 | 11 | 8 | 110 | 25 | 1,775 | 3 | 2,017 |
| 9. Unknown | 333 | 195 | 18 | 15 | 482 | 145 | 6 | 6,469 | 7,663 |
| Total | 2,570 | 4,242 | 583 | 583 | 26,834 | 6,804 | 1,783 | 6,473 | 49,872 |
| Ratio: Total (2000) / Total (1992) | 1.203 | 1.325 | 1.450 | 1.161 | 1.019 | 0.893 | 0.884 | 0.845 | |
| Cancer & noncancer ratios | 1.278 | | | | 0.960 | | | | |
| | | | | | Percent Distribution | | | | |
| 1. Mesothelioma | 99.9 | 0.0 | 0.0 | 0.0 | 0.0 | 0.1 | 0.0 | 0.0 | 100.0 |
| 2. Lung cancer | 0.2 | 99.7 | 0.0 | 0.0 | 0.1 | 0.0 | 0.0 | 0.0 | 100.0 |
| 3. Colon/rectal cancer | 0.5 | 2.0 | 94.8 | 0.7 | 2.0 | 0.0 | 0.0 | 0.0 | 100.0 |
| 4. Other cancer | 0.0 | 4.6 | 0.6 | 93.2 | 1.2 | 0.2 | 0.2 | 0.0 | 100.0 |
| 5. Asbestosis | 0.2 | 2.3 | 0.5 | 0.2 | 96.7 | 0.0 | 0.0 | 0.0 | 100.0 |
| 6. Disputed asbestosis | 0.3 | 2.1 | 0.5 | 0.3 | 96.7 | 0.1 | 0.0 | 0.0 | 100.0 |
| 7. Pleural plaques/thickening | 0.2 | 1.9 | 0.5 | 0.4 | 10.1 | 86.9 | 0.0 | 0.0 | 100.0 |
| 8. Nonasbestos-related disease | 0.6 | 3.6 | 0.5 | 0.4 | 5.5 | 1.2 | 88.0 | 0.1 | 100.0 |
| 9. Unknown | 4.3 | 2.5 | 0.2 | 0.2 | 6.3 | 1.9 | 0.1 | 84.4 | 100.0 |
| Total | 5.2 | 8.5 | 1.2 | 1.2 | 53.8 | 13.6 | 3.6 | 13.0 | 100.0 |

Source: Authors' tabulations of Manville Trust data.

January 1990 through June 1992 as the basis of the initial claim estimates for the 5-year period 1990-1994 was successful.

Table 10.4 shows that the actual number of cancer claims filed in 1990-1994 was 13.1% higher than projected and the actual number of noncancer claims was 0.6% lower than projected. These discrepancies are about half the size of the discrepancies noted earlier for the comparisons of the cancer claims for 1990-1992 obtained from the original 1992 data with corresponding claims obtained from the 2000 data. This suggests that the process that led to the reclassification of "unknown disease" claims in 1990 was not repeated later in the period 1990-1994.

The use of the baseline projections to understand changes that occurred in the period 1995-1999 is most effective when the actual and projected values for the calibration period are the same. We can force equality between these two sets of values by multiplying each projected disease-specific value by the ratio of the actual to projected 1990-1994 values for that disease. This is illustrated in Table 10.4 for 1995-1999 under the heading "Adjusted Difference." For example, the 26.9% adjusted difference for mesothelioma was obtained from the two observed differences (9.6% and 39.0%) as $26.9\% = (139.0/109.6 - 1) \times 100\%$.

The actual total number of claims filed in 1995-1999 was 103.6% higher than projected and 101.1% higher than projected with the 1990-1994 adjustment. The actual number of cancer claims was 45.1% higher than projected and 28.3% higher than projected with the 1990-1994 adjustment. In contrast, the actual number of noncancer claims was 115.1% higher than projected and 116.3% higher than projected with the 1990-1994 adjustment. The noncancer discrepancy represented an additional 88.0% increase in noncancer claims, beyond the 28.3% expected based on the cancer increase.

We also compared actual and projected numbers of cancer and noncancer claims in 1995-1999 for the extreme upper and lower projections in Figures 7.1 and 9.1 (Table 10.5). The actual numbers of claims were larger than the extreme upper projections in all cases. (Entries in this and some of the following tables are rounded. Hence some sums of entries quoted below differ slightly from the sums of the rounded summands in given in the tables.)

The smallest deviations occurred for analysis S2, Model 1, in Chapter 9, which assumed that beyond 1990-1994 each individual disease in the baseline projection would generate a claim – equivalent to an immediate 41.9% increase in claim filing rates in the first-stage calibration. Under this model, the adjusted differences were 21.1% for cancer, 43.9% for noncancer, and 40.6% for total claims.

The next smallest deviations were obtained for analysis S4, Model 2, in Chapter 7, which, along with all of the other models except analysis S2, Model 1, in Chapter 9, assumed that the claim filing rates estimated for 1990-1994 would continue to apply beyond 1990-1994. Under this model, the adjusted differences were 35.9% for cancer, 82.2% for noncancer, and 75.3% for total claims.

**Table 10.4: Actual and Projected Numbers and Types of Qualified Male Claims Filed Against the Manville Trust, 1990-1994 and 1995-1999**

| | Quinquennium | | | | | | |
|---|---|---|---|---|---|---|---|
| | 1990-1994 | | | 1995-1999 | | | 1995-1999 |
| Alleged Disease | Actual | Projected | Difference | Actual | Projected | Difference | Adjusted Difference |
| 1. Mesothelioma | 4,264 | 3,891 | 9.6% | 6,179 | 4,445 | 39.0% | 26.9% |
| 2. Lung cancer | 6,906 | 5,941 | 16.2% | 10,041 | 6,667 | 50.6% | 29.6% |
| 3. Colon/rectal cancer | 944 | 728 | 29.7% | 1,600 | 837 | 91.1% | 47.4% |
| 4. Other cancer | 895 | 946 | -5.4% | 993 | 1,019 | -2.6% | 3.0% |
| 5. Asbestosis | 52,370 | 49,482 | 5.8% | 116,129 | 42,965 | 170.3% | 155.4% |
| 6. Pleural plaques/thickening | 14,040 | 13,374 | 5.0% | 25,481 | 11,880 | 114.5% | 104.3% |
| 7. Nonasbestos-related disease | 1,783 | 9,154 | -80.5% | 62 | 8,118 | -99.2% | -96.1% |
| 8. Unknown | 6,481 | 3,090 | 109.7% | 5 | 2,914 | -99.8% | -99.9% |
| Total | 87,683 | 86,606 | 1.2% | 160,490 | 78,845 | 103.6% | 101.1% |
| Cancer (1-4) | 13,009 | 11,506 | 13.1% | 18,813 | 12,968 | 45.1% | 28.3% |
| Noncancer (5-8) | 74,674 | 75,100 | -0.6% | 141,677 | 65,877 | 115.1% | 116.3% |
| Asbestos-related noncancer (5-6) | 66,410 | 62,856 | 5.7% | 141,610 | 54,844 | 158.2% | 144.4% |
| Nonasbestos-related & unknown (7-8) | 8,264 | 12,244 | -32.5% | 67 | 11,033 | -99.4% | -99.1% |

Source: Stallard and Manton (1994, Tables D and F); and authors' tabulations of Manville Trust data.

**Table 10.5: Actual and Projected Numbers and Types of Qualified Male Claims Filed Against the Manville Trust, 1995-1999 – Baseline Model and Upper and Lower Extremes from Chapters 7 and 9**

| Alleged Disease | Actual | Projected | Difference | Adjusted Difference |
|---|---|---|---|---|
| | | Chapter 8 Baseline Hybrid Model | | |
| Total | 160,490 | 78,845 | 103.6% | 101.1% |
| Cancer | 18,813 | 12,968 | 45.1% | 28.3% |
| Noncancer | 141,677 | 65,877 | 115.1% | 116.3% |
| | | Chapter 9 Analysis S2 Model 1 | | |
| Total | 160,490 | 112,754 | 42.3% | 40.6% |
| Cancer | 18,813 | 13,740 | 36.9% | 21.1% |
| Noncancer | 141,677 | 99,014 | 43.1% | 43.9% |
| | | Chapter 7 Analysis S4 Model 2 | | |
| Total | 160,490 | 90,066 | 78.2% | 75.3% |
| Cancer | 18,813 | 11,276 | 66.8% | 35.9% |
| Noncancer | 141,677 | 78,789 | 79.8% | 82.2% |
| | | Chapter 9 Analysis S5 Model 1 | | |
| Total | 160,490 | 67,350 | 138.3% | 135.4% |
| Cancer | 18,813 | 11,293 | 66.6% | 47.3% |
| Noncancer | 141,677 | 56,057 | 152.7% | 154.2% |
| | | Chapter 7 Analysis S9 Model 3 | | |
| Total | 160,490 | 55,424 | 189.6% | 184.8% |
| Cancer | 18,813 | 7,293 | 158.0% | 110.1% |
| Noncancer | 141,677 | 48,131 | 194.4% | 198.2% |

Source: Stallard and Manton (1994, Tables D, F, and H); and authors' tabulations of Manville Trust data.

All other models exhibited larger adjusted differences, the largest being for analysis S9, Model 3, in Chapter 7 (110.1% for cancer, 198.2% for noncancer, and 184.8% for total claims).

These comparisons indicate that the deviations of actual from projected values were larger than the largest deviations expected under the uncertainty analyses conducted for the Rule 706 Panel. In other words, neither the baseline model nor any of the 40 alternatives evaluated in Chapters 7 and 9 accurately described the numbers and types of claims against the Manville Trust during 1995-1999.

Moreover, because all of the alternative models considered in the uncertainty analyses were forced to exactly reproduce the claim count estimates for the 1990-1994 calibration period, it is highly unlikely that the claim filing process in effect for 1995-1999 can be described as a simple continuation of the process in effect for 1990-1994.

Explanations for the increases in claim filings during the period 1995-1999 must account not only for the increase in cancer claims but also for the additional increase in noncancer claims beyond that accounted for by the increase in cancer claims.

Table 10.6 stratifies the disease comparisons for the baseline hybrid model by age. The panel for total claims shows that the actual exceeded the projected number of claims in 1990-1994 up to age 64, with a reversal thereafter. For 1995-1999, the actual exceeded the projected number of claims at all ages, but the relative excess was substantially larger at younger ages. The age trend was reduced, but not eliminated, by the adjusted differences.

For all diseases, there was instability at the youngest age group, in part due to the inclusion of 651 actual claimants younger than 40 years in 1995-1999 who would have been excluded from the projection because they were younger than 15 years in 1970-1974. Thus, assessments of age trends should be restricted to ages 45-49 and older.

The age trends in the adjusted differences for 1995-1999 are displayed in Figure 10.4, where they fall into three groups: (1) cancers, (2) asbestos-related noncancers, and (3) nonasbestos-related and unknown diseases. The third group had effectively disappeared from the actual claims data. The cancers exhibit a relatively flat pattern over age. Within the asbestos-related noncancers, the adjusted differences for asbestosis exhibited a substantial decline over age while the adjusted differences for pleural plaques were relatively stable. Except for mesothelioma and colon/rectal cancer, the adjusted differences declined for the two oldest age groups.

These patterns were consistent with the interpretation that the increased claims in 1995-1999 were for the less serious diseases, with the largest relative increases occurring for younger claimants. Given that the adjusted differences for the cancers were smaller and occurred relatively uniformly over age, it was informative to consider the changes in the SEER data over the same periods.

## 10.5 Health and Vital Statistics Data, 1990-1999

The estimated national mesothelioma incidence (diagnosis) counts derived from National Cancer Institute's SEER data, displayed for males in Table 10.7 and Figure 10.5, exhibited only modest increases over the period 1990-1999. The total number of deaths increased from 11,030 in 1990-1994 to 11,290 in 1995-1999 (+2.4%), which contrasts with a 44.9% increase in actual mesothelioma claims (Table 10.4). The ratio of Manville Trust claims to SEER diagnoses increased from 38.7% to 54.7%. This ratio may be interpreted as a measure of the propensity to sue the Manville Trust among all mesothelioma cases.

The 2.4% increase in SEER mesothelioma diagnoses was less than the projected 14.2% increase in mesothelioma claims in Table 10.4 (also in Table 8.19). Both increases were consistent with the assumption of a stable biological process underlying the filing of mesothelioma claims. The increase in the ratio of actual claims to diagnoses from 38.7% to 54.7% suggests that

**Table 10.6: Actual and Projected Numbers and Types of Qualified Male Claims Filed Against the Manville Trust, 1990-1994 and 1995-1999, by Age**

| | Quinquennium | | | | | | |
|---|---|---|---|---|---|---|---|
| | 1990-1994 | | | 1995-1999 | | | 1995-1999 |
| Age | Actual[1] | Projected | Difference | Actual[2] | Projected | Difference | Adjusted Difference |
| | | | 1. Mesothelioma | | | | |
| 0-44 | 87 | 47 | 86.0% | 83 | 18 | 351.6% | 142.8% |
| 45-49 | 125 | 90 | 38.4% | 136 | 95 | 43.5% | 3.7% |
| 50-54 | 276 | 214 | 29.1% | 283 | 159 | 78.5% | 38.2% |
| 55-59 | 402 | 332 | 21.0% | 569 | 323 | 76.4% | 45.7% |
| 60-64 | 653 | 556 | 17.5% | 749 | 455 | 64.5% | 40.0% |
| 65-69 | 855 | 765 | 11.8% | 1,018 | 708 | 43.8% | 28.7% |
| 70-74 | 870 | 801 | 8.6% | 1,308 | 926 | 41.3% | 30.1% |
| 75-79 | 609 | 628 | -3.0% | 1,164 | 878 | 32.5% | 36.7% |
| 80-84 | 277 | 311 | -10.8% | 611 | 592 | 3.2% | 15.7% |
| 85-89 | 96 | 125 | -22.8% | 209 | 232 | -9.9% | 16.8% |
| 90+ | 13 | 22 | -40.6% | 49 | 59 | -17.2% | 39.5% |
| | | | 2. Lung Cancer | | | | |
| 0-44 | 74 | 45 | 64.6% | 44 | 28 | 57.5% | -4.3% |
| 45-49 | 179 | 133 | 34.7% | 161 | 185 | -13.0% | -35.4% |
| 50-54 | 364 | 247 | 47.5% | 437 | 201 | 117.1% | 47.2% |
| 55-59 | 696 | 516 | 34.8% | 915 | 554 | 65.1% | 22.5% |
| 60-64 | 1,290 | 1,012 | 27.5% | 1,436 | 840 | 70.9% | 34.0% |
| 65-69 | 1,618 | 1,344 | 20.4% | 2,009 | 1,296 | 55.0% | 28.8% |
| 70-74 | 1,340 | 1,203 | 11.4% | 2,279 | 1,309 | 74.1% | 56.2% |
| 75-79 | 877 | 892 | -1.7% | 1,650 | 1,185 | 39.2% | 41.6% |
| 80-84 | 355 | 408 | -12.9% | 788 | 741 | 6.3% | 22.1% |
| 85-89 | 90 | 123 | -26.8% | 249 | 258 | -3.4% | 31.9% |
| 90+ | 22 | 18 | 22.3% | 73 | 69 | 6.5% | -12.9% |
| | | | 3. Colon/Rectal Cancer | | | | |
| 0-44 | 12 | 3 | 300.4% | 5 | 0.2 | 2438.9% | 534.1% |
| 45-49 | 19 | 15 | 26.8% | 26 | 20 | 30.6% | 3.0% |
| 50-54 | 40 | 15 | 166.9% | 51 | 10 | 388.0% | 82.8% |
| 55-59 | 83 | 57 | 45.8% | 134 | 57 | 136.1% | 62.0% |
| 60-64 | 156 | 109 | 43.3% | 193 | 95 | 104.1% | 42.4% |
| 65-69 | 223 | 190 | 17.5% | 309 | 188 | 64.6% | 40.1% |
| 70-74 | 196 | 151 | 29.9% | 351 | 165 | 112.2% | 63.3% |
| 75-79 | 143 | 108 | 32.5% | 312 | 145 | 115.8% | 62.8% |
| 80-84 | 53 | 66 | -19.6% | 154 | 120 | 28.2% | 59.5% |
| 85-89 | 15 | 13 | 15.5% | 44 | 33 | 33.6% | 15.6% |
| 90+ | 3 | 1 | 200.3% | 21 | 4 | 370.1% | 56.5% |
| | | | 4. Other Cancer | | | | |
| 0-44 | 18 | 6 | 201.7% | 7 | 2 | 297.9% | 31.9% |
| 45-49 | 26 | 15 | 74.3% | 20 | 20 | 1.3% | -41.9% |
| 50-54 | 47 | 25 | 89.1% | 53 | 21 | 152.8% | 33.7% |
| 55-59 | 102 | 95 | 6.9% | 102 | 104 | -2.4% | -8.7% |
| 60-64 | 172 | 175 | -1.7% | 158 | 142 | 10.9% | 12.9% |
| 65-69 | 198 | 229 | -13.5% | 202 | 208 | -2.7% | 12.4% |
| 70-74 | 158 | 193 | -18.2% | 204 | 206 | -1.2% | 20.8% |
| 75-79 | 106 | 125 | -15.5% | 143 | 167 | -14.5% | 1.2% |
| 80-84 | 49 | 64 | -23.0% | 73 | 113 | -35.1% | -15.7% |
| 85-89 | 16 | 18 | -10.6% | 26 | 31 | -15.2% | -5.2% |
| 90+ | 3 | 1 | 201.7% | 5 | 5 | -2.4% | -67.7% |

See footnotes at end of table. (Continued)

**Table 10.6 (Continued)**

| | Quinquennium | | | | | | |
|---|---|---|---|---|---|---|---|
| | 1990-1994 | | | 1995-1999 | | | 1995-1999 |
| Age | Actual[1] | Projected | Difference | Actual[2] | Projected | Difference | Adjusted Difference |
| | | | | 5. Asbestosis | | | |
| 0-44 | 4,190 | 2,891 | 44.9% | 4,066 | 850 | 378.3% | 230.0% |
| 45-49 | 4,671 | 3,457 | 35.1% | 9,777 | 3,340 | 192.7% | 116.7% |
| 50-54 | 5,948 | 4,552 | 30.7% | 14,084 | 2,948 | 377.8% | 265.6% |
| 55-59 | 7,177 | 6,135 | 17.0% | 16,214 | 5,316 | 205.0% | 160.7% |
| 60-64 | 8,819 | 8,299 | 6.3% | 17,711 | 5,787 | 206.0% | 188.0% |
| 65-69 | 9,086 | 9,805 | -7.3% | 19,267 | 8,488 | 127.0% | 144.9% |
| 70-74 | 6,540 | 7,292 | -10.3% | 17,277 | 6,562 | 163.3% | 193.6% |
| 75-79 | 3,910 | 4,470 | -12.5% | 10,877 | 5,310 | 104.8% | 134.1% |
| 80-84 | 1,515 | 1,894 | -20.0% | 4,881 | 3,024 | 61.4% | 101.8% |
| 85-89 | 443 | 604 | -26.7% | 1,573 | 1,046 | 50.5% | 105.3% |
| 90+ | 72 | 83 | -13.1% | 402 | 293 | 37.4% | 58.1% |
| | | | | 6. Pleural Plaques/Thickening | | | |
| 0-44 | 1,006 | 616 | 63.3% | 591 | 207 | 186.2% | 75.2% |
| 45-49 | 1,216 | 945 | 28.6% | 1,465 | 1,118 | 31.1% | 1.9% |
| 50-54 | 1,648 | 1,366 | 20.6% | 2,825 | 945 | 198.9% | 147.8% |
| 55-59 | 2,070 | 1,879 | 10.1% | 3,676 | 1,648 | 123.1% | 102.5% |
| 60-64 | 2,539 | 2,496 | 1.7% | 4,073 | 1,832 | 122.4% | 118.6% |
| 65-69 | 2,590 | 2,657 | -2.5% | 4,342 | 2,343 | 85.3% | 90.1% |
| 70-74 | 1,709 | 1,880 | -9.1% | 4,250 | 1,697 | 150.5% | 175.6% |
| 75-79 | 871 | 1,035 | -15.9% | 2,645 | 1,231 | 114.8% | 155.3% |
| 80-84 | 291 | 376 | -22.7% | 1,187 | 601 | 97.6% | 155.6% |
| 85-89 | 86 | 111 | -22.4% | 340 | 212 | 60.7% | 107.0% |
| 90+ | 16 | 13 | 23.3% | 86 | 47 | 83.3% | 48.6% |
| | | | | 7. Nonasbestos-Related Disease | | | |
| 0-44 | 184 | 507 | -63.7% | 2 | 148 | -98.6% | -96.3% |
| 45-49 | 95 | 517 | -81.6% | 2 | 632 | -99.7% | -98.3% |
| 50-54 | 178 | 730 | -75.6% | 6 | 531 | -98.9% | -95.4% |
| 55-59 | 218 | 931 | -76.6% | 3 | 817 | -99.6% | -98.4% |
| 60-64 | 317 | 1,527 | -79.2% | 14 | 999 | -98.6% | -93.3% |
| 65-69 | 330 | 2,070 | -84.1% | 5 | 1,857 | -99.7% | -98.3% |
| 70-74 | 239 | 1,472 | -83.8% | 12 | 1,265 | -99.1% | -94.2% |
| 75-79 | 138 | 834 | -83.5% | 10 | 1,002 | -99.0% | -94.0% |
| 80-84 | 67 | 423 | -84.2% | 8 | 620 | -98.7% | -91.8% |
| 85-89 | 14 | 115 | -87.7% | 0 | 167 | -100.0% | -100.0% |
| 90+ | 3 | 28 | -89.2% | 0 | 79 | -100.0% | -100.0% |
| | | | | 8. Unknown | | | |
| 0-44 | 694 | 326 | 113.0% | 0 | 92 | -100.0% | -100.0% |
| 45-49 | 461 | 250 | 84.5% | 0 | 168 | -100.0% | -100.0% |
| 50-54 | 765 | 237 | 222.8% | 0 | 234 | -100.0% | -100.0% |
| 55-59 | 1,040 | 306 | 239.7% | 0 | 362 | -100.0% | -100.0% |
| 60-64 | 1,260 | 448 | 181.3% | 1 | 302 | -99.7% | -99.9% |
| 65-69 | 1,046 | 461 | 126.9% | 1 | 483 | -99.8% | -99.9% |
| 70-74 | 567 | 479 | 18.3% | 3 | 439 | -99.3% | -99.4% |
| 75-79 | 388 | 336 | 15.6% | 0 | 471 | -100.0% | -100.0% |
| 80-84 | 202 | 218 | -7.5% | 0 | 294 | -100.0% | -100.0% |
| 85-89 | 47 | 23 | 106.5% | 0 | 39 | -100.0% | -100.0% |
| 90+ | 11 | 6 | 75.9% | 0 | 31 | -100.0% | -100.0% |

See footnotes at end of table. (Continued)

**Table 10.6 (Continued)**

| | Quinquennium | | | | | | |
|---|---|---|---|---|---|---|---|
| | 1990-1994 | | | 1995-1999 | | | 1995-1999 |
| Age | Actual[1] | Projected | Difference | Actual[2] | Projected | Difference | Adjusted Difference |
| | | | | Total | | | |
| 0-44 | 6,266 | 4,441 | 41.1% | 4,798 | 1,344 | 257.0% | 153.0% |
| 45-49 | 6,792 | 5,422 | 25.3% | 11,587 | 5,578 | 107.7% | 65.8% |
| 50-54 | 9,267 | 7,386 | 25.5% | 17,740 | 5,049 | 251.3% | 180.0% |
| 55-59 | 11,786 | 10,251 | 15.0% | 21,613 | 9,182 | 135.4% | 104.7% |
| 60-64 | 15,206 | 14,622 | 4.0% | 24,335 | 10,454 | 132.8% | 123.8% |
| 65-69 | 15,946 | 17,521 | -9.0% | 27,153 | 15,571 | 74.4% | 91.6% |
| 70-74 | 11,619 | 13,471 | -13.7% | 25,684 | 12,570 | 104.3% | 136.9% |
| 75-79 | 7,042 | 8,428 | -16.4% | 16,801 | 10,390 | 61.7% | 93.5% |
| 80-84 | 2,809 | 3,760 | -25.3% | 7,702 | 6,105 | 26.2% | 68.9% |
| 85-89 | 808 | 1,132 | -28.6% | 2,441 | 2,016 | 21.1% | 69.6% |
| 90+ | 143 | 172 | -16.9% | 636 | 587 | 8.4% | 30.5% |

Note 1: Includes pro rata allocation of 513 (of 87,683) claims with unknown age.
Note 2: Includes pro rata allocation of 5 (of 160,490) claims with unknown age.
Source: Authors' projections and tabulations of Manville Trust data.

the primary changes from 1990-1994 to 1995-1999 were increases in (1) the propensity to sue for those men developing mesothelioma who were exposed to Manville asbestos, (2) the pool of men who could document exposure to Manville asbestos, or (3) some combination of these two factors.

Evidence for an increase in the pool of men who could document exposure to Manville asbestos was provided by Carroll et al. (2002, p. 41) who reported that the typical claimant in the 1980s named about 20 defendants but that this number had increased to 60-70 defendants by the mid-1990s. Such increases would require later claimants to identify exposures to specific asbestos products of an increasing number of defendants. From the point of view of any given defendant, it could then appear that the pool of potential claimants was increasing.

Evidence for an increase in the propensity to sue for men developing mesothelioma who were exposed to Manville asbestos, beyond the amount attributable to the increased pool size, is less compelling. Two factors that could have been significant were the increased publicity given to asbestos-related disease lawsuits during the 1990s and the improved ability of plaintiff lawyers to identify cases that could be litigated successfully.

Regardless of which specific combination of the two factors actually occurred, the net result was an increasing propensity to sue among men with mesothelioma.

The corresponding results for lung cancer are displayed in Table 10.8 and Figure 10.6 (using the standard SEER classification of lung and bronchus as a single-disease category). The total number of deaths decreased from 500,869

**Figure 10.4: Adjusted Differences Between Actual and Projected Numbers of Qualified Male Claims Against the Manville Trust, by Age and Disease, 1995-1999 (Source: Authors' Calculations)**

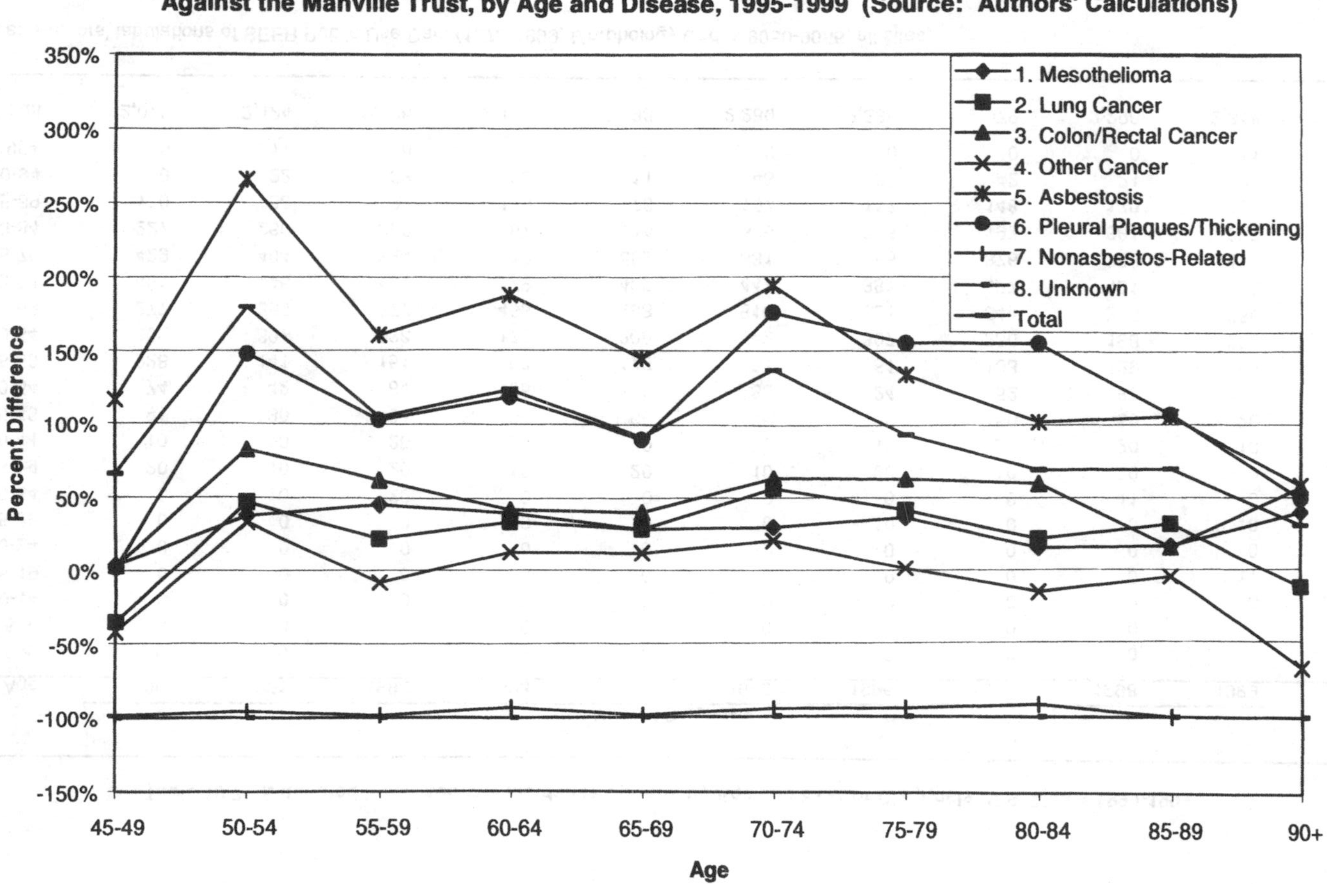

Table 10.7: Estimated Mesothelioma Incidence Counts, by Age and Year of Diagnosis, U.S. Males 1990-1999

| Age | Year of Diagnosis | | | | | | | | | | Total |
|---|---|---|---|---|---|---|---|---|---|---|---|
| | 1990 | 1991 | 1992 | 1993 | 1994 | 1995 | 1996 | 1997 | 1998 | 1999 | |
| 0-4 | 0 | 0 | 0 | 0 | 0 | 0 | 0 | 0 | 0 | 0 | 0 |
| 5-9 | 0 | 0 | 0 | 0 | 0 | 0 | 0 | 0 | 0 | 0 | 0 |
| 10-14 | 0 | 0 | 0 | 0 | 0 | 0 | 0 | 0 | 0 | 0 | 0 |
| 15-19 | 0 | 0 | 0 | 0 | 0 | 0 | 0 | 0 | 0 | 11 | 11 |
| 20-24 | 0 | 0 | 0 | 0 | 0 | 0 | 0 | 0 | 0 | 0 | 0 |
| 25-29 | 0 | 0 | 0 | 0 | 10 | 0 | 10 | 0 | 0 | 10 | 31 |
| 30-34 | 0 | 10 | 20 | 0 | 0 | 0 | 0 | 0 | 41 | 0 | 71 |
| 35-39 | 20 | 10 | 20 | 10 | 20 | 10 | 20 | 10 | 20 | 0 | 141 |
| 40-44 | 10 | 10 | 20 | 20 | 70 | 20 | 10 | 20 | 20 | 10 | 209 |
| 45-49 | 91 | 30 | 101 | 30 | 60 | 50 | 30 | 50 | 50 | 20 | 514 |
| 50-54 | 74 | 42 | 84 | 105 | 146 | 83 | 124 | 52 | 83 | 114 | 907 |
| 55-59 | 228 | 141 | 151 | 97 | 173 | 151 | 151 | 183 | 129 | 193 | 1,596 |
| 60-64 | 209 | 307 | 328 | 175 | 208 | 142 | 307 | 230 | 198 | 253 | 2,358 |
| 65-69 | 277 | 365 | 377 | 455 | 288 | 510 | 321 | 344 | 345 | 256 | 3,539 |
| 70-74 | 401 | 446 | 401 | 479 | 435 | 447 | 381 | 427 | 451 | 530 | 4,397 |
| 75-79 | 429 | 404 | 324 | 410 | 552 | 331 | 386 | 376 | 389 | 445 | 4,045 |
| 80-84 | 227 | 260 | 380 | 167 | 244 | 386 | 395 | 197 | 381 | 337 | 2,974 |
| 85-89 | 110 | 66 | 197 | 120 | 76 | 107 | 117 | 148 | 170 | 137 | 1,247 |
| 90-94 | 0 | 22 | 33 | 33 | 11 | 43 | 32 | 42 | 21 | 21 | 258 |
| 95+ | 0 | 11 | 0 | 0 | 0 | 0 | 0 | 0 | 0 | 11 | 21 |
| Total | 2,075 | 2,124 | 2,436 | 2,101 | 2,293 | 2,280 | 2,284 | 2,079 | 2,298 | 2,348 | 22,319 |

Source: Authors' tabulations of SEER Public Use Data (1973-1999; Morphology Codes 9050-9055, all sites).

**Figure 10.5: Estimated Mesothelioma Incidence Counts, by Age and Year of Diagnosis, U.S. Males Aged 35-94 Years, 1990-1999 (Source: Data from SEER Public Use Data, 1973-1999)**

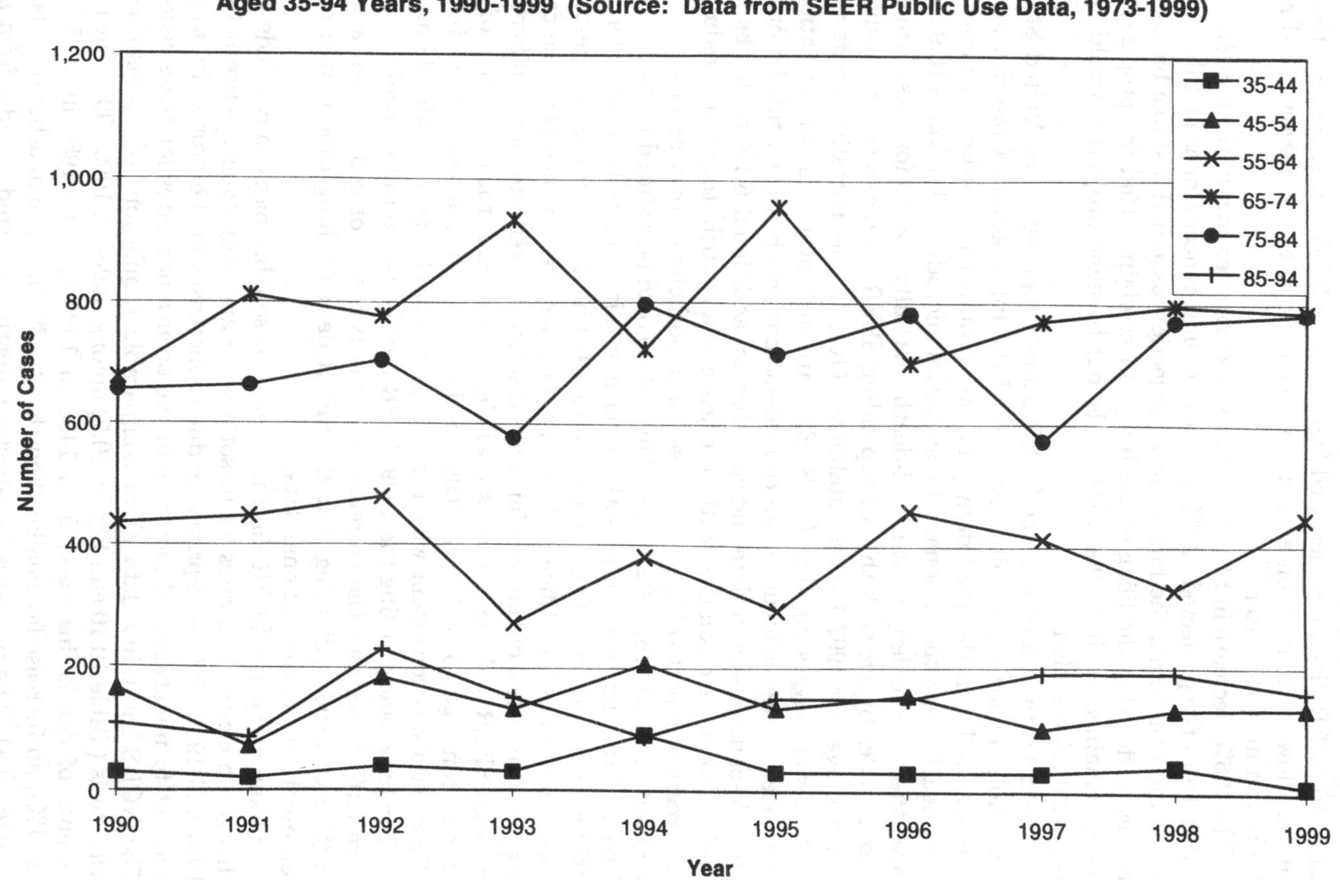

in 1990-1994 to 488,058 in 1995-1999 (−2.6%), which contrasts with a 45.4% increase in actual lung cancer claims (Table 10.4). The ratio of Manville Trust claims to SEER diagnoses increased from 1.4% to 2.1%. These propensities to sue were low in comparison with mesothelioma due to the major causal role of tobacco in lung cancer.

The 2.6% decrease in SEER lung cancer diagnoses was in the opposite direction from the projected 12.2% increase in lung cancer claims in Table 10.4. As for mesothelioma, the lung cancer projection assumed a stable biological process underlying the filing of the lung cancer claims. Had the propensities to sue remained at their 1990-1994 levels, our baseline projection would have been 12-14% too high.

One concern is how well the SEER sample represents the United States as a whole. Following the ninth revision of the International Classification of Diseases (ICD) in the 1999 data year, we obtained tabulations of mesothelioma deaths from the national vital statistics microdata files. Table 10.9 compares the 1999 underlying-cause-of-death mortality counts for mesothelioma among males by age with the corresponding SEER diagnosis counts (using 3-year averages for 1997-1999 for stability). Overall, the mortality counts were 16.1% lower, suggesting that the SEER data may overestimate the national incidence of mesothelioma by an equivalent amount. However, tabulations of multiple-cause-of-death data indicate that an additional 6.1% of deaths had mesothelioma coded on the death certificate as a contributory cause, reducing the excess of reported SEER diagnoses of mesothelioma over reported deaths to about 11% (NCHS, 2002, p. 66). This value could be reduced up to 2% more if part of the difference in mesothelioma mortality counts between 1999 and 2000 were an artifact of the transition in 1999 in the processing of mesothelioma on the death certificate. The very poor survival for mesothelioma cases suggests that most of the remaining 9% discrepancy reflected a real difference between SEER and the nation as a whole. Nonetheless, the correlation of the counts from age 35 to 94 was 0.990, indicating that the age pattern in the SEER data was consistent with that in the national vital statistics data.

For lung cancer (using the 1999 SEER data), the overall mortality counts were 5.9% lower and the correlation of the two sets of counts from age 35 to 94 was 0.996, indicating that the SEER data for lung cancer were more representative of the national data.

In summary, the SEER data provide a reasonable basis to conclude that changes in claim filing rates for mesothelioma and lung cancer between 1990-1994 and 1995-1999 were primarily due to increases in the propensity to sue, and not to unanticipated increases in the occurrence rates for these cancers. The NCHS mortality data allow comparable tabulations of deaths due to asbestosis (Table 10.10 and Figure 10.7) among males in 1990-1999. The total number of deaths increased from 1433 in 1990-1994 to 1950 in 1995-1999 (36.1%), an increase far smaller than the 121.7% increase in asbestosis claims (Table 10.4). The increases in asbestosis deaths occurred broadly from ages 65 to 94. The largest difference from mesothelioma and lung cancer was that

**Table 10.8: Estimated Lung and Bronchus Cancer Incidence Counts, by Age and Year of Diagnosis, U.S. Males 1990-1999**

| | Year of Diagnosis | | | | | | | | | | |
|---|---|---|---|---|---|---|---|---|---|---|---|
| Age | 1990 | 1991 | 1992 | 1993 | 1994 | 1995 | 1996 | 1997 | 1998 | 1999 | Total |
| 0-4 | 0 | 0 | 10 | 10 | 0 | 0 | 0 | 0 | 11 | 0 | 31 |
| 5-9 | 0 | 0 | 0 | 0 | 0 | 0 | 0 | 0 | 0 | 0 | 0 |
| 10-14 | 11 | 0 | 11 | 0 | 0 | 11 | 11 | 0 | 0 | 0 | 42 |
| 15-19 | 22 | 11 | 0 | 11 | 11 | 0 | 11 | 32 | 0 | 0 | 97 |
| 20-24 | 11 | 11 | 32 | 32 | 53 | 0 | 21 | 10 | 42 | 10 | 223 |
| 25-29 | 62 | 82 | 41 | 92 | 62 | 92 | 0 | 92 | 31 | 51 | 605 |
| 30-34 | 224 | 214 | 295 | 204 | 204 | 235 | 194 | 153 | 153 | 133 | 2,009 |
| 35-39 | 793 | 562 | 711 | 543 | 554 | 616 | 587 | 629 | 458 | 450 | 5,903 |
| 40-44 | 1,241 | 1,598 | 1,254 | 1,660 | 1,262 | 1,422 | 1,353 | 1,485 | 1,392 | 1,401 | 14,069 |
| 45-49 | 2,922 | 3,022 | 3,022 | 2,798 | 2,950 | 2,849 | 2,870 | 2,735 | 2,716 | 2,917 | 28,803 |
| 50-54 | 5,259 | 5,677 | 5,399 | 5,370 | 5,318 | 5,685 | 5,038 | 5,145 | 5,415 | 5,205 | 53,511 |
| 55-59 | 9,211 | 8,812 | 8,932 | 8,144 | 8,346 | 8,312 | 8,578 | 8,171 | 8,584 | 8,515 | 85,605 |
| 60-64 | 15,263 | 15,422 | 14,628 | 13,683 | 12,719 | 12,012 | 12,139 | 12,231 | 12,141 | 11,676 | 131,914 |
| 65-69 | 19,882 | 19,730 | 20,351 | 19,983 | 18,177 | 18,560 | 17,435 | 16,839 | 16,405 | 15,163 | 182,525 |
| 70-74 | 18,342 | 17,828 | 19,825 | 19,571 | 19,789 | 19,367 | 19,004 | 19,877 | 19,851 | 18,611 | 192,064 |
| 75-79 | 14,314 | 14,643 | 15,159 | 15,129 | 14,993 | 15,861 | 16,068 | 15,623 | 16,664 | 15,509 | 153,962 |
| 80-84 | 7,731 | 8,830 | 8,484 | 8,707 | 8,850 | 9,107 | 8,989 | 9,248 | 10,400 | 9,907 | 90,252 |
| 85-89 | 3,052 | 3,116 | 3,387 | 3,524 | 3,600 | 3,652 | 3,786 | 3,717 | 4,175 | 4,141 | 36,150 |
| 90-94 | 725 | 897 | 885 | 816 | 1,016 | 750 | 1,085 | 1,006 | 1,049 | 1,162 | 9,390 |
| 95+ | 154 | 153 | 109 | 120 | 205 | 150 | 202 | 254 | 233 | 190 | 1,771 |
| Total | 99,218 | 100,608 | 102,536 | 100,397 | 98,109 | 98,680 | 97,370 | 97,249 | 99,718 | 95,041 | 988,927 |

Source: Authors' tabulations of SEER Public Use Data (1973-1999; ICD-O-2 Codes C340-C349, excluding Types 9050-9055, 9140, and 9590-9989).

**Figure 10.6: Estimated Lung and Bronchus Cancer Incidence Counts, by Age and Year of Diagnosis, U.S. Males Aged 35-94 Years, 1990-1999 (Source: Data from SEER Public Use Data, 1973-1999)**

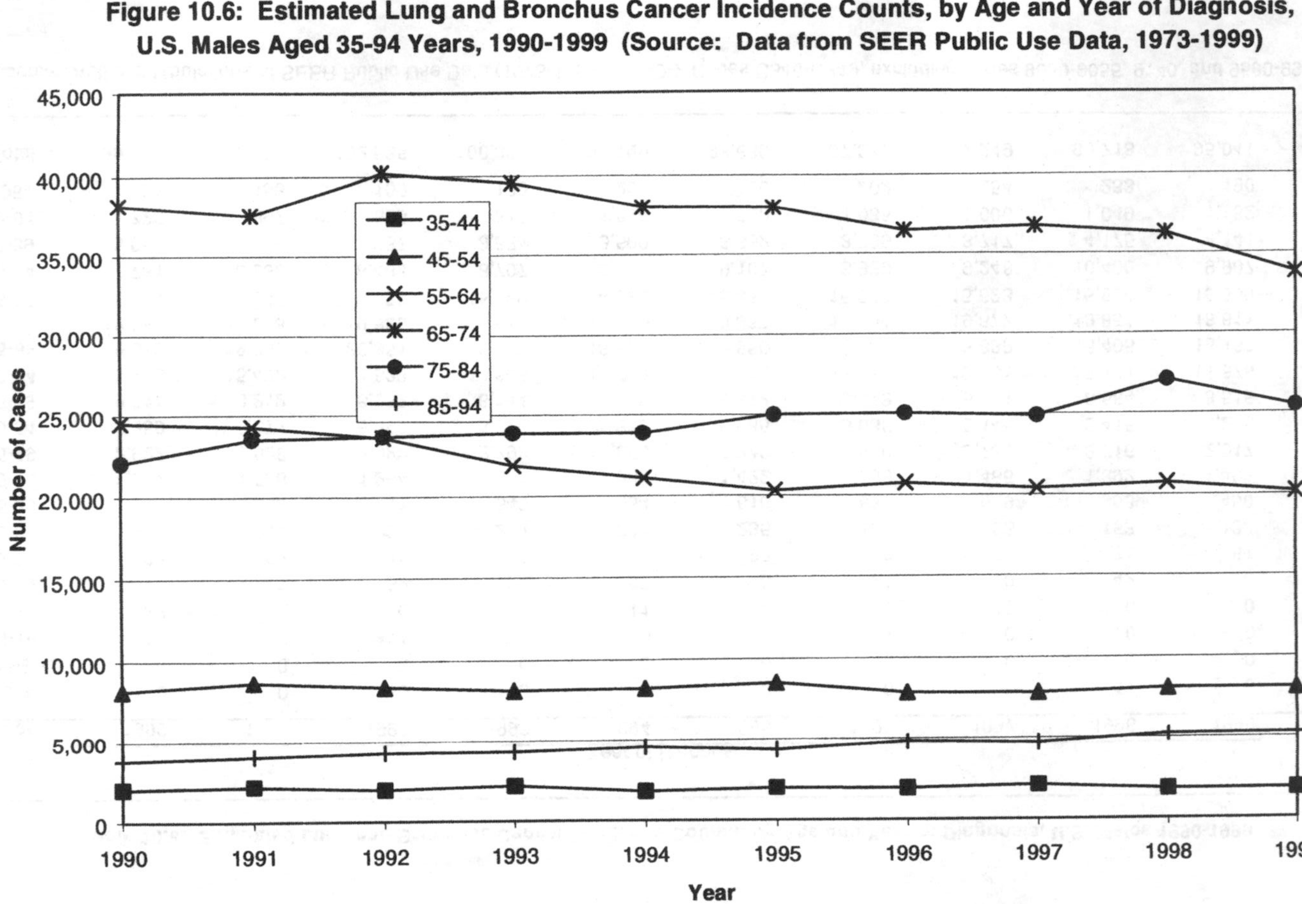

**Table 10.9: Estimates of Mesothelioma and Lung/Bronchus Cancer Incidence Counts from SEER 1997-1999 and Mortality Counts 1999 from NCHS, U.S. Males**

| | Mesothelioma | | | Lung and Bronchus Cancer | | |
|---|---|---|---|---|---|---|
| Age | 1997-1999 Average for SEER Diagnoses | 1999 NCHS Deaths | Ratio: Deaths to Diagnoses | 1999 SEER Diagnoses | 1999 NCHS Deaths | Ratio: Deaths to Diagnoses |
| 0-4 | 0 | 0 | — | 0 | 0 | — |
| 5-9 | 0 | 0 | — | 0 | 2 | — |
| 10-14 | 0 | 0 | — | 0 | 2 | — |
| 15-19 | 4 | 0 | 0.000 | 0 | 5 | — |
| 20-24 | 0 | 0 | — | 10 | 8 | 0.768 |
| 25-29 | 3 | 0 | 0.000 | 51 | 24 | 0.470 |
| 30-34 | 14 | 3 | 0.220 | 133 | 73 | 0.549 |
| 35-39 | 10 | 10 | 0.984 | 450 | 323 | 0.718 |
| 40-44 | 17 | 10 | 0.600 | 1,401 | 1,150 | 0.821 |
| 45-49 | 40 | 35 | 0.869 | 2,917 | 2,344 | 0.803 |
| 50-54 | 83 | 66 | 0.796 | 5,205 | 4,503 | 0.865 |
| 55-59 | 168 | 122 | 0.726 | 8,515 | 7,460 | 0.876 |
| 60-64 | 227 | 168 | 0.740 | 11,676 | 10,543 | 0.903 |
| 65-69 | 315 | 267 | 0.847 | 15,163 | 14,173 | 0.935 |
| 70-74 | 469 | 374 | 0.797 | 18,611 | 17,094 | 0.918 |
| 75-79 | 403 | 389 | 0.965 | 15,509 | 15,453 | 0.996 |
| 80-84 | 305 | 285 | 0.934 | 9,907 | 9,930 | 1.002 |
| 85-89 | 152 | 122 | 0.804 | 4,141 | 4,736 | 1.144 |
| 90-94 | 28 | 27 | 0.957 | 1,162 | 1,314 | 1.131 |
| 95+ | 4 | 3 | 0.852 | 190 | 262 | 1.378 |
| Total | 2,242 | 1,881 | 0.839 | 95,041 | 89,399 | 0.941 |

Source: Authors' tabulations of SEER Public Use Data (1973-1999); and NCHS Public Use Mortality Files (1999; ICD-9 Codes C45 [Mesothelioma] and C34 [Lung and Bronchus]).

the number of asbestosis claims far exceeded the number of deaths. The ratio of Manville Trust claims to NCHS deaths increased from 36.5 in 1990-1994 to 59.6 in 1995-1999. These ratios could not be interpreted as propensities to sue. Instead, the ratios were inverted to show that deaths as fraction of claims declined from 2.7% to 1.7% between 1990-1994 and 1995-1999.

This decline was consistent with a reduction in the average severity of the disease at the time a claim was made, which could have resulted from an expansion of the pool of men who could document exposure to Manville asbestos as well as from an increased propensity to sue within that pool. The fact that the relative increase in numbers of claims for asbestosis was so much larger than for mesothelioma and lung cancer suggested that a reduction in average severity may have been an important component; that is, the diagnosis of either mesothelioma or lung cancer marks a relatively stable point in the disease process, improvements in diagnostic tests notwithstanding. On the other hand, the diagnosis of asbestosis or pleural plaques could have been made in men with a documented history of exposure to asbestos, with minimal or no manifest symptoms.

**Table 10.10: Observed Asbestosis Death Counts, by Age and Year of Death, U.S. Males 1990-1999**

| | Year of Death | | | | | | | | | | |
|---|---|---|---|---|---|---|---|---|---|---|---|
| Age | 1990 | 1991 | 1992 | 1993 | 1994 | 1995 | 1996 | 1997 | 1998 | 1999 | Total |
| 0-4 | 0 | 0 | 1 | 0 | 0 | 0 | 0 | 1 | 0 | 0 | 2 |
| 5-9 | 0 | 0 | 0 | 0 | 0 | 0 | 0 | 0 | 0 | 0 | 0 |
| 10-14 | 0 | 0 | 0 | 0 | 0 | 0 | 0 | 0 | 0 | 0 | 0 |
| 15-19 | 0 | 0 | 0 | 0 | 0 | 0 | 0 | 0 | 0 | 0 | 0 |
| 20-24 | 0 | 0 | 0 | 0 | 0 | 0 | 0 | 0 | 0 | 0 | 0 |
| 25-29 | 0 | 0 | 0 | 0 | 0 | 0 | 0 | 0 | 0 | 0 | 0 |
| 30-34 | 0 | 0 | 0 | 0 | 0 | 0 | 0 | 0 | 0 | 0 | 0 |
| 35-39 | 0 | 0 | 0 | 0 | 0 | 0 | 0 | 0 | 0 | 0 | 0 |
| 40-44 | 1 | 0 | 0 | 0 | 1 | 1 | 0 | 0 | 0 | 0 | 3 |
| 45-49 | 2 | 3 | 0 | 0 | 3 | 0 | 1 | 3 | 0 | 1 | 13 |
| 50-54 | 6 | 6 | 2 | 3 | 3 | 1 | 1 | 3 | 2 | 4 | 31 |
| 55-59 | 8 | 7 | 10 | 9 | 12 | 9 | 4 | 5 | 8 | 9 | 81 |
| 60-64 | 33 | 19 | 22 | 20 | 16 | 21 | 27 | 22 | 26 | 13 | 219 |
| 65-69 | 39 | 35 | 44 | 47 | 50 | 47 | 43 | 35 | 47 | 51 | 438 |
| 70-74 | 59 | 54 | 57 | 59 | 71 | 79 | 82 | 86 | 92 | 83 | 722 |
| 75-79 | 71 | 59 | 65 | 85 | 69 | 81 | 71 | 93 | 94 | 107 | 795 |
| 80-84 | 34 | 44 | 42 | 51 | 58 | 59 | 66 | 77 | 97 | 90 | 618 |
| 85-89 | 20 | 13 | 21 | 30 | 25 | 35 | 39 | 49 | 55 | 56 | 343 |
| 90-94 | 7 | 7 | 6 | 3 | 15 | 8 | 8 | 10 | 13 | 17 | 94 |
| 95+ | 2 | 0 | 1 | 1 | 2 | 1 | 3 | 4 | 5 | 5 | 24 |
| Total | 282 | 247 | 271 | 308 | 325 | 342 | 345 | 388 | 439 | 436 | 3,383 |

Source: Authors' tabulations of NCHS Public Use Mortality Files (1990-1999; ICD-8 Code 501 and ICD-9 Code J61).

**Figure 10.7: Observed Asbestosis Death Counts, by Age and Year of Death, U.S. Males 1990-1999 (Source: Data from NCHS Public Use Mortality Files, 1990-1999)**

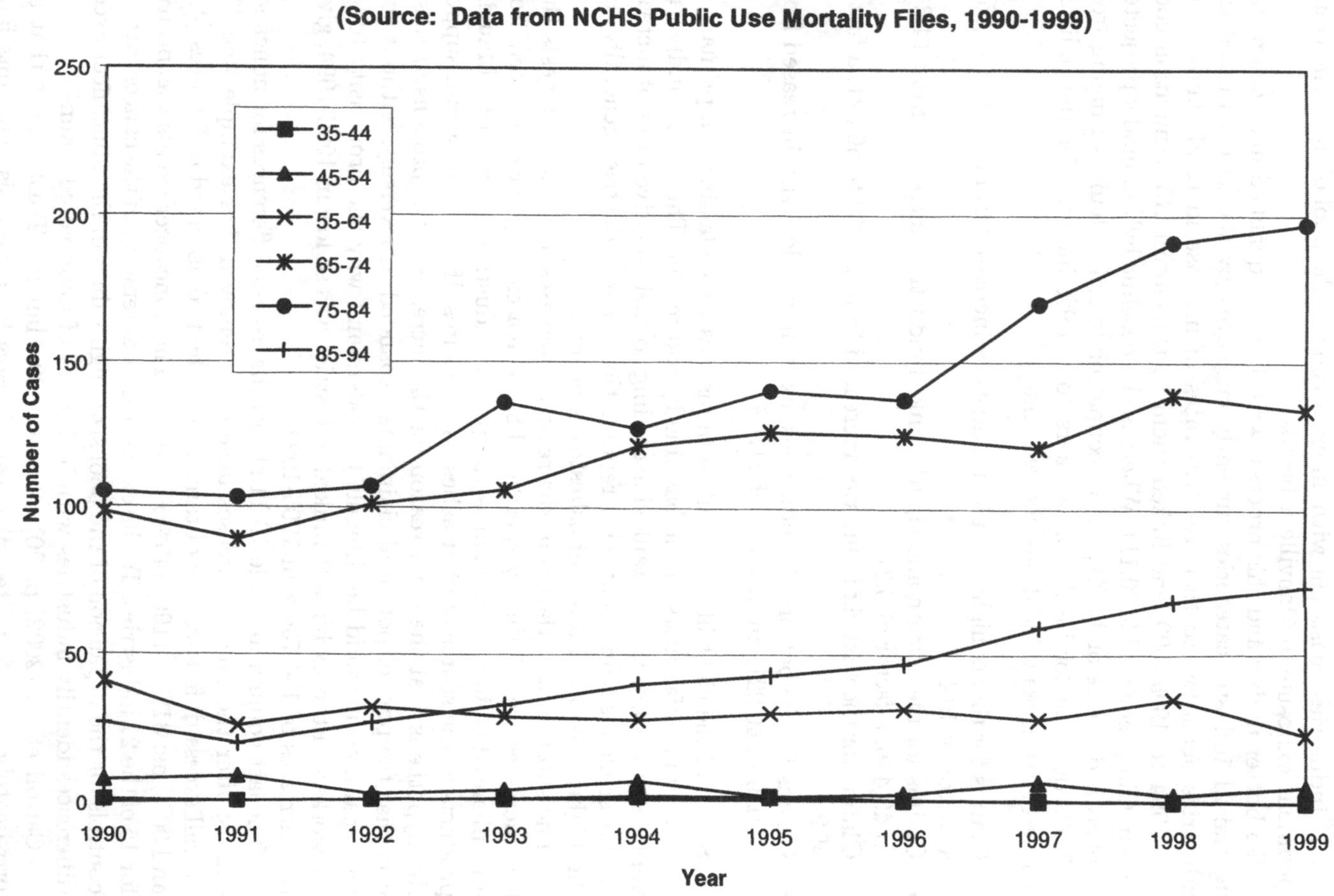

Carroll et al. (2002, p. 47) cited a 2001 Manville Trust report that increasing numbers of claims were from workers exposed to asbestos in "nontraditional" industries, consistent with an expansion of the pool of men who could document exposure to Manville asbestos.

To better understand this process, we tabulated qualified male claims using the 21 industry categories currently employed by the Trust, ranked the industries according to their relative rates of increase in total claims from 1990-1994 to 1995-1999, and linked them to the closest R3 occupation code used in Chapter 8 (Table 10.11). Whereas the baseline hybrid model projected an overall decrease of 17.0%, with occupation specific changes ranging from a 17.9% decrease for insulation workers to a 4.9% increase for the military (Table 8.19), these tabulations showed that:

- Claims for the six highest ranked industries increased from 1352 to 14,052, a factor of 10.4.
- Claims for the seventh and eighth ranked industries increased from 14,676 to 49,436, a factor of 3.37.
- Claims for the next six industries increased from 26,879 to 56,621, a factor of 2.11.
- Claims for the bottom six industries, excluding unknowns, increased from 33,039 to 40,021, an increase of 21.1%.

Seven of the eight highest ranked industries were classified as primary or secondary manufacturers in the baseline hybrid model. The top six industries were also the top six when ranked according to their relative rates of increase in cancer claims, consistent with reports that these industries generally had the highest historical levels of asbestos exposure.

The substantial variation in the relative increases for cancer suggests that the increases were industry-specific. The increases may have resulted from an enhanced ability of claimants in specific industries to identify Manville products as the source of their asbestos exposures. If true, this would support the hypothesis that the main reason for the increase in the propensity to sue was that the pool of potential claimants expanded. Conversely, without such an expansion, it would be difficult to understand why the propensity to sue for workers in the six highest ranked industries was so low in 1990-1994, given the increases of 156-750% in 1995-1999.

The last column in Table 10.11 shows the excess differences for noncancer claims, after the cancer increases have been removed. For example, the 31% overall excess difference was obtained from the two observed differences (45% and 90%) as: $31\% = (190/145 - 1) \times 100\%$. The noncancer excess was positive for 15 of the 21 industries. To the extent that the removal of the cancer increase controls for the expansion of the pool size, high values of the noncancer excess differences identify industries with high rates of unimpaired claims.

Carroll et al. (2002, p. 20) cited a 1999 Manville Trust report that approximately 50% of audited diagnostic X-rays for asbestosis claimants failed to provide evidence of asbestosis on independent review, a figure that was

**Table 10.11: Actual Numbers of Qualified Male Claims Against the Manville Trust, 1990-1994 and 1995-1999, by Industry and Alleged Disease, Ranked According to Relative Rate of Increase in Total Claims from 1990-1994 to 1995-1999**

| Industry Code and Recode | | Alleged Disease: Total | | | Cancer | | | Noncancer | | | |
|---|---|---|---|---|---|---|---|---|---|---|---|
| R3[1] | Description | 1990-1994 | 1995-1999 | Difference | 1990-1994 | 1995-1999 | Difference | 1990-1994 | 1995-1999 | Difference | Excess Difference[2] |
| 6 | Maritime | 325 | 8,559 | 2534% | 147 | 447 | 204% | 178 | 8,112 | 4457% | 1399% |
| 2 | Aerospace/Aviation | 99 | 890 | 799% | 33 | 153 | 364% | 66 | 737 | 1017% | 141% |
| 2 | Textile | 260 | 2,038 | 684% | 40 | 158 | 295% | 220 | 1,880 | 755% | 116% |
| 1 | Asbestos Abatement | 15 | 96 | 540% | 2 | 17 | 750% | 13 | 79 | 508% | -29% |
| 2 | Tire/Rubber Manufacturing | 476 | 1,805 | 279% | 52 | 133 | 156% | 424 | 1,672 | 294% | 54% |
| 1 | Manville Asbestos Products Manufacturing/Mining | 177 | 664 | 275% | 106 | 272 | 157% | 71 | 392 | 452% | 115% |
| 2 | Nonasbestos Products Manufacturing | 3,805 | 14,238 | 274% | 508 | 969 | 91% | 3,297 | 13,269 | 302% | 111% |
| 2 | Iron/Steel/Aluminum/Foundry (Manufacturing) | 10,871 | 35,198 | 224% | 1,274 | 2,860 | 124% | 9,597 | 32,338 | 237% | 50% |
| 6 | Utilities | 2,259 | 6,072 | 169% | 331 | 825 | 149% | 1,928 | 5,247 | 172% | 9% |
| 6 | Chemical | 4,249 | 9,781 | 130% | 472 | 868 | 84% | 3,777 | 8,913 | 136% | 28% |
| 5 | Construction Trades (incl. Insulation Work) | 16,256 | 32,875 | 102% | 2,683 | 4,428 | 65% | 13,573 | 28,447 | 110% | 27% |
| 6 | Petrochemical | 2,448 | 4,892 | 100% | 427 | 674 | 58% | 2,021 | 4,218 | 109% | 32% |
| 6 | Auto Manufacturing | 1,069 | 2,126 | 99% | 184 | 413 | 124% | 885 | 1,713 | 94% | -14% |
| 6 | Longshore | 598 | 875 | 46% | 187 | 133 | -29% | 411 | 742 | 81% | 154% |
| 7 | Military | 2,121 | 2,975 | 40% | 599 | 895 | 49% | 1,522 | 2,080 | 37% | -9% |
| 8 | Building Occupant/Environmental Bystander | 112 | 143 | 28% | 32 | 76 | 138% | 80 | 67 | -16% | -65% |
| 8 | Other | 9,694 | 12,058 | 24% | 1,426 | 1,543 | 8% | 8,268 | 10,515 | 27% | 18% |
| 6 | Railroad | 5,511 | 6,846 | 24% | 626 | 711 | 14% | 4,885 | 6,135 | 26% | 11% |
| 2 | Non-Manville Asbestos Products Manufacturing/ Mining | 1,081 | 1,287 | 19% | 334 | 272 | -19% | 747 | 1,015 | 36% | 67% |
| 4 | Shipyard Construction/Repair | 14,520 | 16,712 | 15% | 2,126 | 2,856 | 34% | 12,394 | 13,856 | 12% | -17% |
| 8 | Unknown | 11,737 | 360 | -97% | 1,420 | 110 | -92% | 10,317 | 250 | -98% | -69% |
| | Total | 87,683 | 160,490 | 83% | 13,009 | 18,813 | 45% | 74,674 | 141,677 | 90% | 31% |

Note 1: Value indicates closest match to R3 recode in Table 8.8.

Note 2: Excess Difference = (Noncancer_1995-1999/Noncancer_1990-1994)/(Cancer_1995-1999/Cancer_1990-1994) – 1.

Source: Author's tabulations of Manville Trust data.

broadly consistent with the relatively greater increases in asbestosis claims compared with mesothelioma and lung cancer. Possible reasons for this audit failure were discussed in Egilman (2002), but the result was clear: A substantial fraction of claims filed in 1995-1999, and beyond, were for men who were functionally unimpaired.

A major part of the motivation for the 2002 TDP was to adjust the disease classification procedures and payment schedule to account for these changes in the mix of claims in a way that maintained the principles of fairness and equity in the 1994 settlement (Weinstein, 2002, p. 41).

## 10.6 Conclusions

The development of reliable forecasts is an iterative process. In this volume, we reviewed the evidentiary basis of models used by the Rule 706 Panel in generating forecasts of future numbers, types, and timings of claims against the Manville Trust. The review covered the primary epidemiological studies that formed the scientific basis for the assumptions of the mathematical models developed in Chapters 6 and 8. The review also covered prior forecasting work by Selikoff (1981), Walker (1982), and others and included detailed assessments and critiques of the assumptions employed therein. In each case, the review yielded insights that resulted in modifications in the next cycle of model development. The Panel's first model, presented in Chapter 6, introduced a range of improvements to the structure of the model employed by Walker (1982), and the Panel's second model, presented in Chapter 8, introduced additional improvements to form a hybrid of Selikoff (1981), Walker (1982), and the Panel's first model.

The Panel preferred the hybrid model because it incorporated all of the assumptions that the Panel and the various reviewers believed to be important. Results of the sensitivity analyses in Chapter 9 were available to the Panel to support their assessment of that model and the recommendations concerning the need to consider the substantial levels of uncertainty surrounding the outcomes of that model.

In this chapter, we critically evaluated the performance of the model by comparing the forecasts for 1990-1994 and 1995-1999 with the actual numbers of claims filed by disease and age group. After adjusting for disease classification changes between the original and updated claim data, we found that the model substantially underpredicted the number of claims in the latter period. The underprediction could be explained by a two-component process involving an increased propensity to sue of up to 28% for cancer, depending on the change in the size of the pool of men with documented exposure to Manville asbestos, with an additional 88% increase in noncancer claims that appeared to be for men with substantially lower disease severity than in the calibration data files. Neither of these findings was anticipated at the time the model was developed. Now, however, it is clear that new applications of the model should allow further changes in the parameters of these two components.

With these adjustments, several types of additional applications could be considered. First, the model could be used to forecast disease types, timings, and costs for future claims against the Manville Trust, based on the experience generated under the 2002 TDP. With substantially more data available for model calibration, the model could be extended to include female claims. Second, with more than 66 companies in bankruptcy, the model could be used to forecast disease types, timings, and costs for bankruptcy trusts created by each company (under Section 524(g) of the U.S. Bankruptcy Code) to set pro rata payout rates meeting established fairness and equity criteria. Third, the model could be used in economic and actuarial evaluations of potential liabilities for future claims against solvent defendants, to ensure that adequate financial reserves are in place to pay the claims when they are filed. Fourth, the model could be used to develop cost estimates associated with comprehensive federal legislative proposals designed to provide fair, equitable, and adequate compensation to persons with asbestos-related injuries.

In each case, the claim filing experience should be continuously monitored and compared with initial and updated forecasts to ensure continued applicability of both the structure of the model and its numerical parameter values.

# References


1. Agency for Toxic Substances and Disease Registry (ATSDR). 1990. Public Health Statement, December.
2. Alleman, J.F., and Mossman, B.T. 1997. Asbestos revisited. *Scientific American* 277(1):70-75.
3. Antman, K.H. 1980. Malignant mesothelioma. *New England Journal of Medicine* 303(4):200-202.
4. Armitage, P., and Doll, R. 1954. The age distribution of cancer and a multi-stage theory of carcinogenesis. *British Journal of Cancer* 8(1):1-12.
5. Armitage, P., and Doll, R. 1957. A two-stage theory of carcinogenesis in relation to the age distribution of human cancer. *British Journal of Cancer* 11(2):161-169.
6. Armitage, P., and Doll, R. 1961. Stochastic models for carcinogenesis. In *Proceedings Fourth Berkeley Symposium on Mathematical Statistics and Probability, Vol. IV, Biology and Problems of Health* (Neyman, J., ed.). University of California Press, Berkeley, pp. 19-38.
7. Associated Press. 2003. Lawyer group pushes limits on asbestos suits. *Wall Street Journal Online*, February 12.
8. Austern, D.T. 2001. *Memorandum: Decision to Reduce Pro Rata Share.* Claims Resolution Management Corporation, Fairfax, VA, June 21.
9. Berry, G. 1981. Mortality of workers certified by pneumoconiosis medical panels as having asbestosis. *British Journal of Medicine* 38:130-137.
10. Berry, G., and Newhouse, M.L. 1983. Mortality of workers manufacturing friction materials using asbestos. *British Journal of Industrial Medicine* 40(1):1-7.
11. Berry, G., Gilson, J.C., Holmes, S., Lewinsohn, H.C., and Roach, S.A. 1979. Asbestosis: A study of dose-response relationships in an asbestos textile factory. *British Journal of Industrial Medicine* 36(2):98-112.
12. Biggs, J.L. 2003. Written testimony on behalf of the Mass Torts Subcommittee of the American Academy of Actuaries, for the National Conference of Insurance Legislators' Committee on Property/Casualty Insurance Hearing on *Proposed Resolution Regarding the Need for Effective Asbestos Reform.* American Academy of Actuaries, Washington, DC, July 10.
13. Blot W.J., and Fraumeni, J.F. 1981. Cancer Among Shipyard Workers. In *Quantification of Occupational Cancer* (Peto, R., and Schneiderman, M., eds.).

Banbury Report 9. Cold Spring Harbor Laboratory, Cold Spring Harbor, NY, pp. 37-49.

14. Blot, W.J., Davies, J.E., Brown, L.M., Nordwall, C.W., Buiatti, E., Ng, A., and Fraumeni, J.F. 1982. Occupation and the high risk of lung cancer in northeast Florida. *Cancer* 50(2):364-371.
15. Blot, W.J., Harrington, J.M., Toledo, A., Hoover, R., Heath, C.W., and Fraumeni, J.F. 1978. Lung cancer after employment in shipyards during World War II. *New England Journal of Medicine* 299(12):620-624.
16. Bocchetta, M., DiResta, I., Powers, A., Fresco, R., Tosolini, A., Testa, J.R., Poss, H.I., Rizzo, P., and Carbone, M. 2000. Human mesothelial cells are unusually susceptible to simian virus 40-mediated transformation and asbestos cocarcinogenicity. *Proceedings of the National Academy of Sciences USA* 97(18):10214-10219.
17. Borow, M., Conston, A., Livornese, L., and Schalet, N. 1973. Mesothelioma following exposure to asbestos: A review of 72 cases. *Chest* 64:641-646.
18. Breslow, N. 1982. Unpublished review of NCI data.
19. Bridbord, K., Decoufle, P., Fraumeni, J.F., Hoel, D.G., Hoover, R.N., Rall, D.P., Saffiotti, U., Schneiderman, M.A., and Upton, A.C. 1978. *Estimates of the Fraction of Cancer in the United States Related to Occupational Factors.* Document submitted by NCI, NIEHS, and NIOSH to the Occupational Safety and Health Administration, September 15.
20. Browne, K., and Smither, W.J. 1983. Asbestos-related mesothelioma: Factors discriminating between pleural and peritoneal sites. *British Journal of Industrial Medicine* 40:145-152.
21. Buckingham, D.A., and Virta, R.L. 2002. Asbestos. In *Historical Statistics for Mineral Commodities in the United States* (Kelly, T., Buckingham, D.A., DiFrancesco, C., Porter, K., Goonan, T., Sznopek, J., Berry, C., and Crane, M., eds.). Available at http://minerals.usgs.gov/minerals/pubs/of01-006/, U.S. Geological Survey, Reston, VA.
22. Burch, P.R.J. 1976. *The Biology of Cancer: A New Approach.* University Park Press, Baltimore, MD.
23. Burnham, C.E. 1982. Unpublished letter to Norman Breslow, June 2.
24. Butel, J.S., and Lednicky, J.A. 1999. Cell and molecular biology of simian virus 40: Implications for human infections and disease. *Journal of the National Cancer Institute* 91(2):119-134.
25. Camus, M., Siemiatycki, J., and Meek, B. 1998. Nonoccupational exposure to chrysotile asbestos and the risk of lung cancer. *New England Journal of Medicine* 338(22):1565-1571.
26. Carroll, S.J., Hensler, D., Abrahamse, A., Gross, J., White, M., Ashwood, S., and Sloss, E. 2002. *Asbestos Litigation Costs and Compensation: An Interim Report.* RAND Institute for Civil Justice, Santa Monica, CA.
27. Churg, A. 1988. Chrysotile, tremolite, and malignant mesothelioma in man. *Chest* 93(3):621-628.
28. Coale, A.J., and Rives, N. 1973. A statistical reconstruction of the Black population of the United States 1880-1970. *Population Index* 39:3-36.
29. Coale, A.J., and Zelnik, M. 1963. *New Estimates of Fertility and Population in the United States.* Princeton University Press, Princeton, NJ.
30. Cochrane, J.C., Webster, I., and Soloman, A. 1980. Prolonged and variable exposure to asbestos fiber. *Current Cancer Research on Occupational and Environmental Carcinogenesis.* DHEW, Washington, DC.

31. Cohen, J.E., Shy, C.M., Checkoway, H., Kupper, L.L., Waldman, G.T., Wilcosky, T.C., and Yoshizawa, C.N. 1984. An analysis of Alexander M. Walker's "Projections of Asbestos-Related Disease 1980-2009." Unpublished manuscript.
32. Cole, P., and Rodu, B. 1996. Declining cancer mortality in the United States. *Cancer* 78(10):2045-2048.
33. Cook, P.J., Doll, R., and Fellingham, S.A. 1969. A mathematical model for the age distribution of cancer in man. *International Journal of Cancer* 4:93-112.
34. Cooke, W.E. 1924. Fibrosis of the lungs due to the inhalation of asbestos dust. *British Medical Journal* 2:147.
35. Cooke, W.E. 1927. Pulmonary asbestosis. *British Medical Journal* 2:1024-1025.
36. Cullen, M.R. 1998. Chrysotile asbestos: Enough is enough. *The Lancet* 351:1377-1378.
37. Daly, A.R., Zupko, A.J., and Hebb, J.L. 1976. *Technological Feasibility and Economic Impact of OSHA Proposed Revision to the Asbestos Standard*, Vol. II. Roy F. Weston Environmental Consultants-Designers, prepared for Asbestos Information Association/North America, Washington, DC.
38. Dement, J. M. 1991. Carcinogenicity of chrysotile asbestos: Evidence from cohort studies. *Annals of the New York Academy of Sciences* 643:15-23.
39. Dement, J.M., Harris, R.L., Symons, M.J., and Shy, C.M. 1982. Estimates of dose-response for respiratory cancer among chrysotile asbestos textile workers. *Annals of Occupational Hygiene* 26(1-4):869-887.
40. Dement, J.M., Harris, R.L., Symons, M.J., and Shy, C.M. 1983a. Exposures and mortality among chrysotile asbestos workers. Part I: Exposure estimates. *American Journal Industrial Medicine* 4(3):399-419.
41. Dement, J.M., Harris, R.L., Symons, M.J., and Shy, C.M. 1983b. Exposures and mortality among chrysotile asbestos workers. Part II: Mortality. *American Journal Industrial Medicine* 4(3):421-433.
42. Doll, R. 1955. Mortality from lung cancer in asbestos workers. *British Journal of Industrial Medicine* 12:81-86.
43. Doll, R., and Peto, R. 1981. The causes of cancer: Quantitative estimates of avoidable risks of cancer in the United States today. *Journal of the National Cancer Institute* 66(6):1191-1308.
44. Edge, J.R. 1979. Incidence of bronchial carcinoma in shipyard workers with pleural plaques. *Annals of the New York Academy of Sciences* 330:289-294.
45. Egilman, D. 2002. Asbestos screenings. *American Journal of Industrial Medicine* 42(2):163.
46. Elmes, P.C., and Simpson, M.J. 1976. The clinical aspects of mesothelioma. *Quarterly Journal of Medicine*, New Series 45:427-449.
47. Enterline, P.E. 1978. Comments on "Estimates of the Fraction of Cancer in the United States Related to Occupational Factors" prepared by NCI, NIEHS, NIOSH. Asbestos Information Association Memorandum, December 14.
48. Enterline, P.E. 1981. Proportion of cancer due to exposure to asbestos. In *Quantification of Occupational Cancer* (Peto, R., and Schneiderman, M., eds.). Banbury Report 9. Cold Spring Harbor Laboratory, Cold Spring Harbor, New York, pp. 19-36.
49. Environmental Protection Agency (EPA). 1986. *Airborne Asbestos Health Assessment Update.* Office of Health and Environmental Assessment, U.S. Environmental Protection Agency, Washington, DC.

50. Felsenthal, S. 1993. In re. *National Gypsum Company, Debtor: Ruling.* United States Bankruptcy Court, Northern District of Texas, Dallas Division, January 29.
51. Finkelstein, M. 1983. Mortality among long-term employees of an Ontario asbestos-cement factory. *British Journal of Industrial Medicine* 40(2):138-144.
52. Finkelstein, M., Kusiak, R., and Suranyi, G. 1981. Mortality among workers receiving compensation for asbestosis in Ontario. *Canadian Medical Association Journal* 125:259-262.
53. Fleiss, J.L. 1981. *Statistical Methods for Rates and Proportions*, 2nd ed. Wiley, New York.
54. Gaensler, E.A. 1992. Asbestos exposure in buildings. *Clinics in Chest Medicine* 13(2):231-242.
55. Garrard, H.G., Gustafson, I.A., and Stovall, M.R. 1992. Medicolegal aspects of asbestos-related diseases: Defendant's attorney's perspective. Chapter 13 in *Pathology of Asbestos-Associated Diseases* (Roggli, V.L., Greenberg, S.D., and Pratt, P.C., eds.). Little, Brown and Co., Boston, pp. 365-381.
56. Gloyne, S.R. 1935. Two cases of squamous carcinoma of the lung occurring in asbestosis. *Tubercle* 17:5-10.
57. Greenberg, S.D. 1992. Benign asbestos-related pleural disease. Chapter 6 in *Pathology of Asbestos-Associated Diseases* (Roggli, V.L., Greenberg, S.D., and Pratt, P.C., eds.). Little, Brown and Co., Boston, pp. 165-188.
58. Greenberg, S.D., and Roggli, V.L. 1992. Other neoplasia. Chapter 8 in *Pathology of Asbestos-Associated Diseases* (Roggli, V.L., Greenberg, S.D., and Pratt, P.C., eds.). Little, Brown and Co., Boston, pp. 211-222.
59. Hahn, W.C., Counter, C.M., Lundberg, A.S., Beijersbergen, R.L., Brooks, M.W., and Weinberg, R.A. 1999. Creation of human tumour cells with defined genetic elements. *Nature* 400:464-468.
60. Hammond, E.C., Selikoff, I.J., and Seidman, H. 1979. Asbestos exposure, cigarette smoking, and death rates. *Annals of the New York Academy of Sciences* 330:473-490.
61. Hemminki, K., Partanen, R., Koskinen, H., Smith, S., Carney, W., and Brandt-Rauf, P.W. 1996. The molecular epidemiology of oncoproteins: Serum p53 protein in patients with asbestosis. *Chest 109*(3 Supplement):22S-26S.
62. Henderson, V.L., and Enterline, P.E. 1979. Asbestos exposure: Factors associated with excess cancer and respiratory disease mortality. *Annals of the New York Academy of Sciences* 330:117-126.
63. Hersch, J. 1992. *Charting the Asbestos Minefield: An Investor's Guide.* Lehman Brothers, New York.
64. Hinds, M.W. 1978. Mesothelioma in the United States: Incidence in the 1970's. *Journal of Occupational Medicine* 20(7):469-471.
65. Hobbs, M.S.T., Woodward, S., Murphy, B., Musk, A.W., and Elder, J.E. 1980. The incidence of pneumoconiosis, mesothelioma, and other respiratory cancer in men engaged in mining and milling crocidolite in Western Australia. In *Biological Effects of Mineral Fibres* (Wagner, J.C., Davis, W., and Crapo, J.D., eds.). International Agency for Research on Cancer, Lyon, France, pp. 615-625.
66. Hogan, M.D., and Hoel, D.G. 1981. Estimated cancer risk associated with occupational asbestos exposure. *Risk Analysis* 1(1):67-76.
67. Jaurand, M.C. 1997. Mechanisms of fiber-induced genotoxicity. *Environmental Health Perspectives* 105(Supplement 5):1073-1084.

68. Kamp, D.W., and Weitzman, S.A. 1999. The molecular basis of asbestos induced lung injury. *Thorax* 54(7):638-652.
69. Keyfitz, N. 1979. Information and allocation: Two uses of the 1980 census. *American Statistician* 33:45-50.
70. Keyfitz, N. 1984. Choice of function for mortality analysis: Effective forecasting depends on a minimum parameter representation. Chapter 10 in *Methodologies for the Collection and Analysis of Mortality Data* (Vallin, J., Pollard, J.H., and Heligman, L., eds.). Ordina Editions, Liege, Belgium.
71. Killian, J.K., Nolan, C.M., Wylie, A.A., Li, T., Vu, T.H., Hoffman, A.R., and Jirtle, R.L. 2001. Divergent evolution in *M6P/IGF2R* imprinting from the Jurassic to the Quaternary. *Human Molecular Genetics* 10(17):1721-1728.
72. Klein, G. 2000. Simian virus 40 and the human mesothelium. *Proceedings of the National Academy of Sciences USA* 97(18):9830-9831.
73. Kolonel, L.N., Hirohata, T., Chappell, B.V., Viola, F.V., and Harris, D.E. 1980. Cancer mortality in a cohort of naval shipyard workers in Hawaii: Early findings. *Journal of the National Cancer Institute* 64:739-743.
74. Lederer, M.E. 1994. Witness statement. In re. *Findley v. Falise*, United States District Court, Eastern District of New York and Southern District of New York, May 5.
75. Lee, R.D., and Carter, L. 1992. Modeling and forecasting U.S. mortality. *Journal of the American Statistical Association* 87:659-675.
76. Leicher, F. 1954. Primary mesothelioma of peritoneum in a case of asbestosis. *Archiv für Gewerbepathologie und Gewerbehygiene* 13:382-392.
77. Lemen, R.A., Dement, J.M., and Wagoner, J.K. 1980. Epidemiology of asbestos-related diseases. *Environmental Health Perspectives* 34:1-11.
78. Levi, I. 1980. *The Enterprise of Knowledge: An Essay on Knowledge, Credal Probability, and Chance.* M.I.T. Press, Cambridge, MA, pp. 431-444.
79. Liddell, F.D., McDonald, A.D., and McDonald, J.C. 1997. The 1891-1920 birth cohort of Quebec chrysotile miners and millers: Development from 1904 and mortality to 1992. *Annals of Occupational Hygiene* 41(1):13-36.
80. Liddell, F.D., McDonald, A.D., and McDonald, J.C. 1998. Dust exposure and lung cancer in Quebec chrysotile miners and millers. *Annals of Occupational Hygiene* 42(1):7-20.
81. Liddell, F.D., McDonald, J.C., and Thomas, D.C. 1977. Methods of cohort analysis: Appraisal by application to asbestos mining. *Journal of the Royal Statistical Society A* 140(Part 4):469-491.
82. Lippmann, M. 1988. Asbestos exposure indices. *Environmental Research* 46:86-106.
83. Lippmann, M. 1990. Effects of fiber characteristics on lung deposition, retention, and disease. *Environmental Health Perspectives* 88:311-317.
84. Longley-Cook, L.H. 1962. An introduction to credibility theory. *Proceedings of the Casualty Actuarial Society* 49:194-221.
85. Lynch, K.M., and Smith, W.A. 1935. Pulmonary asbestosis. Carcinoma of the lung in asbestos-silicosis. *American Journal of Cancer* 24:56-64.
86. Macchiarola, F.J. 1996. The Manville Personal Injury Settlement Trust: Lessons for the future. *Cardozo Law Review* 17(3):583-627.
87. Manton, K.G. 1983. *An Evaluation of Strategies for Forecasting the Implications of Occupational Exposure to Asbestos.* Congressional Research Service, Library of Congress, Washington, DC.

88. Manton, K.G., and Stallard, E. 1979. Maximum likelihood estimation of a stochastic compartment model of cancer latency: Lung cancer mortality among white females in the U.S. *Computers and Biomedical Research* 12:313-325.
89. Manton, K.G., and Stallard, E. 1991. Compartment models of the temporal variation of population lung cancer risks. In *Proceedings of the 13th IMACS World Congress on Computation and Applied Mathematics*, July 22-26, Dublin, Ireland. International Association for Mathematics and Computers in Simulation, Rutgers University, New Brunswick, NJ.
90. Mason, T.J., McKay, F.W., Hoover, R., Blot, W.T., and Fraumeni, J.F. 1975. *Atlas of Cancer Mortality for U.S. Counties: 1950-1969.* DHEW Publication (NIH) 75-780, U.S. Government Printing Office, Washington, DC.
91. McDonald, A.D., and McDonald, J.C. 1980. Malignant mesothelioma in North America. *Cancer* 46(7):1650-1656.
92. McDonald, A.D., Case, B.W., Churg, A., Dufresne, A., Gibbs, G.W., Sebastien, P., and McDonald, J.C. 1997. Mesothelioma in Quebec chrysotile miners and millers: Epidemiology and aetiology. *Annals of Occupational Hygiene* 41(6):707-719.
93. McDonald, A.D., Fry, J.S., Woolley, A.J., and McDonald, J.C. 1983a. Dust exposure and mortality in an American chrysotile textile plant. *British Journal of Industrial Medicine* 40(4):361-367.
94. McDonald, A.D., Fry, J.S., Woolley, A.J., and McDonald, J.C. 1983b. Dust exposure and mortality in an American factory using chrysotile, amosite, and crocidolite in mainly textile manufacturing. *British Journal of Industrial Medicine* 40(4):368-374.
95. McDonald, A.D., Fry, J.S., Woolley, A.J., and McDonald, J.C. 1984. Dust exposure and mortality in an American chrysotile asbestos friction products plant. *British Journal of Industrial Medicine* 41(2):151-157.
96. McDonald, J.C. 1998. Unfinished business: The asbestos textiles mystery. *Annals of Occupational Hygiene* 42(1):3-5.
97. McDonald, J.C., and Liddell, F.D. 1979. Mortality in Canadian miners and millers exposed to chrysotile. *Annals of the New York Academy of Sciences* 330:1-10.
98. McDonald, J.C., and McDonald, A.D. 1981. Mesothelioma as an index of asbestos impact. In *Quantification of Occupational Cancer* (Peto, R., and Schneiderman, M., eds.). Banbury Report 9. Cold Spring Harbor Laboratory, Cold Spring Harbor, New York, pp. 73-85.
99. McDonald, J.C., and McDonald, A.D. 1997. Chrysotile, tremolite and carcinogenicity. *Annals of Occupational Hygiene* 41(6):699-705.
100. McDonald, J.C., Liddell, F.D., Gibbs, G.W., Eyssen, G.E., and McDonald, A.D. 1980. Dust exposure and mortality in chrysotile mining, 1910-75. *British Journal of Industrial Medicine* 37:11-24.
101. Michael, L., and Chissick, S.S. (eds.). 1979. *Asbestos: Properties, Applications, and Hazards.* Wiley, New York.
102. Miettinen, O.S. 1969. Individual matching with multiple controls in the case of all-or-none responses. *Biometrics* 25(2):339-355.
103. Miettinen, O.S. 1972. Standardization of risk ratios. *American Journal of Epidemilogy* 96:383-388.
104. Moolgavkar, S.H., and Venzon, D.J. 1979. Two-event models of carcinogenesis: Incidence curves for childhood and adult tumors. *Mathematical Biosciences* 44:55-77.

105. Mossman, B.T. 1994. Asbestos: Facts and fiction. *Environmental Health Perspectives* 102(5):424-427.
106. Mossman, B.T., Bignon, J., Corn, M., Seaton, A., and Gee, J.B.L. 1990. Asbestos: Scientific developments and implications for public policy. *Science* 247:294-301.
107. Mossman, B.T., and Churg, A. 1998. Mechanisms in the pathogenesis of asbestosis and silicosis. *American Journal of Respiratory and Critical Care Medicine* 157:1666-1680.
108. Murray, H.M. 1907. *Report of the Committee on Compensation for Industrial Diseases.* Minutes of Evidence, CD3946 Her Majesty's Stationery Office, London, UK. (pp. 127-128).
109. Murthy, S.S., and Testa, J.R. 1999. Asbestos, chrosomal deletions, and tumor suppressor gene alterations in human malignant mesothelioma. *Journal of Cellular Physiology* 180:150-157.
110. Muscat, J.E., and Wynder, E.L. 1991. Cigarette smoking, asbestos exposure, and malignant mesothelioma. *Cancer Research* 51(9):2263-2267.
111. National Cancer Institute (NCI). 1996. *Questions and Answers About Asbestos Exposure.* Cancer Facts Sheet 3.21, National Cancer Institute, Bethesda, MD.
112. National Center for Health Statistics. 2002. *Health, United States, 2002: With Chartbook on Trends in the Health of Americans.* National Center for Health Statistics, Hyattsville, MD.
113. National Center for Health Statistics. Each Year 1990 to 1999. *Mortality Data, Underlying Cause-of-Death Public-Use Data Files*, National Center for Health Statistics, Hyattsville, MD.
114. National Research Council (NRC). 1984. *Asbestiform Fibers: Nonoccupational Health Risks.* National Academy Press, Washington, DC.
115. Netherton, N., and Harras, S. American bar association backs limits for asbestos litigation, sets medical criteria. *Class Action Litigation Report* 4(4):140-141, 2003.
116. Newhouse, M.L., and Berry, G. 1976. Predictions of mortality from mesothelioma tumors in asbestos factory workers. *British Journal of Industrial Medicine* 33:147-151.
117. Newhouse, M.L., and Berry, G. 1979. Patterns of mortality in asbestos factory workers in London. *Annals of the New York Academy of Sciences* 330:11-21.
118. Nicholson, W.J. 1981. The role of occupation in the production of cancer. *Risk Analysis* 1(1):77-79.
119. Nicholson, W.J. 1991. Comparative dose-response relationships of asbestos fiber types: Magnitudes and uncertainties. *Annals of the New York Academy of Sciences* 643:78-84.
120. Nicholson, W.J., Perkel, G., Selikoff, I.J., and Seidman, H. 1981a. Cancer from occupational asbestos exposure projections 1980-2000. In *Quantification of Occupational Cancer* (Peto, R., and Schneiderman, M., eds.). Banbury Report 9. Cold Spring Harbor Laboratory, Cold Spring Harbor, NY, pp. 87-111.
121. Nicholson, W.J., Selikoff, I.J., Seidman, H., and Hammond, E.C. 1981b. Mortality experience of asbestos factory workers: Effect of differing intensities of exposure. Unpublished manuscript.
122. Nicholson, W.J., Selikoff, I.J., Seidman, H., Lilis, R., and Formby, P. 1979. Long-term mortality experience of chrysotile miners and millers in Thetford Mines, Quebec. *Annals of the New York Academy of Sciences* 330:11-21.

123. Occupational Safety and Health Administration (OSHA). 1983. Occupational exposure to asbestos: Emergency temporary standard. *Federal Register* 48(215):51,086-51,140.
124. Occupational Safety and Health Administration (OSHA). 1986. Occupational exposure to asbestos, tremolite, anthophyllite, and actinolite: Final rules. *Federal Register* 51(119):22,612-22,790.
125. Occupational Safety and Health Administration (OSHA). 1994. Occupational exposure to asbestos. *Federal Register* 59:40,964-41,162.
126. Passel, J.S., Siegel, J.S., and Robinson, J.G. 1982. *Coverage of the National Population by Age, Sex, and Race: Preliminary Estimates by Demographic Analysis.* Current Population Reports Series P-23, No. 115. U.S. Bureau of the Census, Washington, DC.
127. Peto, J. 1980. The incidence of pleural mesothelioma in chrysotile asbestos textile workers. In *Biological Effects of Mineral Fibres* (Wagner, J.C., Davis, W., and Crapo, J.D., eds.). International Agency for Research on Cancer, Lyon, France, pp. 703-711.
128. Peto, J., and Schneiderman, M. (Eds.). 1981. *Quantification of Occupational Cancer.* Banbury Report 9. Cold Spring Harbor Laboratory, Cold Spring Harbor, New York.
129. Peto, J., Doll, R., Howard, S.V., Kinlen, L.J., and Lewinsohn, H.C. 1977. A mortality study among workers in an English asbestos factory. *British Journal of Industrial Medicine* 34(3):169-173.
130. Peto, J., Henderson, B.E., and Pike, M.C. 1981. Trends in mesothelioma incidence in the United States and the forecast epidemic due to asbestos exposure during World War II. In *Quantification of Occupational Cancer* (Peto, R., and Schneiderman, M., eds.). Banbury Report 9. Cold Spring Harbor Laboratory, Cold Spring Harbor, NY, pp. 51-72.
131. Peto, J., Seidman, H., and Selikoff, I.J. 1982. Mesothelioma mortality in asbestos workers: Implications for models of carcinogenesis and risk assessment. *British Journal of Cancer* 45:124-135.
132. Pilatte, Y., Vivo, C., Renier, A., Kheuang, L., Greffard, A., and Jaurand, M.C. 2000. Absence of SV40 large T-antigen expression in human mesothelioma cell lines. *American Journal of Respiratory Cell and Molecular Biology* 23:788-793.
133. Pollack, E.S., and Horm, J. 1980. Trends in cancer incidence and mortality in the United States, 1969-76. *Journal of the National Cancer Institute* 64:1091-1103.
134. Portier, C.J., and Kopp-Schneider, A. 1991. A multistage model of carcinogenesis incorporating DNA damage and repair. *Risk Analysis* 11(3):535-543.
135. Prentice, R.L., and Breslow, N.E. 1978. Retrospective studies and failure time models. *Biometrika* 65(1):153-158.
136. Proctor, R.N. 1999. *The Nazi War on Cancer.* Princeton University Press, Princeton, NJ.
137. Quinn, J.F., and Burkhauser, R.V. 1994. Retirement and labor force behavior of the elderly. Chapter 3 in *Demography of Aging* (Martin, L.G., and Preston, S.H., eds.). National Academy Press, Washington, DC, pp. 50-101.
138. Rall, D.P. 1994a. Media and science: Harmless dioxin, benign CFCs, and good asbestos. *Environmental Health Perspectives* 102(1):10-11.
139. Rall, D.P. 1994b. Response to Mossman's letter "Asbestos: Facts and fiction." *Environmental Health Perspectives* 102(5):424-427.

140. Resource Planning Corporation (RPC). 1993. *The Manville Personal Injury Settlement Trust Claims Forecast Model.* Resource Planning Corporation, Washington, DC, September 23. Entered into testimony in re. *Findley v. Falise*, United States District Court, Eastern District of New York and Southern District of New York, March 15, 1994.
141. Reynolds, T. 1992. Asbestos-linked cancer rates up less than predicted. *Journal of the National Cancer Institute* 84:560-562.
142. Robeldo, R., and Mossman, B.T. 1999. Cellular and molecular mechanisms of asbestos-induced fibrosis. *Journal of Cellular Physiology* 180:158-166.
143. Roggli, V.L., and Brody, A.R. 1992. Experimental models of asbestos-related diseases. Chapter 10 in *Pathology of Asbestos-Associated Diseases* (Roggli, V.L., Greenberg, S.D., and Pratt, P.C., eds.). Little, Brown and Co., Boston, pp. 257-297.
144. Roggli, V.L., and Coin, P. 1992. Mineralogy of asbestos. Chapter 1 in *Pathology of Asbestos-Associated Diseases* (Roggli, V.L., Greenberg, S.D., and Pratt, P.C., eds.). Little, Brown and Co., Boston, pp. 1-37.
145. Roggli, V.L., and Pratt, P.C. 1992. Asbestosis. Chapter 4 in *Pathology of Asbestos-Associated Diseases* (Roggli, V.L., Greenberg, S.D., and Pratt, P.C., eds.). Little, Brown and Co., Boston, pp. 77-108.
146. Roggli, V.L., Greenberg, S.D., and Pratt, P.C. (eds.) 1992a. *Pathology of Asbestos-Associated Diseases.* Little, Brown and Co., Boston.
147. Roggli, V.L., Kolbeck, J., Sanfilippo, F., and Shelburne, J.D. 1987. Pathology of human mesothelioma: Etiologic and diagnostic consideration. *Pathology Annual* 22(2):91-131.
148. Roggli, V.L., Pratt, P.C., and Brody, A.R. 1992b. Analysis of tissue mineral fiber content. Chapter 11 in *Pathology of Asbestos-Associated Diseases* (Roggli, V.L., Greenberg, S.D., and Pratt, P.C., eds.). Little, Brown and Co., Boston, pp. 299-345.
149. Roggli, V.L., Sanfilippo, F., and Shelburne, J.D. 1992c. Mesothelioma. Chapter 5 in *Pathology of Asbestos-Associated Diseases* (Roggli, V.L., Greenberg, S.D., and Pratt, P.C., eds.). Little, Brown and Co., Boston, pp. 109-164.
150. Rubino, G.F., Piolatto, G., Newhouse, M.L., Scansetti, G., Aresini, G.A., and Murray, R. 1979. Mortality of chrysotile asbestos workers at the Balangero Mine, northern Italy. *British Journal of Industrial Medicine* 36(3):187-194.
151. Seidman, H. 1984. *Short-Term Asbestos Work Exposure and Long-Term Observation.* In: Docket of current rulemaking for revision of the asbestos (dust) standard. U.S. Department of Labor, Occupational Safety and Health Administration, Washington, DC. Available for inspection at U.S. Department of Labor, OSHA Technical Data Center, Francis Perkins Building; Docket No. H033C, Exhibit Nos. 261-A and 261-B.
152. Seidman, H., Selikoff, I.J., and Hammond, E.C. 1979. Short-term asbestos work exposure and long-term observation. *Annals of the New York Academy of Sciences* 330:61-89.
153. Selikoff, I.J. 1978a. Paper presented at meeting of the New York Academy of Sciences, June.
154. Selikoff, I.J. 1978b. Oral testimony at public hearings on OSHA generic cancer policy, June 1.
155. Selikoff, I.J. 1981 (reissued in 1982). *Disability Compensation for Asbestos-Associated Disease in the United States.* Report to the U.S. Dept. of Labor,

Contract No. J-9-M-8-0165. Environmental Sciences Laboratory, Mount Sinai School of Medicine, City University of New York, NY.

156. Selikoff, I.J., and Hammond, E.C. 1978. Asbestos-associated disease in United States shipyards. *CA:A Cancer Journal for Clinicians* 28(2):87-99.
157. Selikoff, I.J., and Seidman, H. 1991. Asbestos-associated deaths among insulation workers in the United States and Canada, 1967-1987. *Annals of the New York Academy of Sciences* 643:1-14.
158. Selikoff, I.J., Churg, J., and Hammond, E.C. 1964. Asbestos exposure and neoplasia. *Journal of the American Medical Association* 188:22-26.
159. Selikoff, I.J., Hammond, E.C., and Seidman, H. 1979. Mortality experience of insulation workers in the United States and Canada, 1943-1976. *Annals of the New York Academy of Sciences* 330:91-116.
160. Selikoff, I.J., Hammond, E.C., and Seidman, H. 1980. Latency of asbestos disease among insulation workers in the United States and Canada. *Cancer* 46:2736-2740.
161. Siegel, J.S. 1974. Estimates of coverage of the population by sex, race, and age in the 1970 Census. *Demography* 11:1-23.
162. Smith, A.H., and Wright, C.C. 1996. Chrysotile asbestos is the main cause of pleural mesothelioma. *American Journal of Industrial Medicine* 30:252-266.
163. Smith, R.J. 1980. Government says cancer rate is increasing. *Science* 209:998-1002.
164. Social Security Administration (SSA). 1992. *Life Tables for the United States Social Security Area 1900-2080.* Actuarial Study No. 107. Social Security Administration, Baltimore, MD.
165. Social Security Administration (SSA). 1996. *Social Security Area Population Projections: 1996.* Actuarial Study No. 110. Social Security Administration, Baltimore, MD.
166. Social Security Administration (SSA). 1999. *The 1999 Technical Panel on Assumptions and Methods: Report to the Social Security Advisory Board.* Social Security Administration, Baltimore, MD.
167. Sparks, J.W. 1931. Pulmonary asbestosis. *Radiology* 17:1249-1257.
168. Spirtas, R., Heineman, E.F., Bernstein, L., Beebe, G.W., Keehn, R.J., Stark, A., Harlow, B.L., and Benichou, J. 1994. Malignant mesothelioma: Attributable risk of asbestos exposure. *Occupational and Environmental Medicine* 51(12):804-811.
169. Stallard, E., and Manton, K.G. 1993. *Estimates and Projections of Asbestos-Related Diseases and Exposures Among Manville Personal Injury Settlement Trust Claimants, 1990-2049.* Center for Demographic Studies, Duke University, Durham, N.C. Entered into testimony in re. *Findley v. Falise*, United States District Court, Eastern District of New York and Southern District of New York, March 15, 1994.
170. Stallard, E., and Manton, K.G. 1994. *Projections of Asbestos-Related Injury Claims Against the Manville Personal Injury Settlement Trust, Males 1990-2049, by Occupation, Date of First Exposure, and Type of Claim.* Center for Demographic Studies, Duke University, Durham, N.C. Entered into testimony in re. *Findley v. Falise*, United States District Court, Eastern District of New York and Southern District of New York, March 15.
171. Stanton, M.F., and Wrench, C. 1972. Mechanisms of mesothelioma induction with asbestos and fibrous glass. *Journal of the National Cancer Institute* 48:797-821.

172. Stayner, L.T., Dankovic, D.A., and Lemen, R.A. 1996. Occupational exposure to chrysotile asbestos and cancer risk: A review of the amphibole hypothesis. *American Journal of Public Health* 86(2):179-186.
173. Strickler, H.D., Rosenberg, P.S., Devesa, S.S., Hertel, J., Fraumeni, J.F., and Goedert, J.J. 1998. Contamination of poliovirus vaccines with simian virus 40 (1955-1963) and subsequent cancer rates. *Journal of the American Medical Association* 279(4):292-295.
174. Supreme Court of the United States. 1997. *Amchem Products, Inc. v. Windsor*, No. 96-270. Opinion delivered by Justice Ginsburg on June 25.
175. Supreme Court of the United States. 1999. *Ortiz v. Fibreboard Corp.*, No. 97-1704. Opinion delivered by Justice Souter on June 23.
176. Surveillance, Epidemiology, and End Results (SEER) Program. 2000. *Public Use CD-ROM (1973-1997).* National Cancer Institute, DCCPS, Cancer Surveillance Research Program, Cancer Statistics Branch, Bethesda, MD, released April 2000, based on the August 1999 submission.
177. Surveillance, Epidemiology, and End Results (SEER) Program. 2002. *Public-Use Data (1973-1999)*. National Cancer Institute, DCCPS, Surveillance Research Program, Cancer Statistics Branch, released April 2002, based on the November 2001 submission.
178. Tagnon, E., Blot, W.J., Stroube, R.B., Day, N.E., Morris, L.E., Peace, B.B., and Fraumeni, J.F. 1980. Mesothelioma associated with the shipbuilding industry in coastal Virginia. *Cancer Research* 40:3875-3879.
179. Tan, W., and Chen, C.W. 1991. Multiple-pathway model of carcinogenesis involving one- and two-stage models. In *Mathematical Population Dynamics* (Arino, O., Axelrod, D., and Kimmel, M., eds.). Marcel Dekker, New York.
180. Virta, R.L. 2001. Asbestos, In *Minerals Yearbook 2001, Vol. I, Metals and Minerals* (U.S. Geological Survey). Available at http://minerals.usgs.gov/minerals/pubs/commodity/asbestos/, U.S. Geological Survey, Reston, VA.
181. Wagner, J.C., Sleggs, C.A., and Marchand, P. 1960. Diffuse pleural mesothelioma and asbestos exposure in the North Western Cape Province. *British Journal of Industrial Medicine* 17:260-271.
182. Walker, A.M. 1982. *Projections of Asbestos-Related Disease 1980-2009.* Epidemiology Resources, Inc., Chestnut Hill, MA.
183. Walker, A.M., Loughlin, J.E., Friedlander, E.R., Rothman, K.J., and Dreyer, N.A. 1983. Projections of asbestos-related disease 1980-2009. *Journal of Occupational Medicine* 25(5):409-425.
184. Weill, H., Hughes, J., and Waggenspack, C. 1979. Influence of dose and fiber type on respiratory malignancy risk in asbestos cement manufacturing. *American Review of Respiratory Diseases* 120(2):345-354.
185. Weinstein, J.B. 1994. In re. *Findley v. Falise: Memorandum, Orders, and Judgment.* United States District Court, Eastern District of New York and Southern District of New York, December 15.
186. Weinstein, J.B. 2002. In re. *Findley v. Trustees of the Manville Personal Injury Settlement Trust: Memorandum and Order.* United States District Court, Eastern District of New York and Southern District of New York, December 27.
187. Weiss, A. 1953. Pleurakrebs bei Lungenasbestose, in vivo morphologisch gesichert. *Medizinische* 1:93-94.

188. Weiss, W., 1983. Heterogeneity in historical cohort studies: A source of bias in assessing lung cancer risk. *Journal of Occupational Medicine* 25:290-294.
189. Whittemore, A., and Keller, J.B. 1978. Quantitative theories of carcinogenesis. *SIAM Review* 20(1):1-30.

# Index